*Reinhard Mahnke, Jevgenijs Kaupužs,
and Ihor Lubashevsky*

Physics of Stochastic Processes

*Reinhard Mahnke, Jevgenijs Kaupužs,
and Ihor Lubashevsky*

Physics of Stochastic Processes

How Randomness Acts in Time

WILEY-VCH Verlag GmbH & Co. KGaA

The Authors

Dr. Reinhard Mahnke
University of Rostock
Institute of Physics
Rostock, Germany
reinhard.mahnke@uni-rostock.de

Dr. Jevgenijs Kaupužs
University of Latvia
Mathematical and Computer Science
Riga, Latvia

Prof. Ihor Lubashevsky
Russian Academy of Sciences
Prokhorov General Physics Institute
Moscow, Russia

All books published by Wiley-VCH are carefully produced. Nevertheless, authors, editors, and publisher do not warrant the information contained in these books, including this book, to be free of errors. Readers are advised to keep in mind that statements, data, illustrations, procedural details or other items may inadvertently be inaccurate.

Library of Congress Card No.: applied for

British Library Cataloguing-in-Publication Data
A catalogue record for this book is available from the British Library.

Bibliographic information published by the Deutsche Nationalbibliothek
Die Deutsche Nationalbibliothek lists this publication in the Deutsche National-bibliografie; detailed bibliographic data are available in the Internet at http://dnb.d-nb.de.

© 2009 WILEY-VCH Verlag GmbH & Co. KGaA, Weinheim

All rights reserved (including those of translation into other languages). No part of this book may be reproduced in any form – by photoprinting, microfilm, or any other means – nor transmitted or translated into a machine language without written permission from the publishers. Registered names, trademarks, etc. used in this book, even when not specifically marked as such, are not to be considered unprotected by law.

Composition Laserwords Private Ltd., Chennai, India
Printing betz-druck GmbH, Darmstadt
Bookbinding Litges & Dopf GmbH, Heppenheim

Printed in the Federal Republic of Germany
Printed on acid-free paper

ISBN: 978-3-527-40840-5

Contents

Preface *XI*

Part I Basic Mathematical Description *1*

1 Fundamental Concepts *3*
1.1 Wiener Process, Adapted Processes and Quadratic Variation *3*
1.2 The Space of Square Integrable Random Variables *8*
1.3 The Ito Integral and the Ito Formula *15*
1.4 The Kolmogorov Differential Equation and the Fokker–Planck Equation *23*
1.5 Special Diffusion Processes *27*
1.6 Exercises *29*

2 Multidimensional Approach *31*
2.1 Bounded Multidimensional Region *31*
2.2 From Chapman–Kolmogorov Equation to Fokker–Planck Description *33*
2.2.1 The Backward Fokker–Planck Equation *35*
2.2.2 Boundary Singularities *37*
2.2.3 The Forward Fokker–Planck Equation *40*
2.2.4 Boundary Relations *43*
2.3 Different Types of Boundaries *44*
2.4 Equivalent Lattice Representation of Random Walks Near the Boundary *45*
2.4.1 Diffusion Tensor Representations *46*
2.4.2 Equivalent Lattice Random Walks *54*
2.4.3 Properties of the Boundary Layer *56*
2.5 Expression for Boundary Singularities *58*

Physics of Stochastic Processes: How Randomness Acts in Time
Reinhard Mahnke, Jevgenijs Kaupužs and Ihor Lubashevsky
Copyright © 2009 WILEY-VCH Verlag GmbH & Co. KGaA, Weinheim
ISBN: 978-3-527-40840-5

2.6	Derivation of Singular Boundary Scaling Properties *61*
2.6.1	Moments of the Walker Distribution and the Generating Function *61*
2.6.2	Master Equation for Lattice Random Walks and its General Solution *62*
2.6.3	Limit of Multiple-Step Random Walks on Small Time Scales *65*
2.6.4	Continuum Limit and a Boundary Model *68*
2.7	Boundary Condition for the Backward Fokker–Planck Equation *69*
2.8	Boundary Condition for the Forward Fokker–Planck Equation *71*
2.9	Concluding Remarks *72*
2.10	Exercises *73*

Part II Physics of Stochastic Processes *75*

3 The Master Equation *77*
- 3.1 Markovian Stochastic Processes *77*
- 3.2 The Master Equation *82*
- 3.3 One-Step Processes in Finite Systems *85*
- 3.4 The First-Passage Time Problem *88*
- 3.5 The Poisson Process in Closed and Open Systems *92*
- 3.6 The Two-Level System *99*
- 3.7 The Three-Level System *105*
- 3.8 Exercises *114*

4 The Fokker–Planck Equation *117*
- 4.1 General Fokker–Planck Equations *117*
- 4.2 Bounded Drift–Diffusion in One Dimension *119*
- 4.3 The Escape Problem and its Solution *123*
- 4.4 Derivation of the Fokker–Planck Equation *127*
- 4.5 Fokker–Planck Dynamics in Finite State Space *128*
- 4.6 Fokker–Planck Dynamics with Coordinate-Dependent Diffusion Coefficient *133*
- 4.7 Alternative Method of Solving the Fokker–Planck Equation *140*
- 4.8 Exercises *142*

5 The Langevin Equation *145*
- 5.1 A System of Many Brownian Particles *145*
- 5.2 A Traditional View of the Langevin Equation *151*
- 5.3 Additive White Noise *152*
- 5.4 Spectral Analysis *157*
- 5.5 Brownian Motion in Three-Dimensional Velocity Space *160*

5.6 Stochastic Differential Equations *166*
5.7 The Standard Wiener Process *168*
5.8 Arithmetic Brownian Motion *173*
5.9 Geometric Brownian Motion *173*
5.10 Exercises *176*

Part III Applications *179*

6 One-Dimensional Diffusion *181*
6.1 Random Walk on a Line and Diffusion: Main Results *181*
6.2 A Drunken Sailor as Random Walker *184*
6.3 Diffusion with Natural Boundaries *186*
6.4 Diffusion in a Finite Interval with Mixed Boundaries *193*
6.5 The Mirror Method and Time Lag *200*
6.6 Maximum Value Distribution *205*
6.7 Summary of Results for Diffusion in a Finite Interval *208*
6.7.1 Reflected Diffusion *208*
6.7.2 Diffusion in a Semi-Open System *209*
6.7.3 Diffusion in an Open System *210*
6.8 Exercises *211*

7 Bounded Drift–Diffusion Motion *213*
7.1 Drift–Diffusion Equation with Natural Boundaries *213*
7.2 Drift–Diffusion Problem with Absorbing and Reflecting Boundaries *215*
7.3 Dimensionless Drift–Diffusion Equation *216*
7.4 Solution in Terms of Orthogonal Eigenfunctions *217*
7.5 First-Passage Time Probability Density *226*
7.6 Cumulative Breakdown Probability *228*
7.7 The Limiting Case for Large Positive Values of the Control Parameter *229*
7.8 A Brief Survey of the Exact Solution *232*
7.8.1 Probability Density *233*
7.8.2 Outflow Probability Density *234*
7.8.3 First Moment of the Outflow Probability Density *234*
7.8.4 Second Moment of the Outflow Probability Density *235*
7.8.5 Outflow Probability *236*
7.9 Relationship to the Sturm–Liouville Theory *238*
7.10 Alternative Method by the Backward Fokker–Planck Equation *240*
7.11 Roots of the Transcendental Equation *249*
7.12 Exercises *251*

8 The Ornstein–Uhlenbeck Process 253
8.1 Definitions and Properties 253
8.2 The Ornstein–Uhlenbeck Process and its Solution 254
8.3 The Ornstein–Uhlenbeck Process with Linear Potential 261
8.4 The Exponential Ornstein–Uhlenbeck Process 266
8.5 Outlook on Econophysics 268
8.6 Exercises 272

9 Nucleation in Supersaturated Vapors 275
9.1 Dynamics of First-Order Phase Transitions in Finite Systems 275
9.2 Condensation of Supersaturated Vapor 277
9.3 The General Multi-Droplet Scenario 286
9.4 Detailed Balance and Free Energy 290
9.5 Relaxation to the Free Energy Minimum 294
9.6 Chemical Potentials 295
9.7 Exercises 296

10 Vehicular Traffic 299
10.1 The Car-Following Theory 299
10.2 The Optimal Velocity Model and its Langevin Approach 302
10.3 Traffic Jam Formation on a Circular Road 316
10.4 Metastability Near Phase Transitions in Traffic Flow 328
10.5 Car Cluster Formation as First-Order Phase Transition 332
10.6 Thermodynamics of Traffic Flow 338
10.7 Exercises 348

11 Noise-Induced Phase Transitions 351
11.1 Equilibrium and Nonequilibrium Phase Transitions 351
11.2 Types of Stochastic Differential Equations 354
11.3 Transformation of Random Variables 358
11.4 Forms of the Fokker–Planck Equation 360
11.5 The Verhulst Model of Third Order 361
11.6 The Genetic Model 364
11.7 Noise-Induced Instability in Geometric Brownian Motion 364
11.8 System Dynamics with Stagnation 367
11.9 Oscillator with Dynamical Traps 369
11.10 Dynamics with Traps in a Chain of Oscillators 372
11.11 Self-Freezing Model for Multi-Lane Traffic 381
11.12 Exercises 385

12   **Many-Particle Systems**   *387*
12.1   Hopping Models with Zero-Range Interaction   *387*
12.2   The Zero-Range Model of Traffic Flow   *389*
12.3   Transition Rates and Phase Separation   *391*
12.4   Metastability   *395*
12.5   Monte Carlo Simulations of the Hopping Model   *400*
12.6   Fundamental Diagram of the Zero-Range Model   *403*
12.7   Polarization Kinetics in Ferroelectrics with Fluctuations   *405*
12.8   Exercises   *409*

Epilogue   *411*

References   *413*

Index   *423*

Preface

A wide variety of systems in nature can be regarded as many-particle ensembles with extremely intricate dynamics of their elements. Numerous examples are known in physics, e.g. gases, fluids, superfluids, electrons and ions in conductors, semiconductors, plasma, nuclear matter in neutron stars, etc. Such macroscopic systems are typically formed of 10^{23}–10^{28} particles, with essentially erratic motion, so a description of the individual elements is really hopeless. However, it is not actually necessary for practical tasks because on the macroscopic level we are dealing only with cumulative effects expressed in macroscopic variables. At this level, details of the individual particle motion are averaged – only the mean characteristics are essential for a description of the system dynamics. The deviation of an individual particle from the mean behavior can then be taken into account, if necessary, in terms of random fluctuations characterized again by some mean parameters. It should be noted that many systems of a nonphysical nature, e.g. fish swarms and bird flocks, vehicle ensembles, pedestrians or stock markets can be regarded (leaving aside social aspects of their behavior) as ensembles of interacting particles.

There are several approaches to tackling many-particle systems. Dealing with a physical object whose dynamics is based on the Newtonian or Schrödinger equation, it is possible to start from the microscopic description and directly write down the corresponding governing equations. Then a rather small part of the system comprising, e.g. one, two, or three particles should be singled out and considered individually. The effect of the other elements on this selected part is taken into account on the average. Roughly speaking, it is in just this way that the notion of a thermal heat bath is introduced – a small part of the system under consideration is singled out and its interaction with the neighboring particles is simulated in terms of stochastic energy exchange with a certain reservoir characterized by some temperature. This approach is the most rigorous and, as a result, the most difficult way of constructing a bridge between the microscopic description dealing with individual particles (atoms, molecules, etc.) and the mesoscopic continuum fields, e.g. density, temperature, and pressure. Typically this bridge is implemented in the form of a partial differential equation or a system of such equations governing the distribution function of the particle or the collection of particles. We

Physics of Stochastic Processes: How Randomness Acts in Time
Reinhard Mahnke, Jevgenijs Kaupužs and Ihor Lubashevsky
Copyright © 2009 WILEY-VCH Verlag GmbH & Co. KGaA, Weinheim
ISBN: 978-3-527-40840-5

should point out that the many-particle ensembles governed by the laws of classical Newtonian mechanics exhibit chaotic dynamics rather than stochastic dynamics. The term chaotic refers to systems whose evolution from the initial conditions is rigorously determined by nonrandom dynamics. Given the initial conditions, the dynamics of such a deterministic system is formally predictable, so it is not stochastic in a rigorous sense. However, if the system trajectories are located inside a given bounded region and are nonperiodic, then their temporal and spatial structure is highly intricate and in fact looks like that of stochastic random paths. Moreover, these trajectories pass all the standard tests for randomness so, for practical purposes, they can be regarded as stochastic. This observation is actually one of the ways to justify introducing a thermal heat bath characterized by *stochastic* energy exchange (between the small part of the system under consideration and the surrounding particles treated as a random reservoir at a particular temperature). The same comments concerning the relationship between chaos and stochasticity should be addressed in time series analysis. Without knowledge of the origin it is practically impossible to distinguish between chaotic and stochastic behavior.

Another way to treat many-particle systems is to construct a collection of microscopic equations governing, e.g. the dynamics of individual particles where, for a given particle, the influence of the other particles is described in terms of both systematic and random forces. The notion of random forces again enables one to derive the corresponding partial differential equations for the distribution function of particles. Indeed the random forces should be introduced in such a way that these governing equations for the distribution function coincide with those obtained via the approach of the previous paragraph. As far as social, ecological, and economic systems are concerned, postulating the appropriate form of the random forces seems to be the only way to construct a mathematical description. This is due to such systems being open. Moreover the behavior of their elements is so intricate and multifactorial that a closed mathematical description is likely to be impossible.

The latter approach is precisely the main topic of this book. It is based on probability theory or, more specifically, on the notion of stochastic processes and the relevant mathematical constructions, which are the subject matter of Chapters 1 and 2 (see the layout of the book shown at the end of this preface, page XVII). On the microscopic level stochastic trajectories of the system motion are the basic elements of the probabilistic description. It is assumed that different stochastic realizations of the random force are independent and also that the motion of particles does not have long-time memory. The notion of stochastic trajectories has a long history, possibly going back to the scientific poem De Rerum Nature (On the Nature of Things, circa 60 BC) by Titus Lucretius Carus. Although very little is known about the Roman philosopher, it seems he described the random motion of dust particles in air. In 1785 Jan Ingenhousz observed the irregular motion of coal dust particles on the surface of alcohol. Then, in 1827, the British botanist Robert Brown also discovered random highly erratic motion of pollen particles floating in water under the microscope. Since that time this phenomenon has been called *Brownian motion*. The generalization of the observed phenomena gave rise to the notion of random walks where the walker dynamics is governed by both regular and stochastic forces.

The first person who proposed a mathematical model for Brownian motion appears to be Thorvald N. Thiele in 1880. This was followed independently by Louis Bachelier in 1900 in his PhD thesis Théorie de la Spéculation devoted to a stochastic analysis of the stock and option markets. He worked out mathematically the idea that the stock market prices are essentially sums of independent, bounded random changes. The results put forward by Bechelier led to a flash of interest in stochastic processes and corresponding probabilistic approaches. However, it was Albert Einstein's independent research into the problem in his 1905 paper that brought the solution to the attention of physicists (see, e.g. *Brownian motion* – Wikipedia, The Free Encyclopedia, 23 October 2007).

The qualitative explanation of Brownian motion as a kinetic phenomenon was put forward by several authors. As mentioned above, it is possible to add random forces to the dynamical laws which were proposed for the first time by the French physicist Paul Langevin. (This resulted in a new mathematical field now known as stochastic differential equations.) The appropriate partial differential equations for the distribution function could then be derived based on the Langevin equation.

It is possible to develop the probabilistic description of a stochastic process in the opposite way – the equations governing the distribution function are postulated and the appropriate Langevin equation is constructed in order to give these equations. This idea was implemented for the first time by Albert Einstein deriving the diffusion equation for Brownian particles in his famous paper Über die von der molekularkinetischen Theorie der Wärme geforderte Bewegung von in ruhenden Flüssigkeiten suspendierten Teilchen published in *Annalen der Physik* (1905). The equation for diffusive motion was then developed by Adriaan Fokker (1914) and later more completely and generally by Max Planck (1918), leading to the transport equation now known as the Fokker–Planck equation. There are also approaches to describing random processes in discrete phase spaces based on ordinary differential equations (e.g. the probability balance law known as the master equation). If a stochastic process develops in discrete space and time the cellular automata models can be used, which form a distinct branch of the theory of stochastic processes. These problems and their mutual interrelationship are considered in Chapters 3–5 which adopt one of the main assumptions in the theory of stochastic processes, the Markovian approximation. According to this approximation, the displacement of a wandering particle on mesoscopic scales can be considered as the result of many small independent identically distributed steps. This reasoning is very close to what is now called a Kramers–Moyal expansion and has been used to derive the Fokker–Planck equation.

To elucidate the main notions of stochastic processes, Chapters 6 to 8 consider in detail some rather simple examples of discrete random walks and continuous Brownian motion. In particular, they touch on the problem of reaching a boundary for the first time. This problem plays an essential role in many physical phenomena such as escaping from a potential well, anomalous diffusion in fractal media, heat diffusion in living tissue, etc.

As mentioned above, the notion of stochastic processes can form the initial mathematical description for objects of a nonphysical nature, e.g. social, ecological, and economic systems. This is a novel branch of science where only the first steps have been taken. It turns out that, in spite of their nonphysical nature, the cooperative phenomena in such systems (for example, self-organization processes in congested traffic or motion of pedestrians and social animals) exhibit a wide variety of properties commonly met in physical systems (for example in gas–liquid phase transitions, spinodal decomposition in solid solutions, ferromagnetic transitions, etc.). So, in some sense, the stochastic description of many-particle ensembles with strong interaction between their elements is of a more general nature than the basic laws of the corresponding mechanical systems.

These questions are considered in Chapters 9 and 10 dealing with the aggregation of particles out of an initially homogeneous situation. This phenomenon is well known in physics, as well as in other branches of the natural sciences and engineering. The formation of bound states as an aggregation process is due to self-organization. The formation of car clusters (jams) at overcritical densities in traffic flow is an analogous phenomenon in the sense that cars can be considered as (strong asymmetrically) interacting particles. The development of traffic jams in vehicular flow is an everyday example of the occurrence of nucleation and aggregation in a system of many point-like cars. Traffic jams are a typical signature of the complex behavior of the many-car system. The master equation approach to stochastic processes can be applied to describe the car-cluster formation on a road in partial analogy to droplet formation in a supersaturated vapor.

This jamming transition is very similar to conventional phase transitions appearing in the study of critical phenomena. Traffic-like collective movements are observed at almost all levels of biological systems. We study the energy balance of motorized particles in a many-car system. New dynamical features, such as steady state motion with energy flux, also appear. This phenomenon is also observed in a system of active Brownian particles with energy take-up and energy dissipation.

The last two Chapters 11 and 12 are devoted to some modern applications in the physics of stochastic processes. First, we consider nonequilibrium phase transition induced by noise or caused by dynamical traps. Probably, the former type of transition can only be described using the Langevin equation with multiplicative noise, that is, stochastic equations for which the intensity of the random forces depends on the system state. During the last few decades it has been demonstrated that the behavior of such systems can be rather complex; in particular, the appearance of new states can be induced by noise as its intensity increases and attains certain critical values. The second type of phase transition seems to be a commonly encountered phenomenon in systems, for example, congested traffic flow, where the human factor is essential. Such transitions are due to the existence of some regions in the corresponding phase space where the system dynamics is stagnated. Following the notions introduced in the theory of Hamiltonian dynamics with complex behavior, these regions are called dynamical traps.

Finally, we turn to the kinetics of many-particle systems. The zero-range process, introduced in 1970 by Frank Spitzer as a system of interacting random walks, serves as a generic model in which rigorous large-scale description of the dynamics for arbitrary initial densities is possible in terms of a hydrodynamic equation for the coarse-grained particle density. It allows one to derive a criterion for phase separation in one-dimensional driven systems of interacting particles, e.g. in traffic flow, as well as to describe nontrivial features of stochastic dynamics like metastability.

Nowadays another aspect which should be taken into account is non-Gaussian behavior; that is, long-tail distributions which are observed in stock market data as well as in transportation theory. In this sense, applied sciences such as sociology and econophysics, biophysics and engineering, consider extreme events in nature and society and deal with effects (like material rupture) which can be investigated only by the probabilistic approach.

In concluding this preface, we would like to underline the spirit in which this book is intended. Here we are in agreement with other authors of books on random processes; in particular, A. J. Chorin and O. H. Hald in *Stochastic Tools in Mathematics and Science* state: 'When you asked alumni graduates from universities in Europe and US moving into nonacademic jobs in society and industry what they actually need in their business, you found that most of them did stochastic things like time series analysis, data processing etc., but that had never appeared in detail in university courses'. So the general aim of the present book is to provide stochastic tools for the multidisciplinary understanding of random events and to illustrate them with many beautiful applications in different disciplines ranging from econophysics to sociology. The central problem under consideration in this book is thus the theoretical modeling of complex systems, that is, many-particle systems with nondeterministic behavior. In contrast to the established classical deterministic approach based on trajectories, we develop and investigate probabilistic dynamics using stochastic tools, such as stochastic differential equations, Fokker–Planck and master equations, to obtain the probability density distribution. The stochastic technique provides an exact and more understandable background to describe complex systems.

The authors have been working for years on the problems to which this monograph is devoted. Nevertheless, the book is also the result of long-standing scientific cooperation with a number of colleagues from all over the world. The authors thank Werner Ebeling, Rudolf Friedrich, Vilnis Frishfelds, Namik Gusein-Zade, Peter Hänggi, Rosemary Harris, Andreas Heuer, Dirk Helbing, Alexander Ignatov, Andris Jakovičs, Holger Kantz,

Boris Kerner, Reinhart Kühne, Kai Nagel, Holger Nobach, Gerd Röpke, Yuri and Michael Romanovsky, Anri Rukhadze, Andreas Schadschneider, Michael Schreckenberg, Gunter M. Schütz, Lutz Schimansky-Geier, Yuki Sugiyama, Steffen Trimper, Peter Wagner and Hans Weber, for fruitful discussions.

Special thanks are due to Friedrich Liese from the Institute of Mathematics at Rostock University for delivering a joint lecture series on *Stochastic Processes* from the mathematical (F. Liese) as well as physical (R. Mahnke) points of view and for preparing Chapter 1 of this book – *Fundamental Concepts*.

The contents of this book took shape over several years, based on research and lectures performed at different locations. One of the recent lecture presentations took place in the summer term of 2007 at Rostock University. The authors have benefited from the contributions of a number of students. We would like to express our gratitude to the active participants, Michael Brüdgam, Matthias Florian, Peter Grünwald, Hannes Hartmann, Julia Hinkel, Bastian Holst, Thomas Kiesel, Susanne Killiches, Knut Klingbeil, Christof Liebe, Daniel Münzner, Ralf Remer, Elisabeth Schöne, Philipp Sperling, Marten Tolk, Andris Voitkans, Norman Wilken, and Mathias Winkel, together with many other students, PhD students and co-workers.

Finally, we would like to acknowledge Andrey Ushakov, a student from Moscow Technical University of Radiophysics, Engineering and Automation, who has contributed to Section 3.7 – *Three-Level System*.

The authors acknowledge support from the Deutsche Forschungsgemeinschaft via grant MA 1508/8.

Rostock, Riga, Moscow *Reinhard Mahnke*
October 2008 *Jevgenijs Kaupužs*
  Ihor Lubashevsky

Preface | **XVII**

Mathematics

Physics

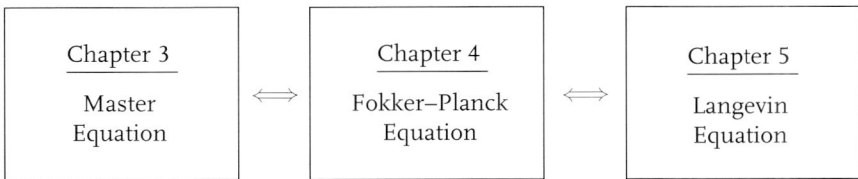

Applications I

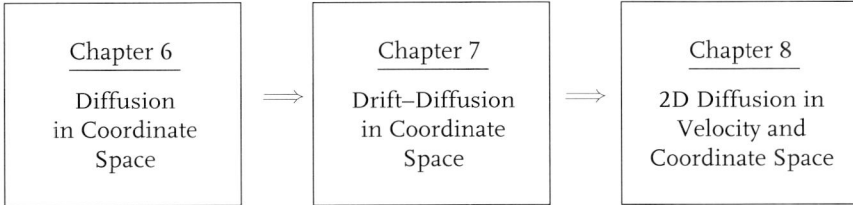

Applications II

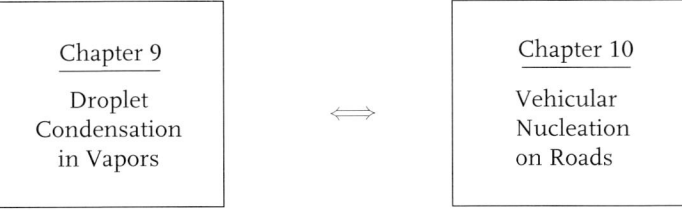

Applications III

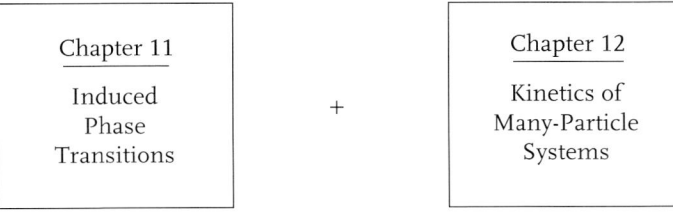

Part I
Basic Mathematical Description

1
Fundamental Concepts

1.1
Wiener Process, Adapted Processes and Quadratic Variation

Stochastic processes represent a fundamental concept used to model the development of a physical or nonphysical system in time. It has turned out that the apparatus of stochastic processes is powerful enough to be applied to many other fields, such as economy, finance, engineering, transportation, biology and medicine.

To start with, we recall that a random variable X is a mapping $X : \Omega \to \mathbb{R}$ that assigns a real value to each elementary event $\omega \in \Omega$. The concrete value $X(\omega)$ is called a realization. It is the value we observe after the experiment has been done. To create a mathematical machine we suppose that a probability space $(\Omega, \mathfrak{F}, \mathbb{P})$ is given. Ω is the set of all elementary events and $\mathfrak{F}$ is the family of events we are interested in. It contains the set of all elementary events Ω and is assumed to be closed with respect to forming the complement and countable intersections and unions of events from this collection of events. Such families of sets or events are called σ-algebras. The character σ indicates that even the union or intersection of countably many sets belongs to $\mathfrak{F}$ as well. For mathematical reasons we have to assume that 'events generated by X', i.e. sets of the type $\{\omega : X(\omega) \in I\}$, where I is an open or closed or semi-open interval, are really events; i.e. such sets are assumed also to belong to $\mathfrak{F}$. Unfortunately the collection of all intervals of the real line is not closed with respect to the operation of union. The smallest collection of subsets of the real line that is a σ-algebra and contains all intervals is called the σ-algebra of Borel sets and will be denoted by $\mathfrak{B}$. It turns out that we have not only $\{\omega : X(\omega) \in I\} \in \mathfrak{F}$ for any interval but even $\{\omega : X(\omega) \in B\} \in \mathfrak{F}$ for every Borel set B. This fact is referred to as the $\mathfrak{F}$-measurability of X.

It turns out that for any random variable X and any continuous or monotone function g the function $Y(\omega) = g(X(\omega))$ is again a random variable. This statement remains true even if we replace g by a function from a larger class of functions, called the family of all measurable functions, to which not only the continuous functions but also the pointwise limit of continuous functions belong. This class of functions is closed with respect to 'almost all' standard manipulations with

Physics of Stochastic Processes: How Randomness Acts in Time
Reinhard Mahnke, Jevgenijs Kaupužs and Ihor Lubashevsky
Copyright © 2009 WILEY-VCH Verlag GmbH & Co. KGaA, Weinheim
ISBN: 978-3-527-40840-5

functions, such as linear combinations and products and finally forming new functions by plugging one function into another function.

The probability measure $\mathbb{P}$ is defined on $\mathfrak{F}$ and it assigns to each event $A \in \mathfrak{F}$ a number $P(A)$ called the probability of A. The mappings $A \mapsto P(A)$ satisfy the axioms of probability theory, i.e. P is a non-negative σ-additive set function on $\mathfrak{F}$ with $P(\Omega) = 1$.

We assume that the reader is familiar with probability theory at an introductory course level and in the following we use basic concepts and results without giving additional motivation or explanation.

Random variables or random vectors are useful concepts to model the random outcome of an experiment. But we have to include the additional variable 'time' when we are going to study random effects which change over time.

Definition 1.1 *By stochastic process we mean a family of random variables $(X_t)_{t \geq 0}$ which are defined on the probability space $(\Omega, \mathfrak{F}, \mathbb{P})$.*

By definition X_t is in fact a function of two variables $X_t(\omega)$. For fixed t this function of ω is a random variable. Otherwise, if we fix ω then we call the function of t defined by $t \mapsto X_t(\omega)$ a realization or a path. This means that the realization of a stochastic process is a function. Therefore stochastic processes are sometimes referred to as random functions. We call a stochastic process continuous if all realizations are continuous functions.

For the construction of a stochastic process, that is, of a suitable probability space, one needs the so-called finite dimensional distributions which are the distributions of random vectors $(X_{t_1}, \ldots, X_{t_n})$, where $t_1 < t_2 < \cdots < t_n$ is any fixed selection. For details of the construction we refer to Øksendal [175].

A fundamental idea of modeling experiments with several random outcomes in both probability theory and mathematical statistics is to start with independent random variables and to create a model by choosing suitable functions of these independent random variables. This fact explains why, in the area of stochastic processes, the particular processes with independent increments play an exceptional role. This, in combination with the fundamental meaning of the normal distribution in probability theory, makes clear the importance of the so-called Wiener process, which will now be defined.

Definition 1.2 *A stochastic process $(W_t)_{t \geq 0}$ is called a standard Wiener process or (briefly) Wiener process if:*

1) $W_0 = 0$,
2) $(W_t)_{t \geq 0}$ has independent increments, i.e. $W_{t_n} - W_{t_{n-1}}, \ldots, W_{t_2} - W_{t_1}, W_{t_1}$ are independent for $t_1 < t_2 < \cdots < t_n$,
3) For all $0 \leq s < t$, $W_t - W_s$ has a normal distribution with expectation $\mathbb{E}(W_t - W_s) = 0$ and variance $\mathbb{V}(W_t - W_s) = t - s$,
4) All paths of $(W_t)_{t \geq 0}$ are continuous.

The Wiener process is also called Brownian motion. This process is named after the biologist Robert Brown whose research dates back to the 1820s. The mathematical theory began with Louis Bachelier (Théorie de la Spéculation, 1900) and later by Albert Einstein (Eine neue Bestimmung der Moleküldimensionen, 1905). Norbert Wiener (1923) was the first to create a firm mathematical basis for Brownian motion.

To study properties of the paths of the Wiener process we use the quadratic variation as a measure of the smoothness of a function.

Definition 1.3 *Let $f : [0, T] \to \mathbb{R}$ be a real function and $\mathfrak{z}_n : a = t_{0,n} < t_{1,n} < \cdots < t_{n,n} = b$, a sequence of partitions with*

$$\delta(\mathfrak{z}_n) := \max_{0 \leq i \leq n-1} (t_{i+1,n} - t_{i,n}) \to 0, \quad \text{as} \quad n \to \infty.$$

If $\lim_{n \to \infty} \sum_{i=0}^{n-1} (f(t_{i+1,n}) - f(t_{i,n}))^2$ exists and is independent of the concrete sequence of partitions then this limit is called the quadratic variation of f and will be denoted by $[f]_T$.

We show that the quadratic variation of a continuously differentiable function is zero.

Lemma 1.1 *If f is differentiable in $[0, T]$ and the derivative $f'(t)$ is continuous then $[f]_T = 0$.*

Proof. Put $C = \sup_{0 \leq t \leq T} |f'(t)|$. Then $|f(t) - f(s)| \leq C|t - s|$ and

$$\sum_{i=0}^{n-1} (f(t_{i+1,n}) - f(t_{i,n}))^2 \leq C^2 \sum_{i=0}^{n-1} (t_{i+1,n} - t_{i,n})^2$$

$$\leq C^2 \delta(\mathfrak{z}_n) T \to_{n \to \infty} 0.$$

If $(X_t)_{0 \leq t \leq T}$ is a stochastic process then the quadratic variation $[X]_T$ is a random variable such that for any sequence of partitions $\mathfrak{z}_n$ with $\delta(\mathfrak{z}_n) \to 0$ it holds for $n \to \infty$

$$\sum_{i=0}^{n-1} (X_{t_{i+1,n}} - X_{t_{i,n}})^2 \to^{\mathbb{P}} [X]_T,$$

where $\to^{\mathbb{P}}$ is the symbol for stochastic convergence. Whether the quadratic variation of a stochastic process does or does not exist depends on the concrete structure of this process and has to be checked in a concrete situation and it is often more useful to deal with the convergence in mean square instead of the stochastic convergence. The relation between the two concepts provides the well known Chebyshev inequality which states that, for any random variables Z_n, Z

$$\mathbb{P}(|Z_n - Z| > \varepsilon) \leq \frac{1}{\varepsilon^2} \mathbb{E}(Z_n - Z)^2.$$

Hence the mean square convergence $\mathbb{E}(Z_n - Z)^2 \to 0$ of Z_n to Z implies the stochastic convergence $\mathbb{P}(|Z_n - Z| > \varepsilon) \to 0$ of Z_n to Z.

Now we are going to calculate the quadratic variation of a Wiener process. To this end we need a well known fact. If V has a normal distribution with expectation μ and variance σ^2 then

$$\mathbb{E}V = \mu, \quad \mathbb{V}(V) = \mathbb{E}(V - \mu)^2 = \sigma^2$$
$$\mathbb{E}(V - \mu)^3 = 0, \quad \mathbb{E}(V - \mu)^4 = 3\sigma^4.$$

If $\mu = 0$ then

$$\mathbb{E}(V^2 - \sigma^2)^2 = \mathbb{E}(V^4 - 2\sigma^2 V^2 + \sigma^4)$$
$$= 3\sigma^4 - \sigma^4 = 2\sigma^4. \tag{1.1}$$

Theorem 1.1 *If $(W_t)_{0 \leq t \leq T}$ is a Wiener process then the quadratic variation*

$$[W]_T = T.$$

Proof. Let $\mathfrak{z}_n$ be a sequence of partitions of $[0, T]$ with $\delta(\mathfrak{z}_n) \to 0$ and put

$$Z_n = \sum_{i=0}^{n-1} (W_{t_{i+1,n}} - W_{t_{i,n}})^2.$$

From the definition of the Wiener process we get that $\mathbb{E}(W_{t_{i+1,n}} - W_{t_{i,n}})^2 = t_{i+1,n} - t_{i,n}$. As the variance of a sum of independent random variables is just the sum of the variances we get from the independent increments

$$\mathbb{E}(Z_n - t)^2 = \mathbb{E}\left(\sum_{i=0}^{n-1} (W_{t_{i+1,n}} - W_{t_{i,n}})^2 - (t_{i+1,n} - t_{i,n})\right)^2$$
$$= \mathbb{V}(Z_n) = \sum_{i=0}^{n-1} \mathbb{V}((W_{t_{i+1,n}} - W_{t_{i,n}})^2)$$
$$= \sum_{i=0}^{n-1} \mathbb{E}((W_{t_{i+1,n}} - W_{t_{i,n}})^2 - (t_{i+1,n} - t_{i,n}))^2$$
$$= 2\sum_{i=0}^{n-1} (t_{i+1,n} - t_{i,n})^2 \leq 2\delta(\mathfrak{z}_n) T \to 0,$$

where for the last equality we have used (1.1).

The statement $[W]_T = T$ is remarkable from different points of view. The exceptional fact is that the quadratic variation of this special stochastic process $(W_t)_{0 \leq t \leq T}$ is a degenerate random variable, it is the deterministic value T. This value is non-zero. Therefore we may conclude from Lemma 1.1 that the paths of

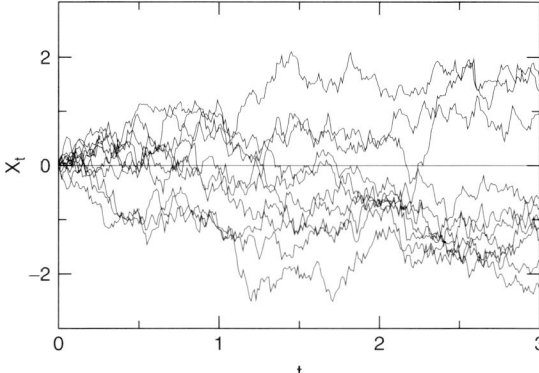

Figure 1.1 Collection of realizations X_t of the special stochastic process $(W_t)_{0 \leq t \leq T}$ named after Norbert Wiener.

a Wiener process cannot be continuously differentiable as otherwise the quadratic variation must be zero. The fact that the quadratic variation is non-zero implies that the absolute value of an increment $W_t - W_s$ cannot be proportional to $t - s$. From here we may conclude that the paths of a Wiener process are continuous but not differentiable and therefore strongly fluctuating. The illustrative picture (see Figure 1.1) of simulated realizations of a Wiener process underlines this statement.

One of the main problems in the theory of stochastic processes is to find mathematical models that describe the evolution of a system in time and can especially be used to predict, of course not without error, the values in the future with the help of information about the process collected from the past. Here and in the sequel by 'the collected information' we mean the family of all events observable up to time t. This collection of events will be denoted by $\mathfrak{F}_t$, where we suppose that $\mathfrak{F}_t$ is a σ-algebra. It is clear that $\mathfrak{F}_s \subseteq \mathfrak{F}_t \subseteq \mathfrak{F}$. Such families of σ-algebras are referred to as a *filtration* and will be denoted by $(\mathfrak{F}_t)_{t \geq 0}$. Each stochastic process $(X_t)_{t \geq 0}$ generates a filtration by the requirement that $\mathfrak{F}_t$ is the smallest σ-algebra that contains all events $\{X_s \in I\}$ where I is any interval and $0 \leq s \leq t$. This filtration will be denoted $\sigma((X_s)_{0 \leq s \leq t})$. We call any stochastic process $(Y_t)_{t \geq 0}$ *adapted* to the filtration $(\mathfrak{F}_t)_{t \geq 0}$ (short $\mathfrak{F}_t$-adapted) if all events that may be constructed by the process up to time t belong to the class of observable events, i.e. already belong to $\mathfrak{F}_t$. The formal mathematical condition is $\sigma((Y_s)_{0 \leq s \leq t}) \subseteq \mathfrak{F}_t$ for every $t \geq 0$. If for any fixed t and any random variable Z all events $\{Z \in I\}$, $I \subseteq \mathbb{R}$, belong to $\mathfrak{F}_t$ and it holds that $\mathbb{E}Z^2 < \infty$ then there are $X_{t_1,n}, \ldots, X_{t_{m_n},n}$, $t_{i,j} \leq t$ and (measurable) functions $f_n(X_{t_1,n}, \ldots, X_{t_{m_n},n})$ such that

$$\mathbb{E}(Z - f_n(X_{t_1,n}, \ldots, X_{t_{m_n},n}))^2 \to 0.$$

We omit the proof which would require additional results from measure theory. We denote by $\mathfrak{P}_t(X)$ the class of all such random variables. $\mathfrak{P}_t(X)$ may be considered as the *past of the process* $(X_t)_{t \geq 0}$.

Example 1.1 Let $(W_t)_{t\geq 0}$ be a Wiener process and $\mathfrak{F}_t = \sigma((Y_s)_{0\leq s\leq t})$. The following processes are $\mathfrak{F}_t$-adapted $X_t = W_t^2$, $X_t = W_{0.5\cdot t}^2 + W_t^4$, $X_t = (W_t^4 / 1 + W_{0.1\cdot t}^2)$. The process W_{t+1} is not $\mathfrak{F}_t$-adapted.

We fix the interval $[0, T]$, set $\mathfrak{F}_t = \sigma((W_s)_{0\leq s\leq t})$ and denote by $\mathfrak{E}_t(W) \subseteq \mathfrak{P}_t(W)$ the collection of all elementary $\mathfrak{F}_t$-adapted processes, that is of all processes that may be written as

$$Y_t = \sum_{i=0}^{n-1} X_{t_i} I_{[t_i, t_{i+1})}(t), \quad X_{t_i} \in \mathfrak{P}_{t_i}(W), \tag{1.2}$$

where $0 = t_0 < t_1 < \cdots < t_n$ and

$$I_{[a,b)}(t) = \begin{cases} 1 & \text{if } a \leq t < b, \\ 0 & \text{if else.} \end{cases}$$

The $\mathfrak{F}_t$-adeptness of the process Y_t follows from the fact that exclusively random variables X_{t_i} with $t_i \leq t$ appear in the sum. The process Y_t is piecewise constant, it has the value X_{t_i} in $[t_i, t_{i+1})$ and jumps at t_i with a height

$$\Delta Y_{t_i} = X_{t_i} - X_{t_{i-1}}.$$

1.2
The Space of Square Integrable Random Variables

By $\mathcal{H}_2$ we denote the space of all random variables X with $\mathbb{E}X^2 < \infty$. Here and in the sequel we identify random variables X and Y that take on different values only with probability zero, i.e. $\mathbb{P}(X \neq Y) = 0$. Set

$$\langle X, Y \rangle := \mathbb{E}(XY).$$

It is not hard to see that $\langle X, Y \rangle$ satisfies all conditions that are imposed on a scalar product, i.e. $\langle X, Y \rangle$ is symmetric in X and Y, it is linear in both X and Y, and it holds that

$$\langle X, X \rangle \geq 0,$$

where the equality is satisfied if and only if $X = 0$.

The norm of a random variable X is given by

$$\|X\| = \sqrt{\mathbb{E}X^2},$$

and the distance of X and Y is the norm of $X - Y$. Recall that a sequence of random variables X_n is said to be convergent in mean square to X if $\mathbb{E}(X_n - X)^2 = 0$. Hence this type of convergence is nothing other than the norm convergence $\lim_{n\to\infty} \|X_n - X\| = 0$. A sequence of random variables $\{X_n\}$ is said to be a Cauchy sequence if

$$\lim_{n,m\to\infty} \|X_n - X_m\| = 0.$$

For a proof of the following theorem we refer to Øksendal [175].

Theorem 1.2 *To each Cauchy sequence $X_n \in \mathcal{H}_2$ there is some $X \in \mathcal{Z}_2$ with*

$$\lim_{n\to\infty} \|X_n - X\| = 0,$$

i.e. the space is complete.

It is clear that $\mathcal{H}_2$ is a linear space. As we have already equipped $\mathcal{H}_2$ with a scalar product we get, together with the completeness, that $\mathcal{H}_2$ is a Hilbert space. This fact allows us to apply methods from the Hilbert space theory to problems of probability theory.

A subset $\mathcal{T} \subseteq \mathcal{H}_2$ is called closed, if every limit X of a sequence $X_n \in \mathcal{T}$ belongs to $\mathcal{T}$ again. If $\mathcal{L} \subseteq \mathcal{H}_2$ is a closed linear subspace of $\mathcal{H}_2$ then there is some element in $\mathcal{L}$ that best approximates X.

Theorem 1.3 *If $\mathcal{L} \subseteq \mathcal{H}_2$ is a closed linear subspace of $\mathcal{H}_2$, then to each $X \in \mathcal{H}_2$ there is a random variable in $\mathcal{L}$, denoted by $\Pi_\mathcal{L} X \in \mathcal{L}$ and called the projection of X on $\mathcal{L}$, such that*

$$\inf_{Y\in\mathcal{L}} \|X - Y\| = \|X - \Pi_\mathcal{L} X\|.$$

Proof. Let $Y_n \in \mathcal{L}$ be a minimum sequence, i.e.

$$\lim_{n\to\infty} \|X - Y_n\| = \inf_{Y\in\mathcal{L}} \|X - Y\|.$$

Then Y_{m_n} is a minimum sequence again. Because

$$\left\|X - \frac{1}{2}(Y_n + Y_{m_n})\right\| \le \frac{1}{2}\|X - Y_n\| + \frac{1}{2}\|X - Y_{m_n}\|$$

$\frac{1}{2}(Y_n + Y_{m_n})$ is also a minimum sequence. Then

$$\lim_{n\to\infty}\left[\frac{1}{2}\|X - Y_n\|^2 + \frac{1}{2}\|X - Y_{m_n}\|^2 - \left\|X - \frac{1}{2}(Y_n + Y_{m_n})\right\|^2\right] = 0.$$

For any random variables U, V it holds that

$$\frac{1}{2}\|U\|^2 + \frac{1}{2}\|V\|^2 - \left\|\frac{1}{2}(U+V)\right\|^2 = \mathbb{E}\left(\frac{1}{2}U^2 + \frac{1}{2}V^2 - \left(\frac{1}{2}(U+V)\right)^2\right)$$

$$= \frac{1}{4}\mathbb{E}(U-V)^2 = \frac{1}{4}\|U - V\|^2.$$

Putting $U = X - Y_n$, $V = X - Y_{m_n}$ we arrive at

$$\frac{1}{2}\|X - Y_n\|^2 + \frac{1}{2}\|X - Y_{m_n}\|^2 - \left\|X - \frac{1}{2}(Y_n + Y_{m_n})\right\|^2$$

$$= \frac{1}{4}\|Y_n - Y_{m_n}\|^2 \to 0.$$

As m_n was an arbitrary sequence we see that Y_n is a Cauchy sequence and converges, by the completeness of $\mathcal{H}_2$, to some random variable $\Pi_\mathcal{L} X$ that belongs to $\mathcal{L}$ since $\mathcal{L}$ is closed by assumption.

Without going into detail we note that the projection $\Pi_\mathcal{L} X$ is uniquely determined in the sense that, for every $Z \in \mathcal{L}$ which also provides a best approximation, it holds that

$$\mathbb{P}(\Pi_\mathcal{L} X \neq Z) = 0. \tag{1.3}$$

The projection $\Pi_\mathcal{L} X$ can be also characterized with the help of conditions imposed on the error $X - \Pi_\mathcal{L} X$.

Corollary 1.1 *It holds that $Y = \Pi_\mathcal{L} X$ if and only if $Y \in \mathcal{L}$ and $Y - X \perp \mathcal{L}$, i.e.*

$$\langle Y - X, Z \rangle = 0 \quad \text{for every } Z \in \mathcal{L}. \tag{1.4}$$

Proof. 1. Assume $Y = \Pi_\mathcal{L} X$. Then $Y \in \mathcal{L}$ by the definition of the projection. We consider

$$g(t) = \left\| (X - Y) - tZ \right\|^2 = \|X - Y\|^2 + t^2 \|Z\|^2 - 2t \langle Y - X, Z \rangle.$$

By the definition of $\Pi_\mathcal{L} X$ the function $g(t)$ attains its minimum at $t = 0$. Hence

$$g'(0) = -2 \langle Y - X, Z \rangle = 0$$

which implies $\langle Y - X, Z \rangle = 0$.

2. If $Y \in \mathcal{L}$ satisfies (1.4) then for every $U \in \mathcal{L}$

$$\|X - U\|^2 = \|X - Y\|^2 + 2 \langle X - Y, Y - U \rangle + \|Y - U\|^2.$$

As $Z = Y - U \in \mathcal{L}$ we see that the middle term vanishes. Hence the right-hand term is minimal if and only if $U = Y$.

The simplest prediction of a random variable X is a constant value. Which value a is the best one? It is easy to see that the function

$$\varphi(a) = \mathbb{E}(X - a)^2$$

attains the minimum at $a_0 = \mathbb{E} X$. Consequently, if $\mathcal{L}$ consists of constant random variables only, then $\Pi_\mathcal{L} X = \mathbb{E} X$. This is the reason why, for any closed linear subspace, we call the projection $\Pi_\mathcal{L} X$ the *conditional expectation* given $\mathcal{L}$. In this case we tacitly assume that all constant random variables are contained in $\mathcal{L}$. As $\mathcal{L}$ is a linear space this is equivalent to the fact that $Z_0 \equiv 1 \in \mathcal{L}$. If this condition is satisfied then we write

$$\mathbb{E}(X|\mathcal{L}) := \Pi_\mathcal{L} X.$$

Choosing $Z = 1$ in (1.4) we get the following.

Conclusion 1.1 *(Iterated expectation)* It holds that

$$\mathbb{E}(\mathbb{E}(X|\mathcal{L})) = \mathbb{E}X. \tag{1.5}$$

The relation (1.4) provides the orthogonal decomposition

$$X = \Pi_\mathcal{L} X + (X - \Pi_\mathcal{L} X). \tag{1.6}$$

Here $\Pi_\mathcal{L} X$ belongs to the subspace $\mathcal{L}$ whereas the error $X - \Pi_\mathcal{L} X$ is perpendicular on $\mathcal{L}$.

The Corollary 1.1 implies that the projection operator $\Pi_\mathcal{L}$ is linear, i.e.

$$\Pi_\mathcal{L}(a_1 X_1 + a_2 X_2) = a_1 \Pi_\mathcal{L}(X_1) + a_2 \Pi_\mathcal{L}(X_2).$$

The relation (1.6) implies

$$\|\Pi_\mathcal{L} X\| \le \|X\|.$$

This inequality yields, in conjunction with the linearity, that $\Pi_\mathcal{L} X$ depends continuously on X. Indeed, $X_n \to X$ implies

$$\|\Pi_\mathcal{L} X_n - \Pi_\mathcal{L} X\| = \|\Pi_\mathcal{L}(X_n - X)\| \le \|X_n - X\| \to 0. \tag{1.7}$$

Now we collect other properties of the conditional expectation that will be used in the sequel.

Lemma 1.2 *If $\mathcal{L}$ is a closed linear subspace of $\mathcal{H}_2$ that contains the constant variables and V is a random variable such that $UV \in \mathcal{L}$ for every $U \in \mathcal{L}$ then*

$$\mathbb{E}(VX|\mathcal{L}) = V\mathbb{E}(X|\mathcal{L}).$$

Proof. The assumption $VU \in \mathcal{L}$ and (1.4) imply

$$0 = \langle X - \mathbb{E}(X|\mathcal{L}), VU \rangle$$
$$= \mathbb{E}(XV - V\mathbb{E}(X|\mathcal{L}))U = \langle XV - V\mathbb{E}(X|\mathcal{L}), U \rangle.$$

The application of Corollary 1.1 completes the proof.

The multiple application of the conditional expectation corresponds to the iterated application of projections.

Lemma 1.3 *If $\mathcal{L}_i$ is a closed linear subspace of $\mathcal{H}_2$ that contains the constant variables and $\mathcal{L}_1 \subseteq \mathcal{L}_2$ then*

$$\mathbb{E}((\mathbb{E}(X|\mathcal{L}_2))|\mathcal{L}_1) = \mathbb{E}(X|\mathcal{L}_1).$$

Proof. Set $R = \mathbb{E}(X|\mathcal{L}_1)$ and $S = \mathbb{E}(X|\mathcal{L}_2)$. Then by Corollary 1.1

$$\langle X - R, U \rangle = 0 \quad \text{for every } U \in \mathcal{L}_1,$$
$$\langle X - S, U \rangle = 0 \quad \text{for every } U \in \mathcal{L}_2.$$

The assumption $\mathcal{L}_1 \subseteq \mathcal{L}_2$ gives

$$\langle S - R, U \rangle = 0 \quad \text{for every } U \in \mathcal{L}_1.$$

Corollary 1.1 completes the proof.

Next we study the relation between the independence of random variables and the conditional expectation.

Lemma 1.4 *If X is independent of every $Z \in \mathcal{L}$ then*

$$\mathbb{E}(X|\mathcal{L}) = \mathbb{E}X.$$

Proof. The required independence implies

$$\mathbb{E}(XZ) = (\mathbb{E}X)(\mathbb{E}Z)$$
$$\mathbb{E}(X - \mathbb{E}X)Z = 0.$$

The statement follows from Corollary 1.4 and the fact that the constant random variable $\mathbb{E}X$ belongs to $\mathcal{L}$.

We say that $\mathcal{L}$ is generated by the random variables $X_1, \ldots, X_n$ if $\mathcal{L}$ consists of all possible functions (not necessarily linear) $h(X_1, \ldots, X_n)$ such that $\mathbb{E}h^2(X_1, \ldots, X_n) < \infty$. Then we write

$$\mathcal{L} = \mathcal{G}(X_1, \ldots, X_n).$$

Suppose the vector $(Y, X_1, \ldots, X_n)$ has the joint density $f(y, x_1, \ldots, x_n)$. Then

$$g(x_1, \ldots, x_n) = \int f(y, x_1, \ldots, x_n) \, dy \tag{1.8}$$

is the marginal density of $(X_1, \ldots, X_n)$ and

$$f(y|x_1, \ldots, x_n) = \frac{f(y, x_1, \ldots, x_n)}{g(x_1, \ldots, x_n)} \tag{1.9}$$

is called the conditional density of of Y given $X_1 = x_1, \ldots, X_n = x_n$.

Theorem 1.4 *Let γ be any function with $\mathbb{E}\gamma^2(Y) < \infty$ and $f(y|x_1, \ldots, x_n)$ be the conditional density of Y given $X_1 = x_1, \ldots, X_n = x_n$. Then*

$$\mathbb{E}(\gamma(Y)|\mathcal{G}(X_1, \ldots, X_n)) = \psi(X_1, \ldots, X_n),$$

where ψ is the so-called regression function that is given by

$$\psi(x_1, \ldots, x_n) = \int_{-\infty}^{+\infty} \gamma(t) f(t|x_1, \ldots, x_n) \, dt. \tag{1.10}$$

Proof. As $\mathcal{G}(X_1, \ldots, X_n)$ consists of all functions $\varphi(X_1, \ldots, X_n)$ it suffices to show that

$$\mathbb{E}(Y - \psi(X_1, \ldots, X_n))^2 \leq \mathbb{E}(Y - \varphi(X_1, \ldots, X_n))^2.$$

It holds that

$$\mathbb{E}(\gamma(Y) - \varphi(X_1,\ldots,X_n))^2$$
$$= \int \cdots \int (\gamma(y) - \varphi(x_1,\ldots,x_n))^2 f(y,x_1,\ldots,x_n)\,dy\,dx_1\cdots dx_n$$
$$= \int \cdots \int (\gamma(y) - \psi)^2 f(y,x_1,\ldots,x_n)\,dy\,dx_1\cdots dx_n$$
$$+ 2 \int \cdots \int (\gamma(y) - \psi)(\psi - \varphi) f(y,x_1,\ldots,x_n)\,dx\,dy_1\cdots dx_n$$
$$+ \int \cdots \int (\varphi - \psi)^2 f(y,x_1,\ldots,x_n)\,dx\,dx_1\cdots dx_n.$$

To calculate the middle term we note that $\varphi - \psi$ does not depend on y. Hence

$$\int \cdots \int (\gamma(x) - \psi)(\psi - \varphi) f(y,x_1,\ldots,x_n)\,dx\,dy_1\cdots dx_n$$
$$= \int \cdots \int \left((\psi - \varphi) \int (\gamma(y) - \psi) f(y|x_1,\ldots,x_n)\,dy \right)$$
$$\times g(x_1,\ldots,x_n)\,dx_1\cdots dx_n$$
$$= 0$$

because of (1.10). Hence

$$\mathbb{E}(\gamma(X) - \varphi(X_1,\ldots,X_n))^2$$
$$= \mathbb{E}(\gamma(X) - \psi(X_1,\ldots,X_n))^2 + \mathbb{E}(\varphi(X_1,\ldots,X_n) - \psi(X_1,\ldots,X_n))^2.$$

The term on the right-hand side becomes minimal if and only if $\varphi(X_1,\ldots,X_n) - \psi(X_1,\ldots,X_n) = 0$ which proves the statement.

Let $(X_t)_{t\geq 0}$ be a stochastic process such that all finite dimensional distributions of $X_{t_1},\ldots,X_{t_n}$ have a density that we will denote by $f_{t_1,\ldots,t_n}(x_1,\ldots,x_n)$, where $t_1 < t_2 < \cdots < t_n$. By

$$f_{t_n|t_1,\ldots,t_{n-1}}(x_n|x_1,\ldots,x_{n-1}) = \frac{f_{t_1,\ldots,t_n}(x_1,\ldots,x_n)}{f_{t_1,\ldots,t_{n-1}}(x_1,\ldots,x_{n-1})} \quad (1.11)$$

we denote the conditional density of X_{t_n} given $X_{t_1} = x_1,\ldots,X_{t_{n-1}} = x_{n-1}$. We call a stochastic process a *Markov process* if the conditional density depends only on the values of the process at the last moment of the past, i.e.

$$f_{t_n|t_1,\ldots,t_{n-1}}(x_n|x_1,\ldots,x_{n-1}) = f_{t_n|t_{n-1}}(x_n|x_{n-1}). \quad (1.12)$$

If $(X_t)_{t\geq 0}$ is a Markov process, then by Theorem 1.4, for every $t_1 < t_2 < \cdots < t_n = t$ and $h > 0$

$$\mathbb{E}(\gamma(X_{t+h})|\mathcal{G}(X_{t_1},\ldots,X_{t_n})) = \mathbb{E}(\gamma(X_{t+h})|\mathcal{G}(X_t)). \quad (1.13)$$

Conversely, if the last condition holds for every γ then

$$\int \gamma(x_n) f_{t_n|n-1}(x_n|x_{n-1})\,dx_n = \int \gamma(x_n) f_{t_n|t_1,\ldots,t_{n-1}}(x_n|x_1,\ldots,x_{n-1})\,dx_n, \quad (1.14)$$

As γ is arbitrary the relation (1.14) yields

$$f_{t_n|n-1}(x_n|x_{n-1}) = f_{t_n|t_1,\ldots,t_{n-1}}(x_n|x_1,\ldots,x_{n-1}).$$

Recall that $\mathfrak{P}_t(X)$ is the smallest closed subspace of $\mathcal{H}_2$ that contains all subspaces $\mathcal{G}(X_{t_1},\ldots,X_{t_m})$, where $t_1 < t_2 < \cdots < t_n$. This means that $\mathfrak{P}_t(X)$ consists of all random variables that are either functions of random variables from the past or a limit of such random variables. Hence by the continuity of the scalar product

$$\gamma(X_{t+h}) - \mathbb{E}(\gamma(X_{t+h})|\mathcal{G}(X_t)) \perp Z, \quad Z \in \mathfrak{P}_t(X),$$

if and only if

$$\gamma(X_{t+h}) - \mathbb{E}(\gamma(X_{t+h})|\mathcal{G}(X_t)) \perp Z, \quad Z \in \mathcal{G}(X_{t_1},\ldots,X_{t_n})$$

for any $t_1 < t_2 < \cdots < t_n \leq t$. As $\mathbb{E}(\gamma(X_{t+h})|\mathcal{G}(X_t)) \in \mathfrak{P}_t(X)$ then from Corollary 1.1 we get the following theorem.

Theorem 1.5 *A stochastic process $(X_t)_{t\geq 0}$ is a Markov process if and only if*

$$\mathbb{E}(\gamma(X_{t+h})|\mathfrak{P}_t(X)) = \mathbb{E}(\gamma(X_{t+h})|\mathcal{G}(X_t))$$

for every function γ with $\mathbb{E}\gamma^2(X_{t+h}) < \infty$. This condition is equivalent to (1.13) for any $t_1 < t_2 < \cdots < t_n = t$.

Now we present a general construction scheme for Markov processes.

Theorem 1.6 *Let $(X_t)_{t\geq 0}$ be a stochastic process and $V(x,t,h)$ for $t,h > 0$, $x \in \mathbb{R}$ a family of random variables such that:*

1) $V(x,t,h)$ *is independent of every* $Z \in \mathfrak{P}_t(X)$ *for every* $t,h > 0$, $x \in \mathbb{R}$
2) $X_{t+h} = V(X_t, t, h).$

Then $(X_t)_{t\geq 0}$ is a Markov process.

Proof. Assume $\mathbb{E}\gamma^2(X_{t+h}) < \infty$ and fix $t_1 < \cdots < t_n = t$. Let $(\Omega, \mathfrak{F}, \mathbb{P})$ be the basic probability space. For fixed $t, h > 0$ the random variable $\gamma(V(x,t,h))$ is a function of x and ω, say $\Gamma(x,\omega)$. Without proof we use the fact that each such function can be approximated by linear combinations of the products of functions $v(x)V(\omega)$ in the sense that, for suitably chosen $v_{i,n}$ and $V_{i,n}$ that are independent of every $Z \in \mathfrak{P}_t(X)$

$$\mathbb{E}\left(\sum_{i=1}^n v_{i,n}(X_t)V_{i,n} - \gamma(V(X_t,t,h))\right)^2 \to 0.$$

In view of Theorem 1.5 and Corollary 1.1 we have to show that

$$\mathbb{E}(\gamma(X_{t+h}) - \mathbb{E}(\gamma(X_{t+h})|\mathcal{L}(X_{t_1},\ldots,X_{t_n})))Z = 0$$

for every $Z \in \mathcal{G}(X_{t_1}, \ldots, X_{t_n})$. Due to the continuity of the projection, see (1.7), it suffices to show that

$$\mathbb{E}\left(\sum_{i=1}^{n} v_{i,n}(X_t) V_{i,n} - \mathbb{E}\left(\sum_{i=1}^{n} v_{i,n}(X_t) V_{i,n} | \mathcal{G}(X_{t_1}, \ldots, X_{t_n})\right)\right) Z = 0. \quad (1.15)$$

To this end we note that $v_{i,n}(X_t) \in \mathcal{G}(X_{t_1}, \ldots, X_{t_n})$. Hence by Lemma 1.2

$$\mathbb{E}(v_{i,n}(X_t) V_{i,n} | \mathcal{G}(X_{t_1}, \ldots, X_{t_n})) = v_{i,n}(X_t) \mathbb{E}(V_{i,n} | \mathcal{G}(X_{t_1}, \ldots, X_{t_n})).$$

Lemma 1.4 and the independence of $V_{i,n}$ of all $X_{t_1}, \ldots, X_{t_n}$ implies

$$\mathbb{E}(V_{i,n} | \mathcal{G}(X_{t_1}, \ldots, X_{t_n})) = \mathbb{E}(V_{i,n}).$$

This yields

$$\mathbb{E}(Z(\mathbb{E}(v_{i,n}(X_t) V_{i,n} | \mathcal{L}(X_{t_1}, \ldots, X_{t_n})))) = \left[\mathbb{E}(Z v_{i,n}(X_t))\right]\left[\mathbb{E}(V_{i,n})\right]. \quad (1.16)$$

Otherwise $V_{i,n}$ is independent of $X_{t_1}, \ldots, X_{t_n}$ and therefore independent of $Z v_{i,n}(X_t)$. This yields

$$\mathbb{E}(V_{i,n} v_{i,n}(X_t) Z) = \left[\mathbb{E}(Z v_{i,n}(X_t))\right]\left[\mathbb{E}(V_{i,n})\right]. \quad (1.17)$$

The relations (1.16) and (1.17) imply (1.15) and thus the statement.

1.3
The Ito Integral and the Ito Formula

The aim of this section is to introduce and study the concept of the Ito integral which is an integral where, instead of the classical Riemann integral, the values of the function to be integrated are not weighted according to the length of the interval from the chosen partition. Instead we weight this values by increments of a Wiener process. A first idea could be to set

$$\int_a^b X_s \, dW_s := \int_a^b X_s W_s' \, ds. \quad (1.18)$$

But we know from the discussion after Theorem 1.1 that the derivative W_s' does not exist. So this fact excludes this method. Ito succeeded in constructing an integral of the above type by starting as a first step with elementary processes and in a second step by extending the integral to a larger class of processes.

Recall that by (1.2) every elementary adapted process $X \in \mathfrak{E}(W)$ can be written as

$$Y_t = \sum_{i=0}^{n-1} X_{t_i} I_{[t_i, t_{i+1})}(t), \quad X_{t_i} \in \mathfrak{P}_{t_i}(W).$$

We set

$$\int_0^T X_s \, dW_s := \sum_{i=0}^{n-1} X_{t_i}(W_{t_{i+1}} - W_{t_i}).$$

A first immediate property of this integral concept is its linearity, i.e.

$$\int_0^T (c_1 X_s^{(1)} + c_2 X_s^{(2)})\, dW_s = c_1 \int_0^T X_s^{(1)}\, dW_s + c_2 \int_0^T X_s^{(2)}\, dW_s.$$

Another property that makes Hilbert space arguments applicable is the so-called isometry property.

Theorem 1.7 *If $X^{(1)}, X^{(2)} \in \mathfrak{E}(W)$ then*

$$\left\langle \int_0^T X_s^{(1)}\, dW_s, \int_0^T X_s^{(2)}\, dW_s \right\rangle = \int_0^T \left\langle X_s^{(1)}, X_s^{(2)} \right\rangle ds. \tag{1.19}$$

Proof. A possible change to a joint refinement shows that the two elementary processes $X_s^{(1)}$ and $X_s^{(2)}$ can be represented about the same partition. Hence

$$Y_t^{(j)} = \sum_{i=0}^{n} X_{t_i}^{(j)} I_{[t_i, t_{i+1})}(t),$$

with some $X_{t_i}^{(j)} \in \mathfrak{P}_{t_i}(W)$. Then

$$\left\langle \int_0^T X_s^{(1)}\, dW_s, \int_0^T X_s^{(2)}\, dW_s \right\rangle$$

$$= \sum_{i,j=0}^{n-1} \mathbb{E}(X_{t_i}^{(1)} X_{t_j}^{(2)} (W_{t_{i+1}} - W_{t_i})(W_{t_{j+1}} - W_{t_j})).$$

Let $i \neq j$ and for example $t_i > t_j$. The independence of the increments implies that $W_{t_{i+1}} - W_{t_i}$ and $X_{t_i}^{(1)} X_{t_j}^{(2)}(W_{t_{j+1}} - W_{t_j})$ are independent. Consequently $\mathbb{E}(W_{t_{i+1}} - W_{t_i}) = 0$ implies that the mixed terms vanish. This yields

$$\left\langle \int_0^T X_s^{(1)}\, dW_s, \int_0^T X_s^{(2)}\, dW_s \right\rangle = \sum_{i=0}^{n-1} \mathbb{E}(X_{t_i}^{(1)} X_{t_i}^{(2)} (W_{t_{i+1}} - W_{t_i})^2).$$

Because of $X_{t_i}^{(1)} X_{t_i}^{(2)} \in \mathfrak{P}_{t_i}(W)$ this random variable from the past is independent of $(W_{t_{i+1}} - W_{t_i})^2$ which implies

$$\mathbb{E}[(W_{t_{i+1}} - W_{t_i})^2 X_{t_i}^{(1)} X_{t_i}^{(2)}] = \mathbb{E}[(W_{t_{i+1}} - W_{t_i})^2] \mathbb{E}[X_{t_i}^{(1)} X_{t_i}^{(2)}]$$

$$= (t_{i+1} - t_i)[\mathbb{E}(X_{t_i}^{(1)} X_{t_i}^{(2)})].$$

Hence

$$\left\langle \int_0^T X_s^{(1)}\, dW_s, \int_0^T X_s^{(2)}\, dW_s \right\rangle = \sum_{i=0}^{n-1} [\mathbb{E}(X_{t_i}^{(1)} X_{t_i}^{(2)})](t_{i+1} - t_i)$$

$$= \int_0^T \left\langle X_s^{(1)}, X_s^{(2)} \right\rangle ds.$$

We denote by $\mathfrak{L}_2(W)$ the set of all $\mathfrak{P}_t(W)$-adapted processes X with

$$\int_0^T \mathbb{E} X_t^2 \, dt < \infty.$$

In the sequel we use the fact that every $X \in \mathfrak{L}_2(W)$ can be approximated by elementary processes $X^{(n)} \in \mathfrak{E}(W)$ in the sense that

$$\lim_{n \to \infty} \int_0^T \mathbb{E}(X_t^{(n)} - X_t)^2 \, dt = 0. \tag{1.20}$$

We refer to Øksendal [175] for a proof.
The relation (1.20) provides

$$\lim_{n,m \to \infty} \int_0^T \mathbb{E}(X_t^{(n)} - X_t^{(m)})^2 \, dt = 0,$$

which, together with the isometry property (1.19), leads to

$$\lim_{n,m \to \infty} \mathbb{E}\left(\int_0^T X_t^{(n)} \, dW_t - \int_0^T X_t^{(m)} \, dW_t\right)^2$$
$$= \lim_{n,m \to \infty} \int_0^T \mathbb{E}(X_t^{(n)} - X_t^{(m)})^2 \, dt = 0.$$

This means that the sequence of random variables $\int_0^T X_t^{(n)} \, dW_t$ is a Cauchy sequence and converges therefore to a random variable that will be denoted by

$$\int_0^T X_t \, dW_t.$$

This random variable is independent of the choice of the approximating sequence $X_t^{(n)}$ and is called the *Ito integral*. The continuity of the scalar product shows that the above isometry property is still valid for the larger class of processes $X \in \mathfrak{L}_2(W)$.

Theorem 1.8 *If $X, Y \in \mathfrak{L}_2(W)$ then*

$$\int_0^T (aX_t + bY_t) \, dW_t = a \int_0^T X_t \, dW_t + b \int_0^T Y_t \, dW_t$$
$$\left\langle \int_0^T X_t \, dW_t, \int_0^T Y_t \, dW_t \right\rangle = \int_0^T \langle X_t, Y_t \rangle \, dt.$$

Letting the upper bound in the integral be variable we may introduce the new stochastic process $\int_0^t X_s \, dW_s$ which has been constructed exclusively with the help of random variables from $\mathfrak{P}_t(W)$. Thus we see that the new process

$$Y_t = \int_0^t X_s \, dW_s \tag{1.21}$$

again belongs to $\mathfrak{L}_2(W)$. This process has an important projection property.

Theorem 1.9 *If $t_1 < t_2$ then Y_t in (1.21) satisfies*

$$\mathbb{E}(Y_{t_2}|\mathfrak{P}_{t_1}(W)) = Y_{t_1} \tag{1.22}$$

$$\mathbb{E} Y_{t_1} = \mathbb{E} Y_{t_2} = 0. \tag{1.23}$$

Proof. By the linearity of the Ito integral and the continuity of the projection we have to prove the statement only for elementary processes of the type $X_t = Z I_{[a,b)}(t)$ where $Z \in \mathfrak{P}_a(W)$. Then

$$Y_t = \int_0^t X_s \, dW_s = Z(W_{b \wedge t} - W_a),$$

where $b \wedge t = \min(b, t)$. This shows that Y_t does not depend on t for $t < a$ and $t > b$. Hence we have only to consider the case $a \leq t_1 < t_2 \leq b$. Then $Y_{t_2} - Y_{t_1} = Z(W_{t_2} - W_{t_1})$ and

$$\mathbb{E}(Y_{t_2} - Y_{t_1}|\mathfrak{P}_{t_1}(W)) = \mathbb{E}(Z(W_{t_2} - W_{t_1})|\mathfrak{P}_{t_1}(W)).$$

As $Z \in \mathfrak{P}_a(W) \subseteq \mathfrak{P}_{t_1}(W)$ we may apply Lemma 1.2 and can take Z out of the conditional expectation

$$\mathbb{E}(Z(W_{t_2} - W_{t_1})|\mathfrak{P}_{t_1}(W)) = Z\mathbb{E}((W_{t_2} - W_{t_1})|\mathfrak{P}_{t_1}(W)).$$

The independence of $W_{t_2} - W_{t_1}$ and the random variables from $\mathfrak{P}_{t_1}(W)$ together with Lemma 1.4 yield

$$\mathbb{E}((W_{t_2} - W_{t_1})|\mathfrak{P}_{t_1}(W)) = 0$$

and therefore

$$\mathbb{E}(Y_{t_2} - Y_{t_1}|\mathfrak{P}_{t_1}(W)) = 0.$$

Because of $Y_{t_1} \in \mathfrak{P}_{t_1}(W)$ we obtain $\mathbb{E}(Y_{t_2}|\mathfrak{V}_{t_1}(W)) = Y_{t_1}$ which is the first statement. The relation (1.5) implies $\mathbb{E} Y_{t_2} = \mathbb{E} Y_{t_1}$ for every $0 \leq t_1 \leq t_2$. As $Y_0 = 0$ we get (1.23).

Stochastic processes that satisfy (1.22) are called martingales in probability theory.

Now we introduce a class of processes that turns out to be useful in order to model the evolution of a time-dependent phenomenon. A stochastic process X is called an *Ito process*, if

$$X_t = X_0 + \int_0^t A_s \, ds + \int_0^t B_s \, dW_s, \tag{1.24}$$

where $A, B \in \mathfrak{L}_2(W)$. It is not hard to show that the quadratic variation of $\int_0^t A_s \, ds$ is zero, so that $\int_0^t A_s \, ds$ is a smooth part of X_t that plays the role of a drift. The second component $\int_0^t B_s \, dW_s$ is irregular as the quadratic variation is

$$[X]_t = \int_0^t B_s^2 \, ds \tag{1.25}$$

which can be easily shown and does not vanish. We also write

$$dX_s = A_s \, ds + B_s \, dW_s. \tag{1.26}$$

instead of (1.24). Ito processes admit the following interpretation. For fixed $h > 0$ the increment $X_{t+h} - X_t$ is approximately given by

$$X_{t+h} - X_t \approx A_t h + B_t(W_{t+h} - W_t). \tag{1.27}$$

The first term $A_t h$ is a drift with a slope which is governed by values from the past. The factors in the product $B_t(W_{t+h} - W_t)$ are independent where $W_{t+h} - W_t$ is normally distributed with expectation zero and variance h. If the values in the past are fixed then $B_t(W_{t+h} - W_t)$ has the variance $B_t^2 h$. This mean that $B_s \, dW_s$ is a diffusion term.

Diffusion processes are special Ito processes. They are characterized by the fact that the drift coefficient A_t as well as the diffusion coefficient B_t only depend on the last state of the process. This means that

$$A_t = a(t, X_t), \quad \text{and} \quad B_t = b(t, X_t),$$

with some $a(t, x)$ and $b(t, x)$. Hence

$$X_t = X_0 + \int_0^t a(s, X_s) \, ds + \int_0^t b(s, X_s) \, dW_s. \tag{1.28}$$

This is an integral equation for X_t, which can formally be written as a differential equation, often used as a basic equation of motion in physics and named after Langevin

$$\dot{X}_t = a(t, X_t) + b(t, X_t) \dot{W}_t. \tag{1.29}$$

The problem is that $\dot{W}_t \equiv dW_t/dt$ does not exist as we have already pointed out by showing that the paths of W_t are not differentiable.

The representation (1.28) raises the question of for which a, b the integral equation has a solution and under which conditions this solution is unique. In the sense of an initial value problem the value X_0 has to be fixed. Necessary and sufficient conditions that guarantee the existence and uniqueness of a solution of this initial value problem can be found in many books, e.g. [30, 57, 91, 104, 175].

Often the starting point X_0 is a deterministic value, say x_0. To indicate the dependence on x_0 we denote the corresponding process by X_{t,x_0}. Hence

$$X_{t,x_0} = x_0 + \int_0^t a(s, X_{s,x}) \, ds + \int_0^t b(s, X_{s,x}) \, dW_s, \tag{1.30}$$

and

$$X_{t+h,x} - X_{t,x} = \int_t^{t+h} a(s, X_{s,x}) \, ds + \int_t^{t+h} b(s, X_{s,x}) \, dW_s.$$

For every fixed x the random variable

$$V(x,t,h) = x + \int_t^{t+h} a(s, X_{s,x})\,ds + \int_t^{t+h} b(s, X_{s,x})\,dW_s$$

is independent of of the random variables from $\mathfrak{P}_t(W)$. If (1.30) has a unique solution then

$$X_{t+h,x_0} = V(X_{t,x_0}, t, h).$$

From Theorem 1.6 we get the Markov property.

Theorem 1.10 *If the equation*

$$X_{t,x} = x + \int_s^t a(\tau, X_{\tau,x})\,d\tau + \int_s^t b(\tau, X_{\tau,x})\,dW_\tau$$

has a unique solution for every x and s then the process starting at x_0 being defined as the solution of

$$X_{t,x_0} = x_0 + \int_0^t a(s, X_{s,x_0})\,ds + \int_0^t b(s, X_{s,x_0})\,dW_s$$

is a Markov process. It is called homogeneous, if a and b are independent of s, hence

$$X_{t,x_0} = x_0 + \int_0^t a(X_{s,x_0})\,ds + \int_0^t b(X_{s,x_0})\,dW_s.$$

The class of Ito processes is closed with respect to the application of smooth functions, i.e. $u(t, X_t)$ is again a Ito process whose drift and diffusion coefficient can be given explicitly.

Theorem 1.11 *(Ito formula)* *Suppose $A, B \in \mathcal{L}_2(W)$ and assume*

$$dX_s = A_s\,ds + B_s\,dW_s.$$

If $u : [0, \infty) \times \mathbb{R} \to \mathbb{R}$ is twice continuously differentiable then

$$du(t, X_t) = \frac{\partial u}{\partial t}(t, X_t)\,dt + \frac{\partial u}{\partial x}(t, X_t)\,dX_t + \frac{1}{2}\frac{\partial^2 u}{\partial x^2}(t, X_t)\cdot(dX_t)^2, \tag{1.31}$$

where $(dX_t)^2 = dX_t \cdot dX_t$ is to be calculated according to the following rules

$$dt \cdot dt = dt \cdot dW_t = dW_t \cdot dt = 0, \tag{1.32}$$

$$dW_t \cdot dW_t = dt. \tag{1.33}$$

Proof. We give only a sketch of the proof. Further details can be found in Øksendal [175] or many other textbooks on stochastic differential equations such as Chorin and Held [30] and Karatzas and Shreve [91].

Suppose $\mathfrak{z}_n = \{t_{0,n}, \ldots, t_{n,n}\}$, $t_{0,n} = 0$, $t_{n,n} = t$ is a sequence of partitions of $[0, t]$ with $\delta(\mathfrak{z}_n) \to 0$. Then

$$u(t, X_t) - u(t, X_0) = \sum_{l=1}^{n} [u(t_{l,n}, X_{t_{l,n}}) - u(t_{l-1,n}, X_{t_{l-1,n}})].$$

and by the Taylor expansion

$$u(t, X_{t_{l,n}}) - u(t, X_{t_{l-1,n}}) = \frac{\partial u}{\partial t}(t_{l,n}, X_{t_{l,n}})(t_{l,n} - t_{l-1,n})$$
$$+ \frac{\partial u}{\partial x}(t_{l,n}, X_{t_{l,n}})(X_{t_{l,n}} - X_{t_{l-1,n}})$$
$$+ \frac{1}{2}\frac{\partial^2 u}{\partial x^2}(t_{l,n}, X_{t_{l,n}})(X_{t_{l,n}} - X_{t_{l-1,n}})^2 + R_{l,n},$$

where

$$\sum_{l=1}^{n} R_{l,n} \xrightarrow{\mathbb{P}} 0$$

can be shown. The sum of the first terms of the above decomposition can be shown to tend to

$$\int_0^t \frac{\partial u}{\partial t}(s, X_s)\,ds,$$

as $n \to \infty$. Similarly, by $dX_s = A_s\,ds + B_s\,dW_s$ the sum of the second terms tends to

$$\int_0^t \frac{\partial u}{\partial x}(s, X_s) A_s\,ds + \int_0^t \frac{\partial u}{\partial x}(s, X_s) B_s\,dW_s.$$

Using $dX_s = A_s\,ds + B_s\,dW_s$ again we see that the sum of the third terms consists of three parts. The first one is

$$\sum_{l=1}^{n} \frac{1}{2}\frac{\partial^2 u}{\partial x^2}(t_{l,n}, X_{t_{l,n}}) A_{t_{l-1,n}}^2 (t_{l,n} - t_{l-1,n})^2. \tag{1.34}$$

Assuming, for simplicity, a boundedness of $\frac{\partial^2 u(t,x)}{\partial x^2} A_t^2$, this sum does not exceed

$$c \sum_{l=1}^{n} (t_{l,n} - t_{l-1,n})^2 \leq c\delta(\mathfrak{z}_n) \cdot t \to 0$$

as $\delta(\mathfrak{z}_n) = \max_{1 \leq l \leq n} |t_{l,n} - t_{l-1,n}| \to 0$. Hence (1.34) tends stochastically to zero. The second part is the mixed term

$$\sum_{l=1}^{n} \frac{\partial^2 u}{\partial x^2}(t_{l,n}, X_{t_{l,n}}) A_{t_{l-1,n}} B_{t_{l-1,n}} (t_{l,n} - t_{l-1,n})(W_{t_{l,n}} - W_{t_{l,n}}).$$

If $\frac{\partial^2 u}{\partial x^2}(t_{l,n}, X_{t_{l,n}}) A_{t_{l-1,n}} B_{t_{l-1,n}}$ is bounded then the expectation of the absolute value can be estimated by

$$c \sum_{l=1}^{n} (t_{l,n} - t_{l-1,n}) \mathbb{E} |W_{t_{l,n}} - W_{t_{l-1,n}}|.$$

Using the inequality $\mathbb{E}|Z| \leq (\mathbb{E}Z^2)^{1/2}$ valid for any random variable Z we get the bound

$$c \sum_{l=1}^{n} (t_{l,n} - t_{l-1,n})(t_{l,n} - t_{l-1,n})^{1/2} \to 0$$

where we used $\max_{1 \leq l \leq n} |t_{l,n} - t_{l-1,n}| \to 0$ again. The sum over the third parts

$$\frac{1}{2} \sum_{l=1}^{n} \frac{\partial^2 u}{\partial x^2}(t_{l,n}, X_{t_{l,n}}) B_{t_{l-1,n}}^2 (W_{t_{l,n}} - W_{t_{l-1,n}})^2$$

does not disappear. By similar arguments that have been used while studying the quadratic variation of the Wiener process one can show that the last sum tends to

$$\frac{1}{2} \int_0^t \frac{\partial^2 u}{\partial x^2}(s, X_s) B_s^2 \, ds,$$

which completes the sketch of the proof.

We now consider special cases. Suppose X_t is a diffusion process already defined by (1.28)

$$X_t = X_0 + \int_0^t a(s, X_s) \, ds + \int_0^t b(s, X_s) \, dW_s \tag{1.35}$$

The transformation rules (1.32) and (1.33) give

$$u(t, X_t) = u(0, X_0)$$
$$+ \int_0^t \left[\frac{\partial u(s, X_s)}{\partial s} \, ds + \frac{\partial u(s, X_s)}{\partial x} a(s, X_s) + \frac{1}{2} \frac{\partial^2 u(s, X_s)}{\partial x^2} b^2(s, X_s) \right] ds$$
$$+ \int_0^t \frac{\partial u(s, X_s)}{\partial x} b(s, X_s) \, dW_s. \tag{1.36}$$

If u depends only on x then

$$u(X_t) = u(X_0)$$
$$+ \int_0^t \left[u'(X_s) a(s, X_s) + \frac{1}{2} u''(X_s) b^2(s, X_s) \right] ds$$
$$+ \int_0^t u'(X_s) b(s, X_s) \, dW_s. \tag{1.37}$$

Corollary 1.2 *If X_t is a solution of $dX_t = a(t, X_t) \, dt + b(t, X_t) \, dW_t$ then*

$$\mathbb{E}(u(X_t) - u(X_0)) = \mathbb{E} \int_0^t \left[u'(X_s) a(s, X_s) + \frac{1}{2} u''(X_s) b^2(s, X_s) \right] ds.$$

Proof. Theorem 1.9 shows that

$$\mathbb{E} \int_0^t \frac{\partial u(X_s)}{\partial x} b(s, X_s) \, dW_s$$

is independent of t and is therefore zero as the expression vanishes for $t = 0$.

To conclude this section we note that the diffusion process X_t in (1.28) reduces to the Wiener process in the special case $a = 0, b = 1$. But in the general case one may replace the probability measure $\mathbb{P}$ by another distribution Q (Girsanov transformation) such that the process X_t becomes a Wiener process with respect to Q.

1.4
The Kolmogorov Differential Equation and the Fokker–Planck Equation

We consider the diffusion process defined by the stochastic differential equation

$$dX_t = a(X_t) \, dt + b(X_t) \, dW_t. \tag{1.38}$$

We know from Theorem 1.10 that this process is a Markov process. As both a and b do not depend on t the process is homogeneous. Let $f(t, x, y)$ be the family of transition densities, i.e. $f(t, x, \cdot)$ is the conditional density of X_t given $X_0 = x$. If the process $X_{t,x}$ starts in $t = 0$ at x then $f(t, x, \cdot)$ is the probability density of $X_{t,x}$. This family of densities satisfies the Chapman–Kolmogorov equation

$$f(s + t, x, y) = \int f(s, x, z) f(t, z, y) \, dz, \quad 0 \leq s, t. \tag{1.39}$$

Let $\mathbb{C}_b$ be the space of all bounded and measurable functions $\mathbb{R}$ and denote by $\mathbb{C}_0^2$ the space of all twice continuously differentiable functions that vanish outside of some finite interval that may depend on the concrete function under consideration. For $u \in \mathbb{C}_b$ we set

$$(T_t u)(x) = \int u(y) f(t, x, y) \, dy$$
$$= \mathbb{E} u(X_{t,x}).$$

It is easy to see that $T_t u \in \mathbb{C}_b$. The Chapman–Kolmogorov equation implies the semigroup property, that is,

$$T_t T_s = T_{s+t}. \tag{1.40}$$

Putting $X_0 = x$ in Corollary 1.2 we get, for any $u \in \mathbb{C}_0^2$,

$$(T_t u)(x) = u(x) + \mathbb{E} \int_0^t \left[u'(X_{s,x}) a(X_{s,x}) + \frac{1}{2} u''(X_{s,x}) b^2(X_{s,x}) \right] ds$$

and therefore

$$\frac{(T_h u)(x) - u(x)}{h} = \mathbb{E} \frac{1}{h} \int_0^h \left[u'(X_{s,x}) a(X_{s,x}) + \frac{1}{2} u''(X_{s,x}) b^2(X_{s,x}) \right] ds.$$

Each diffusion process can be shown to be continuous. Hence $\lim_{s \downarrow 0} X_{s,x} = x$ and

$$\lim_{h \downarrow 0} \frac{1}{h} \int_0^h \left[u'(X_{s,x}) a(X_{s,x}) + \frac{1}{2} u''(X_{s,x}) b^2(X_{s,x}) \right]$$

$$= a(x) \frac{\partial u(x)}{\partial x} + \frac{1}{2} b^2(x) \frac{\partial^2 u(x)}{\partial x^2} = (Au)(x),$$

where A is the differential operator

$$A = a(x) \frac{\partial}{\partial x} + \frac{1}{2} b^2(x) \frac{\partial^2}{\partial x^2}. \tag{1.41}$$

This differential operator is the infinitesimal operator of the semigroup in the sense that

$$(Au)(x) = \lim_{h \downarrow 0} \frac{(T_h u)(x) - u(x)}{h}.$$

Let I be the identical operator. Then we obtain from the semigroup property (1.40) that

$$\lim_{h \downarrow 0} \frac{T_{t+h} u - T_t u}{h} = \lim_{h \downarrow 0} T_t \left(\frac{(T_h - I) u}{h} \right)$$

$$= T_t A u. \tag{1.42}$$

Similarly,

$$\lim_{h \downarrow 0} \frac{T_{t+h} u - T_t u}{h} = \lim_{h \downarrow 0} \left(\frac{(T_h - I)}{h} \right) T_t u$$

$$= A T_t u. \tag{1.43}$$

Thus we have obtained the following result.

Theorem 1.12 *If $X_{t,x}$ is the solution of*

$$dX_{t,x} = a(X_{t,x}) \, dt + b(X_{t,x}) \, dW_t,$$

$$X_{0,x} = x$$

and $u \in \mathbb{C}_0^2$, then

$$u(t, x) = (T_t u)(x)$$

$$= \mathbb{E} u(X_{t,x}) = \int u(y) f(t, x, y) \, dy$$

satisfies the Kolmogorov forward equation

$$\frac{\partial u(t,x)}{\partial t} = (T_t A u)(x)$$
$$= \int \left[a(y) \frac{\partial u(y)}{\partial y} + \frac{1}{2} b^2(y) \frac{\partial^2 u(y)}{\partial y^2} \right] f(t,x,y) \, dy \quad (1.44)$$

and the Kolmogorov backward equation

$$\frac{\partial u(t,x)}{\partial t} = A(T_t u)(x) \quad (1.45)$$
$$= a(x) \frac{\partial u(t,x)}{\partial x} + \frac{1}{2} b^2(x) \frac{\partial^2 u(t,x)}{\partial x^2}.$$

Proof. The statement (1.44) follows from (1.42). Similarly, (1.45) follows from (1.43).

Now we establish differential equations for the transition densities. To this end we apply integration by parts. If $u, v \in \mathbb{C}_0^2$, and both a and b are twice continuously differentiable then

$$\int \left[a(x) \frac{du(x)}{dx} \right] [v(x)] \, dx = -\int \left[\frac{d(a(x)v(x))}{dx} \right] [u(x)] \, dx,$$
$$\int \left[b^2(x) \frac{d^2 u(x)}{dx^2} \right] [v(x)] \, dx = \int \left[\frac{d^2(b^2(x)v(x))}{dx^2} \right] [u(x)] \, dx.$$

The application to (1.44) yields

$$\frac{\partial u(t,x)}{\partial t} = \int \left[a(y) \frac{\partial u(y)}{\partial y} + \frac{1}{2} b^2(y) \frac{\partial^2 u(y)}{\partial y^2} \right] f(t,x,y) \, dy$$
$$= \int \left[-\frac{\partial (a(y) f(t,x,y))}{\partial y} + \frac{1}{2} \frac{\partial^2 (b^2(y) f(t,x,y))}{\partial y^2} \right] u(y) \, dy.$$

Otherwise

$$\frac{\partial u(t,x)}{\partial t} = \frac{\partial}{\partial t} \int u(y) f(t,x,y) \, dy$$
$$= \int u(y) \frac{\partial}{\partial t} f(t,x,y) \, dy. \quad (1.46)$$

Hence for every $u \in \mathbb{C}_0^2$

$$\int \left[\frac{\partial}{\partial t} f(t,x,y) + \frac{\partial (a(y) f(t,x,y))}{\partial y} - \frac{1}{2} \frac{\partial^2 (b^2(y) f(t,x,y))}{\partial y^2} \right] u(y) \, dy = 0. \quad (1.47)$$

Let $u \in C_0^2$ be any probability density with support, e.g.

$$u(t) = ct^2(1-t)^2,$$

where c is determined by

$$\int_0^1 u(t)\,dt = 1.$$

Put for every fixed z

$$u_n(t) = nu(n(t-z)). \tag{1.48}$$

For large n the sequence $u_n(t)$ is concentrated around z. When ψ is twice continuously differentiable we get

$$\int \psi(t) u_n(t)\,dt = \int \psi(t) nu(n(t-z))\,dt$$
$$= \int \psi\left(z + \frac{s}{n}\right) u(s)\,ds \to \int \psi(z) u(s)\,ds = \psi(z).$$

The application of this statement to (1.47) yields the so-called *forward Fokker–Planck equation*

$$\frac{\partial}{\partial t} f(t,x,z) = -\frac{\partial(a(z)f(t,x,z))}{\partial z} + \frac{1}{2}\frac{\partial^2(b^2(z)f(t,x,z))}{\partial z^2}. \tag{1.49}$$

Similarly, the relation (1.45) yields

$$\frac{\partial u(t,x)}{\partial t} = a(y)\frac{\partial u(t,x)}{\partial x} + \frac{1}{2}b^2(x)\frac{\partial^2 u(t,x)}{\partial x^2}$$
$$= a(y)\frac{\partial}{\partial x}\int u(y)f(t,x,y)\,dy + \frac{1}{2}b^2(x)\frac{\partial^2}{\partial x^2}\int u(y)f(t,x,y)\,dy$$
$$= \int u(y)\left[a(y)\frac{\partial f(t,x,y)}{\partial x} + \frac{1}{2}b^2(x)\frac{\partial^2 f(t,x,y)}{\partial x^2}\right] dy.$$

Because of (1.46) we arrive at

$$\int u(y)\left[\frac{\partial}{\partial t}f(t,x,y) - a(y)\frac{\partial f(t,x,y)}{\partial x} - \frac{1}{2}b^2(x)\frac{\partial^2 f(t,x,y)}{\partial x^2}\right] dy = 0.$$

Again by plugging in u_n from (1.48) and by letting $n \to \infty$ we obtain

$$\frac{\partial}{\partial t} f(t,x,y) = a(x)\frac{\partial f(t,x,y)}{\partial x} + \frac{1}{2}b^2(x)\frac{\partial^2 f(t,x,y)}{\partial x^2}, \tag{1.50}$$

which is called the *backward Fokker–Planck equation*.

1.5
Special Diffusion Processes

This section is aimed at presenting special examples of diffusion processes and studying the relation between them.

Example 1.2 If $X_t = W_t$ is the Wiener process then $a = 0$ and $b = 1$ in the stochastic differential equation (1.38). Since $f(t, x, y)$ is the density of $x + W_t$ the family of transition densities is given by

$$f(t, x, y) = \varphi_{0,t}(y - x)$$

where φ_{μ,σ^2} is the density of the normal distribution with parameters μ and σ^2. We see from (1.41) that the infinitesimal operator A is given by

$$A = \frac{1}{2} \frac{\partial^2}{\partial x^2}.$$

Putting

$$u(t, x) = \int u(y) f(t, x, y) \, dy$$

the Kolmogorov backward equation (1.45) reads

$$\frac{\partial u(t, x)}{\partial t} = \frac{1}{2} \frac{\partial^2 u(t, x)}{\partial x^2}.$$

This type of equation is called heat (or pure diffusion, which means without drift) equation in physics. The Fokker–Planck equation has the same form

$$\frac{\partial f(t, x, y)}{\partial t} = \frac{1}{2} \frac{\partial^2 f(t, x, y)}{\partial y^2}.$$

Of course, the above differential equation could also have been directly obtained using the fact that the transition density is, in view of $X_{t,x} = x + W_t$, given by

$$f(t, x, y) = \varphi_{0,t}(y - x) = \frac{1}{\sqrt{2\pi t}} \exp\left\{-\frac{(y - x)^2}{2t}\right\}.$$

Example 1.3 The Ornstein–Uhlenbeck process is defined to be a solution of the following stochastic differential equation

$$dX_t = \mu X_t \, dt + \sigma \, dW_t.$$

To solve this equation we apply the Ito formula to the process $X_t \exp\{\mu t\}$ where we choose $u(t, x) = x \exp\{-\mu t\}$. The formula for $du(t, X_t)$ in Theorem 1.11 (Ito formula) gives

$$\begin{aligned} d(X_t \exp\{-\mu t\}) &= \frac{\partial u(t, X_t)}{\partial t} dt + \frac{\partial u(t, X_t)}{\partial x} dX_t + \frac{1}{2} \frac{\partial^2 u(t, X_t)}{\partial x^2} (dX_t)^2 \\ &= -\mu X_t \exp\{-\mu t\} dt + \exp\{-\mu t\} dX_t \\ &= \exp\{-\mu t\} \sigma \, dW_t. \end{aligned}$$

Hence

$$\exp\{-\mu t\}X_t - X_0 = \sigma \int_0^t \exp\{-\mu s\}\,dW_s$$

$$X_t = X_0 \exp\{\mu t\} + \sigma \int_0^t \exp\{\mu(t-s)\}\,dW_s.$$

The infinitesimal operator reads

$$A = \mu x \frac{\partial}{\partial x} + \frac{1}{2}\sigma^2 \frac{\partial^2}{\partial x^2}.$$

From the definition of the Ito integral one easily concludes that for any nonrandom function h the random variable

$$\int_0^t h(s)\,dW_s$$

has a normal distribution with expectation zero and variance $\int_0^t h^2(s)\,ds$. This means that the distribution of $\sigma \int_0^t \exp\{\mu(t-s)\}\,dW_s$ is a normal distribution with expectation zero and a variance given by

$$\sigma^2 \exp\{2\mu t\} \int_0^t \exp\{-2\mu s\}\,ds = -\frac{\sigma^2}{2\mu} \exp\{2\mu t\} [\exp\{-2\mu t\} - 1]$$

$$= -\frac{\sigma^2}{2\mu}[1 - \exp\{2\mu t\}] \longrightarrow -\frac{\sigma^2}{2\mu} \quad \text{for } t \to \infty$$

if $\mu < 0$. In this case $X_0 \exp\{\mu t\}$ tends to zero. Hence for $\mu < 0$ the one-dimensional marginal distribution of X_t tends to a normal distribution with expectation zero and variance $-(\sigma^2/2\mu)$. One can show that this distribution, when used as an initial distribution of X_0, turns the Ornstein–Uhlenbeck process into a stationary process.

Example 1.4 We consider the geometric Brownian motion that is defined by

$$Y_t = \exp\{\mu t + \sigma W_t\}.$$

Put $X_t = \mu t + \sigma W_t$. We use the Ito formula in Theorem 1.11 with $u(x) = \exp\{x\}$. Hence by (1.32) and (1.33)

$$dY_t = \frac{\partial u(t, X_t)}{\partial t}\,dt + \frac{\partial u(t, X_t)}{\partial x}\,dX_t + \frac{1}{2}\frac{\partial^2 u(t, X_t)}{\partial x^2}(dX_t)^2$$

$$= u(X_t)\,dX_t + \frac{1}{2}u(X_t)(\mu\,dt + \sigma\,dW_t)^2$$

$$= Y_t\,dX_t + \frac{\sigma^2}{2}Y_t\,dt = Y_t\left(\mu + \frac{\sigma^2}{2}\right)dt + \sigma Y_t\,dW_t.$$

In particular, for $\mu = -\sigma^2/2$ we get

$$dY_t = \sigma Y_t \, dW_t.$$

Hence we see from (1.23) that Y_t has the constant expectation $\mathbb{E} Y_0 = 1$.

These special cases of diffusion processes considered in the last three examples will be discussed in more detail in Chapter 6 (Wiener process or Brownian motion from Example 1.2), in Chapter 8 (Ornstein–Uhlenbeck process from Example 1.3) and in Chapter 11 as well as in Section 5.9 (geometric Brownian motion from Example 1.4).

1.6 Exercises

E 1.1 Ito diffusion
Write a computer program using the Euler discretization algorithm of the Ito stochastic differential equation (1.38) to study special cases of Ito diffusion such as the Wiener process, Brownian motion with constant drift, and especially geometric Brownian motion (see Examples 1.2–1.4 in Section 1.5). Start with a simulation of the Wiener process $dX_t = dW_t$ using a discrete time interval Δt and normally distributed random numbers $Z \sim \mathcal{N}(0, 1)$ generated by the Box–Muller and/or the polar method. Check the known properties of the Wiener process by considering the Wiener difference $\Delta W_t = W_{t+\Delta t} - W_t$ over time step $\Delta t = t + \Delta t - t$ in the limit $\Delta t \to 0$.

E 1.2 Brownian paths in higher dimensions
Study Brownian paths (or Wiener trails) in higher dimensions $\mathbb{R}^n (n \geq 2)$ and show that the n-dimensional Brownian motion is isotropic by doing simulations of Brownian paths in $\mathbb{R}^2$.

E 1.3 Hausdorff dimension
The Hausdorff dimension and the box-counting dimension of a Brownian trail in $\mathbb{R}^n (n \geq 2)$ is equal to 2. Try to find the Hausdorff and box dimension for a graph (realization) of Brownian motion in $\mathbb{R}^1$ (one-dimensional case).

E 1.4 Stochastic process with constant drift and diffusion
Find the Fokker–Planck equation for the stochastic process that satisfies the stochastic differential equation $dX_t = -a \, dt + b \, dW_t$, where a and b are constants and $dW_t = W_{t+dt} - W_t$ is the increment of a Wiener process (also called white noise).

E 1.5 Stochastic Ornstein–Uhlenbeck process
Consider Example 1.3 in Section 1.5 (Ornstein–Uhlenbeck process) in more detail and find the solution of the corresponding Fokker–Planck equation related to $du_t = -\mu \, u_t \, dt + \sigma \, dW_t$ with non-negative constants μ, σ and given the initial condition $u_{t=0} = u_0$. Show that the probability density $p(u, t)$ becomes stationary and the so-called fluctuation–dissipation relation holds.

2
Multidimensional Approach

2.1
Bounded Multidimensional Region

As it is known already from Chapter 1 and many textbooks, see e.g. [55,193], Markovian stochastic processes are completely determined by their conditional probabilities which obey the Chapman–Kolmogorov equation. The Kramers–Moyal expansion can be used to determine the Fokker–Planck equation by specifying the drift vector and diffusion matrix based on the assumption of vanishing higher order Kramers–Moyal coefficients.

Usually, the Fokker–Planck equation is derived implicitly assuming that the phase space of the stochastic variables under consideration extends to infinity, so that so-called natural boundary conditions can be applied. If stochastic processes in a finite region of phase space are considered, boundary conditions are introduced a posteriori based on apparent physical arguments leading to the notion of a reflecting barrier, characterized by a vanishing normal component of the probability current, an absorbing barrier, where the probability distribution has to vanish, and boundary conditions at a discontinuity, where probability distributions and the normal components of the probability current have to be continuous. No attempts, so far, have been made to derive the Fokker–Planck equation simultaneously with appropriate boundary conditions from the Chapman–Kolmogorov equation.

It is quite evident that boundaries can strongly influence the stochastic motion of a particle in various ways depending on the microscopic interactions. As an example we mention a boundary formed by a fast diffusion layer. In such a thin layer, particles are able to diffuse in the directions tangential to the boundary on a fast time scale, whereas in the bulk the particle behavior should be accurately described by the Fokker–Planck equation. The theoretical treatment of the particle diffusion requires the formulation of consistent boundary conditions which match the internal Fokker–Planck behavior to the stochastic properties of the boundary layer.

So it would be desirable to have a technique for *deriving* the boundary conditions, referring directly to the way in which the regional boundaries affect the stochastic processes. In this respect, we note the unified formulation of the Fokker–Planck equation for stochastic hybrid systems by Julien Bect [17, 18] devoted to a general

Physics of Stochastic Processes: How Randomness Acts in Time
Reinhard Mahnke, Jevgenijs Kaupužs and Ihor Lubashevsky
Copyright © 2009 WILEY-VCH Verlag GmbH & Co. KGaA, Weinheim
ISBN: 978-3-527-40840-5

description of random processes near boundaries causing deterministic jumps. Boundary conditions for the Fokker–Planck equation which describes coupled transport of photons and electrons are derived in [2]. A series of papers [162, 171, 224] dealing with boundary conditions for the advection–diffusion problem combines the Boltzmann and Fokker–Planck equations and their numerical implementation, and [207] develops diffusion models for molecular transport across membranes via ion channels and wider pores in terms of random walks affected by boundaries with complex properties. In addition, [70] actually constructs the absorbing boundary as a limiting transition of an infinite space with half-spaces having substantially different properties and [242] implements boundary conditions for the Wiener processes in path integrals. Papers [56] and [87] develop a rather sophisticated moment technique for tackling the Fokker–Planck equation with mixed boundary conditions based on a special moment truncation scheme.

In this chapter we extend the method of deriving the Fokker–Planck equation from the Chapman–Kolmogorov equation in such a way that simultaneously consistent boundary conditions can be formulated. Our approach is based on the introduction of physical models for the stochastic behavior close to the boundary. We demonstrate that boundaries break the symmetry of the random forces leading to boundary singularities in the Kramers–Moyal expansion. The cancellation of these singularities yields the appropriate boundary conditions. We explicitly derive the boundary conditions for a reflecting or absorbing barrier and describe the general procedure for the derivation of the boundary conditions for the case of a fast diffusion layer. It should be noted that a similar anomalous effect of the regional boundaries on random processes was analyzed in [117, 178] and [182] by numerical implementation of the Wiener processes near the boundaries. In addition, paper [163] applies the concept of symmetry breakdown caused, however, by external fields, to construct a generalized master equation for the classical and anomalous diffusion processes.

In principle the present approach can be extended to anomalous transport phenomena, e.g. sub- and super-diffusion, which are modeled by fractional diffusion operators. It is well known that the formulation of boundary conditions for these processes is still a challenging problem although several approaches have been developed [9, 109, 136, 219]. The procedure outlined here might be helpful in formulating appropriate boundary conditions for these more complicated processes.

The main subject of the chapter is to derive the boundary conditions for the Fokker–Planck equations, both forward and backward ones, directly from the Chapman–Kolmogorov equation. Also, the Fokker–Planck equations will be obtained because; first, this makes the subject more complete; and second, it demonstrates the relationship between the basic elements in constructing a differential description of Markovian process for internal points and near the boundaries. To do this an M-dimensional region with boundaries is considered. The boundaries are assumed, in addition, to be able to absorb particles or to give rise to fast surface transport. It is demonstrated that the boundaries break down the symmetry of random walks in their vicinity, leading to the boundary singularities in the corresponding kinetic coefficients. Eliminating these singularities we get the desired boundary conditions. As required, the boundary condition for the forward Fokker–Planck equation satisfies mass conservation.

2.2
From Chapman–Kolmogorov Equation to Fokker–Planck Description

We consider the stochastic dynamics of a Markovian system represented as a point $\mathbf{r}$ belonging to a certain domain $\mathbb{Q}$ in the Euclidean M-dimensional space $\mathbb{R}^M$. The domain $\mathbb{Q}$ is assumed to be bounded by a smooth hypersurface Υ. When the detailed information about possible trajectories $\{\mathbf{r}(t)\}$ of the system motion is of minor importance the conditional probability, called also the Green function,

$$G(\mathbf{r}, t | \mathbf{r}_0, t_0) := \mathcal{P}\{\mathbf{r}_0, t_0 \Rightarrow \mathbf{r}, t\}$$

gives us the complete description of system evolution. By definition, the Green function is the probability density of finding the system at the point $\mathbf{r}$ at time t provided it was located at the point $\mathbf{r}_0$ at the initial time t_0.

Since Markovian systems have no memory, the Green function $G(\mathbf{r}, t | \mathbf{r}_0, t_0)$ obeys the integral Chapman–Kolmogorov equation that represents the transition of the system from the initial point $\mathbf{r}_0$ to the terminal one $\mathbf{r}$ within the time interval (t_0, t) as a complex step via an intermediate point $\mathbf{r}_* \in \mathbb{Q}$ at a certain fixed moment of time t_* with succeeding summation over all possible positions of the intermediate point (see, e.g., [55])

$$G(\mathbf{r}, t | \mathbf{r}_0, t_0) = \iiint_\mathbb{Q} d\mathbf{r}_* \, G(\mathbf{r}, t | \mathbf{r}_*, t_*) \, G(\mathbf{r}_*, t_* | \mathbf{r}_0, t_0). \tag{2.1}$$

The time t_* may be chosen arbitrarily between the initial and terminal time moments, $t_* \in [t_0, t]$. Figure 2.1 depicts this equation.

The domain boundary Υ will be regarded as a physical object, and so some individual properties are ascribed to it. In particular, the boundary itself can

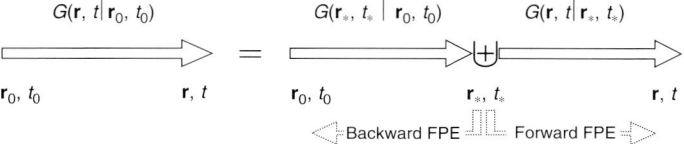

Figure 2.1 Diagram of the Chapman–Kolmogorov equation. The symbol ⊎ denotes summation over the intermediate point $\mathbf{r}_*$ and the arrows illustrate the limiting cases $t_* \to t_0 + 0$ and $t_* \to t - 0$, matching the backward and forward Fokker–Planck equations.

affect the system, for example, trapping it. So the symbol of the triple integral is used in (2.1) to underline this feature and, where appropriate, it should be read as

$$\iiint_{\mathbb{Q}} d\mathbf{r} \ldots = \int_{\mathbb{Q}^+} d\mathbf{r} \ldots + \oint_{\Upsilon} d\mathbf{s} \ldots + \oint_{\Upsilon_{tr}} d\mathbf{s} \ldots$$

where the symbol $\mathbb{Q}^+$ denotes the internal points of the domain $\mathbb{Q}$. The boundary Υ is split from the medium bulk because it can differ essentially from the medium bulk in its properties, and the boundary traps Υ_{tr} are singled out and treated individually for the same reasons. To simplify the notation a similar rule

$$\iint_{\mathbb{Q}} d\mathbf{r} \ldots = \int_{\mathbb{Q}^+} d\mathbf{r} \ldots + \oint_{\Upsilon} d\mathbf{s} \ldots$$

is also adopted. The integrals are split in order to treat the motion of the system inside the internal points $\mathbb{Q}^+$, its possible anomalous transport along the boundary Υ, and the trap effect, individually. Also, according to the probability definition, the equality

$$\iiint_{\mathbb{Q}} d\mathbf{r} \, G(\mathbf{r}, t | \mathbf{r}_0, t_0) = 1 \tag{2.2}$$

holds when the integration runs over all possible states of the system including the boundary traps Υ_{tr}.

In the following a general model for the medium boundary will be studied. Here we paid attention only to the fact that the boundary traps have to be treated individually because the system after being trapped cannot leave the boundary remaining in a trap forever. As a result, if the point $\mathbf{r}_0$ belongs to a trap, then for any internal point $\mathbf{r}$ of the domain $\mathbb{Q}$ the Green function is equal to zero

$$G(\mathbf{r}, t | \mathbf{r}_0, t_0) = 0 \quad \text{for} \quad \mathbf{r}_0 \in \Upsilon_{tr}, \quad \mathbf{r} \in \mathbb{Q}^+.$$

Later, the Green function $G(\mathbf{r}, t | \mathbf{r}_0, t_0)$ for the *internal* initial and terminal points $\mathbf{r}_0, \mathbf{r} \in \mathbb{Q}^+$ will be considered. Therefore, the general Chapman–Kolmogorov equation (2.1) can be reduced by eliminating the integration over the traps, so becoming

$$G(\mathbf{r}, t | \mathbf{r}_0, t_0) = \iint_{\mathbb{Q}} d\mathbf{r}_* \, G(\mathbf{r}, t | \mathbf{r}_*, t_*) \, G(\mathbf{r}_*, t_* | \mathbf{r}_0, t_0). \tag{2.3}$$

In (2.3) this elimination is pointed out by the absence of one integral matching the traps, cf. the general formulation (2.1) of the Chapman–Kolmogorov equation. Within the given integration rule the equality matching identity (2.2) is violated and we have

$$\iint_{\mathbb{Q}} d\mathbf{r} \, G(\mathbf{r}, t | \mathbf{r}_0, t_0) = 1 - \oint_{\Upsilon_{tr}} d\mathbf{s}_{tr} G(\mathbf{s}_{tr}, t | \mathbf{r}_0, t_0) < 1, \tag{2.4}$$

where the symbol $\mathbf{s}_{tr}$ stands for the boundary trap located at the point $\mathbf{s} \in \Upsilon$.

2.2 From Chapman–Kolmogorov Equation to Fokker–Planck Description

In order to obtain the Fokker–Planck equations, two *additional* assumptions must be adopted. The former is the short time confinement, meaning that on small time scales the system cannot jump over long distances, or in terms of the Green function, its first and second moments converge and

$$\lim_{t \to t_0+0} \iiint_{\mathbb{Q}} d\mathbf{r}\, G(\mathbf{r}, t|\mathbf{r}_0, t_0)|\mathbf{r} - \mathbf{r}_0|^p = 0, \quad p = 1, 2. \quad (2.5)$$

The latter is the medium local homogeneity; in other words, the medium where the Markovian process develops, i.e. the domain $\mathbb{Q}$, should be endowed with characteristics being actually some smooth fields determined inside $\mathbb{Q}^+$ or at Υ individually. As a result, the Green function $G(\mathbf{r}, t|\mathbf{r}_0, t_0)$ has to be smooth with respect to all its arguments for $t > t_0$ and $\mathbf{r}, \mathbf{r}_0 \in \mathbb{Q}^+$.

Because the intermediate time t_* entering the Chapman–Kolmogorov equation is any fixed value between the initial and terminal time moments, $t_0 < t_* < t$, there is a freedom to choose it for specific purposes. In particular, the passage to one of the limits $t_* \to t_0 + 0$ or $t_* \to t - 0$ gives rise to either the backward or forward Fokker–Planck equation, respectively (see Figure 2.1).

2.2.1
The Backward Fokker–Planck Equation

To implement the limit $t_* \to t_0 + 0$ let us choose an arbitrary small time scale τ and consider the Chapman–Kolmogorov equation for $t_* = t_0 + \tau$ and an *internal* point $\mathbf{r}_0$. Then, according to the adopted assumptions, the first multiplier $G(\mathbf{r}, t|\mathbf{r}_*, t_*)$ on the right-hand side of (2.3) is a smooth function of both the argument $\mathbf{r}_*$ and t_*, whereas the second one $G(\mathbf{r}_*, t_*|\mathbf{r}_0, t_0)$ exhibits strong variations on small spatial scales. So we can expand the function $G(\mathbf{r}, t|\mathbf{r}_0 + \mathbf{R}, t_0 + \tau)$ in the Taylor series with respect to the variables τ and $\mathbf{R} = \mathbf{r}_* - \mathbf{r}_0$. The required accuracy is the first order in the time step τ and the second order in $\mathbf{R}$ because the characteristic spatial displacement of the system during time τ is of order $\tau^{1/2}$. Within this accuracy it is

$$G(\mathbf{r}, t|\mathbf{r}_0 + \mathbf{R}, t_0 + \tau) = G(\mathbf{r}, t|\mathbf{r}_0, t_0) + \tau \frac{\partial G(\mathbf{r}, t|\mathbf{r}_0, t_0)}{\partial t_0}$$
$$+ \sum_{i=1}^{M} R^i \nabla_i^0 G(\mathbf{r}, t|\mathbf{r}_0, t_0)$$
$$+ \frac{1}{2} \sum_{i,j=1}^{M} R^i R^j \nabla_i^0 \nabla_j^0 G(\mathbf{r}, t|\mathbf{r}_0, t_0), \quad (2.6)$$

where the operator $\nabla_i^0 = \partial/\partial x_0^i$ acts only on the argument $\mathbf{r}_0$ of the Green function. The substitution of expansion (2.6) into the Chapman–Kolmogorov equation (2.3) reduces it to the following

$$-\tau\frac{\partial G(\mathbf{r},t|\mathbf{r}_0,t_0)}{\partial t_0} = -\mathfrak{R}(\mathbf{r}_0,t_0,\tau)\,G(\mathbf{r},t|\mathbf{r}_0,t_0)$$
$$+ \sum_{i=1}^{M} \mathfrak{U}^i(\mathbf{r}_0,t_0,\tau)\nabla_i^0 G(\mathbf{r},t|\mathbf{r}_0,t_0)$$
$$+ \sum_{i,j=1}^{M} \mathfrak{L}^{ij}(\mathbf{r}_0,t_0,\tau)\nabla_i^0\nabla_j^0 G(\mathbf{r},t|\mathbf{r}_0,t_0), \qquad (2.7)$$

where the quantities

$$\mathfrak{R}(\mathbf{r}_0,t_0,\tau) = 1 - \iint_{\mathbb{Q}} d\mathbf{R}\, G(\mathbf{r}_0+\mathbf{R}, t_0+\tau|\mathbf{r}_0,t_0), \qquad (2.8)$$

$$\mathfrak{U}^i(\mathbf{r}_0,t_0,\tau) = \iint_{\mathbb{Q}} d\mathbf{R}\, R^i\, G(\mathbf{r}_0+\mathbf{R}, t_0+\tau|\mathbf{r}_0,t_0), \qquad (2.9)$$

$$\mathfrak{L}^{ij}(\mathbf{r}_0,t_0,\tau) = \frac{1}{2}\iint_{\mathbb{Q}} d\mathbf{R}\, R^i R^j\, G(\mathbf{r}_0+\mathbf{R}, t_0+\tau|\mathbf{r}_0,t_0) \qquad (2.10)$$

have been introduced. Also, the first term on the right-hand side of (2.7) has been assumed to be small and to tend to zero as $\tau \to 0$ which is justified based on the required results.

For an internal point $\mathbf{r}_0$ and, thus, separated from the boundary Υ by a finite distance, the time step τ can be chosen so small that it is possible to construct a neighborhood of the point $\mathbf{r}_0$ with the following properties. First, deviation of the Green function $G(\mathbf{r}_0+\mathbf{R}, t_0+\tau|\mathbf{r}_0,t_0)$ from zero outside this neighborhood is ignorable due the first assumption of short time confinement. Second, inside it the medium can be regarded as the homogeneous space $\mathbb{R}^M$ by virtue of the second assumption on the local homogeneity. In this case actually replicating the proof of the Law of Large Numbers using the generating function notion (see, e.g. [55]) it is possible to demonstrate that quantities (2.9) and (2.10) scale linearly with τ. The difference of quantity (2.8) from zero is ignorable. Therefore, for internal points, we can introduce the drift velocity $v^i(\mathbf{r},t)$ and the diffusion tensor $D^{ij}(\mathbf{r},t)$ by the expressions

$$v^i(\mathbf{r},t) = \lim_{\tau \to +0}\frac{1}{\tau}\int_{\mathbb{Q}^+} d\mathbf{R}\, R^i\, G(\mathbf{r}+\mathbf{R}, t+\tau|\mathbf{r},t), \qquad (2.11)$$

$$D^{ij}(\mathbf{r},t) = \lim_{\tau \to +0}\frac{1}{2\tau}\int_{\mathbb{Q}^+} d\mathbf{R}\, R^i R^j\, G(\mathbf{r}+\mathbf{R}, t+\tau|\mathbf{r},t). \qquad (2.12)$$

Then, for the internal points, the division of (2.7) by τ and the succeeding passage to the limit $\tau \to +0$ yields the *backward Fokker–Planck equation*

$$-\frac{\partial G(\mathbf{r},t|\mathbf{r}_0,t_0)}{\partial t_0} = \widehat{\mathcal{L}}_{\mathrm{FP_B}}\{G(\mathbf{r},t|\mathbf{r}_0,t_0)\}, \qquad (2.13)$$

where the backward Fokker–Planck operator is

$$\widehat{\mathcal{L}}_{\mathrm{FP_B}} := \sum_{i,j=1}^{M} D^{ij}(\mathbf{r}_0, t_0, \tau)\nabla_i^0 \nabla_j^0 + \sum_{i=1}^{M} v^i(\mathbf{r}_0, t_0, \tau)\nabla_i^0. \tag{2.14}$$

We note that the backward Fokker–Planck operator acts on the second spatial argument of the Green function $G(\mathbf{r}, t|\mathbf{r}_0, t_0)$.

This Fokker–Planck equation should be supplemented with the initial condition and the boundary condition. By construction, at the initial time t_0 the system was located at the internal point $\mathbf{r}_0$, so the initial condition just writes the Green function in the form of the Dirac δ-function

$$G(\mathbf{r}, t|\mathbf{r}_0, t_0)\big|_{t=t_0} = \delta(\mathbf{r} - \mathbf{r}_0). \tag{2.15}$$

The boundary condition interrelates the values of the Green function and its derivatives at the internal points adjacent to the domain boundary Υ, that is, values obtained by the continuation $\mathbf{r}_0 \to \mathbf{s}$ from some internal point $\mathbf{r}_0 \in \mathbb{Q}^{M+}$ to a boundary point $\mathbf{s} \in \Upsilon$.

2.2.2
Boundary Singularities

The direct implementation of the passage to the boundary points, however, causes a certain problem. Expansion (2.7) exhibits irregular behavior within the joint passage to limits $\tau \to +0$ and $\mathbf{r}_0 \to \mathbf{s}$. When the former $\tau \to +0$ precedes the latter $\mathbf{r}_0 \to \mathbf{s}$ no boundary conditions are found at all.

In the opposite order, i.e. when the passage $\mathbf{r}_0 \to \mathbf{s}$ is performed first, the kinetic coefficients (2.8)–(2.10) change the scaling type; now they vary with time τ as $\sqrt{\tau}$ at the leading order. The fact is that a path of Markovian system is not smooth at every point and its characteristic variations on small time scales about τ are proportional to $\sqrt{\tau}$. For the internal points of the domain $\mathbb{Q}$ the path deviations in opposite directions are equiprobable within an accuracy of $\sqrt{\tau}$. As a result the coefficient $\mathfrak{U}^i(\mathbf{r}, t, \tau)$ becomes a linear function of the argument τ. In some sense the given anomaly in the Markovian dynamics is hidden at the internal points and reflected only in the linear τ-dependence of the second-order moments $\mathcal{L}^{ij}(\mathbf{r}_0, t_0, \tau)$ of the Green function $G(\mathbf{r}, t|\mathbf{r}_0, t_0)$. The medium boundary Υ breaks down this symmetry because, in particular, it prevents the system from reaching the points on the opposite side. Since the system displacement does not essentially change in amplitude, the terms $\mathfrak{U}^i(\mathbf{r}, t, \tau)$ acquire the root square dependence on the argument τ. In a certain sense the medium boundary reveals this anomaly (Figure 2.2). The succeeding division of expansion (2.7) by τ gives rise to singularities of the type $\tau^{-1/2}$ which will be referred to as *boundary singularities*.

The medium boundary can affect the system dynamics in a more complex way; here, however, we currently confine our speculations only to the effect of its impermeability. The boundary Υ restricts the system motion only in the normal direction. It is quite natural to expect that the boundary singularities quantified

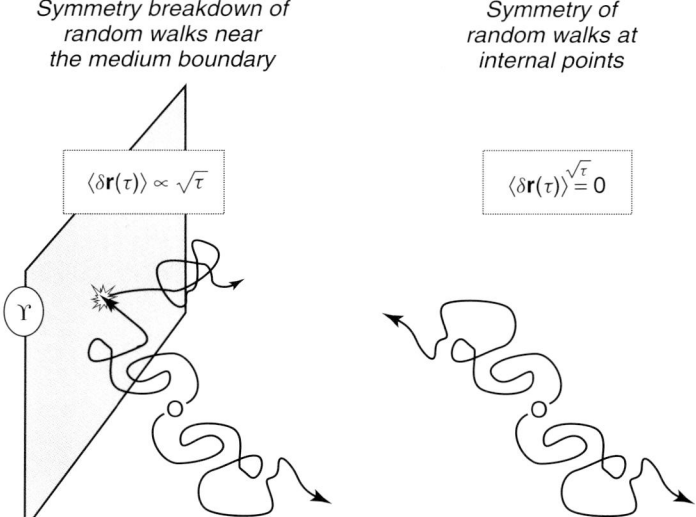

Figure 2.2 The effect of the boundary impermeability on the Markovian system motion. Schematic illustration.

in terms of diverging components $\mathfrak{U}^i(\mathbf{s}, t, \tau)/\tau$ will form a vector object $\mathbf{b}$ that is determined by the mutual effect of two factors. The former is the spatial orientation of the medium boundary Υ described by its unit normal $\mathbf{n}$. The latter is the spatial pattern of random Langevin forces governing the stochastic motion of the given system and is characterized by the diffusion tensor $D^{ij}(\mathbf{r}, t)$. Within a scalar cofactor we have only one possibility to construct the vector $\mathbf{b} = \{b^i\}$ using the two objects,

$$b^i = \sum_{j=1}^{M} D^{ij} n^j \qquad (2.16)$$

or, in the vector form

$$\mathbf{b} = \mathbf{D} \cdot \mathbf{n}. \qquad (2.17)$$

The validity of this construction will be justified in this chapter and $\mathbf{b}$ will be referred to as the *vector of boundary singularities*. To be rigorous it should be noted that, in the general case, the correct expression for the vector of boundary singularities should use the operator D^i_j obtained from the diffusion tensor D^{ij} by lowering one of its indices, namely, $b^i = \sum_j D^i_j n^j$ (these details are discussed in Section 2.4.1). However, dealing with orthonormal bases as is the case at the initial stage of the current consideration, the tensors D^{ij} and D^i_j coincide with each other in the component magnitudes. So, in order not to overload the reader's perception and the mathematical constructions, expressions similar to (2.16) will be used where appropriate.

The notion of the boundary singularity vector immediately enables us to write the desired boundary condition when the medium boundary just confines the system motion. In this case the first and third terms on the right-hand side of expansion (2.7) are absent and the singularity caused by the sequence of transitions $\mathbf{r}_0 \to \mathbf{s}$ and then $\tau \to 0$ takes the form

$$\frac{1}{\sqrt{\tau}} \sum_{i=1}^{M} b^i(\mathbf{s}) \nabla_i^0 \left. G(\mathbf{r},t|\mathbf{r},t_0) \right|_{\mathbf{r}_0 \to \mathbf{s}}$$

$$\to \frac{1}{\sqrt{\tau}} \sum_{i,j=1}^{M} D^{ij}(\mathbf{s},t_0) n^j(\mathbf{s}) \nabla_i^0 \left. G(\mathbf{r},t|\mathbf{r}_0,t_0) \right|_{\mathbf{r}_0 \to \mathbf{s}}.$$

Naturally, for the internal point $\mathbf{r}$ the Green function $G(\mathbf{r},t|\mathbf{r}_0,t_0)$ cannot exhibit any singularity. Therefor the cofactor of the singularity $\tau^{-1/2}$ must be equal to zero, that is

$$\sum_{i,j=1}^{M} D^{ij}(\mathbf{s},t_0) n^j(\mathbf{s}) \nabla_i^0 \left. G(\mathbf{r},t|\mathbf{r}_0,t_0) \right|_{\mathbf{r}_0 \to \mathbf{s}} = 0.$$

It is the well known expression for the boundary condition of the backward Fokker–Planck equation, which is typically obtained in another way, treating the Green function as the concentration of particles spreading over the medium from the point $\mathbf{r}_0$ (see, e.g. [55]).

This chapter is devoted to *deriving* the boundary conditions for the Fokker–Planck equation based on the notion of the boundary singularities. A more general situation will be studied and, in particular, these qualitative speculations will also be justified. The present analysis has demonstrated that the boundary condition for the backward Fokker–Planck equation stems from the boundary singularity terms vanishing in expansion (2.7), that is when $\mathbf{r}_0 \in \Upsilon$

$$- {}^*\mathfrak{R}(\mathbf{r}_0,t_0,\tau) G(\mathbf{r},t|\mathbf{r}_0,t_0)$$
$$+ \sum_{i=1}^{M} {}^*\mathfrak{U}^i(\mathbf{r}_0,t_0,\tau) \nabla_i^0 G(\mathbf{r},t|\mathbf{r}_0,t_0)$$
$$+ \sum_{i,j=1}^{M} {}^*\mathfrak{L}^{ij}(\mathbf{r}_0,t_0,\tau) \nabla_i^0 \nabla_j^0 G(\mathbf{r},t|\mathbf{r}_0,t_0) = 0, \tag{2.18}$$

where the symbol $*$ labels the components of the corresponding kinetic coefficients scaling as $\tau^{1/2}$. It should be pointed out that in (2.18) the argument $\mathbf{r}_0$ is an arbitrary point of a thin layer Υ_τ adjacent to the boundary Υ, which is designated by the symbol $\Subset$. When $\tau \to 0$ its thickness also tends to zero (as $\tau^{1/2}$). However, before passing to the limit $\tau \to 0$ the layer Υ_τ remains volumetric.

Now let us discuss similar problems with respect to the forward Fokker–Planck equation matching the other possibility of passage to the limiting case in the Chapman–Kolmogorov equation (2.3).

2.2.3
The Forward Fokker–Planck Equation

The Chapman–Kolmogorov equation (2.3) also allows for the limit where the intermediate point tends to the terminal one, i.e. $t_* = t - \tau$ with $\tau \to +0$. In this case the former cofactor $G(\mathbf{r}, t | \mathbf{r}_*, t - \tau)$ on the right-hand side of (2.3) exhibits strong variations on small spatial scales, whereas the latter one $G(\mathbf{r}_*, t - \tau | \mathbf{r}_0, t_0)$ becomes a smooth function of the argument $\mathbf{r}_*$. Now, however, applying directly to an expansion similar to that which has been used in deriving the backward Fokker–Planck equation is not appropriate. The fact is that in this way the integration runs over the initial point $\mathbf{r}_*$ of the Green function $G(\mathbf{r}, t | \mathbf{r}_*, t - \tau)$ and coefficients appearing similar to quantities (2.8)–(2.10) have another meaning. In particular, an integral similar to (2.4) can essentially deviate from unity.

To overcome this problem the Pontryagin technique is applied [197]. It is quite similar to the Kramers–Moyal approach (see, e.g. [193]) but is more suitable for tackling the boundary singularity. Let us consider as the first step some arbitrary smooth function $\phi(\mathbf{r})$ determined in the domain $\mathbb{Q}$ and integrate both the sides of the Chapman–Kolmogorov equation (2.3). In this way we get

$$\iint_{\mathbb{Q}} d\mathbf{r} \phi(\mathbf{r}) G(\mathbf{r}, t_* + \tau | \mathbf{r}_0, t_0)$$
$$= \iint_{\mathbb{Q}} \iint_{\mathbb{Q}} d\mathbf{r} \, d\mathbf{r}_* \, \phi(\mathbf{r}) \, G(\mathbf{r}, t_* + \tau | \mathbf{r}_*, t_*) \, G(\mathbf{r}_*, t_* | \mathbf{r}_0, t_0). \tag{2.19}$$

For a rather small time scale τ the Green function $G(\mathbf{r}, t_* + \tau | \mathbf{r}_*, t_*)$ is located practically within some small neighborhood of the point $\mathbf{r}_*$. In this way the function $\phi(\mathbf{r})$ can be expanded in the Taylor series near the point $\mathbf{r}_*$ with respect to the variable $\mathbf{R} = \mathbf{r} - \mathbf{r}_*$

$$\phi(\mathbf{r}) = \phi(\mathbf{r}_*) + \sum_{i=1}^{M} R^i \nabla_i^* \phi(\mathbf{r}_*) + \frac{1}{2} \sum_{i,j=1}^{M} R^i R^j \nabla_i^* \nabla_j^* \phi(\mathbf{r}_*). \tag{2.20}$$

Also, since the Green function $G(\mathbf{r}, t_* + \tau | \mathbf{r}_0, t_0)$ depends smoothly on τ the expansion

$$G(\mathbf{r}, t_* + \tau | \mathbf{r}_0, t_0) = G(\mathbf{r}, t_* | \mathbf{r}_0, t_0) + \tau \frac{\partial G(\mathbf{r}, t_* | \mathbf{r}_0, t_0)}{\partial t_*} \tag{2.21}$$

is also justified for a small value of τ.

Then the substitution of the last two expressions into (2.19) with succeeding integration over $\mathbf{R}$ and the replacement of the dummy variable $\mathbf{r}_*$ by $\mathbf{r}$ as well as t_*

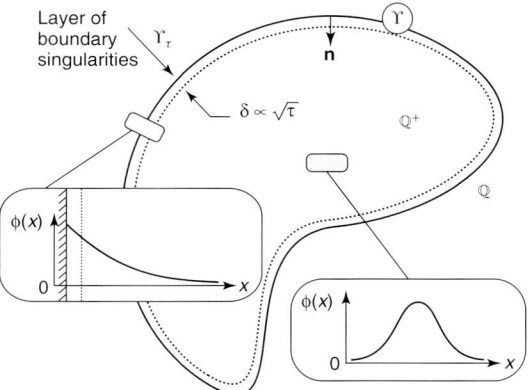

Figure 2.3 Structure of integral (2.22) and division of the region $\mathbb{Q}$ into the layer of boundary singularities and internal points with regular behavior of the kinetic coefficients.

by t, yields

$$\iint_{\mathbb{Q}} d\mathbf{r}\phi(\mathbf{r})\left[\tau\frac{\partial G(\mathbf{r},t|\mathbf{r}_0,t_0)}{\partial t}\right] = \iint_{\mathbb{Q}} d\mathbf{r}\left\{\phi(\mathbf{r})\left[-\mathfrak{R}(\mathbf{r},t,\tau)\,G(\mathbf{r},t|\mathbf{r}_0,t_0)\right]\right.$$
$$+ \sum_{i=1}^{M}\nabla_i\phi(\mathbf{r})\left[\mathfrak{U}^i(\mathbf{r},t,\tau)\,G(\mathbf{r},t|\mathbf{r}_0,t_0)\right]$$
$$\left.+ \sum_{i,j=1}^{M}\nabla_i\nabla_j\phi(\mathbf{r})\left[\mathfrak{L}^{ij}(\mathbf{r},t,\tau)\,G(\mathbf{r},t|\mathbf{r}_0,t_0)\right]\right\}.$$
(2.22)

Here the coefficients $\mathfrak{R}(\mathbf{r},t,\tau)$, $\mathfrak{U}^i(\mathbf{r},t,\tau)$ and $\mathfrak{L}^{ij}(\mathbf{r},t,\tau)$ again exhibit anomalous behavior within a narrow layer Υ_τ adjacent to the medium boundary Υ (Figure 2.3). As should be expected and in accordance with results to be obtained, the thickness of this layer scales with time τ as $\tau^{1/2}$. These coefficients themselves also scale as $\tau^{1/2}$. As a result the corresponding part of integral (2.22) scales as τ. So, after dividing both sides of (2.22) by τ with the following passage to the limit $\tau \to 0$, the contribution to (2.22) caused by integration over this layer remains finite. Therefore, to analyze the properties of the integral relation (2.22) the domain $\mathbb{Q}$ is split into this layer of boundary singularities and an internal part. After passage to the limit $\tau \to 0$ this division allows us to treat the boundary Υ and the internal points $\mathbb{Q}^+$ individually.

Keeping the aforementioned in mind, the integral expression (2.22) is represented as a sum of two terms, the integral over the layer Υ_τ denoted by the formal symbol of surface integral and the integral over the internal part $\mathbb{Q}^+$ of the domain $\mathbb{Q}$

$$\iint_{\mathbb{Q}} d\mathbf{r}\ldots = \oint_{\Upsilon_\tau} d\mathbf{r}\ldots + \int_{\mathbb{Q}^+} d\mathbf{r}\ldots \qquad (2.23)$$

Let us consider the second term first. Inside the region $\mathbb{Q}^+$ the kinetic coefficients $\mathfrak{U}^i(\mathbf{r}, t, \tau)$ and $\mathfrak{L}^{ij}(\mathbf{r}, t, \tau)$ behave in regular way, i.e. they scale as τ according to (2.11) and (2.12), whereas the term $\mathfrak{R}(\mathbf{r}, t, \tau)$ vanishes. So, dividing the corresponding part of the integral relation (2.22) by τ and passing to the limit $\tau \to 0$ we have

$$\int_{\mathbb{Q}^+} d\mathbf{r} \phi(\mathbf{r}) \left[\frac{\partial G(\mathbf{r},t|\mathbf{r}_0,t_0)}{\partial t} \right] = \int_{\mathbb{Q}^+} d\mathbf{r} \Bigg\{ \sum_{i=1}^M \nabla_i \phi(\mathbf{r}) \left[v^i(\mathbf{r},t,\tau) G(\mathbf{r},t|\mathbf{r}_0,t_0) \right]$$
$$+ \sum_{i,j=1}^M \nabla_i \nabla_j \phi(\mathbf{r}) \left[D^{ij}(\mathbf{r},t,\tau) G(\mathbf{r},t|\mathbf{r}_0,t_0) \right] \Bigg\}. $$
$$(2.24)$$

Using the Gauss divergence theorem this integral in turn is split into two surface and volume parts

$$\int_{\mathbb{Q}^+} d\mathbf{r}\ldots = \oint_{\Upsilon} d\mathbf{s}\ldots + \int_{\mathbb{Q}^+} d\mathbf{r}\ldots. \qquad (2.25)$$

The volume integral has the form

$$\int_{\mathbb{Q}^+} d\mathbf{r} \phi(\mathbf{r}) \left[\frac{\partial G(\mathbf{r},t|\mathbf{r}_0,t_0)}{\partial t} \right] = \int_{\mathbb{Q}^+} d\mathbf{r} \phi(\mathbf{r}) \Bigg\{ -\sum_{i=1}^M \nabla_i \left[v^i(\mathbf{r},t,\tau) G(\mathbf{r},t|\mathbf{r}_0,t_0) \right]$$
$$+ \sum_{i,j=1}^M \nabla_i \nabla_j \left[D^{ij}(\mathbf{r},t,\tau) G(\mathbf{r},t|\mathbf{r}_0,t_0) \right] \Bigg\}. \quad (2.26)$$

The latter equality immediately gives rise to the forward Fokker–Planck equation.

Indeed, currently $\phi(\mathbf{r})$ is an arbitrary smooth function and no additional constrain will be imposed on it for the internal points of the domain $\mathbb{Q}$. So applying local variations of $\phi(\mathbf{r})$ at an arbitrary internal point $\mathbf{r}$ (Figure 2.3) we see that the left and right-hand sides of (2.26) should be equal to each other for the points $\mathbf{r} \in \mathbb{Q}^+$ individually, obtaining the forward Fokker–Planck equation

$$\frac{\partial G(\mathbf{r},t|\mathbf{r}_0,t_0)}{\partial t} = \widehat{\mathcal{L}}_{\text{FP}_\text{F}} \{ G(\mathbf{r},t|\mathbf{r}_0,t_0) \} \qquad (2.27)$$

with the forward Fokker–Planck operator

$$\widehat{\mathcal{L}}_{\text{FP}_\text{F}}\{\Diamond\} := \sum_{i=1}^M \nabla_i \left[\sum_{j=1}^M \nabla_j \left(D^{ij}(\mathbf{r},t,\tau) \Diamond \right) - v^i(\mathbf{r},t,\tau) \Diamond \right]. \qquad (2.28)$$

Here the symbol $\Diamond$ stands for a function on which the operator acts. It should be also pointed out that the Fokker–Planck operator acts on the first spatial argument of the Green function.

The forward Fokker–Planck equation can be also written in the conservation form

$$\frac{\partial G(\mathbf{r}, t|\mathbf{r}_0, t_0)}{\partial t} + \sum_{i=1}^{M} \nabla_i J^i \{G(\mathbf{r}, t|\mathbf{r}_0, t_0)\} = 0, \quad (2.29)$$

with the probability flux operator $\widehat{\mathbf{J}} = \{J^i\}_{i=1}^{M}$

$$J^i\{\diamond\} := -\sum_{j=1}^{M} \nabla_j \left(D^{ij}(\mathbf{r}, t, \tau) \diamond \right) + v^i(\mathbf{r}, t, \tau) \diamond. \quad (2.30)$$

The forward Fokker–Planck equation is naturally supplemented by the same initial condition (2.15).

2.2.4
Boundary Relations

Splits (2.23) and (2.25) give rise to two additional terms. The former one is related to the first split and is the integral over the layer Υ_τ of boundary singularities

$$\oint_{\Upsilon_\tau} d\mathbf{r}\, G(\mathbf{s}, t|\mathbf{r}_0, t_0) \Bigg\{ \sum_{i=1}^{M} \nabla_i \phi(\mathbf{s})\, {}^*\mathfrak{L}^i(\mathbf{r}, t, \tau)$$

$$-\phi(\mathbf{s})\, {}^*\mathfrak{R}(\mathbf{r}, t, \tau) + \sum_{i,j=1}^{M} \nabla_i \nabla_j \phi(\mathbf{s})\, {}^*\mathfrak{L}^{ij}(\mathbf{r}, t, \tau) \Bigg\}. \quad (2.31)$$

Here the use of the symbols $d\mathbf{r}$ and $\mathbf{r}$ in the singular components of the kinetic coefficients indicates that, before passing to the limit $\tau \to 0$, we should consider the layer Υ_τ volumetric. The Green function $G(\mathbf{r}, t|\mathbf{r}_0, t_0)$ as well as the test function $\phi(\mathbf{r})$ and its derivatives exhibit minor variations across the layer Υ_τ so their argument $\mathbf{r}$ has been replaced by the corresponding nearest point $\mathbf{s}$ laying on the boundary Υ. The latter term is related to the part of expression (2.24) that remains after integration using the convergence theorem. It can be written in the form

$$\oint_\Upsilon d\mathbf{s} \sum_{i,j=1}^{M} \nabla_j \phi(\mathbf{s}) n^i(\mathbf{s}) \left[D^{ij}(\mathbf{s}, t, \tau)\, G(\mathbf{s}, t|\mathbf{r}_0, t_0) \right]$$

$$= -\oint_\Upsilon d\mathbf{s} \phi(\mathbf{s}) \sum_{i=1}^{M} n^i(\mathbf{s}) J^i \{G(\mathbf{s}, t|\mathbf{r}_0, t_0)\}, \quad (2.32)$$

where $\mathbf{n}(\mathbf{s}) = \{n^i(\mathbf{s})\}$ is the unit normal to the boundary Υ at point $\mathbf{s}$ directed inwards in the domain $\mathbb{Q}$.

Leaping ahead, we note that the appropriate choice of the boundary values of the test function $\phi(\mathbf{s})$ and its derivatives has to fulfill equality (2.31). Then at the next step it gives rise to the required boundary condition for the forward Fokker–Planck equation. Let us demonstrate this for the impermeable boundary using the notion of the boundary singularity vector **b**. Namely, we again assume that, for an internal point **r** located in the vicinity of a boundary point **s**, i.e. $\mathbf{r} \Subset \mathbf{s}$

$$\mathfrak{U}^i(\mathbf{r}, t, \tau) \propto b^i(\mathbf{s}) = \sum_j D^{ij}(\mathbf{s}, t) v^j(\mathbf{s}). \tag{2.33}$$

In this case only the first term in equality (2.31) remains and it is fulfilled when

$$\sum_{ij=1}^{M} D^{ij}(\mathbf{s}, t) n^j(\mathbf{s}) \nabla_i \phi(\mathbf{s}) = 0. \tag{2.34}$$

Equality (2.34) just relates the boundary values of the test function $\phi(\mathbf{s})$ with its derivative along the boundary normal $\mathbf{n}(\mathbf{s})$. So for an arbitrary smooth function $\phi_\Upsilon(\mathbf{s})$ determined at the boundary Υ it is possible to construct the appropriate function $\phi(\mathbf{r})$ determined in the domain $\mathbb{Q}$ and meeting equality (2.34) (see Figure 2.3). So in the given case the left-hand side and, thus, the right-hand side of expression (2.32) become zero. Since the integral on the right-hand side of (2.32) contains an arbitrary function $\phi(\mathbf{s})$ determined at the boundary Υ, the equality

$$\sum_{i=1}^{M} n^i(\mathbf{s}) J^i \{ G(\mathbf{s}, t | \mathbf{r}_0, t_0) \} = 0 \tag{2.35}$$

holds for every point of the boundary Υ individually. Expression (2.35) means that the probability flux in the direction normal to the boundary Υ is equal to zero, which actually reflects its impermeability.

However, to *derive* the boundary conditions for the Fokker–Planck equations, more sophisticated constructions are necessary. Also, in order to take into account other possible properties of the medium boundary its model must be specified.

2.3
Different Types of Boundaries

In this section, to be specific, we consider three typical examples of medium boundaries. They are (*a*) the impermeable boundary, (*b*) the boundary absorbing particles, and (*c*) the boundary with a thin adjacent layer characterized by extremely high values of the kinetic coefficients, the fast diffusion boundary (see Figure 2.4).

The first type matches a medium whose boundary properties are similar to its bulk properties; the boundary points differ from the internal ones only by the absence of medium points on one side. As a result a random walker hopping over the medium points just cannot pass through the boundary.

The second type is similar to the first one except for the fact that the walker can be trapped at the boundary and will not return again to the medium. In this case the corresponding boundary conditions are typically used in describing the first passage time problem or diffusion in solids with fixed boundary values of impurity

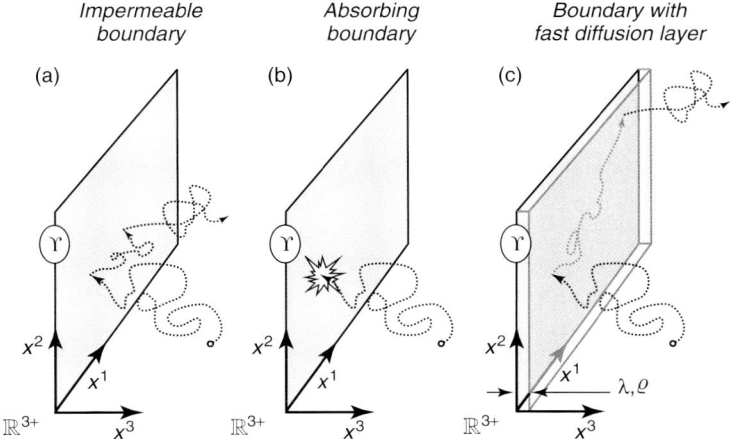

Figure 2.4 Three types of boundaries under consideration.

concentration C_s (see, e.g. [55]). Generally the boundary absorption is described by the rate σC_s, where σ is a certain kinetic coefficient.

The third type of boundaries are very common, for example, in polycrystals or nanoparticle agglomerates. The grain boundaries contain a huge amount of defects and as a result the diffusion coefficient inside the grain boundaries can exceed its value in the crystal bulk by many orders. Therefore, impurity propagation in polycrystals is governed mainly by grain boundary diffusion (for a review see [20] and references therein). In terms of random walks the effect of the fast diffusion layer is reduced to extremely long spatial jumps made by a walker inside it. It is natural to characterize such a boundary layer by its thickness λ about the atomic spacing and the ratio $\varrho \gg 1$ of the diffusion coefficients inside the boundary layer and in the regular crystal lattice.

2.4
Equivalent Lattice Representation of Random Walks Near the Boundary

The derivation of both the forward and backward Fokker–Planck equations, the requires the calculation of three quantities $\mathfrak{R}(\mathbf{r}, t, \tau)$, $\mathfrak{U}^i(\mathbf{r}, t, \tau)$ and $\mathfrak{L}^{ij}(\mathbf{r}, t, \tau)$ specified by expressions (2.8)–(2.10). They are the moments of the system displacement $\mathbf{R}$ during the time τ treated as an arbitrary small value. In order to obtain the desired boundary conditions these quantities should be found in the vicinity of the medium boundary Υ or, more precisely, in its neighborhood Υ_τ of thickness about $(D\tau)^{1/2}$, where D is the characteristic value of the diffusion tensor components. To study the boundary effects it suffices to consider quite a small region wherein the medium and its boundary have practically homogeneous properties and, in addition, the boundary geometry is approximated well by some hyperplane. In this region the system motion will be imitated by random walks on a lattice constructed as follows.

First, the elementary steps of the random walks on it are characterized by a time τ_a such that

$$\tau_a \ll \tau \tag{2.36}$$

and the arrangement of the lattice nodes, i.e. their spacings $\{a_i\}$ and the spatial orientation should again give us the same diffusion tensor **D** as well as the drift field **v** for the internal points on time scales $\tau_a \ll t \ll \tau$. The individual hops of a random walker between the neighboring nodes actually represent a collection of mutually independent Langevin forces governing the random system motion in the given continuum. Second, the boundary Υ is represented as a layer of nodes Υ_0 between which the walker can migrate via elementary hops. In other words, the aforementioned collection of mutually independent Langevin forces has to contain components acting along the boundary Υ and one component moving the walker towards or from Υ. Other characteristics of this effective lattice may be chosen for the sake of convenience. At the final stage we should pass to the limit $\tau_a \to 0$ returning to the continuous description.

2.4.1
Diffusion Tensor Representations

In order to construct the required lattice let us consider Markovian random walks $\{\mathbf{r}(t)\}$ in M-dimensional Euclidean half-space $\mathbb{R}^{M+}$ made of vectors

$$\mathbf{r} = \{x^1, x^2, \ldots, x^M\}$$

such that

$$\mathbf{r} \cdot \mathbf{n} \coloneqq \sum_{i=1}^{M} x^i n^i \geq 0,$$

where $\mathbf{n} = \{n^1, n^2, \ldots, n^M\}$ is a certain unit vector. The boundary of $\mathbb{R}^{M+}$, that is, the hyperplane $\Upsilon = \{\mathbf{r} \cdot \mathbf{n} = 0\}$ perpendicular to the vector $\mathbf{n}$ is, in its turn, the Euclidean space $\mathbb{R}^{M-1}$ of dimension $M - 1$. The half-space $\mathbb{R}^{M+}$ and, correspondingly, the hyperplane Υ are assumed to be homogeneous. The latter means that the local properties of the random walks under consideration have to be independent of position in space; naturally the boundary and internal points are not equivalent. In particular, the diffusion tensor **D** and drift vector **v** are the same at all the internal points of the half-space $\mathbb{R}^{M+}$.

In this case the components of the drift vector and diffusion tensor are determined by the expressions (cf. (2.8)–(2.10))

$$v^i = \frac{1}{\tau} \left\langle \delta X^i(t, \tau) \right\rangle, \tag{2.37}$$

$$D^{ij} = \frac{1}{2\tau} \left\langle \left[\delta X^i(t, \tau) - v^i \tau \right] \left[\delta X^j(t, \tau) - v^j \tau \right] \right\rangle. \tag{2.38}$$

Here the random variable $\delta X^i(t, \tau) := x^i(t+\tau) - x^i(t)$ and $\mathbf{r} = \{x^i\}$ is an arbitrary internal point. The observation time interval τ should be chosen to be small enough so that the length scale $(D\tau)^{1/2}$ is much less than the distance between the point $\mathbf{r}$ and the boundary Υ, i.e. $D\tau \ll (\mathbf{r} \cdot \mathbf{n})^2$, and the triangular brackets $\langle \ldots \rangle$ stand for the average over all the random trajectories passing through the point $\mathbf{r}$ at time t. It should be noted that, due to the space homogeneity, the passage to the limit $\tau \to 0$ can be omitted, which is necessary in the general case.

In the following, nonorthogonal bases will be used. So, keeping in mind the tensor notation (see, e.g. [138]), the upper and lower indices will be distinguished. In these terms $\{x^i\}$ or just x^i, is a vector, whereas, the collection of the basis vectors $\mathbf{e}_i$ is a covector. According to definitions (2.38) and (2.37) the objects D^{ij} and v^i are contravariant tensors. In addition, if the basis $\mathfrak{e}$ has the form $\mathfrak{e} = \mathfrak{e}_\Upsilon \oplus \mathbf{e}$, where $\mathfrak{e}_\Upsilon$ is the basis of the hyperplane Υ and the vector $\mathbf{e}$ does not lie within it, then the Greek letters will label the tensor indices corresponding to the hyperplane Υ to simplify the perception of this fact.

In order to deal with the diffusion tensor in a nonorthogonal basis $\mathfrak{e} = \{\mathbf{e}_i\}$ the metric tensor is also necessary. It is defined as

$$g_{ij} := (\mathbf{e}_i \cdot \mathbf{e}_j) \tag{2.39}$$

and is the kernel of the scalar product of two vectors $\mathbf{r}$ and $\bar{\mathbf{r}}$, namely,

$$(\mathbf{r} \cdot \bar{\mathbf{r}}) := \sum_{i,j=1}^{M} g_{ij} x^i \bar{x}^j. \tag{2.40}$$

For an orthonormal basis the metric tensor $g_{ij} = \delta_{ij}$, where δ_{ij} is the Kronecker delta. The metric tensor g_{ij} defines the conversion of contravariant tensors into covariant ones, in particular,

$$D^{i\cdot}_{\cdot j} = \sum_{k=1}^{M} D^{ik} g_{kj}, \quad D^{\cdot j}_{i\cdot} = \sum_{k=1}^{M} g_{ik} D^{kj}, \tag{2.41}$$

as well as

$$D_{ij} = \sum_{k,p=1}^{M} g_{ik} g_{jp} D^{kp}. \tag{2.42}$$

Due to the diffusion tensor D^{ij} as well as the metric tensor g_{ij} being symmetric, the tensor D_{ij} is also symmetric, whereas the tensors $D^{i\cdot}_{\cdot j}$ and $D^{\cdot i}_{j\cdot}$ are identical and so denoted further as D^i_j. The tensor D^i_j can be regarded as a certain operator $\widehat{\mathcal{D}}$ acting in the space $\mathbb{R}^M$ and the tensor D_{ij} specifies a quadratic form

$$\mathbf{r} \cdot \widehat{\mathcal{D}} \mathbf{r} = \sum_{i,j,k=1}^{M} g_{ij} x^i D^j_k x^k = \sum_{i,j=1}^{M} D_{ij} x^i x^j. \tag{2.43}$$

The quadratic form (2.43) is positive-definite. To demonstrate this, a random variable

$$\delta L = \sum_{p=1}^{M} \left[\delta X^p(t,\tau) - v^p \tau\right] (\mathbf{e}_p \cdot \boldsymbol{\ell}) = \sum_{p,i=1}^{M} \left[\delta X^p(t,\tau) - v^p \tau\right] g_{pi} \ell^i$$

is considered, where $\boldsymbol{\ell} = \sum_{i=1}^{M} \mathbf{e}_i \ell^i$ is an arbitrary vector in the space $\mathbb{R}^M$ and the metric tensor definition (2.39) has been taken into account. From this we we have a chain of equalities

$$0 < \langle [\delta L]^2 \rangle = \sum_{p,p',i,i'=1}^{M} g_{pi} g_{p'i'} l^i l^{i'} \left\langle \left[\delta X^p(t,\tau) - v^p \tau\right]\left[\delta X^{p'}(t,\tau) - v^{p'} \tau\right]\right\rangle$$

$$= \sum_{p,p',i,i'=1}^{M} 2\tau D^{pp'} g_{pi} g_{p'i'} l^i l^{i'} = \sum_{i,i'=1}^{M} 2\tau D_{ii'} l^i l^{i'}$$

$$= \sum_{p,p',i,i'=1}^{M} 2\tau D^{ii'} l_i l_{i'}.$$

So, for any arbitrary vector l^i and covector l_i the inequalities

$$\sum_{i,j=1}^{M} D_{ij} l^i l^j > 0, \qquad \sum_{i,j=1}^{M} D^{ij} l_i l_j > 0 \tag{2.44}$$

hold. The covector and vector representations of the same object are related as $l_i = \sum_{j=1}^{M} g_{ij} l^j$; within orthonormal bases they are identical.

Since the symmetry of the tensor D_{ij} and the quadratic form (2.43) are both positive definite, all the eigenvalues of the operator $\widehat{\mathcal{D}}$ are real positive quantities and its eigenvectors form a basis in the space $\mathbb{R}^M$ which can be chosen to be orthonormal (see, e.g., [53]). In this basis the diffusion tensor takes the diagonal form Therefore, the corresponding eigenvectors and eigenvalues specify the directions and intensity of the mutually independent Langevin forces governing random walks in the medium under consideration. Unfortunately, in the general case where all the eigenvalues are nondegenerate, this basis is unique, so it cannot be used in constructing the desired lattice in the vicinity of the medium boundary Υ because one could find a situation where none of the basis vectors are parallel to the hyperplane Υ. In order to overcome this problem we will construct a special nonorthogonal basis applying to the following statement.

Proposition 2.1 *Let $\mathbb{R}^{M+} = \{\mathbf{r} \cdot \mathbf{n} > 0\}$ be a homogeneous half-space bounded by the hyperplane $\Upsilon = \{\mathbf{r} \cdot \mathbf{n} = 0\}$ and $\mathfrak{e} = \{\mathbf{e}_1, \mathbf{e}_2, \ldots \mathbf{e}_M\}$ be a fixed arbitrary basis of $\mathbb{R}^M$. In this basis the components of the diffusion tensor $\{D^{ij}\}$ as well as the metric tensor $\{g_{ij}\}$ are given. Then there is a basis $\mathfrak{b} = \mathfrak{b}_\Upsilon \oplus \mathbf{b}_M$ with the following properties.*

First, it is composed of a certain orthonormal basis $\mathfrak{b}_\Upsilon$ of the hyperplane Υ and a unit vector $\mathbf{b}_M$ not belonging to Υ that is determined by the expression

$$\mathbf{b}_M = \frac{1}{\omega} \sum_{i,j=1}^{M} \mathbf{e}_i D_j^i n^j. \tag{2.45}$$

Here, according to the construction of the half-space, $\mathbb{R}^{M+}$ $\mathbf{n} = \{n^1, n^2, \ldots n^M\}$ is the unit vector normal to the hyperplane Υ and the normalization factor

$$\omega = \left[\sum_{i,j,p,k=1}^{M} g_{ij} D_p^i D_k^j n^p n^k \right]^{1/2}. \tag{2.46}$$

Second, in the basis $\mathfrak{b}$ the diffusion tensor takes the diagonal form

$$\|D^{ij}\| = \left\| \begin{matrix} \mathcal{D}_1 & 0 & \ldots & 0 \\ 0 & \mathcal{D}_2 & \ldots & 0 \\ \vdots & \vdots & \ddots & \vdots \\ 0 & 0 & \ldots & \mathcal{D}_M \end{matrix} \right\|, \tag{2.47}$$

where all its diagonal components are positive quantities, $\{\mathcal{D}_i > 0\}$, with the value $\mathcal{D}_M$ being given by the expression

$$\mathcal{D}_M = \omega^2 \left[\sum_{i,j=1}^{M} D_{ij} n^i n^j \right]^{-1}. \tag{2.48}$$

Third, let, in addition, the initial basis be of the form $\mathfrak{e} = \mathfrak{e}_\Upsilon \oplus \mathbf{n}$, where $\mathfrak{e}_\Upsilon = \{\mathbf{e}_\alpha\}_1^{M-1}$ is a certain basis of the hyperplane Υ, and $\hat{U}_\Upsilon = \|u_\beta^\alpha\|$ be the transformation of the hyperplane Υ mapping the basis $\mathfrak{b}_\Upsilon$ onto the basis $\mathfrak{e}_\Upsilon$, that is, $\mathfrak{b}_\Upsilon \overset{\hat{u}}{\mapsto} \mathfrak{e}_\Upsilon$. By mapping $\mathbf{b}_M \mapsto \mathbf{n}$ the transformation $\hat{U}_\Upsilon$ is complemented to a certain transformation $\hat{U}$ of the whole space $\mathbb{R}^M$, namely, if $\mathbf{r}$ is an arbitrary vector of the space $\mathbb{R}^M$ with the coordinates specified by its expansion over the bases $\mathfrak{e}$ and $\mathfrak{b}$:

$$\mathbf{r} = \sum_{\gamma=1}^{M-1} \mathbf{e}_\gamma x^\gamma + \mathbf{n} x^M \equiv \sum_{\gamma=1}^{M-1} \mathbf{b}_\gamma \zeta^\gamma + \mathbf{b}_M \zeta^M, \tag{2.49}$$

then its coordinates are related by the expressions

$$\zeta^\alpha = \sum_{\gamma=1}^{M-1} u_\gamma^\alpha \left(x^\gamma - \frac{1}{D^{MM}} D^{\gamma M} x^M \right), \tag{2.50a}$$

$$\zeta^M = \frac{\omega}{D^{MM}} x^M, \tag{2.50b}$$

and for the inverse transformation

$$x^\alpha = \sum_{\gamma=1}^{M-1} \breve{u}_\gamma^\alpha \zeta^\gamma + \frac{1}{\omega} D^{\alpha M} \zeta^M, \tag{2.50c}$$

$$x^M = \frac{D^{MM}}{\omega} \zeta^M. \tag{2.50d}$$

2 Multidimensional Approach

Here $\widehat{U}_\nu^{-1} = \|\check{u}^\alpha_\beta\|$ is the operator inverse to the operator $\hat{U}_\Upsilon$, that is, meeting the identity $\sum_{\gamma=1}^{M-1} \check{u}^\alpha_\gamma u^\gamma_\beta = \delta^\alpha_\beta$. Besides, the equality

$$\sum_{\gamma=1}^{M-1} \check{u}^\alpha_\gamma \check{u}^\beta_\gamma \mathcal{D}_\gamma = D^{\alpha\beta} - \frac{1}{D^{MM}} D^{\alpha M} D^{\beta M} \tag{2.51}$$

holds.

The proof of this proposition requires just formal mathematical manipulation. For the first step, the initial basis $\mathfrak{e}$ of the half-space $\mathbb{R}^{M+}$ is assumed to comprise a certain basis

$$\mathfrak{e}_\Upsilon = \{\mathbf{e}_1, \mathbf{e}_2, \ldots \mathbf{e}_{M-1}\}$$

of the hyperplane Υ and its unit normal $\mathbf{n}$ directed inward $\mathbb{R}^{M+}$, that is, $\mathfrak{e} = \mathfrak{e}_\Upsilon \oplus \mathbf{n}$. Then the results to be obtained will be represented in invariant form where appropriate, enabling us to write the general expressions. The diffusion tensor D^{ij} is assumed to be determined beforehand in the initial basis. Also as before, to simplify reader perception the Greek letters will be used to label tensor indices corresponding to the hyperplane Υ.

Let us consider a new basis $\mathfrak{b} = \mathfrak{b}_\Upsilon \oplus \mathbf{b}_M$ of the same structure except for the last vector $\mathbf{b}_M$; it need not be normal to the hyperplane Υ. A one-to-one map between the two bases, $\mathfrak{e} \Leftrightarrow \mathfrak{b}$, determines a linear transformation $\widehat{\mathcal{U}}$ of the space $\mathbb{R}^M$ mapping, in particular, the hyperplane Υ onto itself. This transformation $\widehat{\mathcal{U}} = \|U^i_j\|$ is specified by the relationship between the basis vectors

$$\mathbf{e}_\alpha = \sum_{\beta=1}^{M-1} \mathbf{b}_\beta u^\beta_\alpha; \quad \mathbf{n} = \sum_{\alpha=1}^{M-1} \mathbf{b}_\alpha \omega^\alpha + \mathbf{b}_M \omega^M. \tag{2.52}$$

Here the tensor u^α_β represents an operator $\widehat{\mathcal{U}}_\Upsilon$ acting in the hyperplane Υ whereas the tensor ω^α (in Υ) and the coefficient $\omega^M \neq 0$ complement it to the operator $\widehat{\mathcal{U}}$, namely,

$$U^\alpha_\beta = u^\alpha_\beta, \quad U^\alpha_M = \omega^\alpha, \tag{2.53a}$$

$$U^M_\beta = 0, \quad U^M_M = \omega^M. \tag{2.53b}$$

According to the rule of tensor transformations (see, e.g. [138]) in the basis $\mathfrak{b}$ the diffusion matrix has the components

$$\tilde{D}^{\alpha\beta} = \sum_{\gamma,\gamma'=1}^{M-1} u^\alpha_\gamma u^\beta_{\gamma'} D^{\gamma\gamma'} + \omega^\alpha \omega^\beta D^{MM} + \sum_{\gamma=1}^{M-1} \left(\omega^\alpha u^\beta_\gamma + \omega^\beta u^\alpha_\gamma\right) D^{\gamma M}, \tag{2.54a}$$

$$\tilde{D}^{\alpha M} = \omega^M \left(\sum_{\gamma=1}^{M-1} u^\alpha_\gamma D^{\gamma M} + \omega^\alpha D^{MM}\right), \tag{2.54b}$$

$$\tilde{D}^{MM} = \left(\omega^M\right)^2 D^{MM}. \tag{2.54c}$$

Correspondingly, an arbitrary vector $x^i = \{x^\alpha, x^M\}$ is converted as

$$\tilde{x}^\alpha = \sum_{\gamma=1}^{M-1} u_\gamma^\alpha x^\gamma + \omega^\alpha x^M, \tag{2.55a}$$

$$\tilde{x}^M = \omega^M x^M. \tag{2.55b}$$

Currently there are no restrictions imposed on the basis $\mathfrak{b}$ (except for its general structure). Now let us choose a specific version of the tensor ω^α that eliminates the off-diagonal elements of the diffusion tensor in the basis $\mathfrak{b}$. By virtue of (2.54b) it is

$$\omega^\alpha = -\frac{1}{D^{MM}} \sum_{\gamma=1}^{M-1} u_\gamma^\alpha D^{\gamma M}. \tag{2.56}$$

Here, division by D^{MM} is possible because, according to definition (2.38), the diagonal elements of diffusion tensor are positive, in particular, $D^{MM} > 0$ except for the case where the system motion along the direction $\mathbf{n}$ is rigorously deterministic. However, setting $D^{MM} \to +0$ the latter case can also be allowed for. The substitution of (2.56) into (2.54a) yields

$$\tilde{D}^{\alpha\beta} = \sum_{\gamma,\gamma'=1}^{M-1} u_\gamma^\alpha u_{\gamma'}^\beta \mathfrak{D}^{\gamma\gamma'}, \tag{2.57}$$

where the object

$$\mathfrak{D}^{\alpha\beta} = D^{\alpha\beta} - \frac{1}{D^{MM}} D^{\alpha M} D^{\beta M} \tag{2.58}$$

is a tensor within the hyperplane Υ because, up to now, the collections of vectors $\mathfrak{e}_\Upsilon$ and $\mathfrak{b}_\Upsilon$ are general bases of this hyperplane.

The tensor $\mathfrak{D}^{\alpha\beta}$ is symmetric and positive definite. The latter property stems directly from inequality (2.44) written for an arbitrary covector l_α of the hyperplane Υ with the component

$$l_M = -\frac{1}{D^{MM}} \sum_{\gamma=1}^{M-1} D^{M\gamma} l_\gamma, \tag{2.59}$$

namely,

$$\sum_{i,j=1}^{M} D^{ij} l_i l_j = \sum_{\alpha,\beta=1}^{M-1} \mathfrak{D}^{\alpha\beta} l_\alpha l_\beta > 0. \tag{2.60}$$

Therefore, the basis $\mathfrak{b}_\Upsilon$ of the hyperplane Υ can be chosen to be an orthonormal one wherein the tensor $\mathfrak{D}^{\alpha\beta}$ takes the diagonal form with the diagonal components being positive values, so $\mathfrak{D}^{\alpha\beta} = \mathfrak{D}_\alpha^\beta = \mathfrak{D}_{\alpha\beta} = \mathcal{D}_\alpha \delta_{\alpha\beta}$ [53]. This basis $\mathfrak{b}_\Upsilon$ is made up of the eigenvectors of the operator $\widehat{\mathfrak{D}} \coloneqq \|\mathfrak{D}_\beta^\alpha\|$ whose eigenvalues are $\{\mathcal{D}_\alpha\}$. For example, in the initial basis $\mathfrak{e}_\Upsilon$ the tensor $\mathfrak{D}_\alpha^\beta$ is related to the tensor $\mathfrak{D}^{\alpha\beta}$ by the expression

$$\mathfrak{D}_\alpha^\beta = \sum_{\gamma=1}^{M-1} g_{\alpha\gamma} \mathfrak{D}^{\gamma\beta}, \quad \text{where} \quad g_{\alpha\beta} \coloneqq (\mathbf{e}_\alpha \cdot \mathbf{e}_\beta)$$

is the metric tensor of the hyperplane Υ.

The choice of the given basis $\mathfrak{b}_\Upsilon$ specifies the transformation matrix $\|u_\beta^\alpha\|$ which, together with expression (2.56), gives us the vector $\mathbf{b}_M$ and the corresponding component $\mathcal{D}_M$ of the diffusion tensor. Namely, first substituting (2.56) into the latter equality of (2.52) and taking into account the former one, we write

$$\mathbf{b}_M \omega_M = \mathbf{n} + \frac{1}{D^{MM}} \sum_{\alpha,\gamma=1}^{M-1} \mathbf{b}_\alpha u_\gamma^\alpha D^{\gamma M} = \mathbf{n} + \frac{1}{D^{MM}} \sum_{\gamma=1}^{M-1} \mathbf{e}_\gamma D^{\gamma M}. \tag{2.61}$$

In the invariant form this expression can be rewritten as

$$\mathbf{b}_M = \frac{1}{\omega} \sum_{i,j=1}^{M} \mathbf{e}_i D^{ij} (\mathbf{e}_j \cdot \mathbf{n}), \tag{2.62}$$

where the normalization factor ω

$$\omega = \left[\sum_{i,j,k,p=1}^{M} D^{ik} D^{jp} (\mathbf{e}_i \cdot \mathbf{e}_j)(\mathbf{e}_k \cdot \mathbf{n})(\mathbf{e}_p \cdot \mathbf{n}) \right]^{1/2} \tag{2.63}$$

is due to the vector $\mathbf{b}_M$ being of unit length. Since the obtained expressions (2.62) and (2.63) are of the tensor form and are scalar in this sense (they do not contain free indices) they hold within any basis, thus proving (2.45) and (2.46).

Second, according to (2.61) and (2.62) the coefficient $\omega_M = \omega / D^{MM}$. So expressions (2.54c) and (2.63) give us the diffusion tensor component $\mathcal{D}_M$ related to the vector $\mathbf{b}_M$ in the basis $\mathfrak{b}$

$$\mathcal{D}_M = \omega^2 \left[\sum_{i,j=1}^{M-1} D^{ij} (\mathbf{e}_i \cdot \mathbf{n})(\mathbf{e}_j \cdot \mathbf{n}) \right]^{-1} \tag{2.64}$$

and written in the invariant form. Formula (2.48) is therefore proved. In addition, expressions (2.55) and (2.56) together with the equality $\omega = \omega_M D^{MM}$ immediately lead to (2.50a) and (2.50b).

Finally, we need the transformation $\widehat{\mathcal{U}}_\Upsilon^{-1} = \|\breve{u}_\beta^\alpha\|$ of the hyperplane Υ that is inverse to the transformation $\widehat{\mathcal{U}}_\Upsilon = \|u_\gamma^\alpha\|$; its components obey the equality

$$\sum_{\gamma=1}^{M-1} \breve{u}_\gamma^\alpha u_\beta^\gamma = \delta_\beta^\alpha. \tag{2.65}$$

This exists due to the transformation $\widehat{\mathcal{U}}_\Upsilon$ being a one-to-one map of the bases $\mathfrak{e}$ and $\mathfrak{b}$. Then the inversion of equalities (2.55) with ω^α given by expression (2.56) yields (2.50c) and (2.50d). Inverting (2.57) and taking into account the tensor $\tilde{D}^{\alpha\beta}$ to have the diagonal form $\mathcal{D}_\alpha \delta_{\alpha\beta}$ in the orthonormal basis $\mathfrak{b}$, we directly obtain

$$\sum_{\gamma=1}^{M-1} \check{u}_\gamma^\alpha \check{u}_\gamma^\beta \mathcal{D}_\gamma = \mathfrak{D}^{\alpha\beta}$$

which, together with (2.58), gives rise to formula (2.51). The Proposition is thus proved.

The following comments on Proposition 2.1 should be made. First, it is worthwhile to note that the basis vector $\mathbf{b}_M$ constructed by (2.45) is actually the vector $\mathbf{b}$ of boundary singularities (2.16) normalized to unity.

Second, for the initial basis $\mathfrak{e}$ of the general form (2.51) persuades us to introduce the surface diffusion tensor

$$\mathfrak{D}^{ij} \doteq D^{ij} - \frac{\sum_{p,k=1}^{M} D_p^i D_k^j n^p n^p}{\sum_{p,k=1}^{M} D_{pk} n^k n^p} \tag{2.66}$$

which describes the system random motion along the hyperplane Υ. Indeed, in a basis $\mathfrak{b}_\Upsilon \oplus \mathbf{n}$ the components of this tensor belonging to the hyperplane Υ coincide with ones given by (2.51) and are equal to zero when one of its indices matches the vector $\mathbf{n}$.

Third, when the initial basis $\mathfrak{e}$ is orthonormal the expressions of Proposition 2.1 can be simplified. Indeed, in this case the metric tensor $g_{ij} \doteq (\mathbf{e}_i \cdot \mathbf{e}_j) = \delta_{ij}$ is the unit matrix and it is possible not to distinguish between the upper and lower tensor indices; in particular, all the components $D_{ij} = D_j^i = D^{ij}$ are identical. If, in addition, the initial basis has the form $\mathfrak{e} = \mathfrak{e}_\Upsilon \oplus \mathbf{n}$ expressions (2.45)–(2.48) become

$$\mathbf{b}_M = \frac{1}{\omega}\left[\sum_{\gamma=1}^{M-1} \mathbf{e}_\gamma D_{\gamma M} + \mathbf{n} D_{MM}\right], \tag{2.67}$$

where the coefficient ω in (2.46) is

$$\omega = \left[\sum_{\gamma=1}^{M-1} D_{\gamma M}^2 + D_{MM}^2\right]^{1/2} \tag{2.68}$$

and the value of $\mathcal{D}_M$ in (2.48) is

$$\mathcal{D}_M = \frac{1}{D_{MM}}\sum_{\gamma=1}^{M-1} D_{\gamma M}^2 + D_{MM}. \tag{2.69}$$

Also, the inverse transformation matrix $\|\check{u}_{\alpha\beta}\|$ coincides with the direct transformation matrix transposed, so $\check{u}_{\alpha\beta} = u_{\beta\alpha}$.

Proposition 2.1 prompts us to use the basis $\mathfrak{b} = \{\mathbf{b}_i\}$ in describing random walks in the half-space $\mathbb{R}^{M+}$. For its *internal* points the continuous random walks are represented as a collection of mutually independent one-dimensional Markovian processes $\{\zeta^i(t)\}$

$$\mathbf{r}(t) = \mathbf{b}_i \zeta^i(t) = \mathbf{b}_i \int_0^t dt' \, \xi^i(t'), \tag{2.70}$$

where the Langevin random forces $\{\xi^i(t)\}$ meet the correlations

$$\left\langle \xi^i(t) \right\rangle = v^i, \tag{2.71}$$

$$\left\langle \xi^i(t) \xi^{i'}(t') \right\rangle = 2\mathcal{D}_i \, \delta_{ii'} \, \delta(t - t'), \tag{2.72}$$

and $\{v^i\}$ are the components of the drift velocity $\mathbf{v} = \mathbf{b}_i v^i$ in the basis $\mathfrak{b}$. As could be shown directly, these random forces lead to expressions (2.37) and (2.38).

2.4.2
Equivalent Lattice Random Walks

The desired lattice is constructed as follows (see also Figure 2.5 for illustration). For the first step a set of nodes $\{\mathbf{a}_\Upsilon\}$ is fixed on the boundary Υ such that

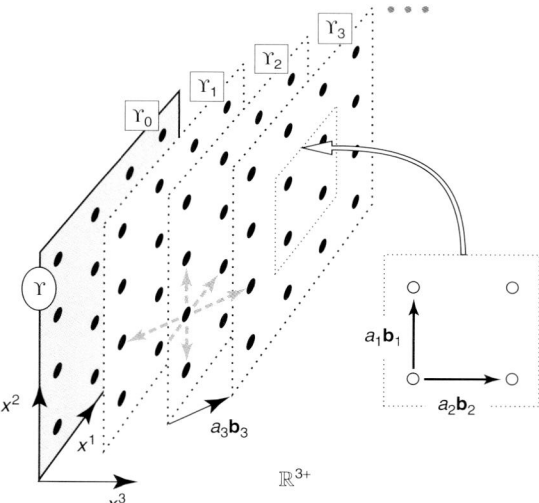

Figure 2.5 The lattice random walks imitating a continuous Markovian process in the half-space R^{3+}. Here Υ is the boundary of R^{3+}, the axes x^1, x^2 are chosen to be directed along the vectors $\mathbf{b}_1$, $\mathbf{b}_2$ of the basis $\mathfrak{b}_\Upsilon$, the axis x^3 is normal to the plane Υ, whereas the basic vector $\mathbf{b}_3$ is not normal to it, in the general case. The values a_1, a_2, a_3 are the lattice spacings and gray arrows show possible hops to the nearest neighbors.

$$\mathbf{a}_\Upsilon(\mathbf{n}_\Upsilon) = \sum_{\alpha=1}^{M-1} \mathbf{b}_\alpha a_\alpha n_\alpha, \tag{2.73}$$

where $\mathbf{n}_\Upsilon = \{n_1, n_2, \ldots, n_{M-1}\}$ is a collection of integers taking values in $\mathbb{Z}$ and the lattice spacings a_α are chosen to be equal to

$$a_\alpha = \sqrt{2\tau_a M \mathcal{D}_\alpha}. \tag{2.74a}$$

Here τ_a is any small time scale meeting inequality (2.36) and being the time step of lattice random walks; a walker hops to one of the nearest neighbors in time τ_a. Such jumps are illustrated by gray arrows in Figure 2.5. These nodes are regarded as the boundary layer Υ_0 of the lattice to be constructed. Then the layer Υ_0 as a whole is shifted inwards in the region $\mathbb{R}^{M+}$ by the vector $a_M \mathbf{b}_M$, where

$$a_M = \sqrt{2\tau_a M \mathcal{D}_M}. \tag{2.74b}$$

Then this new layer Υ_1 in turn is shifted by the same vector $\mathbb{R}^{M+}$, giving rise to the next layer Υ_2 of nodes, and so on. In this way we construct the system of layers $\{\Upsilon_k\}$ making up the desired lattice and random walks on this lattice will imitate the continuous process in the half-space $\mathbb{R}^{M+}$.

Let us now specify the probability of hops from an internal node $\mathbf{n}$ to one of its nearest neighbors $\mathbf{n}'$ along a basis vector $\mathbf{b}_i$ by the expression

$$P_{\mathbf{n}\mathbf{n}'} = \frac{1}{2M} + \frac{\tau_a}{2a_i} v^i \chi_i. \tag{2.75}$$

Here the random value $\chi_i = \pm 1$ takes into accounts the possibility of jumps along the vector $\mathbf{b}_i$ or in the opposite direction. The sequence of such hops with time step τ_a represents equivalently the continuous process quite far from the boundary Υ. Indeed, due to the law of large numbers (see, e.g. [47]) two Markovian processes are identical if, on quite a small time scale, both of them lead to the same mean and mean-square values of the system displacement. By virtue of (2.75) one hop of the walker is characterized by the following mean values of its displacement $\delta \mathbf{r} = \mathbf{b}_i \delta \zeta^i$

$$\sum_{\mathbf{n}'} P_{\mathbf{n}\mathbf{n}'} \delta \zeta^i_{\mathbf{n}\mathbf{n}'} = \tau_a v^i, \tag{2.76}$$

$$\sum_{\mathbf{n}'} P_{\mathbf{n}\mathbf{n}'} \delta \zeta^i_{\mathbf{n}\mathbf{n}'} \delta \zeta^j_{\mathbf{n}\mathbf{n}'} = 2\tau_a \mathcal{D}_i \delta_{ij}, \tag{2.77}$$

where the sums run over all the nearest neighbors $\mathbf{n}'$ of the node $\mathbf{n}$. According to (2.37) and (2.38) the same mean values of the system displacement during the time interval τ_a are given by the continuous random process. Rigorously speaking, the latter mean value and one corresponding to the continuous random process are not identical, but their difference

$$(\mathbf{b}_i \cdot \mathbf{b}_j) v^i v^j \tau_a^2$$

is of the second order in the time scale τ_a, whereas the leading terms are of the first order. So by choosing the time scale τ_a to be sufficiently small, we can make this difference negligible.

2.4.3
Properties of the Boundary Layer

In order to describe the boundary effects on random walks, special properties should be ascribed to the nodes of the boundary layer Υ_0. It is worthwhile to note that it is the place where the model of the medium boundary appears for the first time.

Keeping in mind the boundary types discussed in Section 2.3; first each boundary node is regarded as a unit of two elements, the lattice node itself and a trap. If a walker jumps to a trap it will never return to the lattice nodes. The introduction of traps mimics the absorption effect of medium boundaries. Second, possible fast diffusion inside a thin layer adjacent to medium boundaries is imitated in terms of multiple steps over the boundary nodes during the time interval τ_a. These constructions are illustrated in Figure 2.6.

For the walker located at a certain boundary node the probabilities of hopping to the internal neighboring node, P_l, or being trapped, P_{tr} are specified as

$$P_l = \frac{1 - \sigma_a}{M}, \qquad P_{tr} = \frac{\sigma_a}{M}, \tag{2.78}$$

where the coefficient σ_a quantifies the trapping (absorption) effect. Leaping ahead, we note that the coefficient σ_a can be assumed to be a small value because its magnitude $\sigma_a \to 0$ as $\tau_a \to 0$ within the collection of lattices leading to the equivalent description of the random walks on time scales $\tau_a \ll t \ll \tau$. These probabilities have been chosen to constitute the probability of walker motion along the direction of the basis vector $\mathbf{b}_M$ equal to the same value for the internal points,

$$P_l + P_{tr} = \frac{1}{M}. \tag{2.79}$$

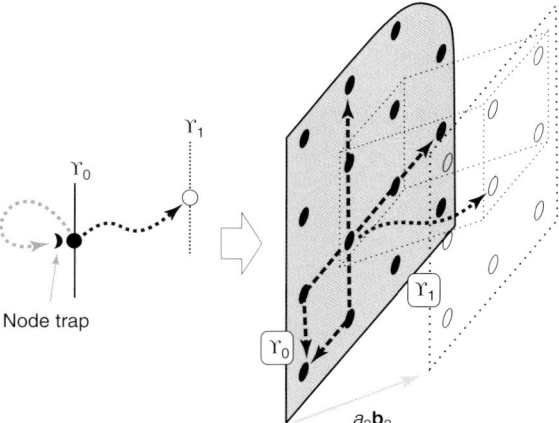

Figure 2.6 Characteristic properties of random walks in the boundary layer Υ_0. The left inset illustrates possible hops from the boundary layer. The main fragment illustrates the walker jumps inside the boundary layer Υ_0 which can be complex and comprises many elementary hops. The latter feature imitates possible fast diffusion inside a certain thin layer adjacent to crystal boundaries.

Therefore the probability of the walker being initially at a boundary node $\mathbf{n}_\Upsilon$ and making a jump within the boundary layer Υ_0 is

$$P_\Upsilon = \frac{M-1}{M}. \tag{2.80}$$

First, let us consider the case where such jumps are the elementary hops to one $\mathbf{n}'_\Upsilon$ of the nearest neighboring nodes in Υ_0. Then, following construction (2.75) its conditional probability is written as

$$P^{(1)}_{\mathbf{n}_\Upsilon \mathbf{n}'_\Upsilon} = \frac{1}{2(M-1)} + \frac{M}{(M-1)} \frac{\tau_a}{2a_\alpha} v^\alpha_\Upsilon \chi_\alpha. \tag{2.81}$$

Here, as before, the value $\chi_\alpha = \pm 1$ is ascribed to the walker hop along the basis vector $\mathbf{b}_\alpha$ or in the opposite direction, v^α_Υ are the components of the drift velocity inside the boundary layer in the basis $\mathfrak{b}_\Upsilon$. It should be noted that regular drift inside the boundary layer and the medium can be different in nature, which is allowed for by the index Υ at the boundary components of the drift velocity. The adopted expression (2.81), as it must, obeys equalities similar to expressions (2.76), (2.77), namely, for the displacement $\delta \mathbf{r}_\Upsilon = \mathbf{b}_\alpha \delta \zeta^\alpha$ along the boundary Υ

$$P_\Upsilon \sum_{\mathbf{m} \in \Upsilon} P^{(1)}_{\mathbf{n}_\Upsilon \mathbf{m}} \delta \zeta^\alpha_{\mathbf{n}_\Upsilon \mathbf{m}} = \tau_a v^\alpha_\Upsilon, \tag{2.82}$$

$$P_\Upsilon \sum_{\mathbf{m} \in \Upsilon} P^{(1)}_{\mathbf{n}_\Upsilon \mathbf{m}} \delta \zeta^\alpha_{\mathbf{n}_\Upsilon \mathbf{m}} \delta \zeta^\beta_{\mathbf{n}_\Upsilon \mathbf{m}} = 2\tau_a \mathcal{D}_\alpha \delta_{\alpha\beta}. \tag{2.83}$$

The fast diffusion inside the boundary layer Υ is imitated by complex jumps made up of g successive elementary hops within the time τ_a. In this case the walker can get not only the nearest neighboring nodes, but also relatively distant ones. The conditional probability of such a g-fold jump from node $\mathbf{n}_\Upsilon$ to node $\mathbf{n}'_\Upsilon$ is given by the expression

$$P^{(g)}_{\mathbf{n}_\Upsilon \mathbf{n}'_\Upsilon} = \sum_{\mathbf{m}_1, \mathbf{m}_2, \ldots, \mathbf{m}_{g-1} \in \Upsilon} P^{(1)}_{\mathbf{n}_\Upsilon \mathbf{m}_1} \times P^{(1)}_{\mathbf{m}_1 \mathbf{m}_2} \times \cdots$$
$$\times P^{(1)}_{\mathbf{m}_{g-2} \mathbf{m}_{g-1}} \times P^{(1)}_{\mathbf{m}_{g-1} \mathbf{n}'_\Upsilon}. \tag{2.84}$$

By virtue of (2.82) and (2.83) the probability function $P^{(g)}_{\mathbf{n}_\Upsilon \mathbf{n}'_\Upsilon}$ of g-fold jumps gives the following values for the first and second moments of the walker displacement $\delta \mathbf{r}_\Upsilon = \sum_{\alpha=1}^{M-1} \mathbf{b}_\alpha \delta \zeta^\alpha$ in the layer Υ_0

$$P_\Upsilon \sum_{\mathbf{m} \in \Upsilon} P^{(g)}_{\mathbf{n}_\Upsilon \mathbf{m}} \delta \zeta^\alpha_{\mathbf{n}_\Upsilon \mathbf{m}} = g\tau_a v^\alpha_\Upsilon, \tag{2.85}$$

$$P_\Upsilon \sum_{\mathbf{m} \in \Upsilon} P^{(g)}_{\mathbf{n}_\Upsilon \mathbf{m}} \delta \zeta^\alpha_{\mathbf{n}_\Upsilon \mathbf{m}} \delta \zeta^\beta_{\mathbf{n}_\Upsilon \mathbf{m}} = 2(g\tau_a) \mathcal{D}_\alpha \delta_{\alpha\beta} + (g\tau_a)^2 \frac{M}{(M-1)} v^\alpha_\Upsilon v^\beta_\Upsilon. \tag{2.86}$$

In expression (2.86) we again have ignored terms of order $g\tau_a^2$ because the displacement of a walker along the boundary Υ caused by its migration inside the layer Υ_0 is considerable only for $g \gg 1$ as will be seen further. In the latter case the conditional probability (2.84) of transition from the node $\mathbf{n}_\Upsilon$ to the node

$$\mathbf{n}'_\Upsilon = \mathbf{n}_\Upsilon + \sum_\alpha \mathbf{b}_\alpha m_\alpha \quad (m_\alpha \text{ are integers}) \tag{2.87}$$

can be approximated by the Gaussian distribution

$$P^{(g)}_{\mathbf{n}_\Upsilon \mathbf{n}'_\Upsilon} = \left(\frac{M-1}{2\pi g}\right)^{\frac{M-1}{2}} \exp\left\{-\frac{(M-1)}{2g}\sum_{\alpha=1}^{M-1}\left[m_\alpha - \frac{gM}{(M-1)}\frac{\tau_a v_\Upsilon^\alpha}{a_\alpha}\right]^2\right\} \tag{2.88}$$

by virtue of the law of large numbers and expressions (2.74a), (2.85), (2.86).

The desired lattice random walks imitating the continuous Markovian process in the vicinity of the medium boundary Υ is therefore constructed.

2.5
Expression for Boundary Singularities

As discussed in Section 2.2.2, the medium boundary Υ breaks down the symmetry of random walks in its vicinity, which is reflected in the anomalous behavior of quantities (2.8)–(2.10) near the boundary Υ. To quantify this effect it is necessary to calculate the given integrals near the boundary Υ for any small time interval τ.

Quantities (2.8)–(2.10) comprise two types of terms which differ in scaling with respect to τ; regular components proportional to τ and anomalous one scaling as $\sqrt{\tau}$. In the present section only the latter terms are under consideration. When deriving the Fokker–Planck equations their division by τ gives rise to the singularity $\tau^{-1/2}$. Their cofactors quantify the influence of the boundary on Markovian processes and by setting them equal to zero we can relate the boundary values of the Green function $G(\mathbf{r}, t|\mathbf{r}_0, t_0)$ to the physical properties of the medium boundaries.

Assuming the time scale τ to be sufficiently small the medium in a certain neighborhood $\mathbb{Q}_\mathbf{s}$ of a boundary point $\mathbf{s} \in \Upsilon$ is treated as a homogeneous continuum with time independent characteristics and the corresponding fragment of the boundary Υ is approximated by a hyperplane. In this case it is natural to choose the coordinate system related to a basis $\mathfrak{e} = \mathfrak{e}_\Upsilon \oplus \mathbf{n}$, which, in particular, reduces the number of the Green function arguments,

$$G(\mathbf{r}, x_0^M|\tau) \coloneqq G(\mathbf{r}, t_0 + \tau|\{\mathbf{0}_\Upsilon, x_0^M\}, t_0).$$

The system origin was located at the hyperplane Υ such that the vector $\mathbf{r}_0 = \{\mathbf{0}_\Upsilon, x_0^M\}$ can have only one component x_0^M determining the distance between the point $\mathbf{r}_0$ and the hyperplane Υ. Then, using the general definitions (2.8)–(2.10) of the quantities $\mathfrak{R}(\mathbf{r}, t, \tau)$, $\mathfrak{U}^i(\mathbf{r}, t, \tau)$, and $\mathfrak{L}^{ij}(\mathbf{r}, t, \tau)$ the anomalous properties of random walks near the boundary Υ are quantified by their singular components $*\mathfrak{U}^i(\tau, x^M)$

and $^*\mathfrak{L}^{ij}(\tau, x^M)$ scaling as $\sqrt{\tau}$. The symbol $*$ is not applied to $\mathfrak{R}(\tau, x^M)$ because it possesses no regular component at all. In other words the desired quantities are determined by the following means

$$\int_{\mathbb{Q}_s} d\tilde{\mathbf{r}}\, G(\tilde{\mathbf{r}}, x^M|\tau) = 1 - \mathfrak{R}(\tau, x^M), \tag{2.89}$$

$$\int_{\mathbb{Q}_s} d\tilde{\mathbf{r}}\, \delta\tilde{x}^i\, G(\tilde{\mathbf{r}}, x^M|\tau) = {}^*\mathfrak{L}^i(\tau, x^M) + O(\tau), \tag{2.90}$$

$$\frac{1}{2}\int_{\mathbb{Q}_s} d\tilde{\mathbf{r}}\, \delta\tilde{x}^i \delta\tilde{x}^j\, G(\tilde{\mathbf{r}}, x^M|\tau) = {}^*\mathfrak{L}^{ij}(\tau, x^M) + O(\tau), \tag{2.91}$$

where $\delta\tilde{x}^\alpha = \tilde{x}^\alpha$ and $\delta\tilde{x}^M = \tilde{x}^M - x^M$.

In order to calculate these boundary singularities we first fix the value τ and introduce a new time scale $\tau_a \ll \tau$. Then the lattice constructed is Section 2.4 and random walks on it are applied to calculate the desired quantities. There are two advantages in using these lattice random walks. First, the choice of the basis $\mathfrak{b} = \mathfrak{b}_\Upsilon \oplus \mathfrak{b}_M$ enables us to simulate the continuous Markovian process as independent random walks along the directions parallel to the hyperplane Υ and along the vector $\mathbf{b}_M$. Second, it becomes possible to ascribe special features to the nodes of the boundary layer and in this way to simulate some physical properties of the medium boundary. In particular, it either can absorb a random walker or cause it to migrate extremely quickly along the boundary within a thin layer. Finally, to restore the continuous description the limit $\tau_a \to 0$ is used.

The implementation of this approach again is based on just mathematical manipulations with the probability function for lattice random walks. So only the final results are stated here and the reader is referred to Section 2.6 for the proof.

Proposition 2.2 *Let us consider a Markovian system in a homogeneous half-space $\mathbb{R}^{M+}$ bounded by a hyperplane Υ and endowed with the basis $\mathfrak{b} = \mathfrak{b}_\Upsilon \oplus \mathfrak{b}_M$ described in Proposition 2.1,*

$$\mathbf{r} = \sum_{\gamma=1}^{M-1} \mathbf{b}_\gamma \varsigma^\gamma + \mathbf{b}_M \varsigma^M.$$

The hyperplane Υ treated as a physical boundary can absorb the system as well as force it to migrate quickly along Υ. The diffusion tensor D^{ij} as well as the drift velocity v^i at the internal points and v^i_Υ at the boundary Υ are assumed to be determined in the basis $\mathfrak{b}$. It should be noted that the boundary drift velocity v^i_Υ is the velocity at which the system would have moved outside the boundary if it had been affected by the same forces.

The continuous motion of the Markovian system is imitated by random walks on the lattice constructed in Section 2.4 with time step τ_a. Finally, the limit $\tau_a \to 0$ is applied.

Then, first, the boundary absorption and fast transport can be characterized by two kinetic coefficients called the surface absorption rate σ and the surface diffusion length l_Υ ascribed directly to the boundary Υ itself, implying that these quantities are independent of the discretization time τ_a.

Second, random walks near the hyperplane Υ exhibit anomalous properties reflected in the following singular mean scaling with the time τ as $\sqrt{\tau}$:

$$\mathfrak{R}_b(\tau, \zeta^M) = D_{MM}^{-1/2} \sigma \cdot \mathcal{K}(\tau, \zeta^M), \tag{2.92}$$

$${}^*\mathfrak{U}_b^M(\tau, \zeta^M) = D_{MM}^{-1/2} \omega \cdot \mathcal{K}(\tau, \zeta^M), \tag{2.93}$$

$${}^*\mathfrak{U}_b^\alpha(\tau, \zeta^M) = D_{MM}^{-1/2} l_\Upsilon v_\Upsilon^\alpha \cdot \mathcal{K}(\tau, \zeta^M), \tag{2.94}$$

$${}^*\mathfrak{L}_b^{\alpha\beta}(\tau, \zeta^M) = D_{MM}^{-1/2} l_\Upsilon \mathcal{D}_\alpha \delta_{\alpha\beta} \cdot \mathcal{K}(\tau, \zeta^M). \tag{2.95}$$

Here the label b denotes that the basis $\mathfrak{b}$ used, ζ^M is the distance between the point $\mathbf{r}$ and the hyperplane Υ measured along the vector $\mathbf{b}_M$, and the function $\mathcal{K}(\tau, \zeta^M)$ is specified by the integral

$$\mathcal{K}(\tau, \zeta^M) = \sqrt{\frac{\tau}{\pi}} \int_0^1 \frac{dz}{\sqrt{z}} \exp\left[-\frac{(\zeta^M)^2}{4\mathcal{D}_M \tau} \frac{1}{z}\right]. \tag{2.96}$$

In order to represent these boundary singularities in the initial basis $\mathfrak{e}$ Proposition 2.1 is again applied. The initial basis has been assumed to be of the form $\mathfrak{e} = \mathfrak{e}_\Upsilon \oplus \mathbf{n}$ with the unit normal $\mathbf{n}$ to the boundary Υ directed inwards into the medium. Let $\hat{U}_\Upsilon^{-1} = \|\breve{u}_\beta^\alpha\|$ be the operator mapping the boundary basis $\mathfrak{e}_\Upsilon$ onto the basis $\mathfrak{b}_\Upsilon$. Then transition from the coordinates $\{\zeta^\alpha\}, \zeta^M$ of a vector $\mathbf{r}$ in the basis $\mathfrak{b}$ to its coordinates $\{x^\alpha\}, x^M$ in the basis $\mathfrak{e}$ is specified by (2.50c) and (2.50d) using the tensor $\breve{u}_\beta^\alpha$ and the diffusion tensor D^{ij} determined in the *initial* basis $\mathfrak{e}$. In the vector form these coordinates are related by (2.49). The quantities ${}^*\mathfrak{U}_b^i(\tau, \zeta^M)$ and ${}^*\mathfrak{L}_b^{ij}(\tau, \zeta^M)$ are obtained by averaging variations of the coordinates ζ^i. So they are a contravariant vector and tensor, respectively, with the latter being proportional to the diffusion tensor written in the basis $\mathfrak{b}$ and reduced to the hyperplane Υ, namely, the tensor $\mathcal{D}_\alpha \delta_{\alpha\beta}$. The value $\mathfrak{R}_b(\tau, \zeta^M)$ is naturally a scalar. Whence it follows directly that

$$\mathfrak{R}(\tau, x^M) = D_{MM}^{-1/2} \sigma \cdot \mathcal{K}(\tau, x^M), \tag{2.97}$$

$${}^*\mathfrak{U}^i(\tau, x^M) = D_{MM}^{-1/2} \left[D^{iM} + l_\Upsilon v_\Upsilon^i\right] \cdot \mathcal{K}(\tau, x^M), \tag{2.98}$$

$${}^*\mathfrak{L}^{\alpha\beta}(\tau, x^M) = D_{MM}^{-1/2} l_\Upsilon \mathfrak{D}^{\alpha\beta} \cdot \mathcal{K}(\tau, x^M). \tag{2.99}$$

Here the coordinate x^M and ζ^M are inter-related by (2.50d) and the boundary diffusion tensor $\mathfrak{D}^{\alpha\beta}$ is specified by (2.66). Formula (2.98) can also be rewritten in the vector form

$${}^*\mathfrak{U}(\tau, x^M) = D_{MM}^{-1/2} [\mathbf{b} + l_\Upsilon \mathbf{v}_\Upsilon] \mathcal{K}(\tau, x^M), \tag{2.100}$$

where the vector $\mathbf{b}$ of boundary singularities is given by (2.16).

2.6
Derivation of Singular Boundary Scaling Properties

The homogeneous half-space $\mathbb{R}^{M+}$ bounded by the hyperplane Υ is under consideration and the lattice described in Section 2.4 is constructed. It is made up of the node layers $\{\Upsilon_i\}$ parallel to the hyperplane Υ with the interplane spacing vector $a_M \mathbf{b}_M$. The individual node arrangement of the layers Υ_i is determined by the vectors of the hyperplane basis $\mathfrak{b}_\Upsilon$ with spacings $\{a_\alpha\}$. In other words, the nodes of this lattice are the points

$$\mathbf{r_n} = \sum_{\alpha=1}^{M-1} n^\alpha \left(a_\alpha \mathbf{b}_\alpha\right) + n a_M \mathbf{b}_M,$$

where $\mathbf{n}$ is the collection of numbers $\{\mathbf{n}_\Upsilon, n\} = \{\{n^\alpha\}, n\}$ taking any integer value, $n^\alpha|_1^{M-1} = 0, \pm 1, \pm 2, \ldots$, except for the last one; it takes only non-negative values $n = 0, 1, 2, \ldots$ In particular, the points $\{\mathbf{r_n}\}_\Upsilon$ with $n = 0$ form the boundary layer Υ_0.

The Markovian process in the half-space $\mathbb{R}^{M+}$ is simulated by random walks on this lattice with the hop probabilities as given in Section 2.4. To find the desired boundary singularities we will analyze evolution of the walker distribution over the given lattice, that is the dynamics of the probability $\mathcal{P}_{t,\mathbf{n}}$ to find the walker at node $\mathbf{n}$ after hop t. Here t is the time measured in jump numbers, that is, in units of the hop duration τ_a. At the initial time $t = 0$ the walker is assumed to be located at a certain internal node $\mathbf{n}_0$. Without lost of generality all the components of the index $\mathbf{n}_0$ can be set equal to zero except for the last one: $\mathbf{n}_0 = \{0, 0, \ldots, 0, n_0\}$.

2.6.1
Moments of the Walker Distribution and the Generating Function

Actually the main purpose here is to find the zeroth, first, and second-order moments of the distribution function $\mathcal{P}_{t,\mathbf{m}}$. The zeroth moment quantifies the trapping effect, whereas the fist and second ones characterize the walker propagation in space. Namely, the following quantities

$$\mathfrak{R}_a(t, n_0) = 1 - \sum_{n=0}^{\infty} \sum_{\mathbf{n}_\Upsilon} \mathcal{P}_{t,\{\mathbf{n}_\Upsilon, n\}}, \tag{2.101}$$

$$\mathfrak{U}_a^i(t, n_0) = \sum_{n=0}^{\infty} \sum_{\mathbf{n}_\Upsilon} \left(n^i - n_0^i\right) \mathcal{P}_{t,\{\mathbf{n}_\Upsilon, n\}}, \tag{2.102}$$

$$\mathfrak{L}_a^{ij}(t, n_0) = \frac{1}{2} \sum_{n=0}^{\infty} \sum_{\mathbf{n}_\Upsilon} \left(n^i - n_0^i\right)\left(n^j - n_0^j\right) \mathcal{P}_{t,\{\mathbf{n}_\Upsilon, n\}} \tag{2.103}$$

have to be calculated. Here the index i is used as a general symbol for one of the indices $\{\alpha\}, M$. In order to do this the generating function and its analogy, written for the boundary nodes only

$$G(s, p, \mathbf{k}_\Upsilon) = \sum_{\substack{t=0 \\ n=0}}^{\infty} \sum_{\mathbf{n}_\Upsilon} e^{-st - p(n-n_0) + i(\mathbf{k}_\Upsilon \cdot \mathbf{n}_\Upsilon)} \mathcal{P}_{t,\{\mathbf{n}_\Upsilon, n\}}, \quad (2.104)$$

$$g(s, \mathbf{k}_\Upsilon) = \sum_{t=0}^{\infty} \sum_{\mathbf{n}_\Upsilon} e^{-st + i(\mathbf{k}_\Upsilon \cdot \mathbf{n}_\Upsilon)} \mathcal{P}_{t,\{\mathbf{n}_\Upsilon, 0\}} = \lim_{p \to \infty} \left[e^{-p n_0} G(s, p, \mathbf{k}_\Upsilon) \right] \quad (2.105)$$

are introduced, where the complex arguments s, p have the positive real parts, Re s, Re $p \geq 0$. It should be noted that the traps are not included in these sums. The discrete Laplace transforms of the desired functions (2.101)–(2.103) are directly related to the generating function. Indeed

$$\mathfrak{R}_a(s, n_0) = \sum_{t=0}^{\infty} e^{-st} \mathfrak{R}_a(t, n_0) = \frac{1}{(1 - e^{-s})} - G(s, 0, \mathbf{0}), \quad (2.106)$$

$$\mathfrak{U}_a^i(s, n_0) = \sum_{t=0}^{\infty} e^{-st} \mathfrak{U}_a^i(t, n_0) = \nabla_i \, G(s, p, \mathbf{k}_\Upsilon) \big|_{p, \mathbf{k}_\Upsilon = 0}, \quad (2.107)$$

$$\mathfrak{L}_a^{ij}(s, n_0) = \sum_{t=0}^{\infty} e^{-st} \mathfrak{L}_a^{ij}(t, n_0) = \frac{1}{2} \nabla_i \nabla_j \, G(s, p, \mathbf{k}_\Upsilon) \big|_{p, \mathbf{k}_\Upsilon = 0}, \quad (2.108)$$

where the operator ∇_i is $\nabla_\alpha = -i \partial_{k^\alpha}$ if the index $i = \alpha$ is one of the indices of the hyperplane Υ and $\nabla_M = -\partial_p$ for the index $i = M$.

2.6.2
Master Equation for Lattice Random Walks and its General Solution

To find the generating function for the discrete random walks under consideration the corresponding master equation is applied. For an internal node $\mathbf{n} = \{\mathbf{n}_\Upsilon, n\}$ with $n \geq 2$ it takes the form

$$\mathcal{P}_{t+1, \mathbf{n}} = {\sum_{\mathbf{m}}}' \mathcal{P}_{t, \mathbf{m}} \, P_{\mathbf{mn}}. \quad (2.109)$$

Here the prime on the sum denotes the index $\mathbf{m}$ running over all the nearest neighbors of the given node $\mathbf{n}$ and according to expression (2.75) the corresponding hop probabilities can be represented as

$$P_{\mathbf{mn}} = \frac{1 + \epsilon_i \chi_i}{2M}, \quad (2.110)$$

where $\epsilon_i = \tau_a v^i M / a_i$ are some small quantities scaling with τ_a as $\epsilon_i \propto \tau_a^{1/2}$ and the value $\chi_i = \pm 1$ stands for hops along the basis vector $\mathbf{b}_i$ or in the opposite direction,

2.6 Derivation of Singular Boundary Scaling Properties

that is, the hop to the node with $m^i = n^i \pm 1$ and $m^j = n^j$ for $j \neq i$. For the nodes of the layer Υ_1 the master equation becomes

$$\mathcal{P}_{t+1,\mathbf{n}} = \sum_{\mathbf{m}}{}' \mathcal{P}_{t,\mathbf{m}} P_{\mathbf{mn}} + \mathcal{P}_{t,\mathbf{n}_b} P_l. \tag{2.111}$$

Here again the prime on the sum has the same meaning except for only internal neighboring nodes being taken into account; $\{\mathbf{n}_b, \mathbf{n}\}$ is the pair of nodes belonging to the boundary layer Υ_0 and the adjacent internal layer Υ_1 that are related to each other via walker hops, and the hop probability P_l is determined by (2.78). The walker distribution function $P_{\mathbf{n}_b,t}$ in the boundary layer obeys the equation

$$\mathcal{P}_{t+1,\mathbf{n}_b} = \sum_{\mathbf{m}_b \in \Upsilon_0} \mathcal{P}_{t,\mathbf{m}_b} P_\Upsilon P^{(g)}_{\mathbf{m}_b \mathbf{n}_b} + \mathcal{P}_{t,\mathbf{n}} P_{\mathbf{nn}_b}. \tag{2.112}$$

We recall that the jumps inside the boundary layer can be complex and comprise, individually, g elementary hops. In this case the multihop probability $P^{(g)}_{\mathbf{m}_b \mathbf{n}_b}$ is determined by formula (2.84). The one-hop probability along the basis vector $\mathbf{b}_\alpha$, provided the walker remains in the boundary layer Υ_0, is

$$P^{(1)}_{\mathbf{m}_b \mathbf{n}_b} = \frac{1 + \epsilon_\alpha^\Upsilon \chi_\alpha}{2(M-1)}, \tag{2.113}$$

where $\epsilon_\alpha^\Upsilon = \tau_a v_\Upsilon^\alpha M / a_\alpha$ is again a small parameter scaling as $\epsilon_\alpha^\Upsilon \propto \tau_a^{1/2}$. The values ϵ_α quantify the asymmetry of hops in the boundary layer Υ_0. In particular, these complex jumps are characterized by the means

$$\langle n^\alpha \rangle_\Upsilon = \sum_{\mathbf{n}_b \in \Upsilon_0} n^\alpha P^{(g)}_{0 \mathbf{n}_b} = \frac{g}{(M-1)} \epsilon_\alpha^\Upsilon, \tag{2.114}$$

$$\langle n^\alpha n^\beta \rangle_\Upsilon = \sum_{\mathbf{n}_b \in \Upsilon_0} n^\alpha n^\beta P^{(g)}_{0 \mathbf{n}_b} = \frac{g}{(M-1)} \delta_{\alpha\beta} + \frac{g(g-1)}{(M-1)^2} \epsilon_\alpha^\Upsilon \epsilon_\beta^\Upsilon. \tag{2.115}$$

Finally, the master equation for the traps is

$$\mathcal{P}^{(\mathrm{tr})}_{t+1,\mathbf{n}_b} = \mathcal{P}^{(\mathrm{tr})}_{t,\mathbf{n}_b} + \mathcal{P}_{t,\mathbf{n}_b} P_{\mathrm{tr}}. \tag{2.116}$$

The hop probabilities P_l, P_{tr}, are given by expressions (2.78) and the kinetic coefficients of walker jumps inside the boundary layer Υ_0 are specified by expressions (2.80), (2.81), and (2.84). At the initial time the walker distribution meets the condition

$$\mathcal{P}_{t=0,\mathbf{n}} = \delta_{\mathbf{nn}_0}. \tag{2.117}$$

To solve this system of equations we substitute (2.109), (2.111), and (2.112) into definition (2.104) of the generating function $G(s, p, \mathbf{k}_\Upsilon)$ and after succeeding mathematical manipulations get the following equation (see the comments about its derivation just after formula (2.120))

$$[e^s - \Phi(p, \mathbf{k}_\Upsilon)] G(s, p, \mathbf{k}_\Upsilon)$$
$$= e^s - e^{pn_0} [\Phi(p, \mathbf{k}_\Upsilon) - \phi(p, \mathbf{k}_\Upsilon)] g(s, \mathbf{k}_\Upsilon) \qquad (2.118)$$

relating the given generating functions $G(s, p, \mathbf{k}_\Upsilon)$ and $g(s, \mathbf{k}_\Upsilon)$ to each other. Here the following functions

$$\Phi(p, \mathbf{k}_\Upsilon) = \frac{1}{M}(\cosh p - \epsilon_M \sinh p) + \frac{1}{M}\sum_{\alpha=1}^{M-1}(\cos k_\alpha + i\epsilon_\alpha \sin k_\alpha), \qquad (2.119)$$

$$\phi(p, \mathbf{k}_\Upsilon) = \frac{(1-\sigma_a)}{M} e^{-p} + \frac{(M-1)}{M}\sum_{\alpha=1}^{M-1} \exp[i(\mathbf{k}_\Upsilon \cdot \mathbf{n}_\Upsilon)] P_{0\mathbf{n}_\Upsilon}^{(g)} \qquad (2.120)$$

have been constructed in deriving (2.118).

The key parts in deriving (2.118) are outlined below. The conversion in (2.104) from $t \to t+1$ leads to the line

$$G(s, p, \mathbf{k}_\Upsilon) = e^{-s}\mathcal{G}(s, p, \mathbf{k}_\Upsilon) + 1, \qquad (2.121)$$

where

$$\mathcal{G}(s, p, \mathbf{k}_\Upsilon) = \sum_{\substack{t=0\\n=0}}^{\infty}\sum_{\mathbf{n}_\Upsilon} e^{-st-p(n-n_0)+i(\mathbf{k}_\Upsilon \cdot \mathbf{n}_\Upsilon)} \mathcal{P}_{t+1,\{\mathbf{n}_\Upsilon,n\}} \qquad (2.122)$$

and the initial condition (2.117) has been taken into account. Equations (2.109), (2.111), and (2.112) relating two succeeding steps of random walks are substituted into the latter expression. As a result the terms in sums (2.109)–(2.117), matching the interlayer hops, split it into two parts

$$\mathcal{G}(s, p, \mathbf{k}_\Upsilon) \Rightarrow \Phi_1(p) G(s, p, \mathbf{k}_\Upsilon) + e^{pn_0}[\phi_1(p) - \Phi_1(p)] g(s, \mathbf{k}_\Upsilon)$$

with the latter summand caused by the fact that the boundary nodes have different properties from the internal ones. In their turn the components of sums (2.109)–(2.117) describing transitions between a given node $\mathbf{n}$ and the nodes of the same layer also split the term $\mathcal{G}(s, p, \mathbf{k}_\Upsilon)$ into two parts

$$\mathcal{G}(s, p, \mathbf{k}_\Upsilon) \Rightarrow \Phi_2(\mathbf{k}_\Upsilon) G(s, p, \mathbf{k}_\Upsilon) + e^{pn_0}[\phi_2(\mathbf{k}_\Upsilon) - \Phi_2(\mathbf{k}_\Upsilon)] g(s, \mathbf{k}_\Upsilon),$$

where the latter summand is due to fast diffusion in the boundary layer. The combination of the two last lines gives (2.118) with $\Phi(p, \mathbf{k}_\Upsilon) = \Phi_1(p) + \Phi_2(\mathbf{k}_\Upsilon)$ and $\phi(p, \mathbf{k}_\Upsilon) = \phi_1(p) + \phi_2(\mathbf{k}_\Upsilon)$.

The generating function $G(s, p, \mathbf{k}_\Upsilon)$ has no singularities in the region Re s, Re $p > 0$. Therefore the left-hand side of (2.118) is equal to zero when $e^s - \Phi(p, \mathbf{k}_\Upsilon) = 0$. Resolving the latter equality with respect to the variable p we obtain a function $p = \omega(s, \mathbf{k}_\Upsilon)$ defined by the equation

$$\Phi[\omega(s, \mathbf{k}_\Upsilon), \mathbf{k}_\Upsilon] = e^s \tag{2.123}$$

which specifies the locus in the space $\{s, p, \mathbf{k}_\Upsilon\}$ where the right-hand side of (2.118) also has to be equal to zero. The latter enables us immediately to write the boundary generating function in the form

$$g(s, \mathbf{k}_\Upsilon) = \frac{\exp[-\omega(s, \mathbf{k}_\Upsilon) n_0]}{1 - e^{-s} \phi[\omega(s, \mathbf{k}_\Upsilon), \mathbf{k}_\Upsilon]}. \tag{2.124}$$

Expressions (2.118) and (2.124) actually solve the problem, giving us the following expression for the generating function

$$G(s, p, \mathbf{k}_\Upsilon) = \frac{1}{1 - e^{-s}} + \frac{1}{[e^s - \Phi(p, \mathbf{k}_\Upsilon)]} \left\{ \frac{[\Phi(p, \mathbf{k}_\Upsilon) - 1]}{[1 - e^{-s}]} + e^{-[\omega(s, \mathbf{k}_\Upsilon) - p] n_0} \frac{[\phi(p, \mathbf{k}_\Upsilon) - \Phi(p, \mathbf{k}_\Upsilon)]}{[1 - e^{-s} \phi(\omega(s, \mathbf{k}_\Upsilon), \mathbf{k}_\Upsilon)]} \right\}, \tag{2.125}$$

where the first summand is the image of the delta function $\mathcal{P}_{t,\mathbf{n}} = \delta_{\mathbf{n}\mathbf{n}_0}$ not contributing to one of the quantities (2.101)–(2.103), the second term is due to random walks over the internal nodes, and the last one is caused by boundary effects. Formula (2.125) specifies the desired generating function in the general form.

2.6.3
Limit of Multiple-Step Random Walks on Small Time Scales

In order to find the Laplace transforms (2.106)–(2.108) it suffices to expand the generating function $G(s, p, \mathbf{k}_\Upsilon)$ into the Taylor series with respect to the arguments p and $\mathbf{k}_\Upsilon$, cutting off the series at the second-order terms. However, in the case under consideration there are additional assumptions essentially simplifying the derivation of the desired results. First, only random walks with many steps are of interest because the hop duration τ_a has been chosen to be much less then the observation time interval τ of the analyzed Markovian process, $\tau_a \ll \tau$. This means that the inequality $s \ll 1$ holds. Second, the time interval τ is regarded as any small value. So only the components of moments (2.101)–(2.103) that are characterized by scaling τ^d with the exponent d not exceeding unity ($d \leq 1$) are to be taken into account. With respect to the generating function $G(s, p, \mathbf{k}_\Upsilon)$ the latter assumption is converted to the statement that all the components of itself and its derivatives calculated at the point $\{\mathbf{k}_\Upsilon = \mathbf{0}, p = 0\}$ that scale with the argument s as s^{-d} and have the exponent d exceeding two ($d > 2$) can be ignored.

At the point $\{\mathbf{k}_\Upsilon = \mathbf{0}, p = 0\}$ according to the definitions (2.119), (2.120) the functions $\Phi(p, \mathbf{k}_\Upsilon)$ and $\phi(p, \mathbf{k}_\Upsilon)$ take the values $\Phi(0, 0) = 1$ and

$$\phi(0,0) = 1 - \frac{\sigma_a}{M}, \tag{2.126}$$

where the coefficient σ_a is considered to be a small parameter, which is justified in the limit $\tau_a \to 0$ as will be seen below. Therefore, in the adopted assumptions, (2.125) for the generating function can be rewritten as

$$G(s, p, \mathbf{k}_\Upsilon) = \frac{1}{s} + \frac{[\Phi(p, \mathbf{k}_\Upsilon) - 1]}{s^2}$$
$$+ e^{-\omega(s,0)n_0} \frac{[\phi(p, \mathbf{k}_\Upsilon) - \Phi(p, \mathbf{k}_\Upsilon)]}{s[s + 1 - \phi[\omega(s,0), 0]]}. \tag{2.127}$$

The expansion of the functions $\Phi(p, \mathbf{k}_\Upsilon)$, $\phi(p, \mathbf{k}_\Upsilon)$ with respect to p and $\mathbf{k}_\Upsilon$ at the required order is

$$\Phi(p, \mathbf{k}_\Upsilon) = 1 - \frac{\epsilon_M p}{M} + \frac{p^2}{2M} + \frac{1}{M} \sum_{\alpha=1}^{M-1} \left(i\epsilon_\alpha k_\alpha - \frac{1}{2} k_\alpha^2 \right) \tag{2.128}$$

and

$$\phi(p, \mathbf{k}_\Upsilon) = 1 - \frac{\sigma_a}{M} - \frac{p}{M} + \frac{p^2}{2M}$$
$$+ \frac{ig}{M} \sum_{\alpha=1}^{M-1} \epsilon_\alpha^\Upsilon k_\alpha - \frac{g}{2M} \sum_{\alpha,\beta=1}^{M-1} k_\alpha k_\beta \left(\delta_{\alpha\beta} + \frac{g-1}{M-1} \epsilon_\alpha^\Upsilon \epsilon_\beta^\Upsilon \right). \tag{2.129}$$

In deriving expression (2.129) equations (2.114) and (2.115) have been used. The substitution of the generating function written in the form (2.127) with approximations (2.128) and (2.129) into relations (2.106)–(2.108) yields

$$\mathfrak{R}_a(s, n_0) = \frac{\sigma_a}{M} \mathcal{K}_a(s, n_0), \tag{2.130}$$

$$\mathfrak{U}_a^M(s, n_0) = \frac{1}{M} \mathcal{K}_a(s, n_0) + \frac{\epsilon_M}{Ms^2}, \tag{2.131}$$

$$\mathfrak{U}_a^\alpha(s, n_0) = \frac{(g-1)\epsilon_\alpha^\Upsilon}{M} \mathcal{K}_a(s, n_0) + \frac{\epsilon_\alpha}{Ms^2}, \tag{2.132}$$

$$\mathfrak{L}_a^{\alpha\beta}(s, n_0) = \frac{(g-1)}{2M} \left[\delta_{\alpha\beta} + \frac{g\epsilon_\alpha^\Upsilon \epsilon_\beta^\Upsilon}{M-1} \right] \mathcal{K}_a(s, n_0) + \frac{\delta_{\alpha\beta}}{2Ms^2}, \tag{2.133}$$

$$\mathfrak{L}_a^{MM}(s, n_0) = \frac{1}{2Ms^2}, \tag{2.134}$$

where the mean $\mathfrak{L}_a^{\alpha M}(s, n_0)$ is equal to zero. Here the function $\mathcal{K}_a(s, n_0)$ is defined by the expression

$$\mathcal{K}_a(s, n_0) = \frac{\exp\left[-\omega(s,\mathbf{0})n_0\right]}{s\left[s + 1 - \phi\left(\omega(s,\mathbf{0}),\mathbf{0}\right)\right]} \qquad (2.135)$$

and we have ignored some insignificant terms where appropriate.

Previously we measured the time t in units of the hop duration τ_a and spatial coordinates $\{\zeta^i\}$ in units of the lattice spacings $\{a_i\}$ within the frame $\mathfrak{b}$. Now let us return to the initial units and deal with the corresponding spatial correlations. To do this, first (2.131)–(2.134) should be multiplied by the spacings a_M and a_α, or their products $a_\alpha a_\beta$ and a_M^2, respectively. Second, the dimensionless Laplace argument s has to be replaced by the product $s\tau_a$, because previously when applying to the discrete Laplace transformation the replacement

$$st \to s\tau_a \cdot \frac{t}{\tau_a}$$

was used obliquely. Third, for the further conversion of the discrete Laplace transformation into a continuous one within the replacement

$$\tau_a \sum_{t/\tau_a=0}^{\infty} \to \int_0^{\infty} dt(\ldots)$$

all the functions (2.130)–(2.134) must be multiplied by the time scale τ_a.

Leaping ahead we note that the absorption coefficient σ_a has to scale with τ_a as $\sigma_a \propto \sqrt{\tau_a}$. As noted before, the coefficients $\{\epsilon_i\}$ also behave in this way. Therefore the observation time interval τ can be chosen to be so small that the solution of (2.123) becomes

$$\omega(s\tau_a, \mathbf{0}) = \sqrt{2Ms\tau_a} \qquad (2.136)$$

and function (2.135) matches a continuous Laplace transform

$$\tau_a \mathcal{K}_a(s\tau_a, n_0) = \sqrt{\frac{M}{2\tau_a}} \mathcal{K}(s, \zeta_0) \qquad (2.137)$$

given by the expression

$$\mathcal{K}(s, \zeta_0^M) = s^{-3/2} \exp\left(-\zeta_0 \sqrt{\frac{s}{\mathcal{D}_M}}\right), \qquad (2.138)$$

with $\zeta_0^M = a_M n_0$ being the distance from the node of the walker's initial position to the medium boundary Υ along the vector $\mathbf{b}_M$.

Indeed, first, if we ignore the second term on the right-hand side of expansion (2.128) the solution of (2.123) for $s\tau_a \ll 1$ and $\mathbf{k}_\Upsilon = \mathbf{0}$ is of the form (2.136). It is justified when $\omega \gg \epsilon_M$, which is equivalent to the condition $s \gg v_M^2/\mathcal{D}_M$ or $\tau \ll \mathcal{D}_M/v_M^2$. Second, according to expansion (2.129) the denominator in expression (2.135) at the leading order is

$$\left[s\tau_a + 1 - \phi\left(\omega(s\tau_a, \mathbf{0}), \mathbf{0}\right)\right] = \frac{\omega(s\tau_a, \mathbf{0})}{M}$$

provided $\omega(s\tau_a, \mathbf{0}) \gg \sigma_a$. Because $\sigma_a \sim \sqrt{\varepsilon\tau_a}$, where ε is some constant, the latter inequality is reduced to the following $s \gg \varepsilon$ and $\tau \ll \varepsilon$. Since the time interval is an arbitrary small value, the two inequalities can be adopted beforehand. So formula (2.136) and (2.137) follow immediately for the spacing a_M given by expression (2.74b).

2.6.4
Continuum Limit and a Boundary Model

To get the final results we analyze the obtained expression in the limit $\tau_a \to 0$. The probability distribution $\mathcal{P}_{t,\mathbf{m}}$ of the lattice random walks can be treated as a discrete implementation of the Green function $G(\mathbf{r}, \mathbf{r}_0, t)$ giving the probability density of finding a walker at the point $\mathbf{r}$ at time t provided it was initially at the point $\mathbf{r}_0$. Using the Green function $G(\mathbf{r}, \mathbf{r}_0, t)$ the means under consideration are written as the following moments

$$\mathfrak{R}(t, \zeta_0) = 1 - \int_{\mathbb{R}^{M+}} d\mathbf{r}\, G(\mathbf{r}, \mathbf{r}_0, t), \tag{2.139}$$

$$\mathfrak{U}_b^i(t, \zeta_0) = \int_{\mathbb{R}^{M+}} d\mathbf{r}(\zeta^i - \zeta_0^i) G(\mathbf{r}, \mathbf{r}_0, t), \tag{2.140}$$

$$\mathfrak{L}_b^{ij}(t, \zeta_0) = \frac{1}{2} \int_{\mathbb{R}^{M+}} d\mathbf{r}(\zeta^i - \zeta_0^i)(\zeta^j - \zeta_0^j) G(\mathbf{r}, \mathbf{r}_0, t), \tag{2.141}$$

and their Laplace transforms can be obtained from the quantities (2.130)–(2.134) in the manner described in the previous subsection. As a result we have

$$\mathfrak{R}(s, \zeta_0) = D_{MM}^{-1/2} \sigma \mathcal{K}(s, \zeta_0), \tag{2.142}$$

$$\mathfrak{U}_b^M(s, \zeta_0) = D_{MM}^{-1/2} \omega \mathcal{K}(s, \zeta_0) + \frac{v^M}{s^2}, \tag{2.143}$$

$$\mathfrak{U}_b^\alpha(s, \zeta_0) = D_{MM}^{-1/2} l_\Upsilon v_\Upsilon^\alpha \mathcal{K}(s, \zeta_0) + \frac{v^\alpha}{s^2}, \tag{2.144}$$

$$\mathfrak{L}_b^{\alpha\beta}(s, \zeta_0) = \left[D_{MM}^{-1/2} l_\Upsilon \mathcal{D}_\alpha \mathcal{K}(s, \zeta_0) + \frac{\mathcal{D}_\alpha}{s^2} \right] \delta_{\alpha\beta}, \tag{2.145}$$

$$\mathfrak{L}_b^{MM}(s, \zeta_0) = \frac{\mathcal{D}_M}{s^2} \tag{2.146}$$

the component $\mathfrak{L}_b^{\alpha M}(s, \zeta_0)$ is equal to zero. Here the following characteristics of the medium boundary, treated as an infinitely thin layer Υ

$$\sigma \doteq \sigma_a \sqrt{\frac{D_{MM}}{2M\tau_a}}, \qquad l_\Upsilon \doteq g\sqrt{\frac{MD_{MM}\tau_a}{2}} \tag{2.147}$$

have been introduced and expression (2.48) has been used. It should be noted that, according to (2.147) the number g of elementary hops forming the long distant jumps of walkers in the boundary layer Υ_0 has to grow with τ_a as $\tau_a^{-1/2}$ in order to retain the effect of boundary fast transport in the limit $\tau_a \to 0$. As a result, the second term in the square brackets of (2.143) scales as $\sqrt{\tau_a}$ because, in turn, the coefficients $\{\epsilon_\alpha^\Upsilon\}$ vary with τ_a as $\sqrt{\tau_a}$. Therefore it vanishes in the limit $\tau_a \to 0$ and the symmetry of the second moments caused by the boundary fast diffusion is restored.

The equality (see, e.g. [60])

$$\int_0^\infty \frac{dt}{\sqrt{\pi t}} \exp\left(-\frac{\zeta_0^2}{4\mathcal{D}_M t} - st\right) = \frac{1}{\sqrt{s}} \exp\left(-\zeta_0 \sqrt{\frac{s}{\mathcal{D}_M}}\right) \tag{2.148}$$

and the Laplace transform of integrals, enable us to represent the inverse Laplace transform $\mathcal{K}(t, \zeta_0)$ of function (2.138) in the integral form

$$\mathcal{K}(t, \zeta_0) = \sqrt{\frac{t}{\pi}} \int_0^1 \frac{dz}{\sqrt{z}} \exp\left(-\frac{\zeta_0^2}{4\mathcal{D}_M t}\frac{1}{z}\right). \tag{2.149}$$

Expression (2.149) together with formula (2.142)–(2.146) proves Proposition 2.2.

2.7
Boundary Condition for the Backward Fokker–Planck Equation

The expressions (2.97)–(2.99) directly lead us to the final results. First, they relate the singular kinetic coefficients to the diffusion tensor and the physical characteristics of the medium boundary. Second, they reduce the problem of canceling the singularities inside a thin layer Υ_τ adjacent to the boundary Υ which, nevertheless, is volumetric before implementing the passage to the limit $\tau \to 0$. Indeed, since all the terms of (2.97)–(2.99) depend on the coordinate x^M in the normal direction via the same function $\mathcal{K}(\tau, x^M)$, the singularities will be canceled at all the points of the layer Υ_τ if it is the case at the boundary Υ. Also, the structure of the function $\mathcal{K}(\tau, x^M)$, namely, expression (2.96) justifies that adopted before the assumption that the characteristic thickness of the layer Υ_τ scales with time as $\tau^{1/2}$.

As shown in Section 2.2.2 the boundary singularities that appear in the expansion of the Chapman–Kolmogorov equation leading to the backward Fokker–Planck equation will vanish if equality (2.18) holds. At first, in order to improve the perception of the results let us consider quite a small neighborhood of the point $\mathbf{s}$ belonging to the boundary Υ wherein it is actually a hyperplane and choose the basis $\mathbf{e}_\Upsilon \oplus \mathbf{n}$ composed of its hyperplane basis $\mathbf{e}_\Upsilon(\mathbf{s})$ and unit normal $\mathbf{n}(\mathbf{s})$ directed inwards the domain $\mathbb{Q}$. Then substituting expressions (2.97)–(2.99) into formula (2.18) we can immediately conclude that at the boundary point $\mathbf{s} \in \Upsilon$ the Green function $G(\mathbf{r}, t | \mathbf{r}_0, t_0)$ with respect to the latter pair of its arguments with $\mathbf{r}_0 \to \mathbf{s}$ has to meet the condition

$$\sum_{i=1}^{M} D^{iM}(\mathbf{s}, t_0) \nabla_i^s G(\mathbf{r}, t|\mathbf{s}, t_0) = \sigma(\mathbf{s}, t_0) G(\mathbf{r}, t|\mathbf{s}, t_0) - l_\Upsilon(\mathbf{s}, t_0)$$

$$\times \left[\sum_{\alpha=1}^{M-1} v_\Upsilon^\alpha(\mathbf{s}, t_0) \nabla_\alpha^s G(\mathbf{r}, t|\mathbf{s}, t_0) \right.$$

$$\left. + \sum_{\alpha,\beta=1}^{M-1} \mathfrak{D}^{\alpha\beta}(\mathbf{s}, t_0) \nabla_\alpha^s \nabla_\beta^s G(\mathbf{r}, t|\mathbf{s}, t_0) \right]. \quad (2.150)$$

We note that the two last terms in expression (2.150) describe the effective motion of the system inside the boundary Υ and have the form of the backward Fokker–Planck operator (2.14) with the diffusion tensor $\mathfrak{D}^{\alpha\beta}$ and drift velocity v_Υ^α whose action is confined to the boundary Υ. In order to rewrite this expression for an orthonormal basis of general orientation we make use of the definition of the boundary singularity vector $\mathbf{b}(\mathbf{s}, t)$, expression (2.16), and take into account (2.66) for the surface tensor diffusion. Then introducing the backward Fokker–Planck operator acting only within the hyperplane Υ

$$\widehat{\ell}_{\text{FP}_B}(\mathbf{s}, t_0)\{\Diamond\} = l_\Upsilon(\mathbf{s}, t_0) \left[\sum_{i,j=1}^{M} \mathfrak{D}^{ij}(\mathbf{s}, t_0) \nabla_i^s \nabla_j^s \Diamond + \sum_{i=1}^{M} v_\Upsilon^i(\mathbf{s}, t_0) \nabla_i^s \Diamond \right], \quad (2.151)$$

where as before the symbol $\Diamond$ stands for a function on which this operator acts. Then in the vector invariant form the boundary condition for the backward Fokker–Planck equation is written as

$$\mathbf{b}(\mathbf{s}, t_0) \cdot \nabla^s G(\mathbf{r}, t|\mathbf{s}, t_0) = \sigma(\mathbf{s}, t_0) G(\mathbf{r}, t|\mathbf{s}, t_0) - \widehat{\ell}_{\text{FP}_B}(\mathbf{s}, t_0)\{G(\mathbf{r}, t|\mathbf{s}, t_0)\}, \quad (2.152)$$

which is the desired formula.

In deriving expression (2.150) the boundary Υ was treated as a hyperplane, so the Euclidian space of dimension $(M-1)$ and its local basis $\mathfrak{e}_\Upsilon$ was used. To write it again in the general form underlining the fact that the operator $\widehat{\ell}_{\text{FP}_B}$ acts in this hyperplane, only the tensor notions of covariant derivatives are used (see, e.g. [138]). In these terms the action of the operator $\widehat{\ell}_{\text{FP}_B}$ on the Green function taken at the boundary Υ can be rewritten as

$$\widehat{\ell}_{\text{FP}_B}(\mathbf{s}, t_0)\{G(\mathbf{r}, t|\mathbf{s}, t_0)\} = l_\Upsilon(\mathbf{s}, t_0) \times \left[\sum_{\alpha=1}^{M-1} v_\Upsilon^\alpha(\mathbf{s}, t_0) G(\mathbf{r}, t|\mathbf{s}, t_0)_{;\alpha} \right.$$

$$\left. + \sum_{\alpha,\beta=1}^{M-1} \mathfrak{D}^{\alpha\beta}(\mathbf{s}, t_0) G(\mathbf{r}, t|\mathbf{s}, t_0)_{;\alpha\beta} \right]. \quad (2.153)$$

In the given case it is simply another form of the corresponding term in expression (2.150). However, for a nonplanar boundary, (2.153) holds allowing for the boundary curvature, whereas expression (2.151) loses the curvature effect. Its analysis goes far beyond the scope of the present chapter, so, here we will just ignore it.

2.8
Boundary Condition for the Forward Fokker–Planck Equation

The boundary conditions are obtained in a similar way. First, we note that the integrand of expression (2.31) is similar to the boundary relation (2.18) within the replacement the test function $\phi(\mathbf{r})$ by the Green function $G(\mathbf{r}, t|\mathbf{r}_0, t_0)$ and the action of the operators at the argument $\mathbf{r}$ instead of $\mathbf{r}_0$. This analogy and the boundary condition (2.152) for the backward Fokker–Planck equation enable us to reduce equality (2.31) to the following

$$\mathbf{b}(\mathbf{s}, t) \cdot \nabla^s \phi(\mathbf{s}) = \sigma(\mathbf{s}, t)\phi(\mathbf{s}) - \widehat{\ell}_{FP_B}(\mathbf{s}, t)\{\phi(\mathbf{s})\} \qquad (2.154)$$

for an arbitrary boundary point $\mathbf{s} \in \Upsilon$. Since the boundary part of the backward Fokker–Planck equation acts only within the boundary Υ, only the left part of expression (2.154) contains the first derivative of the test function $\phi(\mathbf{s})$ in the direction normal to the boundary Υ at the point $\mathbf{s}$. All the other terms are either the boundary value of the function $\phi(\mathbf{s})$ itself or its derivatives along the hyperplane Υ. It justifies the previously adopted statement that, in the vicinity of Υ, the test function $\phi(\mathbf{r})$ can have any boundary value $\phi(\mathbf{s})$.

Then noting that the left-hand side of the condition (2.32) is just the combination

$$\mathbf{b}(\mathbf{s}, t) \cdot \nabla^s G(\mathbf{s}, t|\mathbf{r}_0, t_0),$$

the last term converts expression (2.32) into

$$\oint_\Upsilon d\mathbf{s}\phi(\mathbf{s}) \sum_{i=1}^{M} v^i(\mathbf{s}) J^i \{G(\mathbf{s}, t|\mathbf{r}_0, t_0)\} = -\oint_\Upsilon d\mathbf{s}\phi(\mathbf{s}) \sigma(\mathbf{s}, t) G(\mathbf{s}, t|\mathbf{r}_0, t_0)$$
$$+ \oint_\Upsilon d\mathbf{s} \widehat{\ell}_{FP_B}(\mathbf{s}, t)\{\phi(\mathbf{s})\} G(\mathbf{s}, t|\mathbf{r}_0, t_0). \qquad (2.155)$$

Using the divergence integral theorem for the surfaces the last term in (2.155) is reduced to the form

$$\oint_\Upsilon d\mathbf{s} \widehat{\ell}_{FP_B}(\mathbf{s}, t)\{\phi(\mathbf{s})\} G(\mathbf{s}, t|\mathbf{r}_0, t_0) = \oint_\Upsilon d\mathbf{s}\phi(\mathbf{s}) \widehat{\ell}_{FP_B}(\mathbf{s}, t)\{G(\mathbf{s}, t|\mathbf{r}_0, t_0)\}.$$

Here the operator $\widehat{\ell}_{FP_F}$ is the boundary forward Fokker–Planck equation

$$\widehat{\ell}_{FP_F}(\mathbf{s}, t)\{\diamond\} = \sum_{i=1}^{M} \nabla_i^s \left[\sum_{j=1}^{M} \nabla_j^s \left(l_\Upsilon(\mathbf{s}, t) \mathfrak{D}^{ij}(\mathbf{s}, t) \diamond \right) - l_\Upsilon(\mathbf{s}, t) v_\Upsilon^i(\mathbf{s}, t_0) \diamond, \right] \qquad (2.156)$$

where again the symbol $\Diamond$ stands for the function on which the operator acts. Since the test function $\phi(\mathbf{s})$ takes any arbitrary values at the boundary Υ, equality (2.155) holds for any point on the boundary Υ, so

$$\mathbf{n}(\mathbf{s}) \cdot \widehat{\mathbf{J}}\{G(\mathbf{s}, t|\mathbf{r}_0, t_0)\} = -\sigma(\mathbf{s}, t)\, G(\mathbf{s}, t|\mathbf{r}_0, t_0) + \widehat{\ell}_{\mathrm{FP}_\mathrm{F}}(\mathbf{s}, t)\{G(\mathbf{s}, t|\mathbf{r}_0, t_0)\}, \quad (2.157)$$

which is the desired boundary condition for the forward Fokker–Planck equation. As should be the case, the boundary condition (2.157) can be interpreted in terms of mass conservation; the component of the walker flux normal to the boundary Υ is determined by the surface rate of walker absorption and the rate of fast surface transport, withdrawing the walkers from the given boundary point.

2.9
Concluding Remarks

In this chapter, a technique for *deriving* the boundary conditions for the Fokker–Planck equations based on the Chapman–Kolmogorov integral equation has been developed. The purpose of the chapter (see also [128]) is summarized in Figure 2.7.

The interest to this problem is partly due to the following. It is well known that the Fokker–Planck equations, both forward and backward ones, stem directly from the Chapman–Kolmogorov equation under two additional assumptions; the short time confinement of the corresponding Markovian process and the local homogeneity of the medium. There are quite rigorous methods of deriving them from the integral Chapman–Kolmogorov equation based on expanding the latter

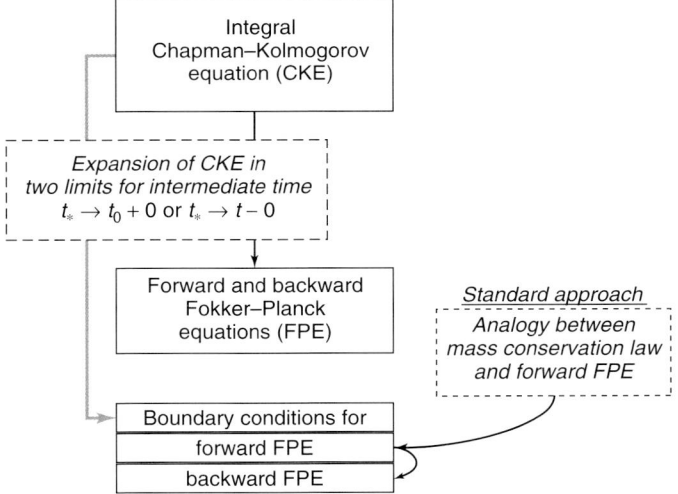

Figure 2.7 Illustration of the main purpose of this chapter concerning derivation of boundary conditions for the forward as well as the backward Fokker–Planck equation.

on short time scales in the possible limits. By contrast, the corresponding boundary conditions are typically postulated applying the physical meaning of the probability flux and the analogy between the forward Fokker–Planck equation and the mass conservation law.

However, such simple arguments can fail when dealing with more complex Markovian processes like sub- or super-diffusion, for which the Fokker–Planck equations with fractional derivatives form the governing equations. In this case it would be appropriate to have a formal technique giving rise to the boundary conditions starting from the general description. However, until now, constructing such a technique has been a challenging problem. This was also the case with respect to the normal Markovian processes in continua.

This chapter has demonstrated how to do this with the normal Markovian processes. The key point is the fact that the medium boundary breaks down the symmetry of random walks near to it. As a result, the coefficients in the corresponding expansion series of the Chapman–Kolmogorov equation are endowed with anomalous features called the boundary singularities. Namely, on short time scales they behave as $(\delta t)^{-1/2}$. Since the probability distribution on macroscopic scales cannot contain such singularities, the corresponding cofactors in the expressions for the boundary singularities should be set equal to zero, leading to the required boundary conditions. In this way we have shown that the boundary conditions of the Fokker–Planck equations also follow directly from the Chapman–Kolmogorov equation supplemented by some rather general assumptions about the properties of the medium boundary. As must be the case, the boundary conditions obtained in this way satisfy mass conservation.

2.10
Exercises

E 2.1 *Unbounded lattice random walk*
Consider lattice random walks on a regular unbounded chain of nodes. Use the Law of Large Numbers and demonstrate that the probability of finding a walker at a given node has the Gaussian form when the number of walker steps tends to infinity.

E 2.2 *Diagonal diffusion tensor I*
Find the orthonormal coordinate system where the following two-dimensional diffusion tensor

$$\{D_{11} = 2, \quad D_{22} = 2, \quad D_{12} = D_{21} = 1\}$$

has the diagonal form, and calculate its components.

E 2.3 *Diagonal diffusion tensor II*
Dealing with the half-space $\mathbb{R}^{2+} = \{x_1, x_2 | x_1 \geq 0\}$ to find the coordinate system (not orthogonal) where the diffusion tensor given in Exercise (2.2) has the diagonal form and one of the basic vectors is parallel to the boundary $x_1 = 0$.

E 2.4 Diffusion with one impermeable boundary

Let us consider the half-space $\mathbb{R}^+ = \{x | x \geq 0\}$ with an impermeable (reflecting) boundary at $x = 0$ and unbiased random walks characterized by the constant diffusion coefficient D. Using the image method, show that the Green function $G(x, x_0|t)$ meets the boundary condition $\partial G/\partial x = 0$ at $x = 0$.

E 2.5 Drift–diffusion with one impermeable boundary

Find the stationary distribution $G_{st}(x) \doteq \lim_{t \to \infty} G(x, x_0|t)$ of random walks in the half-space $\mathbb{R}^+ = \{x | x \geq 0\}$ with the impermeable (reflecting) boundary at $x = 0$ which are characterized by constant diffusion coefficient D and constant drift velocity $-v$ ($v > 0$).

E 2.6 Forward and backward Fokker–Planck equation

Let the Green function $G(x, t | x_0, t_0)$ be introduced as the solution of the forward Fokker–Planck equation for $t > t_0$ and $x, x_0 \in (x_1, x_2)$

$$\frac{\partial G}{\partial t} = \frac{\partial}{\partial x} \left\{ \frac{\partial [D(x,t) G]}{\partial x} - v(x,t) G \right\} \tag{2.158}$$

subject to the following initial and boundary conditions

$$G_{t=0} = \delta(x - x_0), \tag{2.159}$$

$$\left\{ \frac{\partial [D(x,t) G]}{\partial x} - v(x,t) G \right\}_{x=x_1} = \left\{ \frac{\partial [D(x,t) G]}{\partial x} - v(x,t) G \right\}_{x=x_2} = 0. \tag{2.160}$$

Demonstrate directly that this Green function $G(x, t | x_0, t_0)$ also obeys the backward Fokker–Planck equation for $t_0 < t$ and $x, x_0 \in (x_1, x_2)$

$$-\frac{\partial G}{\partial t_0} = D(x_0, t_0) \frac{\partial^2 G}{\partial x_0^2} + v(x_0, t_0) \frac{\partial G}{\partial x_0} \tag{2.161}$$

subject to the initial and boundary conditions

$$G_{t_0=t} = \delta(x - x_0), \tag{2.162}$$

$$\left. \frac{\partial G}{\partial x} \right|_{x=x_1} = \left. \frac{\partial G}{\partial x} \right|_{x=x_2} = 0. \tag{2.163}$$

Prove the same statement going in the opposite direction, that is, from the backward to forward Fokker–Planck equations.

Part II
Physics of Stochastic Processes

3
The Master Equation

3.1
Markovian Stochastic Processes

Stochastic processes occur in many physical descriptions of nature. Historically, the motion of a heavy particle in a fluid of light molecules was the first to be observed. The path of such a *Brownian particle* consists of stochastic displacements due to random collisions. This motion was studied by the Scottish botanist Robert Brown (1773–1858). In 1828 he discovered that the microscopically small particles into which the pollen of plants decay in an aqueous solution are in permanent irregular motion. This type of stochastic process is called *Brownian motion* and can be interpreted as a discrete random walk or continuous diffusion. This topic is considered in textbooks about Statistical Physics [38, 85, 194, 206, 213] and also in many books or monographs about stochastic processes [6, 25, 55, 84, 121, 156, 160, 172, 201, 211, 234].

The intuitive background needed to describe irregular motion completely as a stochastic process is to measure the values $x_1, x_2, \ldots, x_n, \ldots$ at times $t_1, t_2, \ldots, t_n, \ldots$ of a time-dependent random variable $x(t)$ and to assume that a set of joint probability densities, called JPD distributions

$$p_n(x_1, t_1; x_2, t_2; \ldots; x_n, t_n), \quad n = 1, 2, \ldots \tag{3.1}$$

exists. The same can be done by introducing a set of conditional probability densities (called CPD distributions)

$$p_n(x_n, t_n \mid x_{n-1}, t_{n-1}; \ldots; x_1, t_1), \quad n = 2, 3, \ldots \tag{3.2}$$

denoting that, at time t_n, the value x_n can be found, if at previous times $t_{n-1}, \ldots, t_1$ the respective values $x_{n+1}, \ldots x_1$ were present. The relationship between JPD and CPD is given by

$$p_{n+1}(x_1, t_1; \ldots; x_{n+1}, t_{n+1})$$
$$= p_{n+1}(x_{n+1}, t_{n+1} \mid x_n, t_n; \ldots; x_1, t_1) \, p_n(x_1, t_1; \ldots; x_n, t_n). \tag{3.3}$$

This stochastic description in terms of macroscopic variables will be called *mesoscopic*. Why? Typical systems encountered in everyday life like gases, liquids, solids, biological organisms and human or technical objects consist of about 10^{23} interacting units. The macroscopic properties of matter are usually the result of the collective behavior of a large number of atoms and molecules acting under the laws of quantum mechanics. To understand and control these collective macroscopic phenomena a complete knowledge based upon the known fundamental laws of microscopic physics is useless because the problem of interacting particles is far beyond the capabilities of the largest recent, and probably future, computers. The understanding of complex macroscopic systems consisting of many basic particles (in the order of atomic sizes: 10^{-10} m) requires the formulation of new concepts. One method is a stochastic description taking into account statistical behavior. Since the macroscopic features are averages over time of a large number of microscopic interactions, a stochastic description links both approaches together; both the microscopic and the macroscopic, to give probabilistic results. Monographs (recommended for physicists and engineers) devoted to stochastic concepts are mainly written as advanced courses on Statistical Physics like that by Josef Honerkamp [85] and on Statistical Thermodynamics by Werner Ebeling and Igor M. Sokolov [38], or well known textbooks on Stochastic Processes, see e.g. [6] by Vadim S. Anishenko et al., [55] by Crispin W. Gardiner, [84] by Josef Honerkamp and [234] by N. G. van Kampen.

Speaking about a *stochastic process* from the physical point of view we always refer to stochastic variables (random events) changing in time. A realization of a stochastic process is a trajectory $x(t)$ as a function of time. Here we introduce a hierarchy of *probability distributions*

$$p_n(x_1, t_1; x_2, t_2; \ldots; x_n, t_n)\, dx_1\, dx_2 \ldots dx_n, \quad n = 1, 2, \ldots, \tag{3.4}$$

where $p_1(x_1, t_1)\, dx_1$ is known as a time-dependent probability of first order to measure the value x_1 (precisely, the value within $[x_1, x_1 + dx_1]$) at time t_1, $p_2(x_1, t_1; x_2, t_2)$ is the same probability of second order, up to higher-order joint distributions $p_n(x_1, t_1; \ldots; x_n, t_n)\, dx_1\, dx_2 \ldots dx_n$ in order to find, for the stochastic variable, the value x_1 at time t_1, the value x_2 at time t_2 and so on. It is only the knowledge of such an infinite hierarchy of joint probability densities $p_n(x_1, t_1; \ldots; x_n, t_n)$ (expression (3.1)) with $n = 1, 2, \ldots$ which gives us the overall description of the stochastic process.

A stochastic process without any dynamics (like throwing a coin or any game of chance) is called a temporally *uncorrelated process*. It holds that

$$p_2(x_1, t_1; x_2, t_2) = p_1(x_1, t_1)\, p_1(x_2, t_2), \tag{3.5}$$

if random variables at different times are mutually independent. This means that each realization of a random number at time t_2 does not depend on the previous time t_1, that is, the correlation at different times $t_1 \neq t_2$ is zero. Such a stochastic process, where function $p_1(x_1, t_1) \equiv p_1(x)$ is the density of a normal distribution, is called *Gaussian white noise*. The Gaussian white noise with its rapidly varying, highly irregular trajectory is an idealization of a realistic fluctuating quantity. Due to the factorization of all higher-order joint probability densities, the knowledge of the normalized distribution $p_1(x_1, t_1)$ totally describes the process.

Now we are introducing dynamics via correlations between two different time moments. This basic assumption enables us to define the *Markov process*, also called the *Markovian process*, by two quantities totally, namely the first-order $p_1(x_1, t_1)$ and the second-order probability density $p_2(x_1, t_1; x_2, t_2)$, or equivalently by the joint probability $p_1(x_1, t_1)$ and the conditional probability $p_2(x_2, t_2 \mid x_1, t_1)$ of finding the value x_2 at time t_2, given that its value at a previous time t_1 ($t_1 < t_2$) is x_1. Contrary to the uncorrelated processes (3.5) discussed before, Markov processes are characterized by the following temporal relationship

$$p_2(x_1, t_1; x_2, t_2) = p_2(x_2, t_2 \mid x_1, t_1)\, p_1(x_1, t_1). \tag{3.6}$$

The *Markov property*

$$p_n(x_n, t_n \mid x_{n-1}, t_{n-1}; \ldots; x_1, t_1) = p_2(x_n, t_n \mid x_{n-1}, t_{n-1}) \tag{3.7}$$

enables us to calculate all higher-order joint probabilities p_n for $n > 2$. To determine the fundamental equation of stochastic processes of Markov type we start with the third-order distribution ($t_1 < t_2 < t_3$)

$$\begin{aligned} p_3(x_1, t_1; x_2, t_2; x_3, t_3) &= p_3(x_3, t_3 \mid x_2, t_2; x_1, t_1)\, p_2(x_1, t_1; x_2, t_2) \\ &= p_2(x_3, t_3 \mid x_2, t_2)\, p_2(x_2, t_2 \mid x_1, t_1)\, p_1(x_1, t_1) \end{aligned} \tag{3.8}$$

and integrate this identity over x_2 and then divide both sides by $p_1(x_1, t_1)$. We get the following result for the conditional probabilities defining a Markov process

$$p_2(x_3, t_3 \mid x_1, t_1) = \int p_2(x_3, t_3 \mid x_2, t_2)\, p_2(x_2, t_2 \mid x_1, t_1)\, \mathrm{d}x_2, \tag{3.9}$$

called the *Chapman–Kolmogorov equation*.

As already stated the Markov process is uniquely determined by the distribution $p_1(x, t)$ at time t and the conditional probability $p_2(x', t' \mid x, t)$, also called the transition probability from x at t to x' at a later t', to determine the whole hierarchy p_n ($n \geq 3$) by the Markov property (3.7). Also, these two functions cannot be chosen arbitrarily, they have to fulfill two consistency conditions, namely the Chapman–Kolmogorov equation (3.9)

$$p_2(x'', t'' \mid x, t) = \int p_2(x'', t'' \mid x', t')\, p_2(x', t' \mid x, t)\, \mathrm{d}x', \tag{3.10}$$

the Markov relationship (3.6)

$$p_1(x', t') = \int p_2(x', t' \mid x, t)\, p_1(x, t)\, \mathrm{d}x, \tag{3.11}$$

and the normalization condition

$$\int p_1(x', t')\, \mathrm{d}x' = 1. \tag{3.12}$$

The history in a Markov process, given by (3.7), is very short, only one time interval from t to t' is involved. If the trajectory has reached x at time t, the past is forgotten, and it moves toward x' at t' with a probability depending on x, t and x', t' only. The entire information relevant for the future is thus contained in the present. A Markov process is a stochastic process for which the future depends on the past and the present only through the present. It has no memory [201]. In an ordinary case where the space of states x is locally homogeneous it makes sense to transform the Chapman–Kolmogorov equation (3.9) in an equivalent differential equation in the short-time limit $t' = t + \tau$ with small τ tending to zero. The short-time behavior of the transition probability $p_2(\cdot \mid \cdot)$ should be written as series expansion with respect to the time interval τ in the form

$$p_2(x, t + \tau \mid x'', t) = \left[1 - \overline{w}(x, t)\tau\right]\delta(x - x'') + \tau w(x, x'', t) + \mathcal{O}(\tau^2). \qquad (3.13)$$

The new quantity $w(x, x'', t) \geq 0$ is the transition rate, the probability per time unit, for a jump from x'' to $x \neq x''$ at time t. This transition rate w multiplied by the time step τ gives the second term in the series expansion describing transitions from another state x'' to x. The first term (with the delta function) is the probability that no transitions take place during time interval τ. Based on the normalization condition

$$\int p_2(x, t + \tau \mid x'', t)\, dx = 1 \qquad (3.14)$$

it follows that

$$\overline{w}(x, t) = \int w(x'', x, t)\, dx''. \qquad (3.15)$$

The ansatz (3.13) implies that a realization of the random variable after any time interval τ retains the same value with a certain probability or attains a different value with the complementary probability. A typical trajectory $x(t)$ consists of straight lines $x(t) = $ constant, interrupted by jumps. An illustration is presented in Figure 3.1.

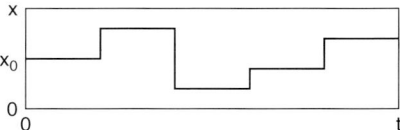

Figure 3.1 Sketch of time evolution of a stochastic one-dimensional variable $x(t)$. The stochastic trajectory consists of pieces of deterministic motion interrupted by jumps.

From Chapman–Kolmogorov equation (3.9) together with (3.13) we get

$$p_2(x, t + \tau \mid x', t') = \int p_2(x, t + \tau \mid x'', t) p_2(x'', t \mid x', t') \, dx''$$

$$= \int \left[1 - \overline{w}(x, t)\tau\right] \delta(x - x'') p_2(x'', t \mid x', t') \, dx''$$

$$+ \int \tau w(x, x'', t) p_2(x'', t \mid x', t') \, dx'' + \mathcal{O}(\tau^2). \tag{3.16}$$

With (3.15) and after taking the short-time limit $\tau \to 0$ one obtains the following differential equation

$$\frac{\partial}{\partial t} p_2(x, t \mid x', t') = \int w(x, x'', t) p_2(x'', t \mid x', t') \, dx''$$

$$- \int w(x'', x, t) p_2(x, t \mid x', t') \, dx''. \tag{3.17}$$

In order to rewrite the derived equation in a well known form using physical concepts, we get after multiplication by $p_1(x', t')$ and integration over x', the differential formulation of the Chapman–Kolmogorov equation

$$\frac{\partial}{\partial t} p_1(x, t) = \int w(x, x', t) p_1(x', t) \, dx' - \int w(x', x, t) p_1(x, t) \, dx' \tag{3.18}$$

called the *master equation* in the (physical) literature.

The name 'master equation' for the above probability balance equation is used in the sense that this differential expression is a general, fundamental or basic equation. For a process which is homogeneous in time the transition rates $w(x, x', t)$ are independent of time t and therefore $w(x, x', t) = w(x, x')$. The short-time transition rates w have to be known from the physical context, often like an intuitive ansatz, or have to be formulated based on a reasonable hypothesis or approximation. One of these is *Fermi's golden rule* originating from microscopic quantum theory [194]. With known transition rates w and given initial distribution $p_1(x, t = 0)$ the master equation (3.18) gives the resulting evolution of the probability p_1 over an infinitely long time period.

The well known master equation can be written in different ways. Besides the continuous formulation with one variable x, the generalization to the multidimensional as well as the discrete case is obvious. Instead of $p_1(x, t)$ with the high-dimensional probability vector $P(\underline{x}, t) \equiv P(x_1, x_2, \ldots, x_n, t)$ we may write the master equation in the discrete form (with summation instead of integration) as

$$\frac{\partial}{\partial t} P(\underline{x}, t) = \sum_{\underline{x}' \neq \underline{x}} \left\{ w(\underline{x}, \underline{x}') P(\underline{x}', t) - w(\underline{x}', \underline{x}) P(\underline{x}, t) \right\}. \tag{3.19}$$

Generalizations of the master equation have been developed by Honerkamp and Breuer [85], Montroll and West [165] and others. To perform stochastic simulations of complex systems like piecewise deterministic Markov processes, a stochastic formulation of fluid dynamics or reaction–diffusion equations; the so-called many-body or multivariate master equation, have been introduced. When describing quantum random systems, the master equation is usually called the Pauli master equation.

3.2
The Master Equation

The basic equation of stochastic Markov processes, called the *master equation* or explicitly the *forward master equation*, is usually written as a gain–loss equation (3.18) for the probabilities $p(x, t)$ in the form

$$\frac{\partial p(x, t)}{\partial t} = \int \{w(x, x')p(x', t) - w(x', x)p(x, t)\} \, dx'. \tag{3.20}$$

This very general equation can be interpreted as local balance for the probability densities which have to fulfill the global normalization condition

$$\int p(x, t) \, dx = 1 \tag{3.21}$$

at each time moment t, also at the beginning for the initial distribution $p(x, t = 0)$. The linear master equation (3.20) with known transition rates per unit time $w(x, x')$ is a so-called Markov evolution equation showing the relaxation from a chosen starting distribution $p(x, t = 0)$ to some final probability distribution $p(x, t \to \infty)$. The linearity of the master equation is based on the assumption that the underlying dynamics is Markovian. The transition probabilities w do not depend on the history of reaching a state, so that the transition rates per unit time are indeed constants for a given temperature or total energy.

If the state space of the stochastic variable is a discrete one, often considering natural numbers within a finite range $0 \leq n \leq N$, the master equation for the time evolution of the probabilities $p(n, t)$ is written as

$$\frac{dp(n, t)}{dt} = \sum_{n' \neq n} \{w(n, n')p(n', t) - w(n', n)p(n, t)\}, \tag{3.22}$$

where $w(n', n) \geq 0$ are rate constants for transitions from n to another $n' \neq n$. Together with the initial probabilities $p(n, t = 0)$ $(n = 0, 1, 2, \ldots, N)$ and the boundary conditions at $n = 0$ and $n = N$ this set of equations governing the time evolution of $p(n, t)$ from the beginning at $t = 0$ to the long-time limit $t \to \infty$ has to be solved. The meaning of both terms is clear. The first (positive) term is the inflow current to state n due to transitions from other states n', and the second (negative) term is the outflow current due to opposite transitions from n to n'.

Now let us define *stationarity*, sometimes called *steady state*, as a time independent distribution $p^{st}(n)$ by the condition $dp(n, t)/dt|_{p=p^{st}} = 0$. Therefore the stationary master equation is given by

3.2 The Master Equation

$$0 = \sum_{n' \neq n} \{w(n, n')p^{st}(n') - w(n', n)p^{st}(n)\}. \tag{3.23}$$

This equation states the obvious fact that, in the stationary or steady-state regime, the sum of all transitions into any state n must be balanced by the sum of all transitions from n into other states n'. Based on the properties of the transition rates per unit time, the probabilities $p(n, t)$ tend in the long-time limit to the uniquely defined stationary distribution $p^{st}(n)$, for which a constant probability flow is possible in open systems. This fundamental property of the master equation may be stated as

$$\lim_{t \to \infty} p(n, t) = p^{st}(n). \tag{3.24}$$

Now we are considering the question of *equilibrium* in a system without external exchange. The condition of equilibrium in closed isolated systems is much stronger than the former condition of stationarity (3.23). Here we require, as an additional constraint, a balance between each pair of states n and n' separately. This so-called *detailed balance* relation is written for the equilibrium distribution $p^{eq}(n)$ as

$$0 = w(n, n')p^{eq}(n') - w(n', n)p^{eq}(n). \tag{3.25}$$

It always holds for one-step processes in one-dimensional systems with closed boundaries considered further in our paper. Of course, each equilibrium state is by definition also stationary. If the initial probability vector $p(n, t = 0)$ is strongly nonequilibrium, many probabilities $p(n, t)$ change rapidly as soon as the evolution starts (short-time regime), and then relax more slowly towards equilibrium (long-time behavior). The final state, called thermodynamic equilibrium, is reached in the limit $t \to \infty$.

Using linear algebra we want to solve the master equation analytically by an expansion in eigenfunctions. This method gives us a general solution of the time-dependent probability vector $p(n, t)$ expressed by eigenvectors and eigenvalues. In a first step we introduce the master equation, written as a set of coupled linear differential equations (3.22), in a compact matrix form

$$\frac{d\mathbf{P}(t)}{dt} = \mathbf{W} \mathbf{P}(t), \tag{3.26}$$

with a probability vector $\mathbf{P}(t) = \{p(n, t) \mid n = 0, \ldots, N\}$ and an undecomposable asymmetric transition matrix $\mathbf{W} = \{W(n, n') \mid n, n' = 0, \ldots, N\}$. The elements of the matrix are given by

$$W(n, n') = w(n, n') - \delta_{n,n'} \sum_{m \neq n} w(m, n) \tag{3.27}$$

and obey the following two properties

$$W(n, n') \geq 0 \quad \text{for } n \neq n', \tag{3.28}$$

$$\sum_{n} W(n, n') = 0 \quad \text{for each } n'. \tag{3.29}$$

As we know from matrix theory [53] there are a number of consequences based on both properties. In particular the transition matrix $\mathbf{W}$ has a single zero eigenvalue whose eigenvector is the equilibrium probability distribution. In general, other eigenvalues can be complex and they always have negative real part. In our special case where the detailed balance (3.25) holds all eigenvalues are real, as discussed further on.

The solution $\mathbf{P}(t)$ of the master equation (3.26) with given initial vector $\mathbf{P}(0)$ may be written formally as

$$\mathbf{P}(t) = \mathbf{P}(0)\,\exp(\mathbf{W}\,t), \tag{3.30}$$

(where $\exp(\mathbf{W}\,t) = \sum_{m=0}^{\infty} (\mathbf{W}\,t)^m/m!$) but this does not help us to find $\mathbf{P}(t)$ explicitly.

The familiar method is to make $\mathbf{W}$ symmetric and thereby diagonalizable and then to construct the solution as superposition of eigenvectors $\mathbf{u}_\lambda$ related to (zero or negative) eigenvalues λ in the form

$$\mathbf{P}(t) = \sum_\lambda c_\lambda \mathbf{u}_\lambda \, e^{\lambda t}. \tag{3.31}$$

with, until now, unknown coefficients c_λ. Using the condition of detailed balance (3.25) we transform the matrix $\mathbf{W} = \{W(n, n')\}$ to a new symmetric transition matrix $\widetilde{\mathbf{W}} = \{\widetilde{W}(n, n')\}$ with elements given by

$$\widetilde{W}(n, n') \stackrel{\text{def}}{=} W(n, n')\sqrt{\frac{p^{\text{eq}}(n')}{p^{\text{eq}}(n)}} = \widetilde{W}(n', n). \tag{3.32}$$

Both matrices $\mathbf{W}$ and $\widetilde{\mathbf{W}}$ have the same eigenvalues λ_i. Due to the symmetry of matrix $\widetilde{\mathbf{W}}$, all eigenvalues are real. They may be labeled in order of decreasing algebraic values, so that $\lambda_0 = 0$ and $\lambda_i < 0$ for $1 \leq i \leq N$. Denoting the normalized eigenvectors by $\mathbf{u}_i$ and $\widetilde{\mathbf{u}}_i$ respectively, defined by the eigenvalue equations

$$\sum_{n'} W(n, n')\, u_i(n') = \lambda_i\, u_i(n); \quad \mathbf{W}\,\mathbf{u}_i = \lambda_i\,\mathbf{u}_i \tag{3.33}$$

$$\sum_{n'} \widetilde{W}(n, n')\, \widetilde{u}_i(n') = \lambda_i\, \widetilde{u}_i(n); \quad \widetilde{\mathbf{W}}\,\widetilde{\mathbf{u}}_i = \lambda_i\,\widetilde{\mathbf{u}}_i \tag{3.34}$$

and related by the transformation $u_i(n) = \sqrt{p^{\text{eq}}(n)}\,\widetilde{u}_i(n)$ to each other, we are ready to construct the time dependent solution of the fundamental master equation (3.26). According to superposition formula (3.31), where the coefficients c_λ are calculated from the initial condition $p(n, 0)$ at $t = 0$, the solution is then

$$p(n, t) = \sqrt{p^{\text{eq}}(n)} \sum_{i=0}^{N} \widetilde{u}_i(n)\, e^{\lambda_i t} \left[\sum_{m=0}^{N} \widetilde{u}_i(m) \frac{p(m, 0)}{\sqrt{p^{\text{eq}}(m)}} \right], \tag{3.35}$$

or

$$p(n, t) = \sum_{i=0}^{N} u_i(n)\, e^{\lambda_i t} \left[\sum_{m=0}^{N} u_i(m) \frac{p(m, 0)}{p^{\text{eq}}(m)} \right]. \tag{3.36}$$

This solution plays a very important role in the stochastic description of Markov processes and can be found in different notations (e.g. as an integral representation) in many textbooks, see e.g. [84, 234].

As time increases to infinity ($t \to \infty$) only the term $i = 0$ in the solution survives and the probabilities tend to equilibrium $\mathbf{P}(t) \to \mathbf{P}^{\text{eq}}$, written as

$$p(n,t) = p^{\text{eq}}(n) + \sum_{i=1}^{N} u_i(n)\, e^{\lambda_i t} \left[\sum_{m=0}^{N} u_i(m) \frac{p(m,0)}{p^{\text{eq}}(m)} \right]. \qquad (3.37)$$

In the long-time limit all remaining modes $c_\lambda \mathbf{u}_\lambda\, e^{\lambda t}$ decay exponentially. In the short-time regime, due to combinations of modes with different signs, there is the possibility of growing and subsequent shrinking of transient states as probability current from initial distribution $\mathbf{P}(0)$ to equilibrium $\mathbf{P}^{\text{eq}}$ via intermediates $\mathbf{P}(t)$ [164].

Master equation dynamics can be studied either by solving the basic equation analytically with implementation of numerical methods or by simulating the stochastic process as a large number of subsequent jumps from state to state with the given transition rates. Both methods have different advantages and disadvantages. One important point is the choice of the appropriate time interval called the numerical integration step or waiting time in simulation techniques. The step size required for a given accuracy is usually smaller when the time t is closer to zero, and can be enlarged as t grows. Therefore, only a numerical algorithm with an adaptive step size should be used. Detailed information about algorithms used to generate a trajectory of a stochastic process described by a master equation can be found in textbooks by Honerkamp [84, 85] or others [140, 149].

3.3
One-Step Processes in Finite Systems

We are speaking about a one-dimensional stochastic process if the state space is characterized by one variable only. Often this discrete variable is a particle number $n \geq 0$ describing the amount of molecules in a box or the size of an aggregate. In chemical physics such aggregation phenomena like formation and/or decay of clusters are of great interest. Examples are the formation of a crystal or glass when cooling a liquid, or the condensation of a droplet out of a supersaturated vapor. To determine the relaxation dynamics of clusters of size n we take a particularly simple Markov process with transitions between neighboring states n and $n' = n \pm 1$. This situation is called a *one-step process*. In biophysics, if the variable n represents the number of living individuals of a particular species, the one-step process is often called *birth-and-death process* used-to investigate problems in population dynamics. The random walk with displacements to the left and right by one step is well known in physics [195] and often plays a role as an introductory example and has been recently revisited and applied to new fields like econophysics [119, 180, 206, 235]. The detailed balance relation (3.25) can be proven for the one-step process, so that in our case the former (see Section 3.2) is completely correct.

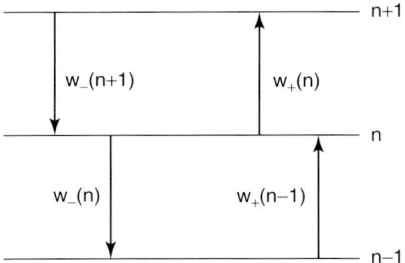

Figure 3.2 Illustration of a one-step process showing the up and down or forward and backward transition probabilities between neighboring states.

Setting the transition rates $w(n, n-1) = w_+(n-1)$, $w(n, n+1) = w_-(n+1)$, and therefore also $w(n+1, n) = w_+(n)$, $w(n-1, n) = w_-(n)$, see Figure 3.2; now the forward master equation (3.22) reads

$$\frac{dp(n,t)}{dt} = w_+(n-1)\, p(n-1,t) + w_-(n+1)\, p(n+1,t)$$
$$- \left[w_+(n) + w_-(n) \right] p(n,t). \tag{3.38}$$

In general, the forward and backward transition rates $w_+(n), w_-(n)$ are nonlinear functions of the random variable n; the physical dimension of $w_\pm$ is one over time (s^{-1}). The master equation is always linear in the unknown probabilities $p(n,t)$ of being at state n at time t. It has to be completed by the boundary conditions. The nonlinearity refers only to the transition coefficients. Further on we will pay attention to particles as aggregates in a closed box or vehicular jams on a circular road. Therefore, in finite systems, the range of the discrete variable n is bounded between 0 and N ($n = 0, 1, 2, \ldots, N$).

The general one-step master equation (3.38) is valid for $n = 1, 2, \ldots, N-1$, but meaningless at the boundaries $n = 0$ and $n = N$. Therefore, we have to add two boundary equations as closure conditions

$$\frac{dp(0,t)}{dt} = w_-(1)\, p(1,t) - w_+(0)\, p(0,t), \tag{3.39}$$

$$\frac{dp(N,t)}{dt} = w_+(N-1)\, p(N-1,t) - w_-(N)\, p(N,t). \tag{3.40}$$

To solve the set of equations we rewrite (3.38) as a balance equation

$$\frac{dp(n,t)}{dt} = J(n+1,t) - J(n,t) \tag{3.41}$$

with the probability current defined by

$$J(n,t) = w_-(n)\, p(n,t) - w_+(n-1)\, p(n-1,t). \tag{3.42}$$

In the stationary regime, remember (3.23), all flows (3.42) have to be independent of n and therefore equal to a constant current of probability: $J(n+1) = J(n) = J$. In open systems the stationary solution is no longer unique, it depends on the current J.

In finite systems with $n = 0, 1, 2, \ldots, N$ one finds a situation with zero flux $J = 0$, which corresponds to the steady state with a detailed balance relationship similar to (3.25). Therefore, the stationary distribution $p^{st}(n)$ fulfills the recurrence relation

$$p^{st}(n) = \frac{w_+(n-1)}{w_-(n)} p^{st}(n-1). \tag{3.43}$$

By applying the iteration successively we get the relation

$$p^{st}(n) = p^{st}(0) \prod_{m=1}^{n} \frac{w_+(m-1)}{w_-(m)}, \tag{3.44}$$

which determines all probabilities $p^{st}(n)$ ($n = 1, 2, \ldots, N$) in terms of the first unknown one $p^{st}(0)$. Taking into account the normalization condition

$$\sum_{n=0}^{N} p^{st}(n) = 1 \quad \text{or} \quad p^{st}(0) + \sum_{n=1}^{N} p^{st}(n) = 1 \tag{3.45}$$

the stationary probability distribution $p^{st}(n)$ in finite systems is finally written as

$$p^{st}(n) = \begin{cases} \dfrac{\prod_{m=1}^{n} \dfrac{w_+(m-1)}{w_-(m)}}{1 + \sum_{k=1}^{N} \prod_{m=1}^{k} \dfrac{w_+(m-1)}{w_-(m)}} & n = 1, 2, \ldots, N \\[2ex] \dfrac{1}{1 + \sum_{k=1}^{N} \prod_{m=1}^{k} \dfrac{w_+(m-1)}{w_-(m)}} & n = 0. \end{cases} \tag{3.46}$$

It is often convenient to write the stationary solution (3.44) in the exponential form

$$p^{st}(n) = p^{st}(0) \exp\{-\Phi(n)\}, \tag{3.47}$$

where, in analogy to physical systems, the function

$$\Phi(n) = \sum_{m=1}^{n} \ln\left(\frac{w_-(m)}{w_+(m-1)}\right) \tag{3.48}$$

is called the potential. An example of a double-well potential $\Phi(n)$ and corresponding bistable stationary probability distribution is shown in Figure 3.3. As we can see, the minimum of the potential corresponds to the probability maximum and vice versa.

The obtained result (3.46) based on the zero-flux relationship (3.43) is a unique solution for the stationary probability distribution in finite systems with closed

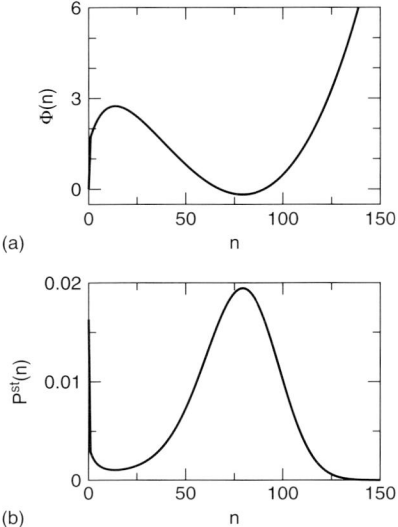

Figure 3.3 An example of a double well potential (a) and the corresponding bistable probability distribution (b) depending on the stochastic variable n.

boundaries. For an isolated system, the stationary solution of the master equation p^{st} is identical with the thermodynamic equilibrium p^{eq}, where the detailed balance holds which, for one-step processes, reads

$$w_-(n)\, p^{eq}(n) = w_+(n-1)\, p^{eq}(n-1). \tag{3.49}$$

The condition of detailed balance states a physical principle. If the distribution p^{eq} is known from equilibrium statistical mechanics and if one of the transition rates is also known (e.g. w_+ by a reasonable ansatz), then (3.49) provides an opportunity to formulate the opposite transition rate w_- in a consistent way. By this procedure the nonequilibrium behavior is adequately described by a sequence of (quasi-) equilibrium states. The relaxation from any initial nonequilibrium distribution tends always to the known final equilibrium. In physical systems the equilibrium distribution is usually represented in an exponential form, see e.g. [167, 213],

$$p^{eq}(n) \propto \exp\left[-\Omega(n)/(k_B T)\right] \tag{3.50}$$

where $\Omega(n)$ is the thermodynamic potential depending on the stochastic variable n, k_B is the Boltzmann constant, and T is the temperature. Equation (3.50) is comparable with (3.47) where $\Phi(n) = \Omega(n)/(k_B T)$.

3.4
The First-Passage Time Problem

In many applications it is important to know the mean time during which the system finds its stable state by overcoming a potential barrier (cf. Figure 3.3) due to stochastic fluctuations. It is closely related to the breakdown phenomena [26, 216]. Particularly, in traffic engineering one speaks about the *traffic breakdown probability* which is the transition rate as inverse quantity of the average breakdown time during which a spontaneous jamming (clustering) of cars appears in an initially homogeneous metastable traffic flow [110–112]. In a more general formulation, it is called the, well known, first-passage problem. The problem is to find the average time during which a stochastic system reaches, for the first time, some given state if started from another state. This time is called the *mean first-passage time*.

For a mathematical formulation of the problem first we derive the backward master equation. Our starting point is the Chapman–Kolmogorov equation (3.10), written for discrete variables

$$p(n,t \mid n',t') = \sum_{n''} p(n,t \mid n'',t'') \, p(n'',t'' \mid n',t'). \tag{3.51}$$

Here $p(n,t \mid n',t')$ represents the conditional probability that the system is in state n at time t if it was in state n' at time moment t', where $t' < t'' < t$. By setting $t'' = t' + \Delta t$ and taking into account the normalization condition (3.14)

$$\sum_{n''} p(n'', t' + \Delta t \mid n', t') = 1, \tag{3.52}$$

using

$$p(n,t \mid n', t' + \Delta t) = \sum_{n''} p(n, t \mid n', t' + \Delta t) p(n'', t' + \Delta t \mid n', t') \tag{3.53}$$

we obtain

$$p(n, t \mid n', t' + \Delta t) - p(n, t \mid n', t')$$
$$= \sum_{n''} p(n'', t' + \Delta t \mid n', t') \left[p(n, t \mid n', t' + \Delta t) - p(n, t \mid n'', t' + \Delta t) \right].$$
$$\tag{3.54}$$

Dividing both sides of (3.54) by Δt and taking the limit $\Delta t \to 0$, we arrive at the *backward master equation*

$$\frac{\partial p(n, t \mid n', t')}{\partial t'} = \sum_{n''} w(n'', n', t') \left\{ p(n, t \mid n', t') - p(n, t \mid n'', t') \right\} \tag{3.55}$$

describing the evolution of the probabilities $p(n, t \mid n', t')$ with respect to the initial time t'. Here

$$w(n'', n', t') = \lim_{\Delta t \to 0} \frac{p(n'', t' + \Delta t \mid n', t')}{\Delta t} \tag{3.56}$$

is the transition rate from state n' to state n'' at time moment t'. An appropriate initial condition for (3.55) is

$$p(n, t' = 0 \mid n', t' = 0) = \delta_{n,n'} \tag{3.57}$$

stating that the system cannot be in two different states simultaneously. For one-step processes, assuming no explicit time dependence of the transition rates $w_+(n) = w(n+1, n, t)$ and $w_-(n) = w(n-1, n, t)$, the backward master equation (3.55) becomes

$$\frac{\partial p(n, t \mid n', t')}{\partial t'} = w_+(n') \left[p(n, t \mid n', t') - p(n, t \mid n'+1, t') \right]$$
$$+ w_-(n') \left[p(n, t \mid n', t') - p(n, t \mid n'-1, t') \right]. \quad (3.58)$$

To study the *first-passage problem*, the backward master equation should be supplied by suitable boundary conditions. Let us assume that the value of the stochastic variable n belongs to the interval $a \le n \le b$ at the initial time $t' = 0$. We consider a reflecting boundary at $n = a$, that is, $w_-(a) = 0$, which means that the system can never reach states with values $n < a$, and an absorbing boundary at $n = b$, that is, $w_-(b+1) = 0$, which means that the system never returns back to $n \in [a; b]$ once it has left this interval. It is often convenient to associate n with the position of a randomly walking particle, assuming that the particle is absorbed at $n = b + 1$.

The question is, how long is the average time till this absorption takes place? The quantity we have to calculate is the *breakdown rate* as inverse of the *mean first-passage time* $\langle T(n) \rangle$ starting from a certain position n inside the interval $[a; b]$ to stick at $b + 1$. Obviously, the reflecting boundary condition $w_-(a) = 0$ in (3.58) can be formally replaced by

$$p(n, t \mid a - 1, t') = p(n, t \mid a, t') \quad (3.59)$$

The absorbing boundary condition for the backward master equation can be written as

$$p(n, t \mid b + 1, t') = 0, \quad (3.60)$$

which states that the transition from state $n = b + 1$ to states $n \le b$ is forbidden.

The probability $G(n, t)$ that at time t the system still has not left the interval $[a; b]$ is given by

$$G(n, t) = \sum_{n'=a}^{b} p(n', t \mid n, 0). \quad (3.61)$$

The function $G(n, t)$ obeys the equation

$$-\frac{\partial}{\partial t} G(n, t) = w_+(n) \left[G(n, t) - G(n+1, t) \right] \quad (3.62)$$
$$+ w_-(n) \left[G(n, t) - G(n-1, t) \right]$$

and the boundary conditions

$$G(a - 1, t) = G(a, t) \quad (3.63)$$
$$G(b + 1, t) = 0, \quad (3.64)$$

as follows from (3.58)–(3.60) and according to (3.61) in view of the fact that probability $p(n, t \mid n', t')$ is a function of the time difference $t - t'$, which means that the derivative with respect to t' is the negative derivative with respect to t.

According to the definition of $G(n, t)$ (3.61) the probability of absorption within an infinitesimal time interval $[t; t + dt]$ is $G(n, t) - G(n, t + dt) = -(\partial G/\partial t)\, dt$, which means that the mean first-passage time $\langle T(n) \rangle$ is

$$\langle T(n) \rangle = -\int_0^\infty t \frac{\partial G}{\partial t}\, dt = \int_0^\infty G(n, t)\, dt, \tag{3.65}$$

where the latter identity is due to the integration by parts. Taking into account (3.65), we obtain just the equation for the mean first-passage time by integration over time in (3.62) and (3.63) to (3.64). It yields the desired equation

$$1 = w_+(n) \left[\langle T(n) \rangle - \langle T(n+1) \rangle \right] + w_-(n) \left[\langle T(n) \rangle - \langle T(n-1) \rangle \right] \tag{3.66}$$

with the boundary conditions

$$\langle T(a-1) \rangle = \langle T(a) \rangle \tag{3.67}$$

$$\langle T(b+1) \rangle = 0. \tag{3.68}$$

To solve (3.66), it is suitable to rewrite it in new variables as

$$w_+(n)\, \phi(n) \left[S(n) - S(n-1) \right] = -1 \tag{3.69}$$

where $\phi(a) = 1$ and

$$\phi(n) = \prod_{m=a+1}^{n} \frac{w_-(m)}{w_+(m)} \tag{3.70}$$

$$S(n) = \frac{\langle T(n+1) \rangle - \langle T(n) \rangle}{\phi(n)}. \tag{3.71}$$

Equation (3.70) holds for $n \in [a+1; b]$ and (3.71) is valid for $n \in [a; b]$ with $S(a-1) = 0$. From this we find immediately

$$\phi(k)\, S(k) = -\phi(k) \sum_{m=a}^{k} \left[w_+(m)\phi(m) \right]^{-1} = \langle T(k+1) \rangle - \langle T(k) \rangle \tag{3.72}$$

The summation in (3.72) from $k = n$ to $k = b$ taking account of the boundary condition (3.68) yields the solution

$$\langle T(n) \rangle = \sum_{k=n}^{b} \phi(k) \sum_{m=a}^{k} \left[w_+(m)\phi(m) \right]^{-1}. \tag{3.73}$$

It is convenient to express the solution (3.73) in terms of the stationary probability distribution $p^{st}(n)$ given by (3.44). It finally yields

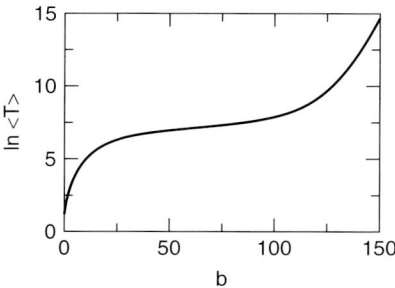

Figure 3.4 Logarithm of the mean first-passage time as a function of state variable. On average the system needs time $\langle T \rangle$ to move from the initial state at $n = 0$ to the absorbing state at $n = b + 1$, corresponding to the stationary probability distribution in Figure 3.3.

$$\langle T(n) \rangle = \sum_{k=n}^{b} \left[w_+(k) p^{\mathrm{st}}(k) \right]^{-1} \sum_{m=a}^{k} p^{\mathrm{st}}(m), \quad (3.74)$$

which allows us to calculate analytically the mean first-passage time $\langle T(n) \rangle$ to reach state $b + 1$ for the first time, when starting at position n, taking into account the forward rate w_+ and the stationary distribution p^{st} defined by (3.44). The mean breakdown rate is given by $1/\langle T(n) \rangle$.

An example of the mean first-passage time, corresponding to the stationary probability distribution in Figure 3.3, is shown in Figure 3.4. The mean first-passage time increases rapidly with changing boundary value b up to $b \approx 15$ due to the necessity to climb up the first hill of the potential in Figure 3.3. Then the increase becomes slower and the $\ln \langle T \rangle$ vs b curve almost has a plateau from $b \approx 30$ to $b \approx 85$. It corresponds to the decreasing part of the potential, therefore the system can pass these states easily in a relatively short time. The mean first-passage time again increases dramatically for larger b values due to the growth of the potential, and in our example it reaches the value of 2.4×10^6 dimensionless time units at $b = 150$. This means that the states with large values of the state variable n will practically never be reached. For comparison, we have $\langle T \rangle = 1.05 \times 10^3$ time units at $b = 50$.

3.5
The Poisson Process in Closed and Open Systems

Until to now we have considered Markov processes in a more general framework without defining the states of the system or the rates for the transitions between these states precisely. The particular case, where the states are characterized by a single particle number n and the rates by a one-step backward transition $w_-(n)$ only, is called *decay process*. A schematic realization of such a stochastic process

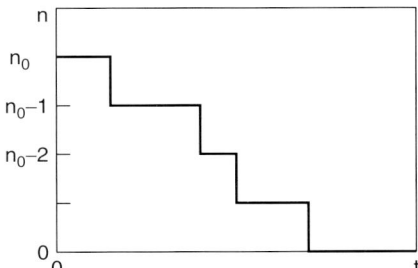

Figure 3.5 Sketch of the realization of a stochastic decay process of Poisson type with shrinking particle number n starting from $n = n_0$ at $t = 0$.

is shown in Figure 3.5 illustrating dissolution or shrinkage of a bound state of n members.

In a first step we present an example of traffic flow considered as a Markov process. We want to investigate the dissolution of a queue of cars standing in front of traffic lights. When the lights switch to green, the first car starts to move. After a certain time interval (waiting time $\tau = $ constant > 0) the next vehicle accelerates to pass the stop line and so on. In our model we consider the decay of traffic congestion without taking into account any influence of external factors, like ramps or intersections, on the driver's behavior. The stochastic variable $n(t)$ is the number of cars which are bounded in the jam at time t. A queue or platoon of n vehicles is also called a car cluster of size n in agreement with the concept of aggregation [141, 144, 203] and traffic flow [146–151].

When the initial jam size is finite, given by the value $n(t = 0) = n_0$, shown in Figure 3.5, the trajectory $n(t) = n_0, n_0 - 1, \ldots, 2, 1, 0$ consists of unit jumps at random times. The jam starting with size n_0 becomes smaller and smaller and dissolves completely. In Figure 3.6 we have shown three different stochastic trajectories to illustrate car cluster dissolution. This one-step stochastic process is a death process only, sometimes called a *Poisson process*.

Defining $p(n, t)$ as the probability of finding a jam of size n at time t, the master equation for the dissolution process reads

$$\frac{\partial}{\partial t} p(n, t) = w_-(n+1) p(n+1, t) - w_-(n) p(n, t) \qquad (3.75)$$

with the decay rate per unit time is assumed to be

$$w(n', n) = w(n-1, n) \equiv w_-(n) = \frac{1}{\tau}. \qquad (3.76)$$

In this approximation the experimentally known waiting time constant τ is a given control parameter in our escape model. It is the reaction time of a driver, usually about 1.5 or 2 seconds, to escape from the jam when the road in front of his car becomes free. Therefore, the transition rate (3.76) is a constant $w_- = 1/\tau$ independent of the jam size n.

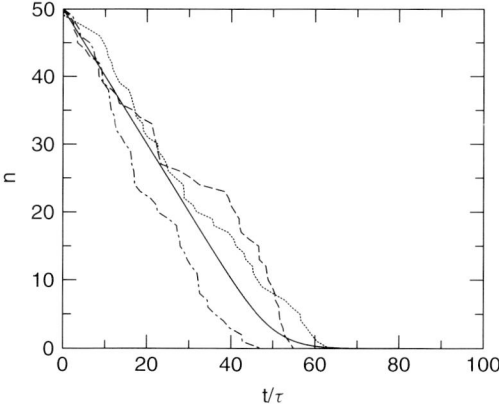

Figure 3.6 Three different stochastic trajectories showing the dissolution of a car queue with the initial length (size) $n_0 = 50$, that is, the cluster size n vs dimensionless time t/τ. The theoretical average value is shown by a smooth solid line.

For the described process of jam shrinkage ($n_0 \geq n \geq 0$), starting with cluster size $n = n_0$ and ending with $n = 0$, we thus obtain the following master equation including boundary conditions (compare (3.38)–(3.40))

$$\frac{\partial}{\partial t}p(n_0, t) = -\frac{1}{\tau}p(n_0, t), \tag{3.77}$$

$$\frac{\partial}{\partial t}p(n, t) = \frac{1}{\tau}\left[p(n+1, t) - p(n, t)\right], \quad n_0 - 1 \geq n > 0, \tag{3.78}$$

$$\frac{\partial}{\partial t}p(0, t) = \frac{1}{\tau}p(1, t) \tag{3.79}$$

and initial probability distribution $p(n, t = 0) = \delta_{n,n_0}$. The delta function means that at the beginning the vehicular queue consists of exactly n_0 cars.

In order to find the explicit expression for the probability distribution $p(n, t)$ we have to solve the set of equations (3.77)–(3.79). This can be done analytically starting with the first equation, getting $p(n_0, t) = \exp(-t/\tau)$ as the exponential decay function, inserting the solution into the next equation for $p(n_0 - 1, t)$, solving it and continue iteratively up to $p(0, t)$. The general solution of the probability $p(n, t)$ of observing a car cluster of size n at time t is

$$p(n, t) = \frac{(t/\tau)^{n_0 - n}}{(n_0 - n)!} e^{-t/\tau}, \quad 0 < n \leq n_0, \tag{3.80}$$

$$p(0, t) = 1 - \sum_{m=0}^{n_0 - 1} \frac{(t/\tau)^m}{m!} e^{-t/\tau}. \tag{3.81}$$

As already mentioned (3.45), the probabilities are always normalized to unity, which can be proven by summation $\sum_{n=0}^{n_0} p(n,t)$ inserting (3.80 and 3.81) to get one. The time evolution of the probability $p(n,t)$ has been calculated from (3.80) and (3.81) for an initial queue length $n_0 = 50$. The result is shown in Figure 3.7 and compared to numerical Monte Carlo simulation experiments.

The average or expectation value $\langle n \rangle$ of the cluster size n is usually given by

$$\langle n \rangle(t) \equiv \sum_{n=0}^{n_0} n\, p(n,t) = \sum_{n=1}^{n_0} n\, p(n,t) \tag{3.82}$$

and can be calculated using the known probabilities (3.80) to get the exact result

$$\langle n \rangle(t) = n_0\, Q(n_0 - 1, t) - \frac{t}{\tau} Q(n_0 - 2, t) \tag{3.83}$$

where $Q(n,t)$ is an abbreviation called the Poisson term

$$Q(n,t) \stackrel{\text{def}}{=} e^{-t/\tau} \sum_{m=0}^{n} \frac{(t/\tau)^m}{m!}. \tag{3.84}$$

The variance or second central moment $\langle\langle n \rangle\rangle(t)$ which measures the fluctuations is given by

$$\langle\langle n \rangle\rangle = \langle (n - \langle n \rangle)^2 \rangle = \langle n^2 \rangle - \langle n \rangle^2 \tag{3.85}$$

and can also be calculated as follows

$$\langle\langle n \rangle\rangle(t) = n_0 \left[n_0\, Q(n_0 - 1, t) - \frac{2t}{\tau} Q(n_0 - 2, t) \right] (1 - Q(n_0 - 1, t))$$
$$+ \left(\frac{t}{\tau}\right)^2 \left[Q(n_0 - 3, t) - Q^2(n_0 - 2, t) \right] + \frac{t}{\tau} Q(n_0 - 2, t). \tag{3.86}$$

In some approximation, where we set $Q(n,t)$ (3.84) to one, the mean value (3.83) reduces to a linearly decreasing function in time

$$\langle n \rangle(t) \approx n_0 - \frac{t}{\tau}, \tag{3.87}$$

whereas the variance (3.86) reduces to a linearly increasing behavior

$$\langle\langle n \rangle\rangle(t) \approx \frac{t}{\tau}. \tag{3.88}$$

In the case of the linear mean value approximation (3.87) the time required for the jam to dissolve totally, is given by

$$t_{\text{end}} = n_0 \tau. \tag{3.89}$$

The exact result (3.85) and the linear approximation (3.87) for the mean value depending on time are shown in Figure 3.8 by solid and dashed lines, respectively. In Figure 3.9 we have shown the same plots for the variance (3.86) and its linearization (3.88).

Equations (3.87) and (3.88), however, do not describe the final stage of dissolution of any finite car cluster. In this case, taking the limit $t \to \infty$ in the time-dependent results (3.80) and (3.81), we have

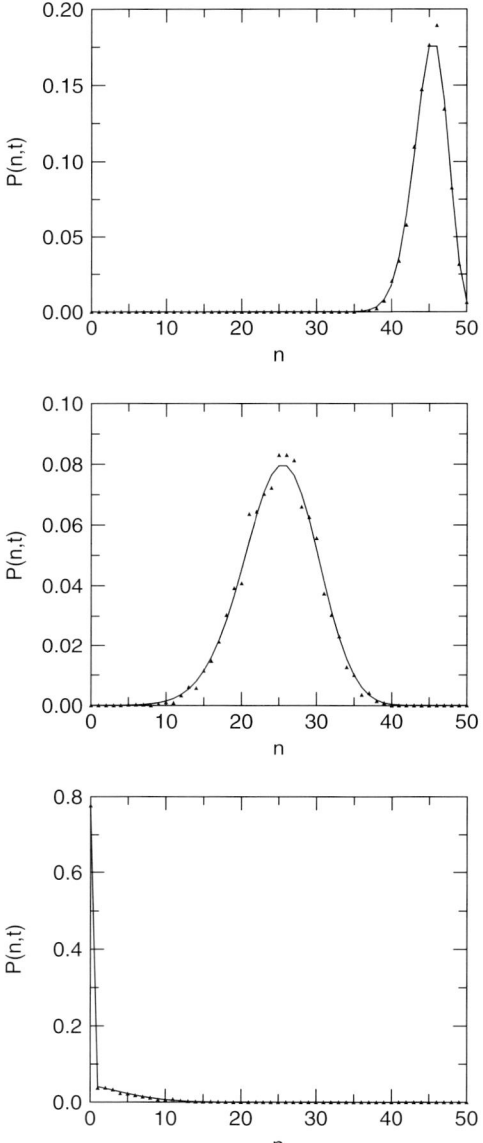

Figure 3.7 Probability distribution $P(n, t)$ at three different times (from the top to the bottom) $t/\tau = 5$, 25, and 55 with the initial condition $P(n, 0) = \delta_{n,50}$. Solid lines show the analytical solution, triangles indicate Monte Carlo results obtained by simulation of 5000 stochastic trajectories. Note that the diagrams have different scales along the probability axis.

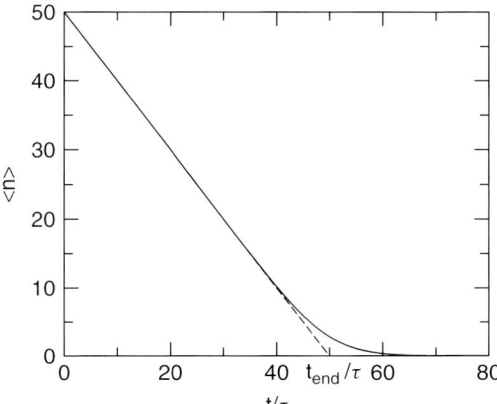

Figure 3.8 The mean value $\langle n \rangle$ of the cluster size depending on the dimensionless time t/τ. The initial size of the cluster is $n_0 = 50$. The exact result is shown by a solid line and the linear approximation by a dashed line.

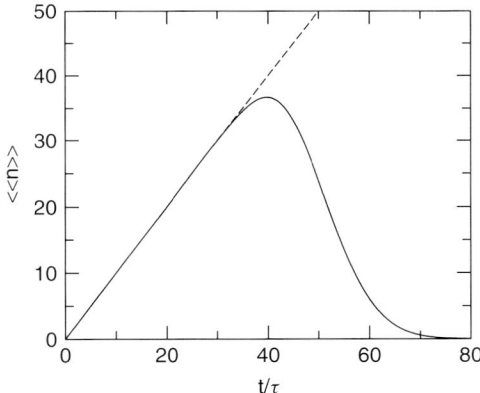

Figure 3.9 The variance $\langle\langle n \rangle\rangle$ depending on the dimensionless time t/τ. The initial size of the cluster is $n_0 = 50$. The exact result is shown by a solid line and the linear approximation by a dashed line.

$$\lim_{t \to \infty} p(n, t) = \delta_{n,0}. \qquad (3.90)$$

If we do not consider the final stage of dissolution of a large cluster, that is, if t is considerably smaller than t_{end} (3.89), then the probability $p(0, t)$ for the cluster to be completely dissolved is very small. This allows us to obtain the correct results for $n > 0$ by the following alternative method.

Let us define the generating function $G(z, t)$ by

$$G(z, t) \stackrel{\text{def}}{=} \sum_n z^n p(n, t). \tag{3.91}$$

According to the situation actually considered, the particular term $p(0, t)$ in this sum is negligible, so that the lower limit of summation may be taken from $n = 1$ instead of $n = 0$. The initial condition corresponding to $p(0, t) = \delta_{n,n_0}$ is represented by

$$G(z, 0) = z^{n_0}. \tag{3.92}$$

The equation for the generating function is obtained if both sides of the master equation (3.78) are multiplied by z^n performing the summation over n afterwards. This yields

$$\frac{\partial}{\partial t} G(z, t) = \frac{1}{\tau} \left(\frac{1}{z} - 1 \right) G(z, t). \tag{3.93}$$

The solution of the partial differential equation (3.93) with respect to the initial condition (3.92) is given by

$$G(z, t) = z^{n_0} \exp\left[\frac{t}{\tau} \left(\frac{1}{z} - 1 \right) \right]. \tag{3.94}$$

The previous result for $p(n, t)$ at $n \geq 1$ (3.80) is obtained from this equation after substitution by (3.91) and expansion of the exponent in z. Starting from (3.94)

$$G(z, t) = z^{n_0} e^{-t/\tau} \exp\left(\frac{t}{\tau} \frac{1}{z} \right) \tag{3.95}$$

the power series is written as follows

$$G(z, t) = \sum_n z^n p(n, t) = z^{n_0} e^{-t/\tau} \sum_m \frac{1}{m!} \left(\frac{t}{\tau z} \right)^m \tag{3.96}$$

$$= e^{-t/\tau} \sum_m \frac{1}{m!} \left(\frac{t}{\tau} \right)^m z^{n_0 - m} \tag{3.97}$$

$$= e^{-t/\tau} \sum_n \frac{1}{(n_0 - n)!} \left(\frac{t}{\tau} \right)^{n_0 - n} z^n \tag{3.98}$$

and therefore by comparison of the same order terms we get the Poisson distribution (3.80)

$$p(n, t) = \frac{(t/\tau)^{n_0 - n}}{(n_0 - n)!} e^{-t/\tau}. \tag{3.99}$$

In order to depict the process of jam shrinkage and to illustrate the developed formalism for the probabilistic description we refer to the graphics Figures 3.6–3.9 shown previously. These are based on numerical calculations simulating the stochastic process $n(t)$ (stochastic trajectories) and illustrating the time-dependent probability distributions $p(n, t)$ and related quantities.

The simple model discussed above can be improved to describe the dissolution of a vehicle queue at a signalized road intersection taking into account the car dynamics of the starting behavior when the red traffic light is switched to green. The quantity we are interested in is a modified detachment probability (3.76) which now depends on the cluster size n. For a long queue the detachment rate $w_-(n)$ has a constant value $1/\tau$ consistent with (3.76). However, due to the time spent in acceleration of the first cars and movement towards the stop line, the detachment rate changes for smaller queues.

3.6
The Two-Level System

In physics one often meets systems which are characterized by two distinct maxima of the probability distribution in the space of all possible states. Such bistable systems can be described by just two states or two levels $+$ and $-$, corresponding to these two maxima. The Ising model of a single spin, fluctuating in an external field, can be considered as a toy example. The spin has two states: spin up or along the field direction ($+$ level), and spin down or opposite to the field direction ($-$ level). It can flip from one state to another and vice versa with certain probabilities per time unit or transition rate, which depend on temperature and interaction strength with the external field. The Ising model has had an enormous impact on modern physics in general and statistical physics in particular, but also on other areas of science, including biology and neuroscience, economics and sociology among others [5, 33, 161, 220].

The Ising model of interacting spins is in fact a nontrivial many-particle system, although in the mean-field approximation it still fits into the simple picture described above. In this case the spin fluctuates in the mean field which is a superposition of the external field and the mean field created by the surrounding spins. In the vicinity of the phase transition or critical point, where the spontaneous ordering of spins occurs in zero external field, the spin system exhibits huge correlated fluctuations, the so-called critical fluctuations, which are not properly captured by the mean-field description. In this case the critical behavior of the Ising model is nontrivial and common for a certain class of models and real systems in accordance with the universality hypothesis known in the theory of critical phenomena [5, 137, 253]. Although exact solutions of one-dimensional and two-dimensional Ising models are well known [15, 161, 177], the critical behavior in the three-dimensional case has been a challenging problem up to the present, and several approaches including renormalization group and numerical methods (see [181] for a review) and, more recently, an alternative analytical method [94] have been proposed.

Returning to the two-level description, also called the Markovian dichotomic system, as a toy example with two states, the dynamics of the system is given by transition rates w_{-+} from $+$ to $-$ and w_{+-} from $-$ to $+$. The states are characterized by the probabilities $p_+(t)$ and $p_-(t)$ that the system at time t occupies the state + and −, respectively. The probabilities obey the normalization condition $p_+ + p_- = 1$. A basic assumption to describe the dynamics is that the switching between two states is a Markov process. This means that the system has no memory, that is, the transition rate from one given state to another is a property of this state and does not depend on the prehistory how it has been reached. The time evolution of $p_+(t)$ and $p_-(t)$ is given by the master equations

$$\frac{d}{dt}p_+ = -w_{-+}p_+ + w_{+-}p_- \tag{3.100}$$

$$\frac{d}{dt}p_- = +w_{-+}p_+ - w_{+-}p_- \tag{3.101}$$

together with initial condition specifying the values of the probabilities $p_+(0)$ and $p_-(0)$ at $t = 0$.

The solution of these differential equations is

$$p_+(t) = \frac{w_{+-}}{w_{+-} + w_{-+}} + \left[p_+(0) - \frac{w_{+-}}{w_{+-} + w_{-+}}\right] e^{-(w_{+-}+w_{-+})t} \tag{3.102}$$

$$p_-(t) = \frac{w_{-+}}{w_{+-} + w_{-+}} + \left[p_-(0) - \frac{w_{-+}}{w_{+-} + w_{-+}}\right] e^{-(w_{+-}+w_{-+})t}. \tag{3.103}$$

In the following we will show two different ways of obtaining the solution. The first one is the direct integration method. Inserting the normalization relation $p_+ = 1 - p_-$ into the first equation of motion, we obtain an inhomogeneous differential equation

$$\frac{d}{dt}p_+ + (w_{+-} + w_{-+})p_+ = w_{+-}, \tag{3.104}$$

which is solved in the standard way. First we consider the homogeneous case

$$\frac{d}{dt}p_+ + (w_{+-} + w_{-+})p_+ = 0 \tag{3.105}$$

and find its solution

$$p_+^{\text{hom}}(t) = C e^{-(w_{+-}+w_{-+})t}. \tag{3.106}$$

In the following we search for a particular solution by variation of constant according to the ansatz

$$p_+^{\text{par}}(t) = C(t) e^{-(w_{+-}+w_{-+})t}. \tag{3.107}$$

By inserting this into the inhomogeneous differential equation we obtain

$$C(t) = w_{+-} \int_0^t e^{+(w_{+-}+w_{-+})s} ds = \frac{w_{+-}}{w_{+-} + w_{-+}} \left[e^{+(w_{+-}+w_{-+})t} - 1\right]. \tag{3.108}$$

Therefore

$$p_+^{\text{par}}(t) = \frac{w_{+-}}{w_{+-} + w_{-+}} \left[1 - e^{-(w_{+-}+w_{-+})t}\right] \tag{3.109}$$

holds and the solution reads

$$\begin{aligned} p_+(t) &= p_+^{\text{par}}(t) + p_+^{\text{hom}}(t) \\ &= \frac{w_{+-}}{w_{+-} + w_{-+}} \left[1 - e^{-(w_{+-}+w_{-+})t}\right] + C e^{-(w_{+-}+w_{-+})t}, \end{aligned} \tag{3.110}$$

The integration constant C is calculated from the initial condition

$$p_+(t=0) = C = p_+(0), \tag{3.111}$$

which finally yields the known result.

The solution can be obtained using another method, so-called diagonalization by eigenstates. It is possible to solve the equations of motion starting from the form

$$\frac{d}{dt}\begin{pmatrix} p_+(t) \\ p_-(t) \end{pmatrix} = \begin{pmatrix} -w_{-+} & w_{+-} \\ w_{-+} & -w_{+-} \end{pmatrix} \begin{pmatrix} p_+(t) \\ p_-(t) \end{pmatrix}, \tag{3.112}$$

or, compare (3.26),

$$\frac{d}{dt} P(t) = W\, P(t), \tag{3.113}$$

where W is the transition matrix, and $P(t)$ is the time-dependent state.

Taking into account the initial condition $P(t=0) = P(0)$ the formal solution reads

$$P(t) = \exp(W\,t)\, P(0) = U(t)\, P(0). \tag{3.114}$$

The general solution $P(t)$, see (3.31), with eigenvalues λ_i and eigenstates u_i is given by

$$P(t) = \sum_i c_i u_i e^{-\lambda_i t}, \tag{3.115}$$

where the coefficients c_i are constants calculated from initial conditions. In our case i has two values: $i = 0, 1$.

The first step consists of the determination of eigenvalues from

$$|W - \lambda E| = \begin{vmatrix} -w_{-+} - \lambda & w_{+-} \\ w_{-+} & -w_{+-} - \lambda \end{vmatrix} = 0. \tag{3.116}$$

The calculation

$$(w_{-+} + \lambda)(w_{+-} + \lambda) - w_{+-} w_{-+} = 0 \tag{3.117}$$

$$\lambda(\lambda + (w_{+-} + w_{-+})) = 0 \tag{3.118}$$

yields the eigenvalues $\lambda_0 = 0$ and $\lambda_1 = -(w_{+-} + w_{-+})$. The term with zero eigenvalue λ_0 in (3.115) represents the stationary solution reached asymptotically as $t \to \infty$. In this case the stationary solution is also the equilibrium one, since the detailed balance (3.25) holds, that is, the condition $dp_+(t)/dt = dp_-(t)/dt = 0$ implies that the probability flux from state $+$ to state $-$ is balanced by the opposite flux. The other term with negative eigenvalue λ_1 describes the relaxation to this stationary solution.

In the second step we calculate the eigenstates $W u_\lambda = \lambda u_\lambda$ from

$$\begin{pmatrix} -w_{-+} & w_{+-} \\ w_{-+} & -w_{+-} \end{pmatrix} \begin{pmatrix} u_+^i \\ u_-^i \end{pmatrix} = \lambda_i \begin{pmatrix} u_+^i \\ u_-^i \end{pmatrix}. \tag{3.119}$$

For $i = 0$ we have

$$\begin{pmatrix} -w_{-+} & w_{+-} \\ w_{-+} & -w_{+-} \end{pmatrix} \begin{pmatrix} u_+^0 \\ u_-^0 \end{pmatrix} = 0 \begin{pmatrix} u_+^0 \\ u_-^0 \end{pmatrix} = \begin{pmatrix} 0 \\ 0 \end{pmatrix}, \tag{3.120}$$

so that the eigenstate u^0 can be written as

$$u_+^0 = \frac{w_{+-}}{w_{-+}} u_-^0. \tag{3.121}$$

For $i = 1$ we have

$$\begin{pmatrix} -w_{-+} & w_{+-} \\ w_{-+} & -w_{+-} \end{pmatrix} \begin{pmatrix} u_+^1 \\ u_-^1 \end{pmatrix} = -(w_{+-} + w_{-+}) \begin{pmatrix} u_+^1 \\ u_-^1 \end{pmatrix} \tag{3.122}$$

and the eigenstate u^1 can be represented as

$$u_+^1 = -u_-^1. \tag{3.123}$$

Putting these eigenstates into the general solution

$$P(t) = c_0 u^0 e^{\lambda_0 t} + c_1 u^1 e^{\lambda_1 t} \tag{3.124}$$

we get

$$P(t) = \begin{pmatrix} p_+(t) \\ p_-(t) \end{pmatrix} = c_0 \begin{pmatrix} \frac{w_{+-}}{w_{-+}} \\ 1 \end{pmatrix} + c_1 \begin{pmatrix} -1 \\ 1 \end{pmatrix} e^{-(w_{+-}+w_{-+})t} \tag{3.125}$$

From the initial condition

$$P(t = 0) = c_0 u^0 + c_1 u^1 = P(0) \tag{3.126}$$

we obtain

$$P(0) = \begin{pmatrix} p_+(0) \\ p_-(0) \end{pmatrix} = c_0 \begin{pmatrix} \frac{w_{+-}}{w_{-+}} \\ 1 \end{pmatrix} + c_1 \begin{pmatrix} -1 \\ 1 \end{pmatrix}. \tag{3.127}$$

Using the initial condition and normalization we finally obtain the previously unknown coefficients

$$c_0 = \frac{w_{-+}}{w_{+-} + w_{-+}}, \tag{3.128}$$

$$c_1 = p_-(0) - c_0 = p_-(0) - \frac{w_{-+}}{w_{+-} + w_{-+}} = -p_+(0) + \frac{w_{+-}}{w_{+-} + w_{-+}} \tag{3.129}$$

to get the known result.

According to (3.114), the solution can be represented by the time evolution matrix U as

$$P(t) = U(t, t_0) P(t_0) \quad \text{with} \quad U(t, t_0) = e^{W(t-t_0)}. \tag{3.130}$$

In our two-state system, setting $t_0 = 0$, we have

$$\begin{pmatrix} p_+(t) \\ p_-(t) \end{pmatrix} = \begin{pmatrix} U_{11} & U_{12} \\ U_{21} & U_{22} \end{pmatrix} \begin{pmatrix} p_+(0) \\ p_-(0) \end{pmatrix} \tag{3.131}$$

Adding up both equations and applying the normalization condition we obtain

$$U = \begin{pmatrix} U_{11} & U_{12} \\ U_{21} & U_{22} \end{pmatrix} = \begin{pmatrix} U_{11} & 1 - U_{22} \\ 1 - U_{11} & U_{22} \end{pmatrix} \tag{3.132}$$

with two unknown matrix elements U_{11} and U_{22}.

Using already known results for $p_+(t)$ and $p_-(t)$ after some mathematical manipulation we get

$$U_{11} = \frac{w_{+-}}{w_{+-} + w_{-+}} + \frac{w_{-+}}{w_{+-} + w_{-+}} e^{-(w_{+-} + w_{-+})t} \tag{3.133}$$

$$U_{12} = \frac{w_{+-}}{w_{+-} + w_{-+}} e^{-(w_{+-} + w_{-+})t} \tag{3.134}$$

$$U_{21} = \frac{w_{-+}}{w_{+-} + w_{-+}} e^{-(w_{+-} + w_{-+})t} \tag{3.135}$$

$$U_{22} = \frac{w_{-+}}{w_{+-} + w_{-+}} + \frac{w_{-+}}{w_{+-} + w_{-+}} e^{-(w_{+-} + w_{-+})t}. \tag{3.136}$$

Up to now we never have specified the transition rates w_{+-} and w_{-+}. They define a particular model. In physical systems like, e.g., the Ising model, the $+$ and $-$ states are characterized by certain energies. Since energy can be defined up to an arbitrary additive constant, we may set the lowest energy equal to zero. Then one state (say $+$) has energy 0 and the other state ($-$) has an energy value $\varepsilon > 0$ called the activation energy.

The transition rates triggered by the heat bath with temperature T are given by the following Arrhenius ansatz

$$w_{-+} = \nu \exp\left\{-\frac{\beta\varepsilon}{2}\right\} \quad \text{(hill up)} \tag{3.137}$$

$$w_{+-} = \nu \exp\left\{+\frac{\beta\varepsilon}{2}\right\} \quad \text{(hill down)}, \tag{3.138}$$

where $\beta = 1/k_B T$ is a parameter proportional to the inverse temperature T, k_B is the Boltzmann constant, and ν is a constant flip rate.

Independent of the initial values $p_+(0), p_-(0)$ the system reaches the long-time limit $p_+(t \to \infty), p_-(t \to \infty)$ (see Figure 3.10) given by

$$p_+(t \to \infty) = \frac{w_{+-}}{w_{+-} + w_{-+}} = \frac{1}{1 + e^{-\beta\varepsilon}}, \tag{3.139}$$

$$p_-(t \to \infty) = \frac{w_{-+}}{w_{+-} + w_{-+}} = \frac{1}{1 + e^{+\beta\varepsilon}}. \tag{3.140}$$

This distribution is shown as function of temperature in Figure 3.11.

An extension of the simple dichotomic spin-flip process coupled to two heat reservoirs is analyzed by Steffen Trimper [228]. While one flip process is triggered by a heat bath at temperature T, the inverse transition is activated by a heat bath at a different temperature T'. The stationary solution of the master equation leads to a generalized Fermi distribution with an effective temperature T_e as the harmonic average of T and T'.

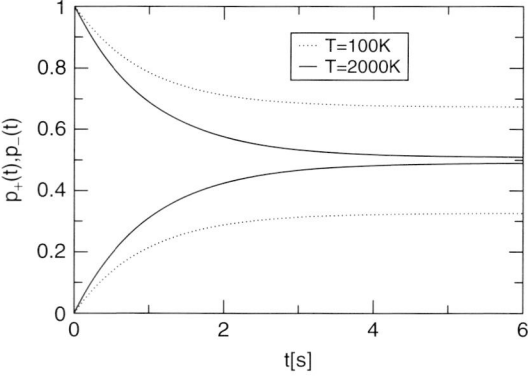

Figure 3.10 The solution of the master equation for different values of temperature T. The results are shown for the following set of parameters: $p_+(t = 0) = 1$, $p_-(t = 0) = 0$, $\nu = 0.5$ s^{-1}, $\varepsilon = 10^{-21}$ Nm, $k_B = 1.3806503 \times 10^{-23}$ m^2 kg s^{-2} K^{-1}.

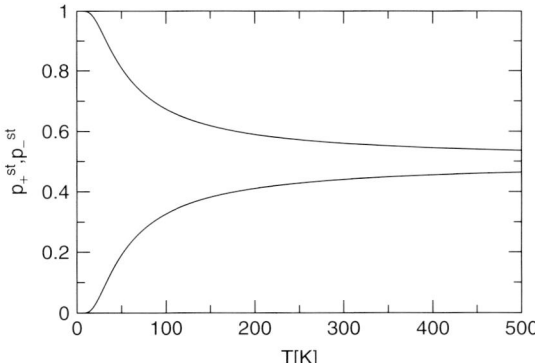

Figure 3.11 Long-time behavior as a function of temperature. Energy difference $\varepsilon = 10^{-21}$ Nm as before.

3.7
The Three-Level System

The foregoing consideration of two-level system can be generalized to a system of three states: 1, 2 and 3. In this case the probability distribution is described by the three-component vector

$$\mathbf{P}(t) = \begin{pmatrix} p_1(t) \\ p_2(t) \\ p_3(t) \end{pmatrix} \qquad (3.141)$$

obeying the master equation (cf. (3.26))

$$\frac{d}{dt}\mathbf{P}(t) = \mathbf{W}\,\mathbf{P}(t), \qquad (3.142)$$

where the transition matrix $\mathbf{W}$ is given by

$$\mathbf{W} = \begin{pmatrix} -(w_{21}+w_{31}) & w_{12} & w_{13} \\ w_{21} & -(w_{12}+w_{32}) & w_{23} \\ w_{31} & w_{32} & -(w_{13}+w_{23}) \end{pmatrix} \qquad (3.143)$$

As distinct from the two-level case, the stationary solution of the three-level system is not necessarily the equilibrium one. It depends on the specific values of the transition rates w_{ij}, and only in a special case is the detailed balance satisfied. In the stationary state the probability flux between two states is constant, but not necessarily zero. For example, if $w_{12} = w_{23} = w_{31} = 0$ holds, then from state 1 it is possible to go only to state 2; from state 2 to state 3; and from state 3 to state 1 with transition rates w_{21}, w_{32}, and w_{13}, respectively. The stationary solution then corresponds to a constant circular flux.

Before searching for the general solution we consider some particular cases. A simple situation is when all $w_{ij} = 0$ except w_{23} and w_{32}. In this case only transitions between states 2 and 3 take place, whereas state 1 is isolated, as illustrated in the following schematic picture

Case I: Isolated state

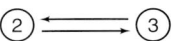

The solution reads

$$p_1(t) = p_1(0) \tag{3.144}$$

$$p_2(t) = \frac{w_{23}}{w_{32} + w_{23}}(1 - p_1(0))$$
$$+ \left[p_2(0) - \frac{w_{23}}{w_{32} + w_{23}}(1 - p_1(0)) \right] e^{-(w_{32}+w_{23})t} \tag{3.145}$$

$$p_3(t) = \frac{w_{32}}{w_{32} + w_{23}}(1 - p_1(0))$$
$$+ \left[p_3(0) - \frac{w_{32}}{w_{32} + w_{23}}(1 - p_1(0)) \right] e^{-(w_{32}+w_{23})t} \tag{3.146}$$

At $p_1(0) = 0$ the probabilities $p_2(t)$ and $p_3(t)$ obey the solution for the two-level system (3.102)–(3.103).

A simple solution exists in the case where all transition rates are equal: $w_{ij} = w$, as shown in the following picture

Case II: Symmetry (equal fluxes)

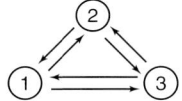

In this case we have

$$p_1(t) = \frac{1}{3} + \left(p_1(0) - \frac{1}{3} \right) e^{-3wt} \tag{3.147}$$

$$p_2(t) = \frac{1}{3} + \left(p_2(0) - \frac{1}{3} \right) e^{-3wt} \tag{3.148}$$

$$p_3(t) = \frac{1}{3} + \left(p_3(0) - \frac{1}{3} \right) e^{-3wt} \tag{3.149}$$

Finally, as a particular situation we consider the already mentioned totally asymmetric case $w_{12} = w_{23} = w_{31} = 0$, represented as

Case III: Asymmetry (circular flux)

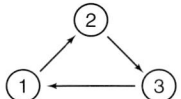

with constant stationary flux. The solution reads

$$\mathbf{P}(t) = \mathbf{P}^{st} + C_1 \mathbf{P}_1 e^{\lambda_1 t} + C_2 \mathbf{P}_2 e^{\lambda_2 t} \tag{3.150}$$

with

$$\lambda_{1,2} = \frac{-A \pm \sqrt{A^2 - 4B}}{2}, \tag{3.151}$$

where

$$A = w_{21} + w_{32} + w_{13} \tag{3.152}$$

$$B = w_{32} w_{13} + w_{21} w_{13} + w_{21} w_{32} \tag{3.153}$$

are constants,

$$\mathbf{P}^{st} = \frac{1}{B} \begin{pmatrix} w_{32} w_{13} \\ w_{21} w_{13} \\ w_{21} w_{32} \end{pmatrix} \tag{3.154}$$

is the stationary solution, and

$$\mathbf{P}_{1,2} = \begin{pmatrix} w_{13} \\ -w_{13} - w_{21} - \lambda_{1,2} \\ w_{21} + \lambda_{1,2} \end{pmatrix} \tag{3.155}$$

are the eigenvectors representing the time-dependent part of the solution with the weight coefficients

$$C_1 = \frac{1}{\lambda_2 - \lambda_1} \left(p_1(0) \frac{w_{21} + \lambda_2}{w_{13}} - p_3(0) - \lambda_2 \frac{w_{32}}{B} \right) \tag{3.156}$$

$$C_2 = \frac{1}{\lambda_1 - \lambda_2} \left(p_1(0) \frac{w_{21} + \lambda_1}{w_{13}} - p_3(0) - \lambda_1 \frac{w_{32}}{B} \right) \tag{3.157}$$

found from the initial condition. As an example, the solution for $w_{21} = w_{13} = 1\,\text{s}^{-1}$ and $w_{32} = 5\,\text{s}^{-1}$ with the initial condition $p_1(0) = 1$, $p_2(0) = p_3(0) = 0$ is shown in Figure 3.12.

In the general case the solution can be written as (3.150)

$$\mathbf{P}(t) = \mathbf{P}^{st} + C_1 \mathbf{P}_1 e^{\lambda_1 t} + C_2 \mathbf{P}_2 e^{\lambda_2 t} \tag{3.158}$$

together with (3.151)

$$\lambda_{1,2} = \frac{-A \pm \sqrt{A^2 - 4B}}{2}, \tag{3.159}$$

where

$$A = w_{12} + w_{13} + w_{21} + w_{23} + w_{31} + w_{32} \tag{3.160}$$

$$B = a + b + c \tag{3.161}$$

$$a = w_{12} w_{23} + w_{13} w_{32} + w_{12} w_{13} \tag{3.162}$$

$$b = w_{13} w_{21} + w_{23} w_{31} + w_{21} w_{23} \tag{3.163}$$

$$c = w_{31} w_{32} + w_{21} w_{32} + w_{12} w_{31} \tag{3.164}$$

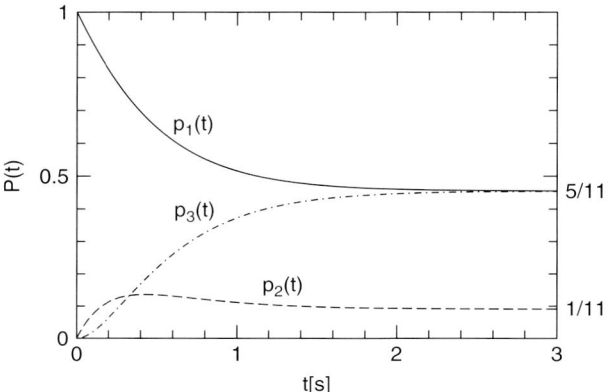

Figure 3.12 The solution of the master equation $\mathbf{P}(t)$ for a three-level system with totally asymmetric transition rates $w_{12} = w_{23} = w_{31} = 0$, $w_{21} = w_{13} = 1\,\mathrm{s}^{-1}$, and $w_{32} = 5\,\mathrm{s}^{-1}$.

and

$$\mathbf{P}^{\mathrm{st}} = \frac{1}{B}\begin{pmatrix} a \\ b \\ c \end{pmatrix} \tag{3.165}$$

$$\mathbf{P}_{1,2} = \begin{pmatrix} w_{13} - w_{12} \\ -w_{13} - w_{21} - w_{31} - \lambda_{1,2} \\ w_{12} + w_{21} + w_{31} + \lambda_{1,2} \end{pmatrix} \tag{3.166}$$

$$C_1 = \frac{1}{\lambda_2 - \lambda_1}\left[\left(p_1(0) - \frac{a}{B}\right)\frac{w_{12} + w_{21} + w_{31} + \lambda_2}{w_{13} - w_{12}} - p_3(0) + \frac{c}{B}\right] \tag{3.167}$$

$$C_2 = \frac{1}{\lambda_1 - \lambda_2}\left[\left(p_1(0) - \frac{a}{B}\right)\frac{w_{12} + w_{21} + w_{31} + \lambda_1}{w_{13} - w_{12}} - p_3(0) + \frac{c}{B}\right] \tag{3.168}$$

The meaning of $\mathbf{P}^{\mathrm{st}}$ (3.165) is the stationary probability distribution. It is the eigenvector of matrix $\mathbf{W}$, which corresponds to the zero eigenvalue $\lambda^{\mathrm{st}} = 0$. Two other eigenvectors $\mathbf{P}_1$ and $\mathbf{P}_2$ correspond to the eigenvalues λ_1 and λ_2, respectively. By definition, the eigenvalues and eigenvectors are determined by the equation

$$\mathbf{W} \cdot \mathbf{P} = \lambda \mathbf{P}. \tag{3.169}$$

This leads to the equation for eigenvalues

$$\det\left(\mathbf{W} - \lambda \mathbf{E}\right) = 0, \tag{3.170}$$

where $\mathbf{E}$ is the unit matrix with components $E_{ij} = \delta_{ij}$.

Note that the condition $\lambda(P_1 + P_2 + P_3) = 0$ is obtained for the components P_i of the eigenvector by summing up all three equations represented in the vectorial form (3.169). This means that $P_1 + P_2 + P_3 = 0$ holds for the eigenvectors with $\lambda \neq 0$. According to (3.170), the system of linear equations for P_i is degenerated, that is, there are no more than two independent equations. A unique solution is obtained by using the probability normalization condition $P_1^{(st)} + P_2^{(st)} + P_3^{(st)} = 1$, as well as the initial condition to find the constants C_1 and C_2.

Equation (3.170) is of the form

$$\begin{vmatrix} -(w_{21} + w_{31} + \lambda) & w_{12} & w_{13} \\ w_{21} & -(w_{12} + w_{32} + \lambda) & w_{23} \\ w_{31} & w_{32} & -(w_{13} + w_{23} + \lambda) \end{vmatrix} = 0. \quad (3.171)$$

Calculating the determinant (3.171) we obtain

$$\lambda[\lambda^2 + \lambda(w_{12} + w_{13} + w_{21} + w_{23} + w_{31} + w_{32}) + (w_{12}w_{13} + w_{13}w_{21} + w_{12}w_{23} + w_{21}w_{23} + w_{12}w_{31} + w_{23}w_{31} + w_{13}w_{32} + w_{21}w_{32} + w_{31}w_{32})] = 0. \quad (3.172)$$

The first root of equation (3.172) is $\lambda^{st} = 0$. It describes the stationary distribution

$$\frac{d\mathbf{P}^{st}}{dt} = 0 \quad (3.173)$$

The other roots satisfy the equation

$$\lambda^2 + A\lambda + B = 0 \quad (3.174)$$

where constants A and B are given by (3.160) and (3.161), respectively.

We have three different cases. In the first case, meeting the inequality $A^2 > 4B$ both of the eigenvalues (λ_1, λ_2) are different real negative numbers. In the second case, when $A^2 = 4B$, these eigenvalues are equal to each other, $\lambda_1 = \lambda_2 = -A/2$. In the third case, when $A^2 < 4B$, the eigenvalues (λ_1, λ_2) are complex numbers $\lambda_1 = (-A + i\sqrt{|D|})/2$ and $\lambda_2 = (-A - i\sqrt{|D|})/2$, where $D = A^2 - 4B$. In this case the system dynamics can exhibit damped oscillation

$$\mathbf{P}(t) = \mathbf{P}^{st} + e^{-At/2}\left[B_1 \sin\left(\sqrt{|D|}t/2\right) + B_2 \cos\left(\sqrt{|D|}t/2\right)\right], \quad (3.175)$$

where the constants B_1 and B_2 are determined from the initial condition.

The discriminant $D = A^2 - 4B$ of polynomial (3.174) is given by the expression

$$\begin{aligned} D = {} & w_{12}^2 + w_{21}^2 + w_{13}^2 + w_{31}^2 + w_{23}^2 + w_{32}^2 \\ & + 2(w_{12}w_{21} + w_{13}w_{31} + w_{23}w_{32} + w_{21}w_{31} + w_{12}w_{32} + w_{13}w_{23}) \\ & - 2(w_{12}w_{23} + w_{13}w_{32} + w_{12}w_{13} + w_{13}w_{21} + w_{23}w_{31} \\ & + w_{21}w_{23} + w_{31}w_{32} + w_{21}w_{32} + w_{12}w_{31}). \end{aligned} \quad (3.176)$$

The stationary solution, defined by (3.173) and given by (3.165), can be found by a more elegant method developed by Gustav Kirchhoff. It is based on some elements of graph theory, as explained in [64, 204]. The Markovian system under consideration is represented by a graph G consisting of vertices and edges. Vertices correspond to system states, whereas edges connect all vertices i and j for which at least one of the transition rates w_{ij} and w_{ji} is nonzero. For nonzero transition rates w_{ij}, the graph G of the three-level system is represented as

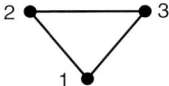

In the following we will assume that the graph G is connected. In principle, it can also consist of unconnected parts. Then Kirchhoff's method can be applied to each part separately. An example is the already considered case with an isolated state where the problem, in fact, reduces to the solution of a two-level system represented by two vertices connected by one edge. The only peculiarity is that the stationary probabilities have to be normalized in such a way that the total probability for the isolated subsystem is that given by the initial condition.

The stationary solution is represented by subgraphs of G, called maximal trees. A maximal tree $T\{G\}$ is a connected subgraph of G such that: (i) all edges of $T\{G\}$ are edges of G; (ii) $T\{G\}$ contains all vertices of G; and (iii) $T\{G\}$ contains no circuits (cyclic sequences of edges). It is easy to realize that one has to drop a certain minimum number of edges of G to exclude circuits and obtain maximal trees. Maximal trees of the graph G for the three-level system are

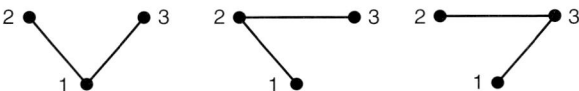

In order to construct the stationary solution of the master equation we need directed maximal trees. For a given state i and tree T the directed tree T_i is obtained from T by directing all its edges towards the vertex i. The superposition of such trees makes up the general solution of the master equation. In the case of the three-level system it is represented as

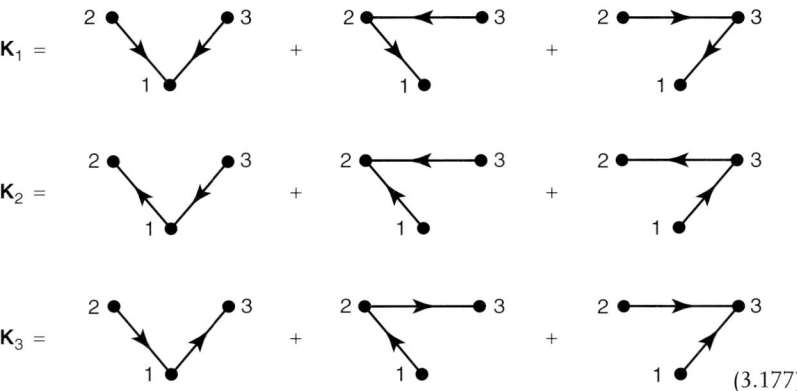

(3.177)

Here the symbol $i \bullet\!\!\leftarrow\!\!\bullet j$ stands for the transition from the state j to the state i with rate w_{ij}. We ascribe to each directed maximal tree T_j a weight $K_i\{T_j\}$ equal to the product of all transition rates w_{ij} corresponding to its edges with the appropriate directions. The value $\mathbf{K}_i$ is defined as the sum of $K_i\{T_j\}$ running over all the maximal directed trees leading to state i

$$\mathbf{K}_1 = \sum_{\text{all } T_j} K_i\{T_j\}. \tag{3.178}$$

According to (3.177), for our three-level system we have

$$\mathbf{K}_1 = w_{12}w_{23} + w_{13}w_{32} + w_{12}w_{13}, \tag{3.179}$$

$$\mathbf{K}_2 = w_{21}w_{13} + w_{23}w_{31} + w_{21}w_{23}, \tag{3.180}$$

$$\mathbf{K}_3 = w_{31}w_{32} + w_{32}w_{21} + w_{31}w_{12}. \tag{3.181}$$

The Kirchhoff formula for the stationary probability distribution p_i^{st} of an N-level system, represented by a connected graph, is given by

$$p_i^{\text{st}} = \frac{\mathbf{K}_i}{\sum_{j=1}^{N} \mathbf{K}_j}. \tag{3.182}$$

We can easily see that this general formula is consistent with (3.165) in the case of the three-level system ($N = 3$), where $\mathbf{K}_i$ are given by (3.179)–(3.181).

In the following we consider the three-level system thermodynamically. For thermodynamic systems the detailed balance typically holds. This means that, for an arbitrary chosen pair of states i, j, the stationary probability flux between them is equal to zero,

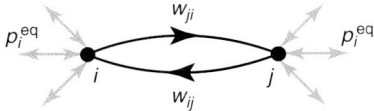

or in mathematical terms

$$J_{ij} := w_{ji} p_i^{\text{eq}} - w_{ij} p_j^{\text{eq}} = 0. \tag{3.183}$$

Following Section 3.2 we use the symbol p_i^{eq} to designate the stationary distribution function p_i^{st} for the systems with detailed balance. As is well known for such a system, the energy H_i specifies the equilibrium distribution via the Boltzmann formula

$$p_i^{\text{eq}} = \frac{1}{Z} \exp\{-\beta H_i\}. \tag{3.184}$$

Let us fix some state, for example, state 1. Then for any state i the detailed balance reads

$$w_{i1} p_1^{eq} = w_{1i} p_i^{eq}. \tag{3.185}$$

By virtue of (3.184) expression (3.185) is rewritten as

$$\frac{w_{1i}}{w_{i1}} = \exp\{-\beta(H_1 - H_i)\}, \tag{3.186}$$

thereby

$$H_i = H_1 + k_B T [\ln w_{1i} - \ln w_{i1}]. \tag{3.187}$$

Expression (3.187) actually enables us to construct the energy H_i using the given transition rates w_{ij}. If H_1 is known we can calculate any H_i. In particular, for the three-level system this construction of the energies H_i is actually based on the following maximal tree

So, on one hand, the energy H_i and correspondingly the distribution of a thermodynamic system can be constructed using one maximal tree only. On the other hand, the Kirchhoff diagram technique deals with all the maximal trees being actually independent. In order to elucidate this seeming contradiction and to propose an approach for constructing the energy H, without applying to a certain fixed maximal tree, we consider the three-level system as an example. The stationary probability flux, for example, along the edge $\{12\}$ can be calculated directly by substituting Kirchhoff's formula (3.182) into (3.183) within the replacement $p^{st} \to p^{eq}$, yielding

$$J_{12}^{eq} = w_{21} p_1^{eq} - w_{12} p_2^{eq} = w_{21} w_{32} w_{13} - w_{12} w_{23} w_{31} = 0 \tag{3.188}$$

The fluxes along $\{13\}$ and $\{23\}$ have the same value equal to zero. Expression (3.188) provides the condition

$$w_{21} w_{32} w_{13} = w_{12} w_{23} w_{31} \tag{3.189}$$

or using Kirchhoff diagrams (3.189) is represented as

$$\text{[diagram]} = 0 \tag{3.190}$$

The following equalities result immediately from (3.189)

$$k_1 := \frac{K_1\{T_j\}}{K_2\{T_j\}} = \frac{w_{12}}{w_{21}} = \frac{w_{32} w_{13}}{w_{23} w_{31}}, \tag{3.191}$$

$$k_2 := \frac{K_1\{T_j\}}{K_3\{T_j\}} = \frac{w_{13}}{w_{31}} = \frac{w_{12}w_{23}}{w_{21}w_{32}}. \tag{3.192}$$

Therefore, to construct the equilibrium distribution only one column in (3.177) may be taken into account. The choice of any number of columns in (3.177) gives the same result

$$p_1^{eq} = \frac{1}{1+k_1+k_2}, \quad p_2^{eq} = \frac{k_1}{1+k_1+k_2}, \quad p_3^{eq} = \frac{k_2}{1+k_1+k_2}. \tag{3.193}$$

So we have demonstrated how to introduce the energy applying to the notion of Kirchhoff's diagrams. By definition, the energy of a state i within a tree T_j is written as

$$H_i\{T_j\} = -k_B T \ln K_i\{T_j\}, \tag{3.194}$$

for example, for state 1 expression (3.194) reads

$$H_1\{T_1\} = -k_B T[\ln w_{12} + \ln w_{13}]. \tag{3.195}$$

For a system with detailed balance the energy $H_i\{T_j\}$ possesses the following property. The difference between the energies of an arbitrarily chosen pair of states $\{i,j\}$ within the same tree T_k is a constant

$$H_i\{T_k\} - H_j\{T_k\} = \text{const.} \tag{3.196}$$

By way of an example, we consider the following two trees directed to state 1 and, similarly, two trees directed to state 2

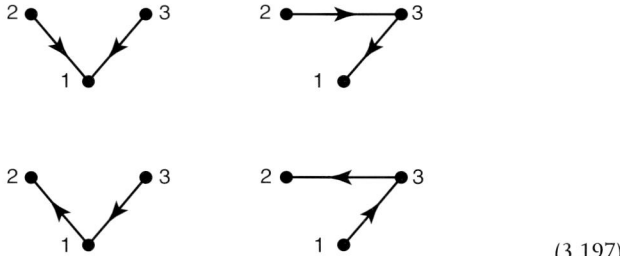

$$(3.197)$$

Using expression (3.194) the corresponding energies are

$$H_1\{T_1\} = -k_B T[\ln w_{12} + \ln w_{13}],$$
$$H_2\{T_1\} = -k_B T[\ln w_{21} + \ln w_{13}],$$
$$H_1\{T_2\} = -k_B T[\ln w_{32} + \ln w_{13}],$$
$$H_2\{T_2\} = -k_B T[\ln w_{31} + \ln w_{32}].$$

The differences between the energies are written as

$$H_1\{T_1\} - H_2\{T_1\} = -k_B T[\ln w_{12} + \ln w_{21}], \quad (3.198)$$

$$H_1\{T_2\} - H_2\{T_2\} = -k_B T[\ln w_{32} + \ln w_{13} + \ln w_{31} + \ln w_{32}]. \quad (3.199)$$

Expressions (3.198) and (3.199) are equal to each other provided the condition (3.189) is fulfilled and, thus,

$$\ln(w_{32} w_{13} w_{21}) = \ln(w_{31} w_{23} w_{12}). \quad (3.200)$$

Now let us construct the energy $\mathbf{H}_i$ of a state i as

$$\mathbf{H}_i = \frac{1}{3} \sum_{k=1}^{3} H_i\{T_k\}. \quad (3.201)$$

Then expression for $H_i\{T_k\}$ can be rewritten as

$$H_i\{T_k\} = \mathbf{H}_i + \Delta H\{T_k\}. \quad (3.202)$$

In (3.202) the former term depends only on the state i. In the general case the latter one should depend on the state i as well as the given tree T_j. However, due to the detailed balance it turns out to depend only on the tree T_j. In fact, for a fixed tree T_k we can write

$$\Delta H_i\{T_k\} - \Delta H_j\{T_k\}$$
$$= H_i\{T_k\} - H_j\{T_k\} - \frac{1}{3}\left[\sum_{j'=1}^{3} H_i\{T_{j'}\} - \sum_{j'=1}^{3} H_j\{T_{j'}\}\right] = 0. \quad (3.203)$$

By virtue of (3.196) using (3.202) expression (3.182) can be rewritten ($\beta = 1/(k_B T)$)

$$p_i^{eq} = \frac{\exp\{-\beta \mathbf{H}_i\}}{\sum_{i=1}^{3} \exp\{-\beta \mathbf{H}_i\}}, \quad (3.204)$$

which has the form of the Boltzmann distribution. We note that these speculations also hold for multilevel systems with detailed balance.

3.8
Exercises

E 3.1 *Radioactive decay*
By analogy with the Poisson process considered in Section 3.5, solve the master equation (3.75) for the radioactive decay process with $w_-(n) = \frac{1}{\tau(n)} = \frac{1}{\tau_0} \cdot n$. With a given initial condition $p(n, t = 0) = \delta_{n,n_0}$ find the analytical time-dependent solution $p(n, t)$ for $n \geq 0$. As in Section 3.5, calculate the first two moments and the variance of the probability distribution.

E 3.2 *Ehrenfest urn model*

Consider the Ehrenfest model of particle diffusion between two boxes. In 1907 Paul and Tatiana Ehrenfest formulated their stochastic urn model [39] for the first time to discuss the H-Theorem investigated by Ludwig Boltzmann. There are totally N particles. Let n be the number of particles in the first (or left) box. Each of them can go over to the second (or right) box with transition rate d_{12}. Similarly, any one of $N - n$ particles in the second box can go over to the first one with the transition rate d_{21}. The probability $p(n, t)$ of finding the system in the state with n particles in the first box at time t is given by the one-step master equation (3.38) with the transition rates

$$w_+(n) = d_{21}(N - n), \tag{3.205}$$

$$w_-(n) = d_{12} n. \tag{3.206}$$

One task is to find the stationary as well as the time-dependent solutions of this master equation with the initial condition $p(n, t = 0) = \delta_{n,n_0}$.

Hint: use the generating function

$$F(s_1, s_2, t) = \sum_{n_i} s_1^{n_1} s_2^{n_2} p(n_1, n_2, t), \quad |s_i| < 1, \tag{3.207}$$

where in our case $p(n_1, n_2, t) \equiv p(n_1, N - n_1, t) \equiv p(n, t)$.

Another task is to obtain the differential equations for the first and the second moments, that is $\langle n \rangle(t)$ and $\langle n^2 \rangle(t)$, of the probability distribution function $p(n, t)$ and solve them. Calculate the variance and estimate its dependence on the number of particles N at $t \to \infty$ (the equilibrium value).

Hint: use the master equation and the definition $\langle n^k \rangle = \sum_n n^k p(n, t)$.

E 3.3 *Schlögl model*

Consider the model of the bistable Schlögl reaction (named after Friedrich Schlögl from Aachen)

$$A + 2X \underset{k_1'}{\overset{k_1}{\rightleftharpoons}} 3X; \quad X \underset{k_2'}{\overset{k_2}{\rightleftharpoons}} F,$$

where A is the raw substance and F is the final product, both having constant concentrations, whereas X is the intermediate substance of interest with varying concentration C. Its dependence on time T is given by the equation

$$\frac{dC}{dT} = k_1 A C^2 - k_1' C^3 - k_2 C + k_2' F. \tag{3.208}$$

The concentration is $C = N/V$, where N is the number of molecules and V is the volume of the system. By introducing elementary volume and time units, V_0 and $t_0 = V_0^2/k_1'$, (3.208) is written in dimensionless variables $X = V_0 C$ and $t = T/t_0$ as

$$\frac{dX}{dt} = -X^3 + aX^2 - bX + c, \tag{3.209}$$

where $a = (k_1 A V_0)/k_1'$, $b = (k_2 V_0^2)/k_1'$, and $c = (k_2' F V_0^3)/k_1'$ are positive dimensionless control parameters. Using the substitution $x = X - a/3$, (3.209) is transformed into the cubic normal form without the quadratic term,

$$\frac{dx}{dt} = -x^3 + \beta x + \gamma \tag{3.210}$$

with two control parameters $\beta = \frac{1}{3}a^2 - b$ and $\gamma = \frac{2}{27}a^3 - \frac{1}{3}ab + c$. Depending on the parameter γ, (3.210) has either one or three stationary solutions.

One task is to find the stationary solutions of (3.210) as functions of the parameter γ and analyze their stability.

Another task is to construct the master equation for the corresponding stochastic model, where $p(N, t)$ is the probability of finding the system in a state with N molecules of substance X at time t.

Hint: consider a finite volume V, where $C = N/V$; estimate the transition rates using the mean-field concentration product ansatz multiplied with the reaction constant.

E 3.4 Stochastic Brusselator

Start a new project called the stochastic Brusselator (named after the city of Brussels, where this model was first discussed by Ilya Prigogine et al.). The model deals with certain idealized autocatalytic reactions. In general, these are chemical reactions in which at least one of the products is also a reactant. We consider the following example

$$R_1 \xrightarrow{k_1} X; \quad R_2 + X \xrightarrow{k_2} Y + F_1; \quad 2X + Y \xrightarrow{k_3} 3X; \quad X \xrightarrow{k_4} F_2,$$

where k_1, k_2, k_3, k_4 are reaction constants; R_1 and R_2 are the raw substances; F_1 and F_2 are final products of the reaction; whereas X and Y are intermediate substances which are of particular interest. Let us denote the concentrations of R_1 and R_2 by r_1 and r_2, and those of X and Y by x and y. In the following it is assumed that the concentrations of the raw substances are so large that their depletion during a considered time of reaction can be neglected, so r_1 and r_2 are constants. We are thus interested only in the temporal variation of the concentrations x and y, which are described by the system of two coupled differential equations

$$\frac{dx}{dt} = k_1 r_1 - k_2 r_2 x + k_3 x^2 y - k_4 x \tag{3.211}$$

$$\frac{dy}{dt} = k_2 r_2 x - k_3 x^2 y. \tag{3.212}$$

One task is to find the fixed-point stationary solution of (3.211)–(3.212), to determine the region of its stability depending on the parameters of the model, and to analyze numerically the solution in the region where the fixed point is unstable.

Another task is to construct the master equation for the corresponding stochastic model, where $p(N_x, N_y, t)$ is the probability of finding the system in a state with N_x molecules of substance X and N_y molecules of substance Y at time t.

Hint: consider a finite volume V, where $x = N_x/V$ and $y = N_y/V$; estimate the transition rates using the mean-field concentration product ansatz multiplied with the reaction constant.

4
The Fokker–Planck Equation

4.1
General Fokker–Planck Equations

One of the fundamental dynamical expressions for Markovian processes is the Fokker–Planck equation [49, 193] in its forward and backward notation. The basic quantity describing the probabilistic evolution for the path from the initial value $\mathbf{r}_0 = \mathbf{r}(t_0)$ to position $\mathbf{r}(t)$ for all times $t \geq t_0$ including $t \to \infty$ is the conditional probability density $p(\mathbf{r}, t \mid \mathbf{r}_0, t_0)$, also called the Green function. Usually the boundary conditions which have to be formulated due to the context of the given problem are essential for the analytical representation of the Fokker–Planck solution which is related to the Sturm–Liouville problem.

The multidimensional forward Fokker–Planck equation (using ∇ known as Nabla notation) consists of drift contributions (given by vector $\mathbf{v} = \{v_i \mid i = 1, \ldots, N\}$) as well as diffusion terms (given by matrix $D = \{d_{ij} \mid i, j = 1, \ldots, N\}$)

$$\frac{\partial p(\mathbf{r}, t \mid \mathbf{r}_0, t_0)}{\partial t} = -\sum_{i=1}^{N} \nabla_i \left[v_i(\mathbf{r}, t) p(\mathbf{r}, t \mid \mathbf{r}_0, t_0) \right]$$
$$+ \sum_{i=1}^{N} \sum_{j=1}^{N} \nabla_i \nabla_j \left[d_{ij}(\mathbf{r}, t) \, p(\mathbf{r}, t \mid \mathbf{r}_0, t_0) \right] \quad (4.1)$$

and can be written as a typical continuity equation

$$\frac{\partial p(\mathbf{r}, t \mid \mathbf{r}_0, t_0)}{\partial t} + \sum_{i=1}^{N} \nabla_i J_i \{p(\mathbf{r}, t \mid \mathbf{r}_0, t_0)\} = 0 \quad (4.2)$$

with the probability flux

$$J_i = v_i(\mathbf{r}, t) p(\mathbf{r}, t \mid \mathbf{r}_0, t_0) - \sum_{j=1}^{N} \nabla_j \left[d_{ij}(\mathbf{r}, t) p(\mathbf{r}, t \mid \mathbf{r}_0, t_0) \right]. \quad (4.3)$$

The continuity equation (4.2) has to be completed by boundary conditions indicating special properties of the flux (4.3) and/or the conditional probability density $p(\mathbf{r}, t \mid \mathbf{r}_0, t_0)$ like reflecting and absorbing barriers.

Physics of Stochastic Processes: How Randomness Acts in Time
Reinhard Mahnke, Jevgenijs Kaupužs and Ihor Lubashevsky
Copyright © 2009 WILEY-VCH Verlag GmbH & Co. KGaA, Weinheim
ISBN: 978-3-527-40840-5

The corresponding multidimensional backward Fokker–Planck equation reads (∇^0 means Nabla derivative with respect to $\mathbf{r}_0$)

$$-\frac{\partial p(\mathbf{r}, t \mid \mathbf{r}_0, t_0)}{\partial t_0} = +\sum_{i=1}^{N} v_i(\mathbf{r}_0, t_0)\, \nabla_i^0 p(\mathbf{r}, t \mid \mathbf{r}_0, t_0)$$

$$+ \sum_{i=1}^{N}\sum_{j=1}^{N} d_{ij}(\mathbf{r}_0, t_0)\, \nabla_i^0 \nabla_j^0 p(\mathbf{r}, t \mid \mathbf{r}_0, t_0). \tag{4.4}$$

The initial condition in both cases is

$$p(\mathbf{r}, t = t_0 \mid \mathbf{r}_0, t_0) = p(\mathbf{r}, t \mid \mathbf{r}_0, t_0 = t) = \delta(\mathbf{r} - \mathbf{r}_0). \tag{4.5}$$

In the following we rewrite the conditional probability $p(\mathbf{r}, t \mid \mathbf{r}_0, t_0)$ for times $t > t_0$ as $p(\{x\}, t \mid \{y\}, s)$ and define the forward and the adjoint backward Fokker–Planck operators $\mathcal{L}_F$ and $\mathcal{L}_B$, respectively. The general Fokker–Planck equations for N variables then read

$$\frac{\partial p(\{x\}, t \mid \{y\}, s)}{\partial t} = \mathcal{L}_F(\{x\}, t)\, p(\{x\}, t \mid \{y\}, s) \quad \text{with}$$

$$\mathcal{L}_F(\{x\}, t) = -\sum_{i=1}^{N} \frac{\partial}{\partial x_i} v_i(\{x\}, t) + \sum_{i=1}^{N}\sum_{j=1}^{N} \frac{\partial^2}{\partial x_i \partial x_j} d_{ij}(\{x\}, t) \tag{4.6}$$

in agreement with (4.1), and

$$\frac{\partial p(\{x\}, t \mid \{y\}, s)}{\partial s} = -\mathcal{L}_B(\{y\}, s)\, p(\{x\}, t \mid \{y\}, s) \quad \text{with}$$

$$\mathcal{L}_B(\{y\}, s) = \sum_{i=1}^{N} v_i(\{y\}, s) \frac{\partial}{\partial y_i} + \sum_{i=1}^{N}\sum_{j=1}^{N} d_{ij}(\{y\}, s) \frac{\partial^2}{\partial y_i \partial y_j} \tag{4.7}$$

in agreement with (4.4).

Finally we consider the mostly investigated one-dimensional time-homogeneous situation with a finite state space I. The forward Fokker–Planck equation (4.6) reduces to

$$\frac{\partial p(x, t \mid y, s)}{\partial t} = \mathcal{L}_F(x)\, p(x, t \mid y, s) \quad \text{with} \quad \mathcal{L}_F(x) = -\frac{\partial}{\partial x} v(x) + \frac{\partial^2}{\partial x^2} d(x), \tag{4.8}$$

whereas the backward equation (4.7) is

$$\frac{\partial p(x, t \mid y, s)}{\partial s} = -\mathcal{L}_B(y)\, p(x, t \mid y, s) \quad \text{with} \quad \mathcal{L}_B(y) = v(y) \frac{\partial}{\partial y} + d(y) \frac{\partial^2}{\partial y^2}. \tag{4.9}$$

Both operators $\mathcal{L}_F$, $\mathcal{L}_B$ are acting on functions in the interval $I \in [a, b]$ subject to appropriate boundary conditions.

To investigate the relationship between both operators we consider the following second-order differential operator $\mathcal{L}$ acting on function $u(x)$

$$\mathcal{L}u(x) \equiv a_0(x)\frac{d^2}{dx^2}u(x) + a_1(x)\frac{d}{dx}u(x) + a_2(x)u(x). \tag{4.10}$$

Then the adjoint operator $\mathcal{L}^\dagger$ defined by

$$\int dx\, v(x)\mathcal{L}u(x) = \int dx\, u(x)\mathcal{L}^\dagger v(x) \tag{4.11}$$

becomes

$$\mathcal{L}^\dagger u(x) \equiv \frac{d^2}{dx^2}\left(a_0(x)u(x)\right) - \frac{d}{dx}\left(a_1(x)u(x)\right) + a_2(x)u(x). \tag{4.12}$$

To find the condition that the operator will be self-adjoint, so that $\mathcal{L}^\dagger = \mathcal{L}$, we figure out from (4.12)

$$\mathcal{L}^\dagger u(x) = a_0(x)\frac{d^2}{dx^2}u(x) + \left(2\frac{da_0(x)}{dx} - a_1(x)\right)\frac{d}{dx}u(x)$$
$$+ \left(\frac{d^2 a_0(x)}{dx^2} - \frac{da_1(x)}{dx} + a_2(x)\right)u(x) \tag{4.13}$$

the required condition

$$\frac{da_0(x)}{dx} = a_1(x) \tag{4.14}$$

that gives

$$\mathcal{L}u(x) = \mathcal{L}^\dagger u(x) = a_0(x)\frac{d^2}{dx^2}u(x) + \frac{da_0(x)}{dx}\frac{d}{dx}u(x) + a_2(x)u(x)$$
$$= \frac{d^2}{dx^2}\left(a_0(x)u(x)\right) - \frac{d}{dx}\left(\frac{da_0(x)}{dx}u(x)\right) + a_2(x)u(x)$$
$$= \frac{d}{dx}\left(a_0(x)\frac{d}{dx}u(x)\right) + a_2(x)u(x). \tag{4.15}$$

This shows the equivalence between the forward and backward Fokker–Planck equation.

4.2
Bounded Drift–Diffusion in One Dimension

Now we are going to consider the one-dimensional drift–diffusion problem in a finite interval with a reflecting (left at position a) and an absorbing (right at position b) boundary.

The forward Fokker–Planck equation reads

$$\frac{\partial p(x,t \mid x_0, t_0)}{\partial t} = \mathcal{L}_F\, p(x,t \mid x_0, t_0)$$
$$\equiv \frac{\partial}{\partial x}\left[-v(x) + \frac{\partial}{\partial x}d(x)\right] p(x,t \mid x_0, t_0) \tag{4.16}$$

with the initial condition (usually the delta distribution)

$$p(x, t = t_0 \mid x_0, t_0) = p_0(x_0) = \delta(x - x_0) \tag{4.17}$$

including the boundary conditions at the left

$$j(x = a, t) \equiv \left[v(x) - \frac{\partial}{\partial x} d(x) \right] p(x, t \mid x_0, t_0) \bigg|_{x=a} = 0 \tag{4.18}$$

and right border

$$p(x, t \mid x_0, t_0)\big|_{x=b} = 0. \tag{4.19}$$

In order to discuss the same physical problem by different means we consider, in analogy to the forward dynamics, the appropriate backward Fokker–Planck equation, given by

$$-\frac{\partial p(x, t \mid x_0, t_0)}{\partial t_0} = \mathcal{L}_B \, p(x, t \mid x_0, t_0)$$

$$\equiv \left[v(x_0) + d(x_0) \frac{\partial}{\partial x_0} \right] \frac{\partial}{\partial x_0} p(x, t \mid x_0, t_0) \tag{4.20}$$

with initial condition (once again the delta distribution)

$$p(x, t \mid x_0, t_0 = t) = p_0(x) = \delta(x - x_0) \tag{4.21}$$

including the reflecting boundary at the left

$$\frac{\partial}{\partial x_0} p(x, t \mid x_0, t_0) \bigg|_{x_0=a} = 0 \tag{4.22}$$

and absorbing at the right border

$$p(x, t \mid x_0, t_0)\big|_{x_0=b} = 0. \tag{4.23}$$

Since we obtain a unique solution of the conditional probability density $p(x, t \mid x_0, t_0)$ by two different ways (forward or backward dynamics, see Figure 4.1) we have to remember the Markov property of the process based on the Chapman–Kolmogorov equation

$$p(x, t \mid x_0, t_0) = \int_a^b dy \, p(x, t \mid y, s) \, p(y, s \mid x_0, t_0). \tag{4.24}$$

After differentiation of both sides with respect to time s in the first step we get (following [84])

$$0 = \frac{\partial}{\partial s} \int_a^b dy \, p(x, t \mid y, s) p(y, s \mid x_0, t_0) \tag{4.25}$$

$$= \int_a^b dy \, p(x, t \mid y, s) \frac{\partial}{\partial s} p(y, s \mid x_0, t_0)$$

$$+ \int_a^b dy \, p(y, s \mid x_0, t_0) \frac{\partial}{\partial s} p(x, t \mid y, s). \tag{4.26}$$

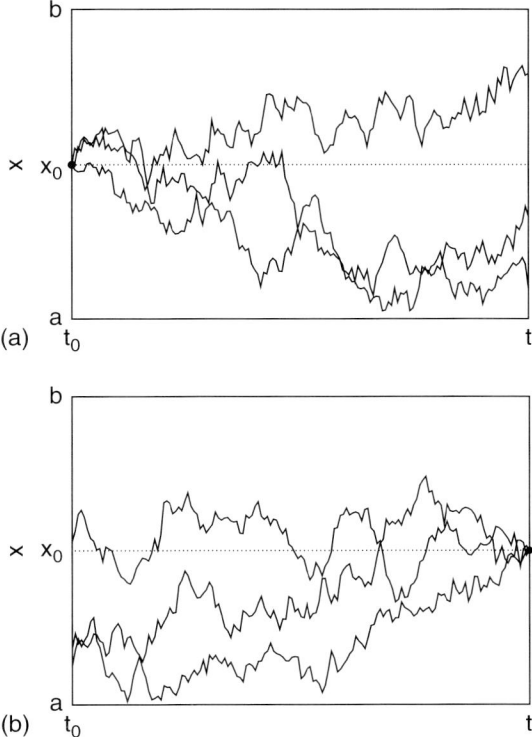

Figure 4.1 Schematic view of forward (a) and backward (b) dynamics showing three different stochastic trajectories $x(t)$ in finite state space $a \leq x \leq b$.

Inserting the corresponding Fokker–Planck operators $\mathcal{L}_\text{F}$ and $\mathcal{L}_\text{B}$

$$0 = \int_a^b dy\, p(x,t \mid y,s)\, \mathcal{L}_\text{F}(y) p(y,s \mid x_0, t_0)$$
$$- \int_a^b dy\, p(y,s \mid x_0, t_0)\, \mathcal{L}_\text{B}(y) p(x,t \mid y,s) \qquad (4.27)$$

we get

$$0 = \int_a^b dy\, p(x,t \mid y,s) \frac{\partial}{\partial y}\left[-v(y) + \frac{\partial}{\partial y} d(y)\right] p(y,s \mid x_0, t_0)$$
$$- \int_a^b dy\, p(y,s \mid x_0, t_0) \left[v(y) + d(y)\frac{\partial}{\partial y}\right]\frac{\partial}{\partial y} p(x,t \mid y,s). \qquad (4.28)$$

4 The Fokker–Planck Equation

Doing several mathematical manipulations using $uv'' - vu'' = (uv' - vu')'$

$$0 = -\int_a^b dy \, \frac{\partial}{\partial y} \left[v(y) \, p(x,t \mid y,s) \, p(y,s \mid x_0, t_0) \right]$$

$$+ \int_a^b dy \, p(x,t \mid y,s) \frac{\partial^2}{\partial y^2} \left[d(y) \, p(y,s \mid x_0, t_0) \right]$$

$$- \int_a^b dy \, p(y,s \mid x_0, t_0) \, d(y) \frac{\partial^2}{\partial y^2} p(x,t \mid y,s)$$

$$= -\int_a^b dy \, \frac{\partial}{\partial y} \left[v(y) \, p(x,t \mid y,s) \, p(y,s \mid x_0, t_0) \right]$$

$$+ \int_a^b dy \, \frac{\partial}{\partial y} \left\{ p(x,t \mid y,s) \frac{\partial}{\partial y} \left[d(y) \, p(y,s \mid x_0, t_0) \right] \right.$$

$$\left. - d(y) \, p(y,s \mid x_0, t_0) \frac{\partial}{\partial y} p(x,t \mid y,s) \right\}$$

$$= \int_a^b dy \, \frac{\partial}{\partial y} \left\{ p(x,t \mid y,s) \left(-v(y) \, p(y,s \mid x_0, t_0) + \frac{\partial}{\partial y} \left[d(y) \, p(y,s \mid x_0, t_0) \right] \right) \right\}$$

$$- \int_a^b dy \, \frac{\partial}{\partial y} \left\{ d(y) \, p(y,s \mid x_0, t_0) \frac{\partial}{\partial y} p(x,t \mid y,s) \right\} \tag{4.29}$$

after integration we arrive at

$$-p(x,t \mid y,s) \left(v(y) p(y,s \mid x_0, t_0) - \frac{\partial}{\partial y} \left[d(y) p(y,s \mid x_0, t_0) \right] \right) \bigg|_{y=a}^{y=b}$$

$$= d(y) \, p(y,s \mid x_0, t_0) \frac{\partial}{\partial y} p(x,t \mid y,s) \bigg|_{y=a}^{y=b} . \tag{4.30}$$

Taking into account forward Fokker–Planck equation with reflecting left boundary

$$j(y = a, t) = v(y = a) \, p(y = a, s \mid x_0, t_0) - \frac{\partial}{\partial y} \left[d(y) \, p(y, s \mid x_0, t_0) \right] \bigg|_{y=a} = 0 \tag{4.31}$$

and absorbing right boundary

$$p(y = b, s \mid x_0, t_0) = 0 \tag{4.32}$$

we get from (4.30)

$$p(x, t \mid y = b, s) d(y = b) \frac{\partial}{\partial y} \left[p(y, s \mid x_0, t_0) \right] \bigg|_{y=b}$$

$$= -p(y = a, s \mid x_0, t_0) d(y = a) \frac{\partial}{\partial y} p(x, t \mid y, s) \bigg|_{y=a} = 0 \tag{4.33}$$

the corresponding conditions for the backward Fokker–Planck equation, e.g. at the left reflecting wall

$$\left.\frac{\partial}{\partial y} p(x, t \mid y, s)\right|_{y=a} = 0 \qquad (4.34)$$

and at the right absorbing border

$$p(x, t \mid y = b, s) = 0. \qquad (4.35)$$

Similar calculations can be done for other boundaries such as two reflecting or two open (absorbing) ones.

To summarize to this point we state that both Fokker–Planck dynamics with the corresponding boundary conditions are equivalent and give the same results.

In the following we consider the typical one-dimensional exit problem of a Brownian particle (drift–diffusion dynamics) from a bounded domain, whose boundary is usually reflecting, except for an absorbing window. Investigating stochastic dynamics in higher dimensions, the average life-time or the mean first passage time (MFPT) increases indefinitely as the absorbing part of the, usually reflecting, surface shrinks to zero. This so-called narrow escape problem becomes important for different kinds of channels such as capillary outflow. Important investigations by A. Singer et al. [217] derive the leading order term in the expansion of the mean first-passage time $\langle t \rangle$ of a Brownian particle with diffusion coefficient D escaping from a general domain of volume $|V|$ to an elliptical hole of large semi-axis a that is much smaller than $|V|^{1/3}$

$$\langle t \rangle \sim \frac{|V|}{2\pi a D} K(e), \qquad (4.36)$$

where e is the eccentricity of the ellipse, and $K(x)$ is the complete elliptic integral of the first kind. In the special case of a small circular hole of radius a the result

$$\langle t \rangle \sim \frac{|V|}{4aD} \qquad (4.37)$$

was already known by Baron Rayleigh published in his famous book *The Theory of Sound* (1877, first edition). In [217] the Rayleigh formula has been extended by derivation of the second-order term and the error estimate for a ball of radius R with a circular hole (radius a) in the boundary

$$\langle t \rangle = \frac{|V|}{4aD} \left[1 + \frac{a}{R} \log \frac{R}{a} + \mathcal{O}\left(\frac{a}{R}\right) \right]. \qquad (4.38)$$

4.3
The Escape Problem and its Solution

Define a new quantity $G(t, x_0)$ based on the solution $p(x, t \mid x_0, t_0)$ as the probability of finding $x(t)$ starting from x_0 at initial time t_0, still in the finite interval $a \leq x < b$ by

$$G(t, x_0) = \int_a^b p(x, t \mid x_0, t_0) \, dx, \quad (4.39)$$

which is related to the probability current density $\mathcal{P}(t, x_0)$ of leaving the system for the first time at the right border $x = b$

$$\mathcal{P}(t, x_0) = -\frac{\partial}{\partial t} G(t, x_0). \quad (4.40)$$

This equation is a global conservation law stating that $x(t)$ is either in the interval $a \leq x < b$ (volume) or is passing the absorbing boundary $x = b$ (surface).

The outflow function $\mathcal{P}(t, x_0) \, dt$ is the first-passage time probability that $x(t)$ passes the value $x = b$ (right boundary) for the first time in time interval $(t, t + dt)$ after starting from $x(t = t_0) = x_0$. In a paper published in 1951 by Siegert [215] the so-called first passage time problem has been treated analytically for the case when the reflecting boundary is going to infinity to become a natural one, that is, with $a \to -\infty$.

Defining all moments by

$$\langle t_n(x_0 \to b) \rangle = \int_0^\infty t^n \mathcal{P}(t, x_0) \, dt \quad (4.41)$$

the zeroth moment is obviously normalization. The first moment is called the mean first passage time (MFPT) whereas the inverse is known as the escape or breakdown rate. The mean first-passage time tells us how long it takes on average to move from $x_0 \in [a, b)$ inside the interval to the open boundary at b taking into account the reflecting wall at a.

Using the above definition we get for the first moment

$$\langle t_1(x_0 \to b) \rangle = \int_0^\infty t \mathcal{P}(t, x_0) \, dt = -\int_0^\infty t \frac{\partial}{\partial t} G(t, x_0)$$
$$dt = \int_0^\infty G(t, x_0) \, dt \quad (4.42)$$

and also higher moments ($n \geq 1$) with similar integration by parts

$$\langle t_n(x_0 \to b) \rangle = n \int_0^\infty t^{n-1} G(t, x_0) \, dt. \quad (4.43)$$

Now we are looking for an alternative way to calculate the moments (4.41) directly to avoid the knowledge of the basic function $p(x, t \mid x_0, t_0)$ as a solution of the Fokker–Planck dynamics. Starting with the backward Fokker–Planck equation and exchanging the time derivation from t_0 to t

$$-\frac{\partial p(x, t \mid x_0, t_0)}{\partial t_0} = \frac{\partial p(x, t \mid x_0, t_0)}{\partial t} = \mathcal{L}_B \, p(x, t \mid x_0, t_0) \quad (4.44)$$

and integrating over x we get a partial differential equation for $G(t, x_0)$

$$\frac{\partial}{\partial t} G(t, x_0) = \left[v(x_0) \frac{\partial}{\partial x_0} + d(x_0) \frac{\partial^2}{\partial x_0^2} \right] G(t, x_0) \quad (4.45)$$

together with $G(t = t_0, x_0) = 1$ and

$$\frac{\partial}{\partial x_0} G(t, x_0)\bigg|_{x_0=a} = 0; \quad G(t, x_0)\bigg|_{x_0=b} = 0. \tag{4.46}$$

Either this equation can be solved to get the moments from $G(t, x_0)$ or we can derive a hierarchic set of equations for the moments. After multiplication of (4.45) by nt^{n-1} and integration over time t from $t_0 = 0$ to infinity we get

$$\left[v(x_0)\frac{\partial}{\partial x_0} + d(x_0)\frac{\partial^2}{\partial x_0^2}\right]\int_0^\infty nt^{n-1} G(t, x_0)\, dt$$

$$= \int_0^\infty nt^{n-1}\frac{\partial G(t, x_0)}{\partial t}\, dt = -n(n-1)\int_0^\infty t^{n-2} G(t, x_0)\, dt. \tag{4.47}$$

The result is for all $n \geq 1$

$$\left[v(x_0)\frac{d}{dx_0} + d(x_0)\frac{d^2}{dx_0^2}\right]\langle t_n(x_0 \to b)\rangle = -n\langle t_{n-1}(x_0 \to b)\rangle \tag{4.48}$$

with closure condition $\langle t_0(x_0 \to b)\rangle = 1$ and

$$\frac{d}{dx_0}\langle t_n(x_0 \to b)\rangle\bigg|_{x_0=a} = 0; \quad \langle t_n(x_0 \to b)\rangle\bigg|_{x_0=b} = 0. \tag{4.49}$$

To get the mean first-passage time (MFPT, $n = 1$) the following equation has to be solved

$$\left[v(x_0)\frac{d}{dx_0} + d(x_0)\frac{d^2}{dx_0^2}\right]\langle t_1(x_0 \to b)\rangle = -1 \tag{4.50}$$

which can be done directly in simple cases.

In the following we consider the so-called drift–diffusion motion with constant drift $v(x_0) = v$ as well as constant diffusion $d(x_0) = D$. The first moment $\langle t_1(x_0 \to b)\rangle$ (mean first passage time) has to be calculated from the inhomogeneous differential equation

$$\left[\frac{d^2}{dx_0^2} + \frac{v}{D}\frac{d}{dx_0}\right]\langle t_1(x_0 \to b)\rangle = -\frac{1}{D} \tag{4.51}$$

as superposition of the homogeneous and the particular solution

$$\langle t_1(x_0 \to b)\rangle = \langle t_1\rangle^{\text{hom}} + \langle t_1\rangle^{\text{par}} = C_1 + C_2 e^{-\frac{v}{D}x_0} - \frac{1}{v}x_0. \tag{4.52}$$

Taking into account the boundary conditions (4.49) the previously unknown coefficients C_1 and C_2 become

$$C_1 = \frac{b}{v} + \frac{D}{v^2}e^{-\frac{v}{D}(b-a)}; \quad C_2 = -\frac{D}{v^2}e^{-\frac{v}{D}a} \tag{4.53}$$

which finally gives the following solution

$$\langle t_1(x_0 \to b) \rangle = \frac{b - x_0}{v} + \frac{D}{v^2} \left(e^{-\frac{v}{D}(b-a)} - e^{-\frac{v}{D}(x_0 - a)} \right) \qquad (4.54)$$

including the special situation starting most far away at left wall (put x_0 at the left boundary a)

$$\langle t_1(x_0 = a \to b) \rangle = \frac{b - a}{v} + \frac{D}{v^2} \left(e^{-\frac{v}{D}(b-a)} - 1 \right). \qquad (4.55)$$

The special case $v = 0$ indicates pure diffusion which follows as the limit from above

$$\lim_{v \to 0} \langle t_1(x_0 = a \to b) \rangle = \frac{1}{2} \frac{(b-a)^2}{D}. \qquad (4.56)$$

To make expressions easier in order to draw the solution (4.54) as a function of the drift parameter we introduce the dimensionless quantities

$$y = \frac{x - a}{b - a}; \quad T = \frac{D}{(b-a)^2} t; \quad \Omega = \frac{v}{D}(b - a) \qquad (4.57)$$

in order to get, instead of $\langle t_1(x_0 \to b) \rangle$ (4.54), the dimensionless first-passage time $\langle T_1(y_0 \to 1) \rangle$ including the limiting case $\Omega = 0$ as

$$\langle T_1(y_0 \to 1) \rangle (\Omega) = \begin{cases} \frac{1}{\Omega}(1 - y_0) + \frac{1}{\Omega^2}\left(e^{-\Omega} - e^{-\Omega y_0}\right) & \Omega \neq 0 \\ \frac{1}{2}(1 - y_0^2) & \Omega = 0. \end{cases} \qquad (4.58)$$

The results are shown in Figure 4.2 for the case $y_0 = 0$. The full line gives the complete solution (4.58) including the diffusion result $1/2$ for $\Omega = 0$. If the drift

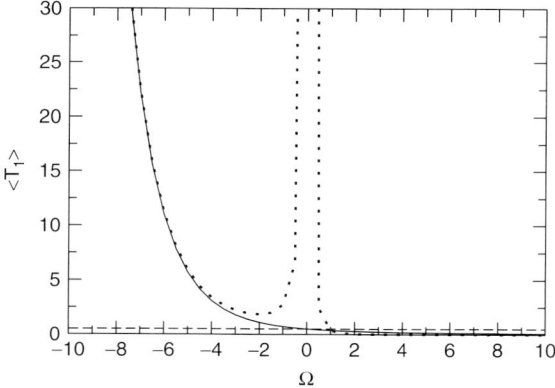

Figure 4.2 Dimensionless mean first-passage time $\langle T_1(y_0 \to 1) \rangle$ (4.58) with initial position $y_0 = 0$ (at the left reflecting wall) depending on the drift force Ω (full curve). The dotted curve shows the approximation by $\Omega^{-2} e^{-\Omega}$; the dashed line shows the correct value $1/2$ valid for $\Omega = 0$ (pure diffusion) only.

is directed to the right ($\Omega > 0$) the system reaches the absorbing border very quickly with $\langle T_1(0 \to 1) \rangle \to 0$ at $\Omega \to +\infty$. In the opposite case (negative drift) the probability of reaching the right border is shrinking and the mean first passage time increases very quickly with $\langle T_1(0 \to 1) \rangle \to \infty$ at $\Omega \to -\infty$. The second term in (4.58) $e^{-\Omega}/\Omega^2$ (dotted curve) is a good approximation in the latter extreme case. It fails around $\Omega \approx 0$, where this term diverges. The dashed straight line indicates the value $1/2$ which is correct for pure diffusion only ($\Omega = 0$).

4.4
Derivation of the Fokker–Planck Equation

The master equation and also the Fokker–Planck equation are both useful for describing the time development of the probability density function $p(x, t)$ for a continuous variable x.

In the following we want to discuss the one-dimensional case in detail. The Fokker–Planck equation follows from the master equation (3.20)

$$\frac{\partial p(x,t)}{\partial t} = \int_{-\infty}^{+\infty} \left\{ w(x, x', t) p(x', t) - w(x', x, t) p(x, t) \right\} dx' \tag{4.59}$$

due to the Kramers–Moyal expansion where only the first two leading terms are retained. As distinct from (3.20), here we allow a more general case, that is where the transition frequencies depend on time t. The derivation can be found in many textbooks, see e.g. [55]. By introducing the quantity $f(y, x, t) = w(x + y, x, t)$, the master equation (4.59) can be written as

$$\frac{\partial p(x,t)}{\partial t} = \int_{-\infty}^{+\infty} \left\{ f(y, x - y, t) p(x - y, t) - f(y, x, t) p(x, t) \right\} dy. \tag{4.60}$$

It is assumed that $f(y, x - y, t)$ is a smooth function with respect to y. The basic idea is to expand the quantity $f(y, x - y, t) p(x - y, t)$ in a Taylor series around $y = 0$, which yields the Kramers–Moyal expansion

$$\frac{\partial p(x,t)}{\partial t} = \sum_{n=1}^{\infty} \frac{(-1)^n}{n!} \frac{\partial^n}{\partial x^n} \left[\alpha_n(x, t) p(x, t) \right], \tag{4.61}$$

where

$$\alpha_n(x, t) = \int_{-\infty}^{+\infty} y^n f(y, x, t) \, dy = \int_{-\infty}^{+\infty} (x' - x)^n w(x', x, t) \, dx' \tag{4.62}$$

are the nth-order moments of the transition frequencies $w(x', x, t)$. Retaining only the first two expansion terms in (4.61) one obtains the well known Fokker–Planck equation in forward notation

$$\frac{\partial p(x,t)}{\partial t} = -\frac{\partial}{\partial x} \left[\alpha_1(x, t) p(x, t) \right] + \frac{1}{2} \frac{\partial^2}{\partial x^2} \left[\alpha_2(x, t) p(x, t) \right]. \tag{4.63}$$

The first term in (4.63) is called the drift term and the second one is called the diffusion or fluctuation term. This is due to the analogy with a drift–diffusion equation where the first derivative describes the drift of the probability profile without changing its form, whereas the second one describes the pure diffusion effect. In fact, (4.63) is a drift–diffusion equation for the probability $p(x, t)$. The diffusion or effluence of the probability distribution profile occurs due to the stochastic fluctuations, therefore the second term in (4.63) is also called the fluctuation term. More explicitly, (4.63) is called the *forward Fokker–Planck equation* to distinguish it from the *backward Fokker–Planck equation* which describes the evolution of the conditional probability $p(x, t \mid x', t')$ with respect to the initial time t'.

Like the backward master equation considered in Section 3.4, the backward Fokker–Planck equation is useful for studying the first-passage problem, which has been already presented in the previous section. The derivation in both cases is similar. Namely, we can write (3.54) for the continuous variable x

$$p(x, t \mid x', t' + \Delta t) - p(x, t \mid x', t') \qquad (4.64)$$
$$= \int p(x'', t' + \Delta t \mid x', t') \left[p(x, t \mid x', t' + \Delta t) - p(x, t \mid x'', t' + \Delta t) \right] dx''$$

and take the limit $\Delta t \to 0$. In the same approximation which has been used to obtain (4.63), the probability $p(x, t \mid x'', t')$ can be expanded in a Taylor series around $x'' = x'$ retaining the first two terms only. It yields the backward Fokker–Planck equation

$$\frac{\partial p(x, t \mid x', t')}{\partial t'} = -\alpha_1(x', t') \frac{\partial p(x, t \mid x', t')}{\partial x'} - \frac{1}{2} \alpha_2(x', t') \frac{\partial^2 p(x, t \mid x', t')}{\partial x'^2}, \qquad (4.65)$$

where the drift coefficient $\alpha_1(x', t')$ and the diffusion coefficient $\alpha_2(x', t')$ are the moments of the transition frequency (4.62) also entering the forward Fokker–Planck equation.

The connection to the stochastic differential equation calculus is given by the following *Langevin equation*

$$dx(t) = \alpha_1(x, t) \, dt + \sqrt{\alpha_2(x, t)} \, dW(t) \qquad (4.66)$$

The term $dW(t) = W(t + dt) - W(t)$ is called the Wiener increment noise term with the properties $\langle W(t) \rangle = 0$ and $\langle dW(t)^2 \rangle = dt$.

4.5
Fokker–Planck Dynamics in Finite State Space

We consider the general one-dimensional forward Fokker–Planck dynamics (4.63) with time-homogeneous coefficients $D_1(x) (= \alpha_1;$ drift$)$ and $D_2(x) (= \alpha_2/2;$ diffusion$)$

$$\frac{\partial p(x, t)}{\partial t} = -\frac{\partial}{\partial x} \left[D_1(x) \, p(x, t) \right] + \frac{\partial^2}{\partial x^2} \left[D_2(x) \, p(x, t) \right] \qquad (4.67)$$

on a finite state space interval $a < x < b$. The properties of the boundaries (closed at $x = a$ and open at $x = b$) are fixed. In the following comparison, presented in Table 4.1, we show the similarities between forward and backward dynamics in detail and indicate that both methods of solving the problem are equivalent and therefore the solutions are identical.

Coming back to the basic equation (4.67) in the first step we change the general diffusion term given by function $D_2(x)$ to an expression with a constant value D using the transformation

$$p(x,t)\,dx = w(\bar{x},t)\,d\bar{x} \quad \text{with} \quad \bar{x} = \bar{x}(x). \tag{4.68}$$

Since we have $p(x,t) = w(\bar{x},t)\,d\bar{x}/dx \equiv \bar{x}'w(\bar{x},t)$ the time derivation becomes

$$\frac{\partial p(x,t)}{\partial t} = \frac{\partial}{\partial t}\left[\bar{x}'w\right] = \bar{x}'\frac{\partial w}{\partial t}, \tag{4.69}$$

whereas the coordinate derivative changes to

$$\frac{\partial p(x,t)}{\partial x} = \frac{\partial}{\partial x}\left[\bar{x}'w\right] = \bar{x}''w + \bar{x}'\frac{\partial w}{\partial x} = \bar{x}''w + (\bar{x}')^2\frac{\partial w}{\partial \bar{x}}. \tag{4.70}$$

According to (4.67) we rewrite the Fokker–Planck equation as follows

$$\frac{\partial p(x,t)}{\partial t} = -\frac{\partial}{\partial x}\left[D_1(x)p(x,t) - \frac{\partial}{\partial x}\left(D_2(x)\,p(x,t)\right)\right] \tag{4.71}$$

$$= -\frac{\partial}{\partial x}\left[\left(D_1(x) - D_2'(x)\right)p(x,t) - D_2(x)\frac{\partial p(x,t)}{\partial x}\right] \tag{4.72}$$

and exchange the derivatives by (4.69) and (4.70) to get the transformed Fokker–Planck equation

$$\frac{\partial w}{\partial t} = \frac{\partial}{\partial \bar{x}}\left[\left(-D_1\bar{x}' + D_2'\bar{x}' + D_2\bar{x}''\right)w + D_2\,(\bar{x}')^2\frac{\partial w}{\partial \bar{x}}\right]. \tag{4.73}$$

Here we set

$$\widetilde{D_2}(\bar{x}) = D_2(x)\,(\bar{x}')^2 = D = \text{const.} \tag{4.74}$$

and obtain the previously unknown coordinate transformation $\bar{x} = \bar{x}(x)$ as

$$\bar{x}(x) = \int^x \sqrt{\frac{D}{D_2(x')}}\,dx' \tag{4.75}$$

including a constant which has to be determined according to the scaled boundaries $\bar{x}(x = a) = \bar{a}$ and $\bar{x}(x = b) = \bar{b}$.

Defining the new drift coefficient by

$$-\widetilde{D_1}(\bar{x}) = -D_1(x(\bar{x}))\,\bar{x}' + D_2'(x)\,\bar{x}' + D_2\,\bar{x}'' \tag{4.76}$$

and taking into account $\widetilde{D_2}' = D_2'\,(\bar{x}')^2 + 2D_2\,\bar{x}'\bar{x}'' = 0$, finally the transformed Fokker–Planck equation reads

Table 4.1 Comparison between forward and backward Fokker–Planck dynamics.

Forward dynamics $p(x, t \mid x_0, t_0)$	Backward dynamics $p(x, t \mid x_0, t_0)$
Forward Fokker–Planck equation: $\frac{\partial}{\partial t} p = -\frac{\partial}{\partial x}[D_1 p] + \frac{\partial^2}{\partial x^2}[D_2 p]$	Backward Fokker–Planck equation: $-\frac{\partial}{\partial t_0} p = \frac{\partial}{\partial t} p = D_1 \frac{\partial}{\partial x_0} p + D_2 \frac{\partial^2}{\partial x_0^2} p$
Initial condition: $p(x, t = t_0 \mid x_0, t_0) = \delta(x - x_0)$	Initial condition: $p(x, t \mid x_0, t_0 = t) = \delta(x - x_0)$
Closed boundary (reflecting at $x = a$): $j(x = a, t) = \left(D_1 p - D_2 \frac{\partial}{\partial x} p\right)\bigg|_{x=a} = 0$	Closed boundary (reflecting at $x_0 = a$): $\frac{\partial}{\partial x_0} p \bigg|_{x_0 = a} = 0$
Open boundary (absorbing at $x = b$): $p(x = b, t \mid x_0, t_0) = 0$ Solution $p = p_F(x, t \mid x_0, t_0)$ identical $p = p_F = p_B$	Open boundary (absorbing at $x_0 = b$): $p(x, t \mid x_0 = b, t_0) = 0$ Solution $p = p_B(x, t \mid x_0, t_0)$ identical $p = p_F = p_B$
Survival probability: $G(t, x_0) = \int_a^b p(x, t \mid x_0, t_0)\, dx$	Survival probability: $G(t, x_0) = \int_a^b p(x, t \mid x_0, t_0)\, dx$ or directly from $\frac{\partial}{\partial t} G = D_1 \frac{\partial}{\partial x_0} G + D_2 \frac{\partial^2}{\partial x_0^2} G$ with $G(t = t_0, x_0) = 1$ and $\frac{\partial}{\partial x_0} G \bigg|_{x=a} = 0$; $G(t, x_0 = b) = 0$.
Outflow probability density: $\mathcal{P}(t, x_0 \to b) = j(x = b, t)$	Outflow probability density: $\mathcal{P}(t, x_0 \to b) = -\frac{d}{dt} G(t, x_0)$
Cumulative outflow from $t_0 = 0$ to t_{obs}: $W(t \leq t_{\text{obs}}, x_0 \to b) = \int_0^{t_{\text{obs}}} \mathcal{P}(t, x_0 \to b)\, dt$	Cumulative outflow from $t_0 = 0$ to t_{obs}: $W(t \leq t_{\text{obs}}, x_0 \to b) = 1 - G(t_{\text{obs}}, x_0)$
Moments: $\langle t_n(x_0 \to b)\rangle = \int_0^\infty t^n \mathcal{P}(t, x_0 \to b)\, dt$	Moments: $\langle t_n(x_0 \to b)\rangle = n \int_0^\infty t^{n-1} G(t, x_0)\, dt$
First moment (MFPT with $x_0 = a$): $\langle t_1(a \to b)\rangle = \int_0^\infty t\, \mathcal{P}(t, a \to b)\, dt$	First moment (MFPT with $x_0 = a$): $\langle t_1(a \to b)\rangle = \int_0^\infty G(t, x_0 = a)\, dt$ or directly from $D_2 \frac{d^2 \langle t_1 \rangle}{dx_0^2} + D_1 \frac{d\langle t_1 \rangle}{dx_0} = -1$ with $\frac{d\langle t_1 \rangle}{dx_0}\bigg|_{x_0=a} = 0$ and $\langle t_1 \rangle\big|_{x_0=b} = 0$.

$$\frac{\partial w(\bar{x},t)}{\partial t} = -\frac{\partial}{\partial \bar{x}}\left(\widetilde{D}_1(\bar{x})\,w(\bar{x},t)\right) + D\,\frac{\partial^2 w(\bar{x},t)}{\partial \bar{x}^2}. \tag{4.77}$$

with

$$\widetilde{D}_1(\bar{x}) = D_1(x(\bar{x}))\,\frac{d\bar{x}}{dx} + D_2(x(\bar{x}))\,\frac{d^2\bar{x}}{dx^2}. \tag{4.78}$$

In the next step we introduce dimensionless variables, coordinate $y \in [0,1]$ and time T, via

$$y = y(\bar{x}) = \frac{\bar{x} - \bar{a}}{\bar{b} - \bar{a}} \quad \text{and} \quad T = T(t) = \frac{D}{(\bar{b} - \bar{a})^2}\,t \tag{4.79}$$

together with a new probability density $P(y,T)$ by

$$w(\bar{x},t)\,d\bar{x} = P(y,T)\,dy. \tag{4.80}$$

The resulting Fokker–Planck equation is

$$\frac{\partial P(y,T)}{\partial T} = \left(-\frac{\partial}{\partial y}A(y) + \frac{\partial^2}{\partial y^2}\right)P(y,T) \tag{4.81}$$

with drift function $A(y) = \dfrac{\bar{b} - \bar{a}}{D}\,\widetilde{D}_1(y)$.

In the next step we transform this equation to a differential equation of the Sturm–Liouville type (without first derivative) by using the substitution

$$P(y,T) = e^{-\Phi(y)/2}\,Q(y,T). \tag{4.82}$$

According to (4.82), the terms in the Fokker–Planck equation (4.81) are transformed as follows:

$$\frac{\partial P}{\partial y} = e^{-\Phi(y)/2}\left[-\frac{1}{2}\Phi' Q + \frac{\partial Q}{\partial y}\right], \tag{4.83}$$

$$\frac{\partial^2 P}{\partial y^2} = e^{-\Phi(y)/2}\left[\frac{\partial^2 Q}{\partial y^2} - \Phi'\frac{\partial Q}{\partial y} + \frac{1}{4}\Phi'^2 Q - \frac{1}{2}\Phi'' Q\right], \tag{4.84}$$

$$\frac{\partial (AP)}{\partial y} = e^{-\Phi(y)/2}\left[-\frac{1}{2}\Phi' A Q + A' Q + A\frac{\partial Q}{\partial y}\right]. \tag{4.85}$$

Inserting these relations into (4.81) yields

$$\frac{\partial Q(y,T)}{\partial T} = \left(\frac{1}{2}\Phi' A - A' + \frac{1}{4}\Phi'^2 - \frac{1}{2}\Phi''\right)Q - (A + \Phi')\frac{\partial Q}{\partial y} + \frac{\partial^2 Q}{\partial y^2}. \tag{4.86}$$

In order to cancel the term with the first derivative, one has to choose $\Phi' = -A$ or $\Phi = -\int^y A(y')\,dy'$, which finally leads to the equation of the desired form

$$\frac{\partial Q}{\partial T} = -A_1(y)Q + \frac{\partial^2 Q}{\partial y^2} \tag{4.87}$$

with $A_1(y) = \tfrac{1}{4}A^2 + \tfrac{1}{2}A'$.

Now we separate the variables by setting

$$Q(y, T) = e^{-\lambda T} \psi(y) \tag{4.88}$$

which leads to the equation

$$\frac{d^2 \psi}{dy^2} + (\lambda - A_1(y)) \psi = 0 \tag{4.89}$$

for the y-dependent function $\psi(y)$, which can be written also as

$$\mathcal{L}\psi + \lambda \psi = 0, \tag{4.90}$$

where

$$\mathcal{L} = \frac{d^2}{dy^2} - A_1(y) \tag{4.91}$$

is the Sturm–Liouville operator. Equation (4.90) has to be completed with appropriate boundary conditions

$$\alpha_1 \psi + \beta_1 \frac{d\psi}{dy} = 0: \quad y = a, \tag{4.92}$$

$$\alpha_2 \psi + \beta_2 \frac{d\psi}{dy} = 0: \quad y = b, \tag{4.93}$$

where α_1, β_1, α_2, and β_2 are constants. Equation (4.89) together with the boundary conditions (4.92) and (4.93) represent the Sturm–Liouville problem. Nontrivial particular solutions, which are called eigenfunctions, are possible only at certain values of λ which are called eigenvalues of the problem. The eigenvalues and eigenfunctions of the Sturm–Liouville problem depend on the specific form of $A_1(y)$ and boundary conditions. However, the eigenvalues are always real if $A(y)$ is a real function, since the Sturm–Liouville operator $\mathcal{L}$ is self-adjoint. This means that for any two functions $g(y)$ and $u(y)$, which fulfill the boundary conditions (4.92) and (4.93), we have

$$(g, \mathcal{L}u) = (\mathcal{L}g, u), \tag{4.94}$$

where $(\cdot, \cdot)$ denotes the scalar product. For scalar functions this is defined by an integral

$$(g, u) = \int_a^b g(y') u(y') \, dy'. \tag{4.95}$$

Hence, the condition (4.94) for the operator (4.91) becomes

$$(g, \mathcal{L}u) - (\mathcal{L}g, u) = \int_a^b \left[g \frac{d^2 u}{dy^2} - u \frac{d^2 g}{dy^2} \right] dy$$

$$= \int_a^b \frac{d}{dy} \left[g \frac{du}{dy} - u \frac{dg}{dy} \right] dy = \left(g \frac{du}{dy} - u \frac{dg}{dy} \right) \bigg|_a^b = 0. \tag{4.96}$$

It obviously is fulfilled according to (4.92) and (4.93), which proves that the Sturm–Liouville operator is self-adjoint.

If $A(y)$ is real, then $\mathcal{L} = \mathcal{L}^*$ holds, and therefore the equation which is complex conjugated to (4.90) reads

$$\mathcal{L}\psi^* + \lambda^*\psi^* = 0. \tag{4.97}$$

By setting $g = \psi^*$ and $u = \psi$ in (4.94) we obtain

$$(\psi^*, \mathcal{L}\psi) = (\mathcal{L}\psi^*, \psi). \tag{4.98}$$

According to (4.90) and (4.97) it reduces to

$$\lambda^*(\psi^*, \psi) = \lambda(\psi^*, \psi) \tag{4.99}$$

or $\lambda^* = \lambda$, which means that the eigenvalue λ is real, as we have mentioned already.

4.6
Fokker–Planck Dynamics with Coordinate-Dependent Diffusion Coefficient

Now we are going to investigate a simple example: the evolution of a particle concentration profile $c(x, t)$ with a diffusion coefficient D which linearly depends on coordinate x as $D(x) = D_0(1 + gx)$. This task was first solved by Martin [159], but we are following our pathway outlined in Section 4.5 to get the solution for the probability density $p(x, t)$. Our result received in a different way from that obtained by Martin for zero initial value, is also a generalization for arbitrary initial conditions.

The corresponding diffusion equation reads

$$\begin{aligned}\frac{\partial p(x,t)}{\partial t} &= \frac{\partial}{\partial x}\left(D(x)\frac{\partial p(x,t)}{\partial x}\right) \\ &= D_0 g \frac{\partial p(x,t)}{\partial x} + D_0(1 + gx)\frac{\partial^2 p(x,t)}{\partial x^2}.\end{aligned} \tag{4.100}$$

This equation is completed by the reflecting boundary condition at $x = -1/g$ (vanishing flux) and natural boundary condition at $x = \infty$ (vanishing flux and concentration) as well as the initial condition $p(x, t = 0) = \delta(x - x_0)$.

In the following we will show how this equation is treated according to the scheme described in the previous section. First we rewrite it in the form (4.67), that is,

$$\frac{\partial p(x,t)}{\partial t} = -\frac{\partial}{\partial x}\left[D_1(x)\, p(x,t)\right] + \frac{\partial^2}{\partial x^2}\left[D_2(x)\, p(x,t)\right], \tag{4.101}$$

where $D_1(x) = D_0 g$ and $D_2(x) = D_0(1 + gx)$. Further on, we make the coordinate transformation (4.75)

$$\overline{x}(x) = \int^x \sqrt{\frac{D_0}{D_2(x')}}\, dx' = \int^x \frac{dx'}{\sqrt{1 + gx'}} = \frac{2}{g}\sqrt{1 + gx} \tag{4.102}$$

and the transformation $p(x, t) = w(\overline{x}, t) \cdot d\overline{x}/dx = w(\overline{x}, t) \cdot 2/(g\overline{x})$ to the density function $w(\overline{x}, t)$ in the new coordinate space in order to obtain the equation with constant diffusion coefficient

$$\frac{\partial w(\bar{x}, t)}{\partial t} = -\frac{\partial}{\partial \bar{x}} \left[\frac{D_0}{\bar{x}} w(\bar{x}, t) \right] + D_0 \frac{\partial^2 w(\bar{x}, t)}{\partial \bar{x}^2} \qquad (4.103)$$

in accordance with (4.77) and (4.78). The boundary values of the new coordinate $\bar{x}$ are $\bar{a} = \bar{x}(x = -1/g) = 0$ and $\bar{b} = \infty$. The constant D_0 can be removed by changing the variables $T = tD_0$ and $P(\bar{x}, T) = w(\bar{x}, t)$. This yields an equation of the form (4.81) with $A(\bar{x}) = 1/\bar{x}$:

$$\frac{\partial P(\bar{x}, T)}{\partial T} = -\frac{\partial}{\partial \bar{x}} \left(\frac{1}{\bar{x}} P(\bar{x}, T) \right) + \frac{\partial^2 P(\bar{x}, T)}{\partial \bar{x}^2}. \qquad (4.104)$$

In the next step we remove the first derivative by changing the variable as

$$P(\bar{x}, T) = e^{-\Phi(\bar{x})/2} Q(\bar{x}, T), \qquad (4.105)$$

where

$$\Phi(\bar{x}) = -\int^{\bar{x}} \frac{dy}{y} = -\ln \bar{x} \qquad (4.106)$$

holds for $\bar{x} > 0$. This leads to an equation of the form (4.87)

$$\frac{\partial Q}{\partial T} = \frac{1}{(2\bar{x})^2} Q + \frac{\partial^2 Q}{\partial \bar{x}^2}, \qquad (4.107)$$

where the coefficient at Q is set to $-A_1(\bar{x}) = -\frac{1}{4}A^2(\bar{x}) - \frac{1}{2} dA(\bar{x})/d\bar{x} = 1/(2\bar{x})^2$ according to $A(\bar{x}) = 1/\bar{x}$.

The boundary conditions are obtained by transforming into the new variables the condition of zero flux,

$$j = D_0 g\, p(x, t) - \frac{\partial}{\partial x} \left[D_0(1 + gx)\, p(x, t) \right] = 0, \qquad (4.108)$$

at the boundary $x = -1/g$. The resulting condition is that

$$\frac{1}{\sqrt{\bar{x}}} \left[\frac{1}{2} Q(\bar{x}, T) - \bar{x} \frac{\partial Q(\bar{x}, T)}{\partial \bar{x}} \right] \qquad (4.109)$$

must vanish at $\bar{x} \to 0$ as well as at $\bar{x} \to \infty$.

The separation of variables (cf. (4.88))

$$Q(\bar{x}, T) = e^{-\lambda^2 T} \Psi(\bar{x}) \qquad (4.110)$$

leads finally to the Sturm–Liouville problem

$$\frac{d^2 \Psi}{d\bar{x}^2} + \left(\lambda^2 + \frac{1}{(2\bar{x})^2} \right) \Psi = 0 \qquad (4.111)$$

with the boundary condition at $\bar{x} = 0$

$$\frac{1}{2} \Psi(\bar{x}) = \bar{x} \left. \frac{\partial \Psi(\bar{x})}{\partial \bar{x}} \right|_{\bar{x}=0}, \qquad (4.112)$$

4.6 Fokker–Planck Dynamics with Coordinate-Dependent Diffusion Coefficient

following from the vanishing of (4.109), and $\lim_{\bar{x}\to\infty} \Psi(\bar{x}) = 0$ according to the physical meaning of the natural boundary condition. The condition (4.112) corresponds to $\Psi(0) = 0$, as will be seen from the following solution. In this case the boundary conditions take the standard form (4.92)–(4.93).

To solve our diffusion problem, it is more suitable to return to the form (4.104) containing the first derivative with respect to coordinate $\bar{x}$, since we then obtain the known Bessel equation instead of (4.111). The boundary (vanishing flux) condition at $\bar{x} = 0$ then reads

$$P(\bar{x}, T) = \bar{x} \left.\frac{\partial P(\bar{x}, T)}{\partial \bar{x}}\right|_{\bar{x}=0}, \tag{4.113}$$

and the initial condition $c(x, t = 0) = \delta(x - x_0)$ is

$$\frac{2P(\bar{x}, T = 0)}{g\bar{x}} = \delta\left(\frac{g}{4}\left(\bar{x}^2 - \bar{x}_0^2\right)\right) \tag{4.114}$$

or

$$P(\bar{x}, T = 0) = \frac{\bar{x}}{\bar{x}_0}\delta(\bar{x} - \bar{x}_0). \tag{4.115}$$

Here we have used the relation $\delta(f(x)) = \frac{1}{|f'(x_0)|}\delta(x - x_0)$ with x_0 as a root of $f(x)$.

To solve this new problem, we now try the separation ansatz $P(\bar{x}, T) = \chi(T)\psi(\bar{x})$. Inserting it into (4.104) and dividing by $\chi(T)\psi(\bar{x})$ we find

$$\frac{1}{\chi(T)}\frac{\partial \chi(T)}{\partial T} = \frac{1}{\bar{x}^2} - \frac{1}{\bar{x}\psi(\bar{x})}\frac{\partial \psi(\bar{x})}{\partial \bar{x}} + \frac{1}{\psi(\bar{x})}\frac{\partial^2 \psi(\bar{x})}{\partial \bar{x}^2}. \tag{4.116}$$

Since both sides depend only on different variables, they have to be constant. Choosing $-\lambda^2$ as the separation constant we get a simple relation for the temporal evolution

$$\frac{\partial \chi(T)}{\partial T} = -\lambda^2 \chi(T) \tag{4.117}$$

having a solution

$$\chi(T) = e^{-\lambda^2 T}. \tag{4.118}$$

The general solution contains an arbitrary normalization constant, which is not important here and has been set to one.

The spatial solution has to satisfy a more complicated differential equation

$$\bar{x}^2 \frac{\partial^2 \psi(\bar{x})}{\partial \bar{x}^2} - \bar{x}\frac{\partial \psi(\bar{x})}{\partial \bar{x}} + \left(1 + \lambda^2 \bar{x}^2\right)\psi(\bar{x}) = 0. \tag{4.119}$$

This equation is quite similar to the Bessel equation

$$x^2 y''(x) + x y'(x) + (n^2 + x^2) y(x) = 0. \tag{4.120}$$

We can transform our equation into this form by a change of variables:

$$y = \lambda \bar{x}, \quad \psi(\bar{x}) = y\phi(y). \tag{4.121}$$

Now we compute

$$\left(1 + \lambda^2 \bar{x}^2\right) \psi(\bar{x}) = y\phi(y) + y^3 \phi(y),$$

$$\bar{x}\frac{\partial \psi(\bar{x})}{\partial \bar{x}} = y\frac{\partial}{\partial y}\left(y\phi(y)\right) = y\phi(y) + y^2 \frac{\partial \phi(y)}{\partial y},$$

$$\bar{x}^2 \frac{\partial^2 \psi(\bar{x})}{\partial \bar{x}^2} = y^2 \frac{\partial^2}{\partial y^2}\left(y\phi(y)\right) = 2y^2 \frac{\partial \phi(y)}{\partial y} + y^3 \frac{\partial^2 \phi(y)}{\partial y^2}.$$

By inserting this into (4.119) we obtain the Bessel equation

$$y^2 \frac{\partial^2 \phi(y)}{\partial y^2} + y\frac{\partial \phi(y)}{\partial y} + y^2 \phi(y) = 0. \tag{4.122}$$

The solution is given by a linear combination of Bessel functions $J_n(y)$ and $Y_n(y)$ of zeroth order $n = 0$:

$$\phi(y) = \frac{1}{\lambda}\left(c_1 J_0(y) + c_2 Y_0(y)\right). \tag{4.123}$$

Written in terms of $\bar{x}$ and $\psi(\bar{x})$, this reads as

$$\psi(\bar{x}) = c_1 \bar{x} J_0(\lambda \bar{x}) + c_2 \bar{x} Y_0(\lambda \bar{x}). \tag{4.124}$$

This solution has to fit to our boundary condition

$$P(\bar{x}, T) = \bar{x}\frac{\partial P(\bar{x}, T)}{\partial \bar{x}}\bigg|_{\bar{x}=0} \Rightarrow \psi(\bar{x}) = \bar{x}\frac{\partial \psi(\bar{x})}{\partial \bar{x}}\bigg|_{\bar{x}=0}. \tag{4.125}$$

Inserting our solution into this equation we obtain

$$0 = \bar{x}\left(c_1 J_0'(\lambda \bar{x}) + c_2 Y_0'(\lambda \bar{x})\right)_{\bar{x}\to 0}. \tag{4.126}$$

It can be shown that the Bessel functions can be approximated in the limit $x \to 0$ by

$$\lim_{\lambda x \to 0} J_0'(\lambda x) = 0, \quad \lim_{\lambda x \to 0} Y_0'(\lambda x) = \frac{2}{\pi x}, \tag{4.127}$$

therefore we find that the constant c_2 is zero. In conclusion, the particular solution for each λ reads

$$P_\lambda(\bar{x}, T) = \chi_\lambda(T)\psi_\lambda(\bar{x}) = c_\lambda e^{-\lambda^2 T}\bar{x} J_0(\lambda \bar{x}). \tag{4.128}$$

The general solution can be written as a linear combination of the solutions for different λ. Since we have a continuous set of eigenvalues λ, we have to formulate the linear combinations with an integral

$$P(\bar{x}, T) = \int_0^\infty d\lambda \, C(\lambda) e^{-\lambda^2 T} \bar{x} J_0(\lambda \bar{x}), \tag{4.129}$$

where the weight function $C(\lambda)$ is determined by the initial condition (4.115). Consequently, in our example we have to solve the equation

4.6 Fokker–Planck Dynamics with Coordinate-Dependent Diffusion Coefficient

$$\frac{\bar{x}}{\bar{x}_0}\delta(\bar{x}-\bar{x}_0) = \int_0^\infty d\lambda\, C(\lambda)\bar{x}J_0(\lambda\bar{x}). \tag{4.130}$$

For this purpose we intend to use the orthogonality of the Bessel functions. It can be shown that the Bessel functions are orthogonal when we consider the following scalar product

$$(f,g) = \int_0^\infty f(x)g(x)x\, dx. \tag{4.131}$$

Thus, the corresponding relation reads

$$\left(J_0(\lambda\bar{x}), J_0(\kappa\bar{x})\right) = \lambda \int_0^\infty J_0(\lambda\bar{x})J_0(\kappa\bar{x})\,\bar{x}\,d\bar{x} = \delta(\lambda-\kappa). \tag{4.132}$$

Therefore, we multiply both sides of (4.130) by $\kappa J_0(\kappa\bar{x})$ and integrate over $\bar{x}$, which yields

$$\kappa\int_0^\infty J_0(\kappa\bar{x})\frac{\bar{x}}{\bar{x}_0}\delta(\bar{x}-\bar{x}_0)\,d\bar{x} = \kappa\int_0^\infty d\bar{x}\, J_0(\kappa\bar{x})\int_0^\infty d\lambda\, C(\lambda)\bar{x}J_0(\lambda\bar{x})$$

$$= \int_0^\infty d\lambda\, C(\lambda)\int_0^\infty \kappa J_0(\kappa\bar{x})J_0(\lambda\bar{x})\,\bar{x}\,d\bar{x}$$

and we get

$$\kappa J_0(\kappa\bar{x}_0)\frac{\bar{x}_0}{\bar{x}_0} = \int_0^\infty d\lambda\, C(\lambda)\delta(\lambda-\kappa),$$
$$C(\kappa) = \kappa J_0(\kappa\bar{x}_0). \tag{4.133}$$

According to (4.133) and (4.129) and using an integral taken from [60], the final solution reads

$$P(\bar{x}, T) = \int_0^\infty d\lambda\, \left(\lambda J_0(\lambda\bar{x}_0)\right)\left(\bar{x}J_0(\lambda\bar{x})\,e^{-\lambda^2 T}\right)$$

$$= \bar{x}\int_0^\infty d\lambda\, \lambda J_0(\lambda\bar{x}_0)J_0(\lambda\bar{x})e^{-\lambda^2 T}$$

$$= \bar{x}\left(\frac{1}{2T}\exp\left\{-\frac{\bar{x}^2+\bar{x}_0^2}{4T}\right\}I_0\left(\frac{\bar{x}_0\bar{x}}{2T}\right)\right), \tag{4.134}$$

where I_0 is the zeroth-order Bessel function of complex argument given by the general relation $I_n(x) = i^{-n}J_n(ix)$, or for $n=0$

$$I_0(x) = J_0(ix) = \sum_{k=0}^\infty \frac{(x/2)^{2k}}{(k!)^2} = 1 + \left(\frac{x}{2}\right) + \cdots. \tag{4.135}$$

Replacing $\bar{x} = \frac{2}{g}\sqrt{1+gx}$, $T = D_0 t$ and $p(x,t) = P(\bar{x},T)\cdot 2/(g\bar{x})$, we find the concentration c in variables of x, t as final result

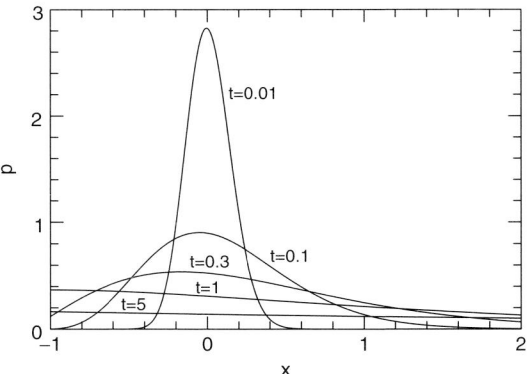

Figure 4.3 Plot of the probability density profile $p(x,t)$ for different times moments starting from an initial delta-like distribution showing the evolution of the distribution (4.136). Parameters: $x_0 = 0$, $D = 1$ m² s⁻¹, $g = 1$ m⁻¹.

$$p(x,t) = \frac{1}{gD_0 t} \exp\left\{-\frac{2 + g(x + x_0)}{g^2 D_0 t}\right\} I_0\left(\frac{2\sqrt{1+gx}\sqrt{1+gx_0}}{g^2 D_0 t}\right). \quad (4.136)$$

This resulting distribution is in agreement with that one found in [159] for the special case of zero initial value x_0. Its temporal development is shown in Figure 4.3.

For a well-founded discussion we want to know the properties of the moments and therefore we have to compute the first moments of the concentration

$$\langle x^n \rangle = \int_{-1/g}^{\infty} x^n p(x,t)\, dx. \quad (4.137)$$

For reasons of simplicity we perform a linear transformation $y = 1 + gx$ such that the lower boundary equals to zero. Inserting our solution for the probability density, we find the following integral

$$\langle x^n \rangle = \frac{1}{gD_0 t} \exp\left\{-\frac{1+gx_0}{g^2 D_0 t}\right\} \quad (4.138)$$

$$\times \int_0^{\infty} \frac{(y-1)^n}{g^n} \exp\left\{-\frac{y}{g^2 D_0 t}\right\} I_0\left(2\frac{\sqrt{1+gx_0}}{g^2 D_0 t}\sqrt{y}\right) \frac{dy}{g}.$$

These integrals can be solved analytically with the help of appropriate integral tables. In [60] we can find the following formula

$$\int_0^{\infty} x^{\mu - \frac{1}{2}} e^{-\alpha x} I_{2\nu}(2\beta\sqrt{x})\, dx = \frac{\Gamma\left(\mu + \nu + \frac{1}{2}\right)}{\beta \alpha^\mu \Gamma(2\nu + 1)} e^{\frac{\beta^2}{2\alpha}} M_{-\mu,\nu}\left(\frac{\beta^2}{\alpha}\right),$$

$$\text{for } \Re\left\{\mu + \nu + \frac{1}{2}\right\} > 0. \quad (4.139)$$

Since we want to compute the zeroth, first, and second moment, we are interested in the parameters $\mu = \frac{1}{2}, \frac{3}{2}, \frac{5}{2}$ and $\nu = 0$. The functions $M_{-\mu,\nu}$ in these cases read:

$$M_{-\frac{1}{2},0}(x) = \sqrt{x} e^{\frac{1}{2}z}, \quad \Gamma(1) = 1 \tag{4.140}$$

$$M_{-\frac{3}{2},0}(x) = \sqrt{x}(1+x) e^{\frac{1}{2}z}, \quad \Gamma(2) = 1 \tag{4.141}$$

$$M_{-\frac{5}{2},0}(x) = \frac{1}{2}\sqrt{x}(2 + 4x + x^2) e^{\frac{1}{2}z}, \quad \Gamma(3) = 2. \tag{4.142}$$

Therefore we can use the following formulas

$$\int_0^\infty e^{-\alpha x} I_0(2\beta\sqrt{x}) \, dx = \frac{1}{\alpha} e^{\frac{\beta^2}{\alpha}}, \tag{4.143}$$

$$\int_0^\infty x e^{-\alpha x} I_0(2\beta\sqrt{x}) \, dx = \frac{1}{\alpha^2}\left(1 + \frac{\beta^2}{\alpha}\right) e^{\frac{\beta^2}{\alpha}}, \tag{4.144}$$

$$\int_0^\infty x^2 e^{-\alpha x} I_0(2\beta\sqrt{x}) \, dx = \frac{1}{\alpha^3}\left(2 + 4\frac{\beta^2}{\alpha} + \frac{\beta^4}{\alpha^2}\right) e^{\frac{\beta^2}{\alpha}}. \tag{4.145}$$

In our calculations the parameters α and β are given by

$$\alpha = \frac{1}{g^2 D_0 t}, \quad \beta = \frac{\sqrt{1+gx_0}}{g^2 D_0 t}. \tag{4.146}$$

The zeroth moment is due to the normalization of the distribution. Namely, the integral of the probability density over the total space gives us one:

$$\langle x^0 \rangle = \int_{-1/g}^\infty p(x,t) \, dx = \frac{1}{gD_0 t} e^{-\frac{1+gx_0}{g^2 D_0 t}} \int_0^\infty e^{-\alpha y} I_0(2\beta\sqrt{y}) \frac{dy}{g}$$

$$= \frac{1}{g^2 D_0 t} e^{-\frac{1+gx_0}{g^2 D_0 t}} \frac{1}{\alpha} e^{\frac{\beta^2}{\alpha}} = 1. \tag{4.147}$$

The first moment describes the mean value of the distribution

$$\langle x \rangle = \int_{-1/g}^\infty x p(x,t) \, dx$$

$$= \frac{1}{gD_0 t} e^{-\frac{1+gx_0}{g^2 D_0 t}} \int_0^\infty \frac{y-1}{g} e^{-\alpha y} I_0(2\beta\sqrt{y}) \frac{dy}{g}$$

$$= \frac{1}{g^3 D_0 t} e^{-\frac{1+gx_0}{g^2 D_0 t}} \int_0^\infty y e^{-\alpha y} I_0(2\beta\sqrt{y}) \, dy - \frac{1}{g}\langle x^0 \rangle$$

$$= \frac{1}{g^3 D_0 t} e^{-\frac{1+gx_0}{g^2 D_0 t}} \frac{1}{\alpha^2}\left(1 + \frac{\beta^2}{\alpha}\right) e^{\frac{\beta^2}{\alpha}} - \frac{1}{g}$$

$$= gD_0 t + x_0. \tag{4.148}$$

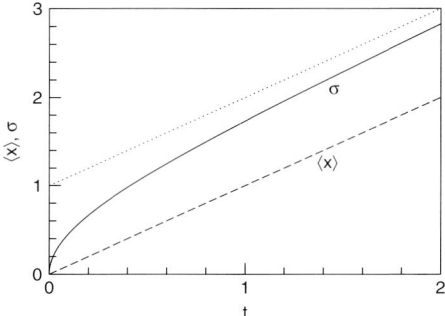

Figure 4.4 The mean value $\langle x \rangle$ (dashed line) and the standard deviation σ (solid line) of the distribution $p(x,t)$ depending on time. The dotted lines shows the linear asymptotics of σ. Parameters: $x_0 = 0$, $D = 1$ m² s⁻¹, $g = 1$m⁻¹.

Hence, the mean value drifts with a constant velocity $v_{\text{drift}} = gD_0$ to the area with a higher diffusion parameter.

The second moment can be computed analogously

$$\langle x^2 \rangle = \int_{-1/g}^{\infty} x^2 p(x,t)\,dx$$

$$= \frac{1}{gD_0 t} e^{-\frac{1+gx_0}{g^2 D_0 t}} \int_0^{\infty} \frac{(y-1)^2}{g^2} e^{-\alpha y} I_0(2\beta\sqrt{y}) \frac{dy}{g}$$

$$= \frac{1}{g^4 D_0 t} e^{-\frac{1+gx_0}{g^2 D_0 t}} \int_0^{\infty} y^2 e^{-\alpha y} I_0(2\beta\sqrt{y})\,dy - \frac{2}{g}\left(\langle x \rangle + \frac{1}{g}\right) + \frac{1}{g^2}\langle x^0 \rangle$$

$$= \frac{1}{g^4 D_0 t} e^{-\frac{1+gx_0}{g^2 D_0 t}} \frac{1}{\alpha^3} \left(2 + 4\frac{\beta^2}{\alpha} + \frac{\beta^4}{\alpha^2}\right) e^{\frac{\beta^2}{\alpha}} - 2\frac{1}{g^2}(g^2 D_0 t + gx_0 + 1) + \frac{1}{g^2}$$

$$= 2D_0^2 g^2 t^2 + 2D_0 t(1 + 2gx_0) + x_0^2. \tag{4.149}$$

Thus, the variance is given by

$$\sigma^2 = \langle x^2 \rangle - \langle x \rangle^2 = D_0^2 g^2 t^2 + 2D_0 t(1 + gx_0). \tag{4.150}$$

Hence we find that, for small times, the standard deviation σ grows as the square root of the time, $\sigma \approx \sqrt{2D_0(1+gx_0)} \cdot \sqrt{t}$, whereas for large times the growth is linear, $\sigma \approx gD_0 t + x_0 + 1/g = \langle x \rangle + 1/g$. The time-dependence of the first moment $\langle x \rangle$, and of σ is illustrated in Figure 4.4.

4.7
Alternative Method of Solving the Fokker–Planck Equation

Coming back to the diffusion equation (4.100) we want finally to present an alternative method of solving the Fokker–Planck equation via the Laplace transformation

4.7 Alternative Method of Solving the Fokker–Planck Equation

to get the long-time behavior as an approximative solution. Applying linear transformation of the coordinate $y = 1 + gx$ and scaling the time $T = D_0 g^2 t$ the probability transforms by $P(y, T)\, dy = p(x, t)\, dx$ and we end up with the following problem, given by the Fokker–Planck equation for $y \in (0, \infty)$

$$\frac{\partial P(y, T)}{\partial T} = \frac{\partial}{\partial y}\left(y \frac{\partial P(y, T)}{\partial y} \right) \tag{4.151}$$

subject to the boundary conditions

$$y\, \partial_y P \big|_{y \to +0} = 0, \quad P\big|_{y \to \infty} \to 0 \tag{4.152}$$

and the initial delta distribution at time $T = 0$

$$P\big|_{T=0} = \delta(y - y_0), \quad y_0 > 0. \tag{4.153}$$

To solve it we make use of the Laplace transformation with respect to time and space.

The space–time Laplace transform of P is defined by

$$F(p, s) = \widehat{\mathfrak{L}}\, P := \int_0^\infty \int_0^\infty dy\, dT\, P(y, T) \exp(-sT - py). \tag{4.154}$$

Applying this Laplace transformation, the Fokker–Planck equation becomes

$$sF = -p \frac{\partial}{\partial p}(pF) + e^{-py_0}. \tag{4.155}$$

To obtain this result we have used the following identities

$$\widehat{\mathfrak{L}}\frac{\partial P}{\partial T} := sF - \exp(-py_0), \tag{4.156}$$

$$\widehat{\mathfrak{L}}\frac{\partial}{\partial y}\left(y \frac{\partial P}{\partial y} \right) := -p \frac{\partial}{\partial p}(pF). \tag{4.157}$$

The solution of (4.155) is found in the form

$$F(p, s) = \frac{A(p)}{p} \exp\left(\frac{s}{p} \right),$$

whence we obtain the equation for $A(p)$

$$A' = p \exp\left(-\frac{s}{p} - py_0 \right).$$

Since $A(0) = 0$ holds, we have

$$A(p) = \int_0^p \frac{dq}{q} \exp\left(-\frac{s}{q} - qx_0 \right),$$

and the desired solution is

$$F(p, s) = \frac{1}{p} \int_0^p \frac{dq}{q} \exp\left[s\left(\frac{1}{p} - \frac{1}{q} \right) - qy_0 \right]. \tag{4.158}$$

Changing the variable
$$q = p\left(1 + \gamma \frac{p}{s}\right)^{-1}$$
the integral (4.158) is converted to
$$F(p,s) = \int_0^\infty \frac{d\gamma}{s + p\gamma} \exp\left(-\gamma - \gamma_0 \frac{sp}{s + \gamma p}\right). \quad (4.159)$$

The inverse transformation from the Laplace space $F(p,s)$ to the original variables $P(y, T)$ becomes easy in the long-time limit. As time increases, the influence of the initial system position y_0 becomes negligible. This limit is described by expression (4.159) where we set $y_0 = 0$. In this case formula (4.159) reads

$$F(p,s) = \int_0^\infty \frac{d\gamma}{s + p\gamma} \exp(-\gamma) = \int_0^\infty \int_0^\infty dy\, dT\, P(y, T) \exp(-sT - py) \quad (4.160)$$

Taking into account the identity
$$\int_0^\infty dt \exp(-st - \xi t) = \frac{1}{s + \xi} \quad (4.161)$$
from (4.160) we have
$$\int_0^\infty dx P(t, x) \exp(-px) = \int_0^\infty dx \exp\left[-x(1 + pt)\right]$$
$$= \frac{1}{1 + pt} = \frac{1}{t}\frac{1}{t^{-1} + p}. \quad (4.162)$$

By using again (4.161), the distribution function can be written as
$$P(y, T) = \frac{1}{T} \exp\left(-\frac{y}{T}\right). \quad (4.163)$$

Going back to the original variables $p(x, t)$, we have
$$p(x, t) = \frac{1}{D_0 g t} \exp\left(-\frac{1 + gx}{D_0 g^2 t}\right). \quad (4.164)$$

It should be noted that expressions (4.136) and (4.164) coincide according to (4.135) for $1 + gx_0 = 0$ as should be the case for the initial position of the delta distribution located at the boundary $x = x_0 = -1/g$.

4.8
Exercises

E 4.1 Fokker–Planck equation and Maxwell distribution
Consider the Fokker–Planck equation (4.67) in the one-dimensional velocity space of particles, where the variable x represents the velocity v, that is $x \to v$. Here we assume $D_1(v) = -\gamma v$ and $D_2 = \gamma k_B T/m$, where γ is the friction coefficient, k_B is the Boltzmann constant, T is the temperature, and m is the mass of a particle. The task is to obtain the stationary solution and compare it with the Maxwell distribution.

E 4.2 Hermitean Fokker–Planck operator

Transform the Fokker–Planck equation for the velocity distribution function $p(v, t)$ considered in the previous exercise into the equation for $W(v, t) = e^{\phi(v)/2} p(v, t)$. Find the function $\phi(v)$ for which the resulting Fokker–Planck operator is Hermitean.

E 4.3 Fokker–Planck equation into the Schrödinger equation

Transform the Fokker–Planck equation (4.67) into the form

$$\frac{\partial p}{\partial t} = \mathcal{L}_{\text{FP}} p(x, t) \quad \text{with} \quad \mathcal{L}_{\text{FP}} = D \frac{\partial^2}{\partial x^2} - V(x) \tag{4.165}$$

and then further into the Schrödinger equation, for a particle in the potential $V(x)$, by introducing the imaginary Schrödinger time $t_{\text{Schrödinger}} = -i\hbar t$. Find the relation between the diffusion coefficient D in (4.165) and the particle mass in the Schrödinger equation.

E 4.4 Fokker–Planck equation with harmonic potential

Consider the Fokker–Planck equation (4.165) for the harmonic potential $V(x) = (1/2)\, m\omega_0^2 x^2$. Transform it into the Schrödinger equation and find the eigenvalues and the eigenfunctions of the Fokker–Planck operator.

5
The Langevin Equation

5.1
A System of Many Brownian Particles

Stochastic differential equations, which in physics are often denoted by Langevin equations, represent a tool for the modeling of quasi-continuous diffusion processes. The range of their applications is very wide, including physical problems like the motion of Brownian particles, the movement of grains in a granular media, the motion of vehicles in traffic flow or of animals in a biological system, the problems of economics and financial mathematics, the description of chemical reactions, and numerous optimization problems, etc.

This chapter is devoted to the Langevin equation, first published by Paul Langevin in 1908, to describe the Brownian motion of passive (force-free) or active particles [6, 34, 38, 64, 211]. Active or self-driven motion of motorized particles (also called Brownian agents) is based on the assumption that the friction coefficient of the ordinary Langevin equation is velocity dependent, representing the take-up of energy from the environment and its conversion into kinetic energy. Before we are going to discuss this in more detail in this Section, at first we introduce the Langevin equation as follows.

A general Langevin equation in multidimensional case with the state vector $\mathbf{q} = (q_1, q_2, \ldots, q_n)$, the deterministic force $\mathbf{F} = (F_1, F_2, \ldots, F_n)$, and the Langevin force as an additive white noise $\xi = (\xi_1, \xi_2, \ldots, \xi_n)$ reads [64]

$$\frac{dq_i}{dt} = F_i(\mathbf{q}) + \hat{\xi}_i(t) \tag{5.1}$$

where

$$\langle \hat{\xi}_i(t) \rangle = 0, \tag{5.2}$$

$$\langle \hat{\xi}_i(t)\hat{\xi}_j(t') \rangle = Q_{ij}\delta(t - t'). \tag{5.3}$$

The corresponding Fokker–Planck equation for the probability density function $p(q_1, q_2, \ldots, q_n, t) \equiv p(\mathbf{q}, t)$ is (cf. (4.1))

$$\frac{\partial p(\mathbf{q},t)}{\partial t} = -\nabla_\mathbf{q}\{\mathbf{F}\,p(\mathbf{q},t)\} + \frac{1}{2}\sum_{ij} Q_{ij}\frac{\partial^2 p(\mathbf{q},t)}{\partial q_i \partial q_j}, \tag{5.4}$$

where

$$\nabla_\mathbf{q}\{\mathbf{F}\,p(\mathbf{q},t)\} = \sum_i \frac{\partial}{\partial q_i}\{F_i\,p(\mathbf{q},t)\} \tag{5.5}$$

denotes the divergence in the n-dimensional state space. The Fokker-Planck equation (5.4) can be written as a continuity equation (cf. (4.2))

$$\frac{\partial p(\mathbf{q},t)}{\partial t} + \nabla_\mathbf{q}\mathbf{J}(\mathbf{q},t) = 0, \tag{5.6}$$

where $\mathbf{J}(\mathbf{q},t)$ is the probability flux vector with components

$$J_i(\mathbf{q},t) = F_i\,p(\mathbf{q},t) - \sum_j Q_{ij}\frac{\partial p(\mathbf{q},t)}{\partial q_j}. \tag{5.7}$$

In the case where the Langevin force has only diagonal terms, so $Q_{ij} = 2D_i\delta_{ij}$ holds, the Langevin equation can be represented as

$$\frac{dq_i}{dt} = F_i(\mathbf{q}) + \sqrt{2D_i}\,\xi_i(t), \tag{5.8}$$

where $\xi_i(t) = \hat{\xi}_i(t)/\sqrt{2D_i}$ is the standard Gaussian white noise with the property

$$\langle \xi_i(t)\xi_j(t')\rangle = \delta_{ij}\delta(t-t'), \tag{5.9}$$

and D_i has the meaning of the diffusion coefficient for the variable q_i.

This equation has many applications in physics, chemistry, engineering and social sciences [6, 34, 38, 64, 211]. In particular, the system of N Brownian particles moving in d-dimensional coordinate space of a given volume V is described by the Langevin equations

$$\frac{d\mathbf{r}_\alpha}{dt} = \mathbf{v}_\alpha, \quad \alpha = 1,2,\ldots,N \tag{5.10}$$

$$\frac{d\mathbf{v}_\alpha}{dt} = -\gamma_\alpha \mathbf{v}_\alpha - \frac{1}{m_\alpha}\nabla_\alpha U(\mathbf{r}_1,\ldots,\mathbf{r}_N) + \sqrt{2B_\alpha}\,\xi_\alpha(t), \tag{5.11}$$

which correspond to (5.8). In this case the state vector has $2dN$ components comprising d components of the coordinate vector $\mathbf{r}_\alpha$ and d components of the velocity vector $\mathbf{v}_\alpha$ of each particle with mass m_α. The particles are numbered by the index α. In this case $\xi_\alpha(t)$ also is a d-dimensional vector, whereas γ_α and B_α are scalar quantities – the friction coefficient and the diffusion coefficient in the velocity space. The noise has only diagonal terms, which in our notation means

$$\langle \xi_\alpha(t)\rangle = 0, \tag{5.12}$$

$$\langle \xi_{\alpha,i}(t)\xi_{\beta,j}(t')\rangle = \delta_{\alpha\beta}\delta_{ij}\delta(t-t'), \tag{5.13}$$

where the indices α and β refer to particles, whereas i and j (from 1 to d) refer to the vector components of a given particle.

The operator ∇_α represents the gradient in the space of $\mathbf{r}_\alpha$. It gives the force $\mathbf{F}_\alpha$ acting on the particle number α,

$$-\nabla_\alpha U(\mathbf{r}_1, \ldots, \mathbf{r}_N) = \mathbf{F}_\alpha(\mathbf{r}_1, \ldots, \mathbf{r}_N). \tag{5.14}$$

In this notation the system of equations (5.10)–(5.11) becomes

$$\frac{d\mathbf{r}_\alpha}{dt} = \mathbf{v}_\alpha, \quad \alpha = 1, 2, \ldots, N \tag{5.15}$$

$$\frac{d\mathbf{v}_\alpha}{dt} = -\gamma_\alpha \mathbf{v}_\alpha + \frac{1}{m_\alpha} \mathbf{F}_\alpha(\mathbf{r}_1, \ldots, \mathbf{r}_N) + \sqrt{2B_\alpha}\, \xi_\alpha(t). \tag{5.16}$$

The corresponding Fokker–Planck equation for the probability density function $p_N = p_N(\mathbf{r}_1, \mathbf{v}_1, \ldots, \mathbf{r}_N, \mathbf{v}_N, t)$ consistent with the particular case of (5.4) where the double sum over i and j contains only the diagonal terms, reads

$$\frac{\partial p_N}{\partial t} = -\sum_\alpha \frac{\partial}{\partial \mathbf{r}_\alpha}(\mathbf{v}_\alpha p_N) - \sum_\alpha \frac{\partial}{\partial \mathbf{v}_\alpha}\left(\frac{1}{m_\alpha} \mathbf{F}_\alpha p_N\right)$$
$$+ \sum_\alpha \frac{\partial}{\partial \mathbf{v}_\alpha}(\gamma_\alpha \mathbf{v}_\alpha p_N) + \sum_\alpha B_\alpha \frac{\partial^2 p_N}{\partial \mathbf{v}_\alpha^2}. \tag{5.17}$$

Here $\partial/\partial \mathbf{r}_\alpha$ and $\partial/\partial \mathbf{v}_\alpha$ are the nabla or the divergence operators acting in the space of $\mathbf{r}_\alpha$ and $\mathbf{v}_\alpha$, respectively. By analogy, the second-order derivative in the last sum represents the Laplace operator.

In the thermodynamic equilibrium, which corresponds to the stationary solution $\partial p_N/\partial t = 0$ of (5.17), the probability density is consistent with the Boltzmann–Gibbs distribution

$$p_N = \frac{1}{Z(T)} e^{-H_N(\mathbf{r}_1, \mathbf{v}_1, \ldots, \mathbf{r}_N, \mathbf{v}_N)/(k_B T)}. \tag{5.18}$$

Here H is the Hamiltonian of the system and $Z(T)$ is the partition function, which depends on the thermal energy $k_B T$, where k_B is the Boltzmann constant and T is the temperature. The Hamiltonian is given by

$$H = \sum_{\alpha=1}^{N} m_\alpha \frac{\mathbf{v}_\alpha^2}{2} + U(\mathbf{r}_1, \ldots, \mathbf{r}_N). \tag{5.19}$$

Hence, the diffusion coefficient in the velocity space B_α should obey the Einstein relation, also called the fluctuation-dissipation relationship

$$B_\alpha = \frac{\gamma_\alpha k_B T}{m_\alpha}. \tag{5.20}$$

In the following we consider a system of Brownian particles with the potential

$$U(\mathbf{r}_1, \ldots, \mathbf{r}_N) = \sum_{\alpha < \beta} m_\alpha m_\beta\, u(|\mathbf{r}_\alpha - \mathbf{r}_\beta|). \tag{5.21}$$

Here $u(r)$ is the potential of interaction between two particles. In particular, the case with gravitational potential in two dimensions has been widely studied by P. H. Chavanis in [27].

In the overdamped limit $\gamma_\alpha \to +\infty$, the inertia of the particles in (5.11) can be neglected [27], which means that we can solve this equation with respect to $\mathbf{v}_\alpha$ by setting $d\mathbf{v}_\alpha/dt = 0$ and substitute the result into (5.10). It yields

$$\frac{d\mathbf{r}_\alpha}{dt} = -\mu_\alpha \nabla_\alpha (\mathbf{r}_1, \ldots, \mathbf{r}_N) + \sqrt{2D_\alpha}\,\xi_\alpha(t), \tag{5.22}$$

where

$$\mu_\alpha = \frac{1}{\gamma_\alpha m_\alpha} \tag{5.23}$$

is the mobility and

$$D_\alpha = \frac{B_\alpha}{\gamma_\alpha^2} \tag{5.24}$$

is the spatial diffusion coefficient of the particle number α. From (5.23), (5.24), and (5.20) we obtain the well known Einstein relation between the spatial diffusion coefficient and mobility, that is,

$$D_\alpha = \mu_\alpha k_B T. \tag{5.25}$$

The Fokker–Planck equation corresponding to (5.22)

$$\frac{\partial p_N}{\partial t} = \sum_{\alpha=1}^N \frac{\partial}{\partial \mathbf{r}_\alpha}\left[D_\alpha \frac{\partial p_N}{\partial \mathbf{r}_\alpha} + \mu_\alpha p_N \frac{\partial U}{\partial \mathbf{r}_\alpha}\right] \tag{5.26}$$

describes the time evolution of the probability density function $p_N = p_N(\mathbf{r}_1, \ldots, \mathbf{r}_N, t)$ in the (Nd)-dimensional coordinate space of N particles.

The systems with long-range interaction like, e.g., the N-star model in astrophysics, demonstrate a particularly interesting behavior [27]. They have a complex thermodynamics and present phase transitions between 'gaseous' and 'clustered' states. In particular, the model of self-gravitating Brownian particles is of interest [27]. In this model, the particles interact gravitationally, but they also experience a friction force and a stochastic force. The latter mimics a coupling with a thermal bath of nongravitational origin. This is a dissipative system, therefore a description by canonical ensemble with given temperature is appropriate. The dynamics of such a model is described by a set of N coupled Langevin equations which we have already discussed with the gravitational potential

$$u(r) = -\frac{1}{d-2}\frac{G}{r^{d-2}} \quad \text{for} \quad d \neq 2, \tag{5.27}$$

$$u(r) = G \ln r \quad \text{for} \quad d = 2, \tag{5.28}$$

where G is the gravity constant. The two-dimensional case $d = 2$ is special. An interesting and, generally difficult, problem consists in determining the effective diffusion coefficient $D(T)$ of a particle of the system. For the self-gravitating gas of Brownian particles in two dimensions it can be calculated exactly [27],

$$D(T) = \frac{K_B T}{\gamma m}\left(1 - \frac{T_c}{T}\right), \qquad (5.29)$$

where T_c is the critical temperature

$$K_B T_c = (N-1)\frac{Gm^2}{4}. \qquad (5.30)$$

For $T \gg T_c$ the self-gravity becomes negligible, and we recover the Einstein relation (5.25). This corresponds to a diffusive motion of particles which is slightly modified by self-gravity. The diffusion coefficient vanishes at $T = T_c$ and becomes negative at $T < T_c$. The latter implies the collapse. In that case the system forms a Dirac peak containing the whole mass in a finite time.

Interacting random walkers, described by coupled Langevin equations, are also studied in soft-matter physics in order to compute the transport properties of interacting particles. Examples are supercooled liquids and colloids in solution. In these examples, however, the interaction potential is short-range and the system is homogeneous at equilibrium.

In other applications, one considers systems with short-range next-neighbor interactions, such as one-dimensional ring chains with Toda or Morse potential, but with velocity-dependent friction coefficient $\gamma(v)$ [38, 211]. The system of coupled Langevin equations ($i = 1, 2, \ldots, N$) then reads

$$\frac{dx_i}{dt} = v_i, \qquad (5.31)$$

$$m\frac{dv_i}{dt} = -\frac{\partial U}{\partial x_i} - m\gamma(v_i)v_i + \sqrt{2B}\xi_i(t). \qquad (5.32)$$

In this case N particles are numbered by the index i and the periodic boundary conditions $x_{i+N} = x_i + L$ are assumed, where L is the length of the chain. The derivative $-\partial U/\partial x_i$ of the system's potential energy U represents the interaction force, whereas $-m\gamma(v_i)v_i$ represents the friction force acting on the ith particle. In the particular case studied in [28] we have

$$\gamma(v) = \gamma_0 + \gamma_1(v), \qquad (5.33)$$

where the constant part γ_0 describes the viscous friction between the particle and the surrounding heat bath, whereas the nonlinear coefficient

$$\gamma_1(v) = -\frac{q}{\kappa + v^2} < 0 \qquad (5.34)$$

is introduced to model the active Brownian particles. A positive parameter $q > 0$ describes the energy flux from the external reservoir into the depots carried by the particles. The parameter κ is connected to the internal dissipation and the conversion of the energy. Introducing a new parameter

$$\mu = \frac{q}{\gamma_0} - \kappa, \qquad (5.35)$$

Equation (5.34) can be written as

$$\gamma(v) = \gamma_0 \left(1 - \frac{\kappa + \mu}{\kappa + v^2}\right) = \gamma_0 \frac{v^2 - \mu}{\kappa + v^2}. \tag{5.36}$$

Here μ plays the role of a bifurcation parameter, since $\gamma(v) = 0$ holds if $v = \pm\sqrt{\mu}$. For $\mu < 0$ the friction coefficient $\gamma(v)$ is always positive and thus leads to damping of the particle motion. At $\mu > 0$, the friction coefficient becomes negative for small velocities $v^2 < \mu$, generating so-called active motion.

The interaction potential (or potential energy) is given by

$$U = \sum_i u_i(\Delta x_i), \tag{5.37}$$

where $\Delta x_i = x_{i+1} - x_i$ are the distances between nearest neighbors. In the following we assume that the pair interaction potential is given by $u_i(\Delta x_i) = u(\Delta x_i)$ for all i. A particular example is the Morse potential

$$u(\Delta x) = \frac{a}{2b}\left[e^{-b(\Delta x - \sigma)} - 1\right]^2 - \frac{a}{2b}, \tag{5.38}$$

which is qualitatively similar to the well known Lennard–Jones potential

$$u(\Delta x) = \frac{a}{2b}\left[\left(\frac{\Delta x}{\sigma}\right)^{-12} - 2\left(\frac{\Delta x}{\sigma}\right)^{-6}\right]. \tag{5.39}$$

The parameters are chosen such that the minimum $-a/(2b)$ of the potentials (5.38) and (5.39) is located at $\Delta x = \sigma$. The Morse potential can be considered as a generalization of the Toda potential

$$u(\Delta x) = \frac{a}{b}\left[e^{-b(\Delta x - \sigma)} - 1\right] + a(\Delta x - \sigma) - \frac{a}{2b}. \tag{5.40}$$

The latter is widely appreciated, since it yields an exactly solvable nonlinear model.

The models with short-range interaction and negative velocity-dependent friction coefficient, describe active particles and show interesting features like clustering. There exists a critical density interval, where both equidistant and cluster configurations correspond to local minima of the potential energy [28].

Soliton-like oscillations are observed in the Toda chains with negative friction [37]. A conservative chain from N elements possesses $N + 1$ different modes of oscillation. They differ from one another in shape, amplitude, frequency and phase shift, between the oscillations of neighboring particles. In the active chain each of these modes may generate the corresponding attractor [38].

Various interesting examples based on Langevin dynamics are considered in [212, 226] such as stochastic thermodynamics and nonequilibrium steady states.

5.2
A Traditional View of the Langevin Equation

The Langevin equation describes the dynamics of a system in the presence of an interaction with the environment. For simplicity here we consider the one-dimensional case, where the state of the system is characterized by a scalar quantity $x(t)$ which depends on time t. The time evolution is described by the Langevin equation

$$\frac{dx}{dt} = f(x) + \psi(x)\xi(t) \tag{5.41}$$

together with the initial condition

$$x(t=0) = x_0. \tag{5.42}$$

Here the dynamics of the system itself is given by the deterministic force $f(x)$, whereas the interaction with the environment is represented by the stochastic or Langevin force $\psi(x)\xi(t)$, where $\psi(x)$ is the noise intensity. If the latter is constant then the Langevin force represents an additive noise. The intensity $\psi(x)$ may depend on x in general. In this case we deal with the so-called multiplicative noise. In the classical case, $\xi(t)$ is the Gaussian white noise, representing random and normally distributed fluctuations, which are completely uncorrelated for different times.

It is important to notice, however, that another kind of noise $\xi(t)$ may also be of interest. For example, the Markovian dichotomous noise represents a stochastic process of switching between two discrete values. This type of noise is frequently used for modeling various phenomena in biology, physics, and chemistry. States of the dichotomous process can be associated, e.g. with two different levels of external stimuli, the presence or absence of an external perturbation, etc. It is interesting to mention that a combination of dichotomous and white noise can lead to a bimodal probability distribution even in a system with a single-well potential [35] $\phi(x) = \alpha x^2$ or linear force $f(x) = -d\phi/dx$. Thus, the noise can significantly change the behavior of a system. In this sense we can speak about noise-induced phase transitions. The latter topic will be discussed in detail in Chapter 11.

Here we give some general statements only and recommend the monograph by Werner Horsthemke and Rene Lefever [86] to check for further information concerning noise-induced transitions. First we notice that the phase transition point is shifted depending on the noise intensity, which is a usual phenomenon. It is a general feature of nonlinear systems subject to multiplicative external noise. The shift in the bifurcation diagram is not too surprising when one thinks about it. The external noise can induce even deeper and far less intuitive modifications in the macroscopic behavior of nonlinear systems [86]. Nonequilibrium systems are, by their very nature, closely dependent on their environment. A question therefore arises as to how the nonequilibrium and the environmental randomness interact. Can this interaction lead to drastic changes in the macroscopic behavior of the system even outside the neighborhood of a deterministic instability point? In other words, is it possible that the external noise modifies bifurcation diagrams in a more profound way than by just a shift in the parameter space? The basic question can be formulated as follows: do nonlinear systems always adjust their macroscopic behavior to the average properties of the environment, or can one find situations in which the system responds to the randomness of the environment in a certain more active way displaying, for instance, behavior forbidden under deterministic external conditions? The answer to these questions is indeed positive. It turns out that even extremely rapid totally random external noise can considerably alter the macroscopic behavior of nonlinear systems. It can induce new transition phenomena, known as noise-induced phase transitions, which are quite unexpected from the usual phenomenological description.

In the following we turn back to the traditional approach with white noise added to the deterministic drift.

5.3
Additive White Noise

Historically, the Langevin equation has been designed to describe Brownian motion, assuming $\psi(x) = \sigma$ in (5.41) as a constant. This is the usual case of the Langevin equation with the additive noise

$$\frac{dx}{dt} = f(x) + \sigma \xi(t). \tag{5.43}$$

In general, $\xi(t)$ is a randomly fluctuating quantity. Traditionally it is white noise, which has the following properties

$$\langle \xi(t) \rangle = 0, \tag{5.44}$$

$$\langle \xi(t)\xi(t') \rangle = \delta(t - t'). \tag{5.45}$$

Equation (5.43) can be formulated as a stochastic differential equation with the initial condition (5.42). It is the conventional form of writing used in the mathematical literature, that is

$$dx(t) = f(x(t))\,dt + \sigma\,dW(t); \quad x(t = 0) = x_0, \tag{5.46}$$

where $W(t)$ is the standard Wiener process with the following properties

$$\langle W(t) \rangle = 0, \tag{5.47}$$

$$\langle W(t)W(t') \rangle = \min(t, t'). \tag{5.48}$$

For the increments of the Wiener process $dW(t) = W(t + dt) - W(t)$ at $dt \to 0$ we have

$$\langle dW(t) \rangle = 0, \tag{5.49}$$

$$\langle dW(t)\, dW(t') \rangle = \begin{cases} dt, & t' = t \\ 0, & t' \neq t \end{cases} \tag{5.50}$$

The formal relation between the Wiener process and the Langevin force is given by

$$\xi(t) = \frac{dW(t)}{dt} \iff W(t) = \int_0^t \xi(s)\, ds. \tag{5.51}$$

The stochastic differential equations and this formal relation will be discussed in more detail in Section 5.6. Here we would like to mention that the formal solution of (5.46) is

$$x(t) = x_0 + \int_0^t f(x(s))\, ds + \sigma W(t). \tag{5.52}$$

This, however, is only a different formulation of the problem by rewriting the stochastic differential equation (5.46) as an integral equation (5.52). Since the right-hand side of (5.52) contains the unknown function $x(s)$, it cannot serve as a solution in practical applications.

The probability density distribution $p(x, t)$ for the variable x at time t is given by the following Fokker–Planck equation which corresponds to (5.43) or (5.46), respectively,

$$\frac{\partial}{\partial t} p(x, t) = -\frac{\partial}{\partial x}\left[f(x) p(x, t)\right] + \frac{\sigma^2}{2} \frac{\partial^2 p(x, t)}{\partial x^2} \tag{5.53}$$

with the initial condition

$$p(x, t = 0) = \delta(x - x_0). \tag{5.54}$$

The averages over the ensemble of stochastic realizations, like the mean value $\langle x(t) \rangle$ and the correlation function $\langle x(t) x(t') \rangle$, can be expressed in terms of the probability distribution functions as

$$\langle x(t) \rangle = \int_{-\infty}^{\infty} x p(x, t)\, dx, \tag{5.55}$$

$$\langle x(t) x(t') \rangle = \int_{-\infty}^{\infty} \int_{-\infty}^{\infty} xy\, p(x, t; y, t')\, dx\, dy. \tag{5.56}$$

Here $p(x, t; y, t')$ is the joint probability density for the two times introduced in Section 3.1.

Returning to the Langevin equation (5.43), first let us consider the dynamics without fluctuations, which is given by the equation with $\sigma = 0$,

$$\frac{dx}{dt} = f(x). \tag{5.57}$$

The force can be represented as

$$f(x) = -\frac{d\phi(x)}{dx}, \tag{5.58}$$

where $\phi(x)$ is the potential. A simple classical example is the double-well potential

$$\phi(x) = \frac{\alpha}{2}x^2 + \frac{\beta}{4}x^4, \tag{5.59}$$

where $\beta > 0$. It has one minimum if $\alpha > 0$ and two minima if $\alpha < 0$. The corresponding force is

$$f(x) = -\alpha x - \beta x^3. \tag{5.60}$$

The stationary solutions of (5.57) are the roots of the equation $f(x) = 0$ or the extremum points of the potential $\phi(x)$. They are given by

$$x(\alpha + \beta x^2) = 0. \tag{5.61}$$

One root always is $x_0 = 0$. At $\alpha \geq 0$ this is the only real solution. At $\alpha < 0$, two other real solutions appear $x_{1,2} = \pm\sqrt{-\alpha/\beta}$ corresponding to two minima of the potential. The solution $x_0 = 0$ corresponds to the only minimum of the potential at $\alpha > 0$, which is changed to the maximum at $\alpha < 0$. Minimum of $\phi(x)$ always corresponds to a stable solution and the maximum to an unstable solution of (5.57), as follows from the stability analysis considering small deviations from the extremum point. These solutions, depending on the parameter α, represent the so-called supercritical bifurcation diagram shown in Figure 5.1. It is called supercritical, since the stable branches merge continuously at the bifurcation point $\alpha = 0$.

A bifurcation diagram of an other kind emerges for the potential

$$\phi(x) = \frac{\alpha}{2}x^2 + \frac{\beta}{4}x^4 + \frac{\gamma}{6}x^6 \tag{5.62}$$

with $\beta < 0$ and $\gamma > 0$. It corresponds to

$$f(x) = -\alpha x - \beta x^3 - \gamma x^5. \tag{5.63}$$

In this case the equation $f(x) = 0$ has five roots, some of which may be complex. One solution is $x_0 = 0$. The other four roots are given by

$$x_{1,2,3,4} = \pm\sqrt{-\frac{\beta}{2\gamma} \pm \sqrt{\left(\frac{\beta}{2\gamma}\right)^2 - \frac{\alpha}{\gamma}}}. \tag{5.64}$$

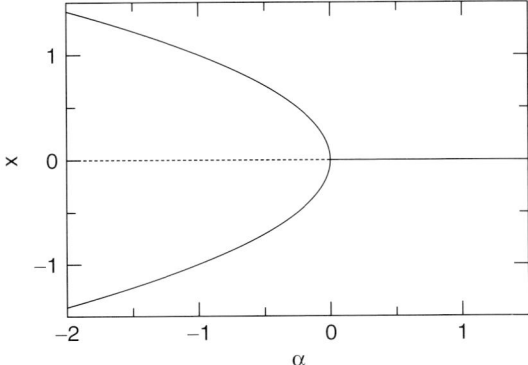

Figure 5.1 The supercritical bifurcation diagram for the potential (5.59) at $\beta = 1$. Stable solutions of (5.57) depending on the parameter α are shown by solid lines and the unstable one by a dashed line.

Only the real solutions have a physical meaning. Also, the solutions corresponding to the minima of the potential are stable, whereas those representing the maxima are unstable. At $\alpha > \beta^2/(4\gamma)$ the only real solution is $x_0 = 0$. All five solutions are real within $0 \leq \alpha \leq \beta^2/(4\gamma)$. Three of them, including $x_0 = 0$, are stable and correspond to three minima of $\phi(x)$. The other two roots represent two local maxima in between. At $\alpha = 0$, the minimum at $x = 0$ transforms into the maximum and two other maxima disappear. Thus, at $\alpha < 0$ there are two stable solutions and one unstable solution $x_0 = 0$. The corresponding so-called subcritical bifurcation diagram is shown in Figure 5.2.

As distinct from the supercritical bifurcation diagram in Figure 5.1, here the stable nonzero branches start at certain nonzero x values at $\alpha = \beta^2/(4\gamma)$, where the

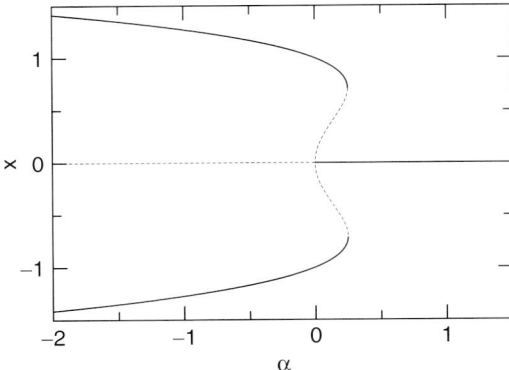

Figure 5.2 The subcritical bifurcation diagram for the potential (5.62) at $\beta = -1$ and $\gamma = 1$. Stable solutions of (5.57) depending on the parameter α are shown by solid lines, and the unstable ones by dashed lines.

$x_0 = 0$ branch is still stable. Therefore, the system cannot switch to these nonzero branches if the initial x value is near zero. In the deterministic dynamics it first happens with a jump only at $\alpha = 0$ when α is decreased. If α is increased, starting from negative values, then a jump from one of the nonzero stable solutions to the zero solution occurs at $\alpha = \beta^2/(4\gamma) > 0$. In other words, hysteresis is observed.

The behavior of the dynamical system in the case of supercritical as well as subcritical bifurcation is essentially changed by the noise included in the Langevin equation (5.43). Due to this noise, the system with potential (5.59) can be randomly switched between two stable states $x_{1,2} = \pm\sqrt{-\alpha/\beta}$ at $\alpha < 0$, which is never possible in the deterministic dynamics. Similarly, in the system with potential (5.62), the noise enables a switching between three stable states within $0 \le \alpha \le \beta^2/(4\gamma)$, or between two stable branches of the bifurcation diagram in Figure 5.2 at $\alpha < 0$. Considering an ensemble of different stochastic realizations of the process $\xi(t)$, the Langevin equation (5.43) allows one to calculate the probability density $p(x, t)$ which has a certain value depending on x at time t. The stationary probability density $p^{st}(x) = \lim_{t \to \infty} p(x, t)$ is given by the stationary solution of the corresponding Fokker–Planck equation (5.53), so

$$p^{st}(x) = \frac{e^{-2\phi(x)/\sigma^2}}{\int_{-\infty}^{\infty} e^{-2\phi(x)/\sigma^2}\, dx}. \qquad (5.65)$$

The stable branches of the bifurcation diagrams in Figures 5.1 and 5.2 correspond to maxima of the stationary probability distribution function $p^{st}(x)$, whereas the unstable branches correspond to its local minima. It can be seen in Figures 5.3 and 5.4, where the stationary probability density is shown for different values of α, corresponding to the bifurcation diagrams in Figures 5.1 and 5.2, respectively. Because of the symmetry of the considered potentials, the probability distribution function is symmetrical with respect to $x = 0$ in these examples.

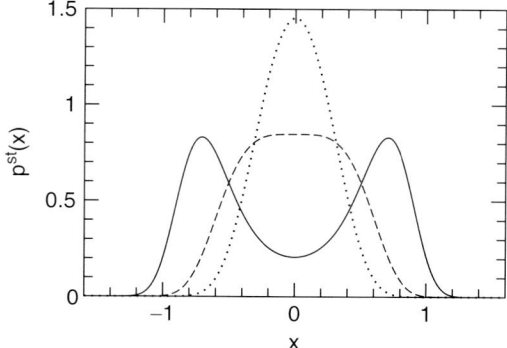

Figure 5.3 The stationary probability density $p^{st}(x)$ for the potential (5.59) with $\beta = 1$ at $\alpha = 0.5$ (dotted curve), $\alpha = 0$ (dashed curve), and $\alpha = -0.5$ (solid curve).

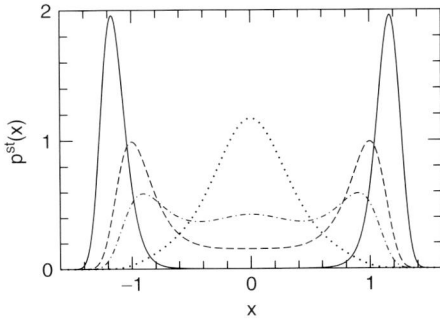

Figure 5.4 The stationary probability density $p^{st}(x)$ for the potential (5.62) with $\beta = -1$ and $\gamma = 1$ at $\alpha = 0.5$ (dotted curve), $\alpha = 0.15$ (dot-dashed curve), $\alpha = 0$ (dashed curve), and $\alpha = -0.5$ (solid curve).

5.4
Spectral Analysis

Fourier or spectral analysis is a powerful tool used to analyze the solution of the Langevin equation [174, 194]. As an example, here we apply spectral analysis to the time evolution of a vector $\mathbf{v}(t)$. Specifically, $\mathbf{v}(t)$ can be the velocity of a Brownian particle moving in three-dimensional space. Its Fourier representation as an infinite sum is

$$\mathbf{v}(t) = \mathbf{v}(t+T) = \sum_{k=-\infty}^{\infty} \mathbf{a}_k \cos\left(\frac{2\pi k t}{T}\right) + \sum_{k=-\infty}^{\infty} \mathbf{b}_k \sin\left(\frac{2\pi k t}{T}\right). \tag{5.66}$$

Here $\mathbf{a}_k$ and $\mathbf{b}_k$ are the Fourier coefficients given by

$$\mathbf{a}_k = \frac{1}{T} \int_0^T \mathbf{v}(t) \cos\left(\frac{2\pi k t}{T}\right) dt \tag{5.67}$$

$$\mathbf{b}_k = \frac{1}{T} \int_0^T \mathbf{v}(t) \sin\left(\frac{2\pi k t}{T}\right) dt. \tag{5.68}$$

Using the well known Euler formulas

$$e^{ix} = \cos(x) + i \sin(x), \tag{5.69}$$

and

$$\cos(x) = \frac{e^{ix} + e^{-ix}}{2}, \quad \sin(x) = \frac{e^{ix} - e^{-ix}}{2i}, \tag{5.70}$$

the transformation (5.66) can be represented by complex Fourier amplitudes $\tilde{\mathbf{v}}(\omega)$ as follows

$$\mathbf{v}(t) = \mathbf{v}(t+T) = \frac{1}{T} \sum_{\omega} \tilde{\mathbf{v}}(\omega) e^{i\omega t}, \tag{5.71}$$

where the summation runs over a set of discrete frequencies $\omega = 2\pi k/T$ with $k = 0, \pm1, \pm2, \ldots$. The inverse transformation reads

$$\tilde{\mathbf{v}}(\omega) = \int_0^T \mathbf{v}(t) e^{-i\omega t}\, dt. \tag{5.72}$$

Equation (5.71) can be viewed as an expansion in the basis of orthogonal wave functions $e^{i\omega t}$, which satisfies the periodic boundary conditions and has the orthogonality property

$$\frac{1}{T}\int_0^T e^{i\omega t} e^{-i\omega' t}\, dt = \delta_{\omega,\omega'}. \tag{5.73}$$

Note that the term $\omega = 0$ in (5.71) represents the constant, that is, the time-independent contribution. If, in general, $\lim_{T\to\infty}\langle \mathbf{v}(t) \rangle$ is a constant, it is just $\tilde{\mathbf{v}}(0)/T$. In this case we have

$$\mathbf{v}(t) - \langle \mathbf{v} \rangle = \frac{1}{T}\sum_{\omega \neq 0} \tilde{\mathbf{v}}(\omega) e^{i\omega t}. \tag{5.74}$$

If the period T tends to infinity ($T \to \infty$), the discrete sum over $\omega \neq 0$ may be replaced by an integral. This substitution in (5.74) yields

$$\mathbf{v}(t) - \langle \mathbf{v} \rangle = \frac{1}{2\pi}\int_{-\infty}^{\infty} \tilde{\mathbf{v}}(\omega) e^{i\omega t}\, d\omega. \tag{5.75}$$

For any given time difference $t' - t = \Delta t$, the correlation function in the stationary regime at $T \to \infty$ is given by the components v_i of the velocity vector $\mathbf{v}$ as

$$\varphi_{ij}(\Delta t) = \langle (v_i(t) - \langle v_i \rangle)(v_j(t + \Delta t) - \langle v_j \rangle) \rangle \tag{5.76}$$

and depends only on the absolute value of Δt. Due to the periodic boundary conditions the correlation function (5.76) can be defined within $\Delta t \in [-T/2, T/2]$.

The spectral density then is given by the Fourier transformation

$$S_{jl}(\omega) = \lim_{T\to\infty} \int_{-T/2}^{T/2} \varphi_{jl}(\Delta t) e^{-i\omega \Delta t}\, d(\Delta t). \tag{5.77}$$

The inverse transformation reads

$$\varphi_{jl}(\Delta t) = \lim_{T\to\infty} \frac{1}{T} \sum_{\omega} S_{jl}(\omega) e^{i\omega \Delta t} = \int_{-\infty}^{\infty} \frac{d\omega}{2\pi} S_{jl}(\omega) e^{i\omega \Delta t}. \tag{5.78}$$

The spectral density (5.77) can be expressed in terms of $\tilde{v}_i(\omega)$, where $\tilde{v}_i(\omega)$ are the components of the vector $\tilde{\mathbf{v}}(\omega)$. Since $S_{jl}(\omega)$ is independent of time, a formal averaging over time t in (5.77) does not change the result. Thus we can write

$$S_{jl}(\omega) = \lim_{T\to\infty} \frac{1}{T} \int_0^T \int_{-T/2}^{T/2} \varphi_{jl}(\Delta t) e^{-i\omega \Delta t} \, dt \, d(\Delta t). \tag{5.79}$$

Inserting (5.76) and (5.74), we obtain

$$S_{jl}(\omega) = \tag{5.80}$$

$$\lim_{T\to\infty} \frac{1}{T^3} \sum_{\omega_1 \neq 0} \sum_{\omega_2 \neq 0} \langle \tilde{v}_j(\omega_1) \tilde{v}_l(\omega_2) \rangle \int_0^T \int_{-T/2}^{T/2} e^{i(\omega_1+\omega_2)t} e^{i(\omega_2-\omega)\Delta t} \, dt \, d(\Delta t).$$

Taking into account the orthogonality property (5.73), only the terms with $\omega_2 = \omega$ and $\omega_1 = -\omega$ remain after the integration. Thus, for $\omega \neq 0$, we have

$$S_{jl}(\omega) = \lim_{T\to\infty} \frac{1}{T} \langle \tilde{v}_j(-\omega) \tilde{v}_l(\omega) \rangle. \tag{5.81}$$

Since the velocity **v** is real, we have $\tilde{v}_j(-\omega) = \tilde{v}_j^*(\omega)$ according to (5.71), and therefore

$$S_{jj}(\omega) = \lim_{T\to\infty} \frac{1}{T} \langle \tilde{v}_j^*(\omega) \tilde{v}_j(\omega) \rangle = \lim_{T\to\infty} \frac{1}{T} \langle |\tilde{v}_j(\omega)|^2 \rangle. \tag{5.82}$$

The relation between the spectral density given by (5.82) and (5.77) (at $j = l$) is known as the Wiener–Khinchin theorem.

As a simple example, where the spectral density can be easily calculated, we consider an exponentially decaying correlation function

$$\varphi(\Delta t) = A e^{-|\Delta t|/\tau}. \tag{5.83}$$

In this case the correlation function is a scalar quantity corresponding to a one-dimensional motion, therefore the indices are omitted. The spectral density calculated from (5.77) is

$$S(\omega) = \int_{-\infty}^{\infty} \varphi(\Delta t) e^{-i\omega \Delta t} \, d(\Delta t) = \int_{-\infty}^{\infty} A e^{-|\Delta t|/\tau} e^{-i\omega \Delta t} \, d(\Delta t)$$

$$= A \left[\int_{-\infty}^{0} e^{(1-i\omega\tau)\Delta t/\tau} \, d(\Delta t) + \int_0^{\infty} e^{-(1+i\omega\tau)\Delta t/\tau} \, d(\Delta t) \right]$$

$$= A \left[\frac{\tau}{1 - i\omega\tau} + \frac{\tau}{1 + i\omega\tau} \right] = \frac{2A\tau}{1 + (\omega\tau)^2}. \tag{5.84}$$

In the limit $\tau \to 0$ and $A \to \infty$ at a constant $A\tau$, the correlation function (5.83) is proportional to the delta function and corresponds to the white noise. According to (5.84), the Fourier spectrum of the white noise, obtained in this limit, is independent of the frequency ω. In other words, like the white light, it contains the whole uniform spectrum of frequencies. At a finite value of the parameter τ, which can be interpreted as a correlation time, the Fourier spectrum has a smooth cut-off at $\omega \approx 1/\tau$. This spectrum corresponds to colored noise. For $1/f$ noise in vehicular traffic, see [251].

5.5
Brownian Motion in Three-Dimensional Velocity Space

Here we will show how the techniques of correlation functions and Fourier analysis, introduced in Section 5.4, are applied to a specific problem of Brownian motion. Consider first the deterministic motion of a Brownian particle with initial velocity $\mathbf{v}(t=0) = \mathbf{v}_0$ in a medium (liquid) with friction. Here the velocity is a three-dimensional vector. Its time evolution is described by the equation

$$\frac{d\mathbf{v}(t)}{dt} = -\gamma \mathbf{v}(t), \tag{5.85}$$

where γ is the friction coefficient. The solution reads

$$\mathbf{v}(t) = \mathbf{v}_0 e^{-\gamma t}. \tag{5.86}$$

Thus, in this simple model, the particle reduces its velocity asymptotically to zero due to friction. This equation, however, does not completely describe the motion of a particle in a liquid. One needs to take into account the randomness caused by stochastic collisions with liquid molecules, which never allow the velocity to relax to zero. This effect is described by the Langevin equation

$$\frac{d\mathbf{v}(t)}{dt} = -\gamma \mathbf{v}(t) + \sqrt{2B}\,\xi(t), \tag{5.87}$$

where (5.85) is completed by a stochastic (Langevin) force $\sqrt{2B}\,\xi(t)$. Here B is the diffusion coefficient in the velocity space and $\xi(t)$ is a three-dimensional vector with components $\xi_i(t)$, representing a stochastic process. The actual Brownian motion in the space of velocity $\mathbf{v}$ and coordinate $\mathbf{x}$ is known as the Ornstein–Uhlenbeck process. For the case where the velocity and the coordinate are scalar (one-dimensional) quantities this process will be discussed, based on the Fokker–Planck equation, in Chapter 8.

The stochastic force should have the following properties.

1. Each component of the stochastic force has zero mean value

$$\langle \xi_i(t) \rangle_{\mathbf{v}_0} = 0, \tag{5.88}$$

where the symbol $\mathbf{v}_0$ indicates that only those stochastic realizations are considered for which $\mathbf{v}(t=0) = \mathbf{v}_0$. This means that the stochastic force has no influence on the averaged motion.

2. The Langevin force is the Gaussian stochastic process, which means that all higher order correlation functions reduce to the two-time correlation function $\langle \xi_i(t_1)\xi_j(t_2) \rangle_{\mathbf{v}_0}$ according to

$$\langle \xi(t_1)\xi(t_2)\cdots\xi(t_{2n}) \rangle_{\mathbf{v}_0} = \sum_{\text{all pairings}} \langle \xi(t_i)\xi(t_j) \rangle_{\mathbf{v}_0} \cdots \langle \xi(t_k)\xi(t_l) \rangle_{\mathbf{v}_0}. \tag{5.89}$$

Like the first moment (5.88), all odd-order moments are zero.

3. The $\langle \xi_i(t)\xi_j(t')\rangle_{v_0}$ function is δ-correlated in time

$$\langle \xi_i(t)\xi_j(t')\rangle_{v_0} = \delta_{ij}\delta(t-t'). \qquad (5.90)$$

Also, this formula implies that different components are uncorrelated or statistically independent.

4. The stochastic process for the velocity $\mathbf{v}(t)$ of the Brownian particle is statistically independent of the stochastic force $\sqrt{2B}\,\xi(t')$ for $t' > t$, so that $\mathbf{v}(t)$ at a given time is independent of the stochastic force in the future:

$$\langle \mathbf{v}(t)\xi(t')\rangle_{v_0} = 0 \quad \text{for} \quad t' > t. \qquad (5.91)$$

The velocity $\mathbf{v}(t)$, naturally, will be affected by $\xi(t')$ at $t' < t$.

In the following, we consider two different ways to get the solution of the Langevin equation (5.87): by direct integration and harmonic analysis. Direct integration yields a formal solution for each specific realization of the stochastic process $\xi(t)$,

$$\mathbf{v}(t) = \mathbf{v}_0 e^{-\gamma t} + e^{-\gamma t}\int_0^t e^{\gamma t'}\sqrt{2B}\,\xi(t')\,\mathrm{d}t', \qquad (5.92)$$

as can be verified by inserting (5.92) into (5.87). This solution allows us to calculate moments of the velocity distribution for the ensemble of all stochastic realizations with given initial velocity $\mathbf{v}_0$. The first moment is

$$\langle \mathbf{v}(t)\rangle_{v_0} = \mathbf{v}_0 e^{-\gamma t} + e^{-\gamma t}\int_0^t e^{\gamma t'}\sqrt{2B}\,\langle \xi(t')\rangle_{v_0}\,\mathrm{d}t'. \qquad (5.93)$$

The last term vanishes, since the Langevin force has zero mean value, as discussed above. Thus we have

$$\langle \mathbf{v}(t)\rangle_{v_0} = \mathbf{v}_0 e^{-\gamma t}. \qquad (5.94)$$

The correlation function $\langle \mathbf{v}(t)\mathbf{v}(t')\rangle_{v_0}$ for velocities at different times can also be calculated in this way. Alternatively, the correlation function can be defined for deviations from the mean values as $\langle (\mathbf{v}(t) - \langle \mathbf{v}(t)\rangle)(\mathbf{v}(t') - \langle \mathbf{v}(t')\rangle)\rangle_{v_0}$ by analogy with (5.76). Both definitions are equivalent for long times, where the mean velocity $\langle \mathbf{v}(t)\rangle_{v_0}$ tends to zero. For definiteness we assume that $t' > t$ holds. Then for any velocity component we have

$$\begin{aligned}\langle v_i(t)v_i(t')\rangle_{v_0} &= v_{i,0}^2\, e^{-\gamma(t'+t)} + 2Be^{-\gamma(t'+t)}\int_0^t\int_0^{t'} e^{\gamma(s'+s)}\langle \xi_i(s)\xi_i(s')\rangle\,\mathrm{d}s\,\mathrm{d}s' \\ &= v_{i,0}^2\, e^{-\gamma(t'+t)} + 2Be^{-\gamma(t'+t)}\int_0^t e^{\gamma(s+s)}\,\mathrm{d}s \\ &= v_{i,0}^2\, e^{-\gamma(t'+t)} + \frac{B}{\gamma}\left(e^{-\gamma(t'-t)} - e^{-\gamma(t'+t)}\right).\end{aligned} \qquad (5.95)$$

By using the definition of scalar product, the correlation function $\langle \mathbf{v}(t)\mathbf{v}(t')\rangle_{\mathbf{v}_0}$ is easily calculated from (5.95) as

$$\langle \mathbf{v}(t)\mathbf{v}(t')\rangle_{\mathbf{v}_0} = \sum_i \langle v_i(t)v_i(t')\rangle_{\mathbf{v}_0}. \tag{5.96}$$

The second moment for each velocity component is obtained from (5.95) by setting $t' = t$, so

$$\langle v_i^2(t)\rangle_{\mathbf{v}_0} = v_{i,0}^2 \, e^{-2\gamma t} + \frac{B}{\gamma}\left(1 - e^{-2\gamma t}\right). \tag{5.97}$$

As an illustrative example, the mean value for one of the velocity components $v_i = v_x$ and the variance $\langle v_x^2\rangle_{\mathbf{v}_0} - \langle v_x\rangle_{\mathbf{v}_0}^2$, calculated from (5.94) and (5.97), are shown in Figure 5.5. The theoretical curve for the mean value (dashed line) is compared with one specific realization of the process (fluctuating curve). The variance (solid line) shows the magnitude of the stochastic fluctuations.

Apart from the mean values, the probability density $p(v_x, v_y, v_z, t)$ in the three-dimensional velocity space is also of interest. Taking into account that the velocity components in (5.87) are not coupled, their probability distributions are independent, and we have

$$p(v_x, v_y, v_z, t) = p(v_x, t)\, p(v_y, t)\, p(v_z, t), \tag{5.98}$$

where $p(v_x, t)$, $p(v_y, t)$, and $p(v_z, t)$ are the probability densities for one component. The latter ones can be calculated by solving the corresponding Fokker–Planck equation for the one-dimensional problem, considered in detail in Chapter 8. Here we only report the result (cf. (8.34))

$$p(v_i, t) = \frac{1}{\sqrt{2\pi\sigma^2(t)}} \exp\left[-\frac{(v_i - v_{i,0}\exp[-\gamma t])^2}{2\sigma^2(t)}\right], \tag{5.99}$$

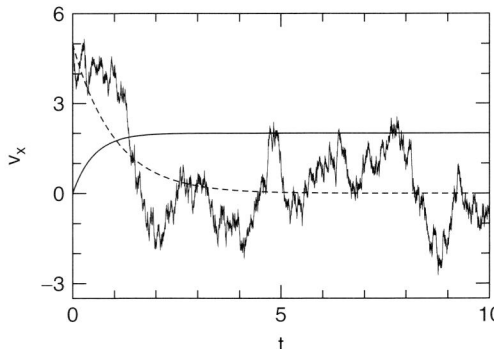

Figure 5.5 A stochastic trajectory (fluctuating curve) and the mean value (dashed curve) of the velocity component v_x, measured in ms^{-1}, depending on the time t measured in seconds. The initial condition is $v_{x,0} = 5$ ms^{-1} and the values of the parameters are $\gamma = 1$ s^{-1} and $B = 2$ m^{2}s^{-3}. The solid curve shows the temporal behavior of the variance $\langle v_x^2\rangle_{\mathbf{v}_0} - \langle v_x\rangle_{\mathbf{v}_0}^2$.

where $i = x, y, z$ denotes the ith component of vector $\mathbf{v}$ and

$$\sigma^2(t) = \langle v_i^2 \rangle - \langle v_i \rangle^2 = \frac{B}{\gamma}(1 - \exp[-2\gamma t]) \qquad (5.100)$$

is the variance consistent with (5.94) and (5.97).

For large times t the initial state (velocity $\mathbf{v}_0$) is forgotten and the final equilibrium state is given by

$$\lim_{t \to \infty} \langle v_i^2(t) \rangle_{\mathbf{v}_0} = B/\gamma. \qquad (5.101)$$

On the other hand, it is well known that

$$\langle v_i^2 \rangle = \frac{k_B T}{m} \qquad (5.102)$$

holds in the equilibrium of a classical system. Comparing (5.101) and (5.102) we arrive to the relation

$$\frac{B}{\gamma} = \frac{k_B T}{m} \qquad (5.103)$$

known as the Einstein formula. It relates the macroscopic quantity (friction coefficient) γ, which describes the dissipation of the momentum, to the microscopic quantity (diffusion coefficient) B, which describes the stochastic force.

The Langevin equation (5.87) can be also solved by the Fourier transformation method. According to (5.71), we have

$$\frac{d\mathbf{v}(t)}{dt} = \frac{1}{T} \sum_\omega \frac{d}{dt}\left(\tilde{\mathbf{v}}(\omega) e^{i\omega t}\right) = \frac{1}{T} \sum_\omega i\omega \tilde{\mathbf{v}}(\omega) e^{i\omega t}. \qquad (5.104)$$

This, in fact, is the Fourier expansion of $d\mathbf{v}(t)/dt$ with the coefficients $i\omega\tilde{\mathbf{v}}(\omega)$. The latter ones thus represent the Fourier transform of $d\mathbf{v}(t)/dt$. Hence, (5.87) in Fourier representation reads

$$(i\omega + \gamma)\tilde{\mathbf{v}}(\omega) = \sqrt{2B}\,\tilde{\xi}(\omega), \qquad (5.105)$$

where $\tilde{\xi}(\omega)$ is the Fourier transform of the noise $\xi(t)$. This is a simple algebraic equation yielding

$$\tilde{\mathbf{v}}(\omega) = \frac{\sqrt{2B}\,\tilde{\xi}(\omega)}{i\omega + \gamma}. \qquad (5.106)$$

By analogy with $S_{jl}(\omega)$ given by (5.77), we will now consider the spectral density for the velocity $S_{v_j v_l}(\omega) \equiv S_{jl}(\omega)$, as well as for the noise $S_{\xi_j \xi_l}(\omega)$. These spectral densities are the Fourier transforms of the corresponding correlation functions for the velocity (5.76) and for the noise (5.90). Following (5.82), the diagonal terms (in the stationary regime at $T \to \infty$) are given by $S_{v_j v_j}(\omega) = \langle |\tilde{v}_j(\omega)|^2 \rangle$ and $S_{\xi_j \xi_j}(\omega) = \langle |\tilde{\xi}_j(\omega)|^2 \rangle$. The nondiagonal case is consistent with (5.81), that is $S_{v_j v_l}(\omega) = \langle \tilde{v}_j(-\omega)\tilde{v}_l(\omega) \rangle$ and $S_{\xi_j \xi_l}(\omega) = \langle \tilde{\xi}_j(-\omega)\tilde{\xi}_l(\omega) \rangle$. The spectral density for the coordinate, defined as $S_{x_j x_l}(\omega) = \langle \tilde{x}_j(-\omega)\tilde{x}_l(\omega) \rangle$, can also be calculated. In the latter case, however, the corresponding correlation function $\langle(x_j(t) - \langle x_j\rangle)(x_l(t + \Delta t) - \langle x_l\rangle)\rangle$ diverges in the long-time limit at $j = l$. Hence, the spectral density of the noise term is

$$S_{\xi_j\xi_l}(\omega) = \langle\tilde{\xi}_j(-\omega)\tilde{\xi}_l(\omega)\rangle = \int_{-\infty}^{\infty} \langle\xi_j(t)\xi_l(t + \Delta t)\rangle e^{-i\omega\Delta t}\,d(\Delta t)$$

$$= \int_{-\infty}^{\infty} \delta_{jl}\,\delta(\Delta t)\,e^{-i\omega\Delta t}\,d(\Delta t) = \delta_{jl}. \tag{5.107}$$

The spectral density of the velocity can easily be calculated from (5.106) and (5.107),

$$S_{v_j v_l}(\omega) = \langle\tilde{v}_j(-\omega)\tilde{v}_l(\omega)\rangle = \frac{2B\langle\tilde{\xi}_j(-\omega)\tilde{\xi}_l(\omega)\rangle}{\omega^2 + \gamma^2} = \frac{2B\,\delta_{jl}}{\omega^2 + \gamma^2}. \tag{5.108}$$

The spectral density of the coordinate is calculated using the relation $\mathbf{x}(t)/dt = \mathbf{v}(t)$. According to

$$\mathbf{x}(t) = \frac{1}{T}\sum_{\omega}\tilde{\mathbf{x}}(\omega)e^{i\omega t}, \tag{5.109}$$

$$\frac{d\mathbf{x}(t)}{dt} = \frac{1}{T}\sum_{\omega} i\omega\tilde{\mathbf{x}}(\omega)e^{i\omega t}, \tag{5.110}$$

the Fourier transform of $\mathbf{x}(t)$ is $\tilde{\mathbf{x}}(\omega)$, whereas that of $d\mathbf{x}(t)/dt$ is $i\omega\tilde{\mathbf{x}}(\omega)$. The latter is equal to $\tilde{\mathbf{v}}(\omega)$, as consistent with $d\mathbf{x}(t)/dt = \mathbf{v}(t)$, that is,

$$\tilde{\mathbf{v}}(\omega) = i\omega\tilde{\mathbf{x}}(\omega). \tag{5.111}$$

Using this relation, we obtain

$$S_{x_j x_l}(\omega) = \langle\tilde{x}_j(-\omega)\tilde{x}_l(\omega)\rangle = \frac{1}{\omega^2}\langle\tilde{v}_j(-\omega)\tilde{v}_l(\omega)\rangle = \frac{1}{\omega^2}\frac{2B\,\delta_{jl}}{\omega^2 + \gamma^2}. \tag{5.112}$$

The second moment of a given velocity component can be calculated from (5.78). Taking into account the definition of the correlation function (5.76), its Fourier representation (5.78), and the fact that the mean velocity is zero at $t \to \infty$, (5.108) yields in the large-time limit:

$$\langle v_j^2\rangle = \int_{-\infty}^{\infty}\frac{d\omega}{2\pi}S_{v_j v_j}(\omega) = \frac{1}{\pi}\int_{-\infty}^{\infty}\frac{B\,d\omega}{\omega^2 + \gamma^2} = \frac{B}{\gamma}, \tag{5.113}$$

as well as

$$\lim_{t\to\infty}\langle v_j(t)v_j(t + \Delta t)\rangle = \int_{-\infty}^{\infty}\frac{d\omega}{2\pi}S_{v_j v_j}(\omega)e^{i\omega\Delta t} = \frac{1}{\pi}\int_{-\infty}^{\infty}\frac{B\,e^{i\omega\Delta t}\,d\omega}{\omega^2 + \gamma^2}$$

$$= \frac{B}{\gamma}e^{-\gamma|\Delta t|} = \langle v_j^2\rangle e^{-\gamma|\Delta t|}. \tag{5.114}$$

The result (5.113) is the same as (5.101), obtained earlier by the method of direct integration of the Langevin equation. The integral in (5.114) is calculated by making a suitable contour in the complex ω plane and using the residue theorem. For $\Delta t > 0$ we need only to calculate the residue at $\omega = i\gamma$, whereas for $\Delta t < 0$, the residue at $\omega = -i\gamma$. Equation (5.114) is consistent with the relation between the exponential correlation function and its spectral density given by (5.84). The correlation function (5.114) corresponds to the long-time limit $t \to \infty$ at a given positive $\Delta t = t' - t$ in the formula (5.95) obtained by the method of direct integration for arbitrary t.

The decay of the correlation function $\langle v_j(t) v_j(t + \Delta t) \rangle$ is illustrated in Figure 5.6, comparing analytical and simulation results. The latter ones are obtained by a time-averaging over one stochastic trajectory, which is equivalent to an ensemble-averaging.

The diffusion coefficient is related to the spectral density of the stochastic force $\sqrt{2B}\,\xi(t)$ via

$$\langle |\sqrt{2B}\,\tilde{\xi}(\omega)|^2 \rangle = 2B, \tag{5.115}$$

as follows from (5.107) at $j = l$. The Einstein formula (5.103) thus can be written as

$$\langle |\sqrt{2B}\,\tilde{\xi}(\omega)|^2 \rangle = 2\gamma k_B T/m. \tag{5.116}$$

This equation represents a relation between the friction coefficient and the spectral density of the noise source. Hence, it is a special form of the fluctuation–dissipation theorem.

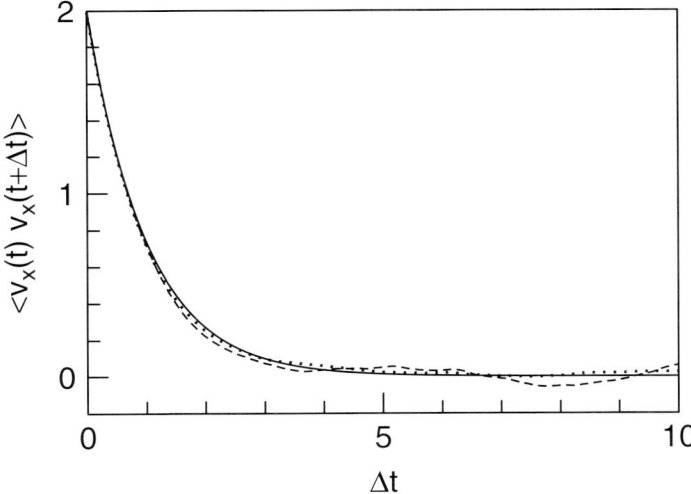

Figure 5.6 The correlation function $\langle v_x(t) v_x(t + \Delta t) \rangle$ (at $t \to \infty$) for the velocity component $v_i = v_x$, measured in ms^{-1}, depending on the time difference Δt given in seconds at $\gamma = 1\,\text{s}^{-1}$ and $B = 2\,\text{m}^2\text{s}^{-3}$. The solid curve shows the analytical result (5.114). The dashed and dotted curves represent estimates obtained by a time-averaging over $t \in [100\,\text{s}, 2000\,\text{s}]$ and $t \in [100\,\text{s}, 9000\,\text{s}]$, respectively, for one stochastic trajectory.

5.6
Stochastic Differential Equations

As a starting point we consider the one-dimensional stochastic differential equation for the variable x, which can be, e.g. the coordinate of a particle performing random walk in one dimension. The motion of the particle is described by the stochastic differential equation (SDE)

$$dx(t) = a(x)\,dt + b_\eta(x)\,dW(t), \tag{5.117}$$

where $a(x)$ and $b_\eta(x)$ are given functions of x, $dx(t) = x(t+dt) - x(t)$ is the increment of x in the time interval from t to $t+dt$, whereas $dW(t) = W(t+dt) - W(t)$ is the increment of the standard Wiener process having the properties $\langle dW(t) \rangle = 0$ and $\langle (dW(t))^2 \rangle = dt$. Later on, the Wiener process will be discussed in detail. This equation (5.117), written in the form (5.41)

$$\frac{dx}{dt} = a(x) + b_\eta(x)\,\xi(t), \tag{5.118}$$

is known as Langevin equation [61, 234]. The Langevin force, formally $\xi(t) = dW(t)/dt$, has to be understood as a fluctuating quantity having the Gaussian distribution

$$p(\xi(t)) = \sqrt{\frac{dt}{2\pi}} \exp\left[-\frac{dt}{2}\xi^2\right] \tag{5.119}$$

with $\langle \xi(t) \rangle = 0$ and $\langle \xi(t)\xi(t') \rangle = \delta(t-t')$. According to the formal substitution $\xi(t) = dW(t)/dt$ we should have the variance which diverges as $\langle \xi(t)^2 \rangle = 1/dt \to \infty$ at $dt \to 0$. The above incorrect substitution $dW(t) = \xi(t)\,dt$, however, represents only a formal way of writing and has no rigorous mathematical meaning, since stochastic trajectories are not differentiable.

An important peculiarity of the stochastic differential equations (5.117) and of the Langevin equation (5.118) is that their solution essentially depends on how the coefficient $b_\eta(x)$ at the noise term is defined. Namely, it is important whether this coefficient is determined at $x = x(t)$, $x = x(t+dt)$, or at x in some intermediate time moment. The parameter η is introduced to distinguish between these cases. Different possibilities can be chosen according to

$$b_\eta(x) = b(x(t+\eta\,dt)). \tag{5.120}$$

The case $\eta = 0$, when the coefficient b is determined at the left border of the integration interval $[t, t+dt]$, is called the Ito stochastic process. In the case of the Stratonovich process, where $\eta = 1/2$, it is determined in the middle of the interval. Finally, if $b_\eta(x)$ is determined at the right border $t+dt$, which corresponds to $\eta = 1$, then we are dealing with the Hänggi–Klimontovich process.

Alternatively, one can define the coefficient $b_{\eta'}(x)$ as

$$b_{\eta'}(x) = b((1-\eta')x(t) + \eta'\,x(t+dt)). \tag{5.121}$$

The two definitions (5.120) and (5.121) are identical at $\eta = \eta' = 0$ and $\eta = \eta' = 1$. For an arbitrary stochastic trajectory $x(t)$, however, the relationship between η and η' is different if $0 < \eta < 1$.

The solution of the stochastic differential equation of Ito type (Ito-SDE)

$$dx(t) = a[x(t)] \, dt + b[x(t)] \, dW(t) \tag{5.122}$$

is represented by the Ito stochastic integral

$$x(t) = x(t_0) + \int_{t_0}^{t} a[x(t')] \, dt' + \int_{t_0}^{t} b[x(t')] \, dW(t'). \tag{5.123}$$

Equation (5.122) thus has a unique solution (5.123) which is a Markov process.

The probability density $p(x, t)$ of finding the particle at a position x at time t is given by the Fokker–Planck equation with general η

$$\frac{\partial}{\partial t} p = \frac{\partial}{\partial x} \left\{ -a(x)p + \frac{1}{2} b(x)^{2\eta} \frac{\partial}{\partial x} \left[b(x)^{2(1-\eta)} p \right] \right\}. \tag{5.124}$$

The stationary solution of (5.124) reads

$$p_{\text{st}}(x) = \frac{C}{b(x)^{2(1-\eta)}} \exp \left[2 \int^{x} dy \frac{a(y)}{b(y)^2} \right] \tag{5.125}$$

with integration constant C given by the normalization condition $\int p_{\text{st}}(x) \, dx = 1$.

In the case of Ito stochastic calculus (integration at left border), the stochastic differential equation (5.122) in typical notations is written as

$$dx(t) = a(x) \, dt + b(x) \, dW(t), \tag{5.126}$$

and the corresponding Fokker–Planck equation reads

$$\frac{\partial p}{\partial t} = \frac{\partial}{\partial x} \left\{ -a(x)p + \frac{1}{2} \frac{\partial}{\partial x} \left[b(x)^2 p \right] \right\} \tag{5.127}$$

$$= -\frac{\partial}{\partial x} \left[a(x)p \right] + \frac{1}{2} \frac{\partial^2}{\partial x^2} \left[b(x)^2 p \right]. \tag{5.128}$$

To distinguish it from the Ito-SDE, the Stratonovich-SDE (integration in the middle) in these notations is written using the special symbol $\circ$

$$dx(t) = a(x) \, dt + b(x) \circ dW(t). \tag{5.129}$$

The corresponding Fokker–Planck equation is

$$\frac{\partial p}{\partial t} = \frac{\partial}{\partial x} \left\{ -a(x)p + \frac{1}{2} b(x) \frac{\partial}{\partial x} \left[b(x) p \right] \right\}. \tag{5.130}$$

One has to take into account that deviations from the usual differentiation rules take place at $\eta \neq 1/2$. It is important when making a transformation of variable $y = g(x) \iff x = g^{-1}(y)$. The transformed Langevin equation then reads

$$\frac{dy}{dt} = \tilde{a}(y) + \tilde{b}_\eta(y) \xi(t) \tag{5.131}$$

or

$$dy = \tilde{a}(y)\,dt + \tilde{b}_\eta(y)\,dW(t) \tag{5.132}$$

with coefficients

$$\tilde{a}(y) = g'(x)\,a(x) + \left(\tfrac{1}{2} - \eta\right) g''(x)\,b(x)^2, \tag{5.133}$$

$$\tilde{b}(y) = g'(x)\,b(x), \tag{5.134}$$

where $g' = dg/dx$.

In the follwing we want to consider some special cases in detail starting with a driftless process.

5.7
The Standard Wiener Process

The stochastic equation of motion driven by Gaussian white noise (compare Example 1.2 in Section 1.5) reads

$$dx(t) = dW(t). \tag{5.135}$$

As pointed out in Chapter 1, (5.135) cannot be written as $dx(t) = \xi(t)\,dt$, since $dW(t)/dt \neq \xi(t)$, as already explained in connection with (5.117)–(5.119).

The solution of (5.135) is given by the initial condition $x(t=0) = x_0$ for the state variable x, and the properties of the standard Wiener process $W(t)$ are stated by

1. $W(t=0) \equiv W_0 = 0$ almost surely;
2. $W(t)$ is a process with independent increments
   $dW(t) = W(t + dt) - W(t)$;
3. $W(t) - W(s)$ is normally distributed $N(0, t-s)$
   with mean 0 and variance $t - s$ ($0 \leq s < t$).

Integrating (5.135) we obtain

$$x(t) = x_0 + W(t). \tag{5.136}$$

The moments of variable $x(t)$ are

$$\begin{aligned} \langle x(t) \rangle &= x_0 & \text{since} & & \langle W(t) \rangle &= 0 \\ \langle x(t)^2 \rangle &= x_0^2 + t & \text{since} & & \langle W(t)^2 \rangle &= t. \end{aligned} \tag{5.137}$$

The variance is linearly increasing in time

$$\langle x(t)^2 \rangle - \langle x(t) \rangle^2 = t, \tag{5.138}$$

and the covariance or correlation function for different times (as extension of the variance) is given by

$$\langle x(t)x(t') \rangle - \langle x(t) \rangle \langle x(t') \rangle = \min\{t, t'\}. \tag{5.139}$$

5.7 The Standard Wiener Process

The Wiener process $W(t)$, which is the solution of (5.135) with $x_0 = 0$, is a Gaussian process normally distributed as $N(0, t)$, thus

$$P(a \leq W(t) \leq b) = \frac{1}{\sqrt{2\pi t}} \int_a^b e^{-\frac{x^2}{2t}} \, dx. \tag{5.140}$$

According to the definition of the Wiener process, its increments are independent. This means that each following step is independent of the previous history. In other words, every Wiener process is a Markov process. In a precise mathematical notation: almost every trajectory of the Wiener process is differentiable almost nowhere [25, 61, 201].

Because $(W(t+h) - W(t))/\sqrt{h}$ is normally distributed as $N(0, 1)$, it follows

$$P(|W(t+h) - W(t)| < a) = P\left(\frac{|W(t+h) - W(t)|}{\sqrt{h}} < \frac{a}{\sqrt{h}}\right)$$

$$= \frac{1}{\sqrt{2\pi}} \int_{-a/\sqrt{h}}^{a/\sqrt{h}} e^{-x^2/2} \, dx \tag{5.141}$$

$$\leq \frac{1}{\sqrt{2\pi}} 2 \frac{a}{\sqrt{h}} = \frac{2a}{\sqrt{2\pi h}}. \tag{5.142}$$

The so-called strong law of large numbers is stated as

$$\frac{W(t)}{t} \to 0 \quad \text{for} \quad t \to \infty. \tag{5.143}$$

The Wiener process has a scaling property which states that if $W(t)$ is a Wiener process, then the time-scaled process $\widetilde{W}(t)$ defined by

$$\widetilde{W}(t) = t \, W(1/t) \quad \text{with} \quad \widetilde{W}(0) = 0 \tag{5.144}$$

is also a Wiener process.

Often the Wiener process, with finite stopping time T, is discussed. The most fundamental one is the first passage time $T(b)$ which is defined by

$$T(b) = \min\{t \geq 0; x(t) = b\} \tag{5.145}$$

as minimal time when the trajectory hits for the first time T the value b starting from x_0.

We shall first obtain the probability density function of $T(b)$ by an heuristic argument, called the reflection principle, using a shadow path [91].

Let us denote by $P[T(b) < t]$ the probability that for a given time moment t the relation $T(b) < t$ holds. It is the probability that the value b has been reached within the time interval $[0, t]$. Let $P[T(b) < t, x(t) > b]$ be the probability that two conditions $T(b) < t$ and $x(t) > b$ are satisfied simultaneously at time moment t. Further on, we have

$$P[x(t) > b] = P[T(b) < t, x(t) > b] + P[T(b) > t, x(t) > b]$$

$$= P[T(b) < t, x(t) > b] \tag{5.146}$$

for the probability $P[x(t) > b]$ that $x(t)$ exceeds b at time moment t, since $P[T(b) > t, x(t) > b] = 0$ holds according to the definition of $T(b)$. The reflection principle tells us that, after reaching the border $x = b$, each trajectory of Wiener process has a shadow path which is a reflection with respect to the line $x(t) = b$. Due to the symmetry of random walk, the probability for a given set of stochastic trajectories passing $x = b$ is the same as for the corresponding set of shadow trajectories. This shadow symmetry is reflected in the relation

$$P[T(b) < t, x(t) > b] = P[T(b) < t, x(t) < b] \tag{5.147}$$

According to (5.146) and (5.147) we have

$$P[T(b) < t, x(t) > b] = P[x(t) > b] \tag{5.148}$$
$$P[T(b) < t, x(t) < b] = P[x(t) > b] \tag{5.149}$$

Summing these two equations, we obtain

$$P[T(b) < t] = 2P[x(t) > b]. \tag{5.150}$$

Using the known result (5.140), leads to the relation

$$P[T(b) < t] = \sqrt{\frac{2}{\pi}} \int_{bt^{-1/2}}^{\infty} e^{-z^2/2} \, dz \tag{5.151}$$

for the cumulative probability $P[T(b) < t]$ that a trajectory has reached the border $x = b$ before the time moment t when starting at $x_0 = 0$. For the general value of $x_0 < b$ one has to replace b with $b - x_0$.

The probability density $\mathcal{P}(T; b)$ of the first passage time of the Wiener process starting at $x_0 = 0$ (see Figure 5.7) is obtained from (5.151) via differentiation with respect to time t,

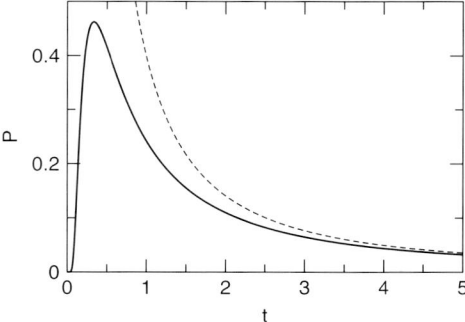

Figure 5.7 The first-passage time probability density $\mathcal{P}(t; b)$ for a standard Wiener process starting at $x_0 = 0$ with absorbing boundary $b = 1$ (solid line). For comparison, the dashed curve shows the decreasing prefactor ($\sim t^{-3/2}$) of the exponential function.

$$\mathcal{P}(T; b) = \frac{b}{\sqrt{2\pi T^3}} \exp\left\{-\frac{b^2}{2T}\right\}. \tag{5.152}$$

The mean first passage time $\langle T \rangle$ in this case is infinite, that is, the integral $\int_0^\infty \mathcal{P}(T; b)\, dT$ diverges.

There are special cases under consideration which are often discussed in literature, see e.g. [25, 61, 91, 201]:

1. The Wiener process on a half–line $[0, \infty]$. One has to specify what happens when the origin is reached.
   (a) Absorbing boundary at $x = 0$.
   (b) Instantaneous reflection at $x = 0$.
2. The Wiener process on a finite interval $[a, b]$.
   (a) Reflection at both endpoints (doubly reflected Wiener process).
   (b) Absorption at both endpoints.
   (c) Mixed boundaries.

In the following we shall briefly discuss the numerical realization of stochastic trajectories. The analytical properties of the Wiener process help us to calculate the increment $\Delta W(t)$ numerically from standard normally distributed random numbers $Z \sim N(0, 1)$ via

$$\Delta W(t) = Z \sqrt{\Delta t}. \tag{5.153}$$

We would like to mention three different algorithms used to obtain normally distributed random numbers Z. All of them produce the transformation to generate $Z \sim N(0, 1)$ from uniformly distributed random numbers $U_i \in (0, 1)$.

The transformation by the *Polar method* has three steps:

1. Generate two uniformly distributed random numbers U_1 and U_2.
2. Define $V_i = 2U_i - 1$.
3. Check the condition that $W = V_1^2 + V_2^2 < 1$.
4. If 'yes' create $Z = V_1 \sqrt{-2 \log(W)/W}$.
5. If 'no' generate new random numbers (go to 1) and check this condition again.

For the same realization the *Box–Muller method* proposes the necessary transformation after the generation of $U_1, U_2 \in (0, 1)$ by one step only

$$Z = \sqrt{-2 \ln U_1} \cos(2\pi U_2). \tag{5.154}$$

The difference between both described methods is insignificant and connected only with the simulation time.

The third strategy used to calculate variable Z is based on theoretical explanations. It is well-known that the uniform distribution in the interval $[0, 1]$

$$p_{\text{uniform}}(x) = \left\{ \begin{array}{lll} 0 & : & x < 0 \\ 1 & : & 0 < x < 1 \\ 0 & : & x > 1 \end{array} \right\} \tag{5.155}$$

has the mean value

$$\mu = \int_{-\infty}^{+\infty} x\, p_{\text{uniform}}(x)\, dx = \int_0^1 x\, dx = \frac{1}{2} \tag{5.156}$$

and the variance

$$\sigma^2 = \int_{-\infty}^{+\infty} (x-\mu)^2\, p_{\text{uniform}}(x)\, dx = \int_0^1 \left(x - \frac{1}{2}\right)^2 dx = \frac{1}{12}. \tag{5.157}$$

Therefore, the new variable

$$Z = \sum_{i=1}^{12} \left(U_i - \frac{1}{2}\right) \tag{5.158}$$

has to be approximately the normal distributed one with parameters

$$\langle Z \rangle = 0 \quad \text{and} \quad \langle Z^2 \rangle - \langle Z \rangle^2 = 1. \tag{5.159}$$

This method does not require the calculation of any function such as logarithm, square root, or cosine. However, it is necessary to generate twelve uniformly distributed random numbers to calculate Z for one time step $[t, t+\Delta t]$. The Box–Muller and Polar methods need for it only two random numbers.

As an example, ten numerically simulated stochastic trajectories of the Wiener process starting at $x_0 = 2$ are shown in Figure 5.8. Here we show also the mean value $\langle x(t) \rangle$ and the variance $\langle x(t)^2 \rangle - \langle x(t) \rangle^2$, approximately evaluated from the ensemble of ten trajectories, and compare them with the theoretical results (5.137) and (5.138).

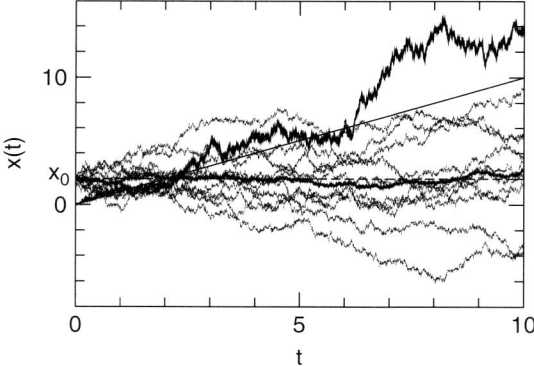

Figure 5.8 Sample of ten stochastic trajectories starting at $x_0 = 2$ (dotted zig-zag lines), the mean value, and the variance of the Wiener process depending on time t. The theoretical mean value is depicted by the dashed line, and the variance by the solid straight line. The corresponding numerical estimates are shown as bold curves.

5.8
Arithmetic Brownian Motion

The standard Brownian motion is defined as the constant drift function together with the white noise already known from the Wiener process

$$dx(t) = a\, dt + b\, dW(t) \tag{5.160}$$

together with following initial conditions $x(t=0) = x_0$ and $W(t=0) = 0$.

Simple integration gives the solution

$$x(t) = x_0 + at + b\, W(t). \tag{5.161}$$

The probability distribution function thus is Gaussian, that is:

$$p(x,t) = \frac{1}{\sqrt{2\pi b^2 t}} \exp\left(-\frac{(x - x_0 - at)^2}{2b^2 t}\right). \tag{5.162}$$

It is a solution of the Fokker–Planck equation (5.124) with $\eta = 0$

$$\frac{\partial}{\partial t} p(x,t) = -a \frac{\partial}{\partial x} p(x,t) + \frac{b^2}{2} \frac{\partial^2}{\partial x^2} p(x,t). \tag{5.163}$$

This equation will be discussed in some detail in Chapters 6 and 7.
From (5.162) we directly obtain the first two moments

$$\langle x(t) \rangle = x_0 + at \tag{5.164}$$

$$\langle x(t)^2 \rangle = \langle x(t) \rangle^2 + b^2 t. \tag{5.165}$$

An example of numerical simulation including ten stochastic trajectories with evaluation of the mean and variance is illustrated in Figure 5.9.

5.9
Geometric Brownian Motion

The stochastic differential equation with linear drift and a multiplicative noise term is called geometric Brownian motion. It has wide applicability in financial modeling and is given in Ito notation by

$$dx(t) = a\,x(t)\, dt + b\,x(t)\, dW(t) \tag{5.166}$$

with typical initial conditions $x(t=0) = x_0 > 0$ and $W(t=0) = 0$. It is a special case of the Ito-SDE (5.126) with $a(x) = ax$ and $b(x) = bx$.
In the following we will use the transformation

$$y(x) = \ln\left(\frac{x}{x_0}\right), \tag{5.167}$$

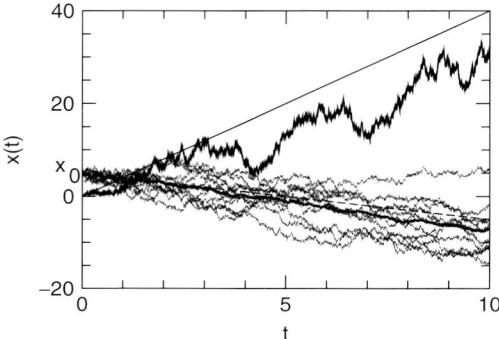

Figure 5.9 Sample of ten stochastic trajectories starting at $x_0 = 5$ (dotted random zig-zag lines), the mean value, and the variance of the arithmetic Brownian motion with parameters $a = 1$ and $b = 2$ depending on time t. The theoretical mean value is depicted by a dashed line, and the variance by the solid straight line. The corresponding numerical estimates are shown as solid curves.

where $x_0 = x(t = 0)$. According to (5.132) we have

$$dy(t) = \tilde{a}(y)\, dt + \tilde{b}(y)\, dW \tag{5.168}$$

with

$$\tilde{a}(y) = \frac{dy}{dx} a(x) + \frac{1}{2} \frac{d^2 y}{dx^2} b(x)^2 \tag{5.169}$$

$$\tilde{b}(y) = \frac{dy}{dx} b(x). \tag{5.170}$$

Here

$$y' \equiv \frac{dy}{dx} = \frac{x_0}{x} \frac{1}{x_0} = \frac{1}{x}; \quad y'' = -\frac{1}{x^2},$$

and hence we have $\tilde{a}(y) = a - b^2/2$ and $\tilde{b}(y) = b$. Consequently, the stochastic differential equation for the transformed variable $y(t)$ reads

$$dy(t) = d \ln x(t) = \left(a - \frac{b^2}{2}\right) dt + b\, dW(t), \tag{5.171}$$

Integrating both sides the solution results in

$$x(t) = x_0 \exp\left\{\left(a - \frac{b^2}{2}\right) t + b\, W(t)\right\} \tag{5.172}$$

The probability density for the variable $y(t)$ is given by the Gaussian distribution

$$\tilde{p}(y, t) = \frac{1}{\sqrt{2\pi b^2 t}} \exp\left[-\frac{(y - (a - b^2/2)t)^2}{2b^2 t}\right]. \tag{5.173}$$

The probability distribution $p(x,t)$ for the original variable is easily calculated according to

$$p(x,t) = \tilde{p}(y,t)\frac{dy}{dx} = \frac{1}{x}\tilde{p}(\ln[x/x_0],t). \tag{5.174}$$

It yields the log-normal distribution

$$p(x,t) = \frac{1}{\sqrt{2\pi b^2 t}}\frac{1}{x}\exp\left[-\frac{(\ln[x/x_0]-(a-b^2/2)t)^2}{2b^2 t}\right]. \tag{5.175}$$

This distribution is a solution of the following Fokker–Planck equation

$$\frac{\partial}{\partial t}p(x,t) = -\frac{\partial}{\partial x}[ax\,p(x,t)] + \frac{1}{2}\frac{\partial^2}{\partial x^2}\left[b^2 x^2\,p(x,t)\right]$$

$$= (b^2-a)\,p + (2b^2-a)\,x\frac{\partial p}{\partial x} + \frac{1}{2}(bx)^2\frac{\partial^2 p}{\partial x^2}. \tag{5.176}$$

The mean value (first moment) of (5.175) is calculated as follows

$$\langle x(t) \rangle = \int_0^\infty x p(x,t)\,dx = \int_0^\infty x\tilde{p}(y,t)\frac{dy}{dx}\,dx = x_0\int_{-\infty}^\infty e^y \tilde{p}(y,t)\,dy$$

$$= \frac{x_0}{\sqrt{2\pi b^2 t}}\int_{-\infty}^\infty \exp\left(y - \frac{(y-[a-b^2/2]t)^2}{2b^2 t}\right)dy$$

$$= \frac{x_0 e^{at}}{\sqrt{2\pi b^2 t}}\int_{-\infty}^\infty \exp\left(-\frac{(y-[a+b^2/2]t)^2}{2b^2 t}\right)dy$$

$$= x_0 e^{at}\frac{1}{\sqrt{2\pi b^2 t}}\int_{-\infty}^\infty \exp\left(-\frac{z^2}{2b^2 t}\right)dz = x_0 e^{at}. \tag{5.177}$$

It increases exponentially

$$\langle x(t) \rangle = x_0\, e^{at}. \tag{5.178}$$

The mean square value (second moment) is calculated in a similar way

$$\langle x(t)^2 \rangle = \int_0^\infty x^2 p(x,t)\,dx = \int_0^\infty x^2 \tilde{p}(y,t)\frac{dy}{dx}\,dx = x_0^2 \int_{-\infty}^\infty e^{2y}\tilde{p}(y,t)\,dy$$

$$= \frac{x_0^2}{\sqrt{2\pi b^2 t}}\int_{-\infty}^\infty \exp\left(2y - \frac{(y-[a-b^2/2]t)^2}{2b^2 t}\right)dy$$

$$= \frac{x_0^2 e^{(2a+b^2)t}}{\sqrt{2\pi b^2 t}}\int_{-\infty}^\infty \exp\left(-\frac{(y-[a+3b^2/2]t)^2}{2b^2 t}\right)dy$$

$$= \frac{x_0^2 e^{(2a+b^2)t}}{\sqrt{2\pi b^2 t}}\int_{-\infty}^\infty \exp\left(-\frac{z^2}{2b^2 t}\right)dz = x_0^2 e^{(2a+b^2)t}. \tag{5.179}$$

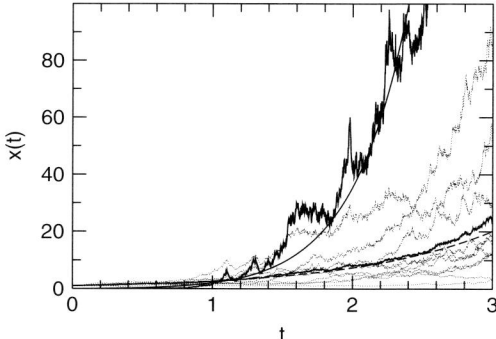

Figure 5.10 Sample of ten stochastic trajectories starting at $x_0 = 1$ (dotted random zig-zag lines), the mean value, and the variance of the geometric Brownian motion with parameters $a = 1$ and $b = 0.5$ depending on time t. The theoretical mean value is depicted by a dashed line, and the variance by the solid straight line. The corresponding numerical estimates are shown as solid curves.

Thus we have

$$\langle x(t)^2 \rangle = x_0^2 \, e^{2(a+b^2/2)\,t} \tag{5.180}$$

giving the variance

$$\langle x(t)^2 \rangle - \langle x(t) \rangle^2 = x_0^2 \, e^{2at} \left[e^{b^2 t} - 1 \right]. \tag{5.181}$$

Similarly, as in the case of the arithmetic Brownian motion, we have shown a numerical realization of ten stochastic trajectories with the evaluation of the mean and variance Figure 5.10. In this case, at $a = 1 > 0$, typical stochastic trajectories show exponential growth in time. This agrees with the formulas (5.178) and (5.181).

5.10
Exercises

E 5.1 *Brownian particle*
Verify that function **v**(t), given by (5.92), is the solution of the Langevin equation (5.87) for the velocity of a Brownian particle; first, by performing the integration in (5.87); and second, by inserting (5.92) into (5.87).

E 5.2 *Noisy harmonic oscillator*
Consider an oscillator with random frequency modulation. The position x and velocity v of the oscillator is described by the complex variable $z = x + iv$. This obeys the Langevin equation of motion [160]

$$\frac{dz}{dt} = i(\omega + \varepsilon \xi(t))z, \qquad (5.182)$$

where $\xi(t)$ is the Gaussian white noise with amplitude given by the parameter ε. Find the analytical solution in the absence of noise at $\varepsilon = 0$. Analyze the behavior of the oscillator depending on the noise amplitude and compare it with the solution without noise.

E 5.3 *Geometric Brownian motion*

Study the geometric Brownian motion (5.166) as the special case of an Ito process in more detail. Fix the initial condition $x(t = 0) = x_0 > 0$, but change the control parameters $a \geq 0$ and $b \geq 0$ to distinguish three cases: (i) $a < b^2/2$; (ii) $a = b^2/2$; (iii) $a > b^2/2$.

Plot the analytical solution, given by the probability density distribution (5.175), for the three cases at different times and compare with your simulation results based on the numerical realizations of a sample of stochastic trajectories, see Figure 5.10.

E 5.4 *Financial market*

Create a project about econophysics. Study important models for buyers and sellers as market participants.

Follow the spirit of 'Stochastic Processes. From Physics to Finance' by Wolfgang Paul and Jörg Baschnagel [180] to find out about modeling the financial market. Discuss the Black–Scholes equation mathematically and compare it with the geometric Brownian motion approach.

Follow Appendix F of [180] to transform the Fokker–Planck equation (5.176) into an ordinary diffusion equation to finally obtain the result (5.175) using inverse transformations.

Part III
Applications

6
One-Dimensional Diffusion

6.1
Random Walk on a Line and Diffusion: Main Results

In an earlier chapter we have discussed in some detail the Brownian motion of a particle as a stochastic process. Another closely related problem is the *random walk*. In distinction to the Brownian motion, where the randomness appears as a continuous Wiener process, the random walk proceeds by discrete steps. It is described by the diffusion equation in the continuum limit. As will be seen from the following examples, the rules that control the random walk are very simple. However, as often occurs in physical models, the consequences of simple rules are far from elementary. Random walks are interesting in themselves, as they provide a basis for the understanding of a wide range of phenomena and require the use of many mathematical techniques to solve the related problems [180, 195]. Random walks introduce us to the concept of *scale invariance* (as they look the same on all scales) and *universality*. The latter means that the general features of the statistical behavior are independent of the microscopic details.

The concept of the *random walk*, also called *drunkard's walk*, was introduced into science by Karl Pearson in a letter to Nature in 1905 [180]:

> A man starts from a point 0 and walks l yards in a straight line: he then turns through any angle whatever and walks another l yards in a straight line. He repeats this process n times. I require the probability that after these n stretches he is at a distance between r and $r + \delta r$ from the starting point 0.

The drunkard takes a series of steps of equal length away from the last point but each at a random angle. The solution to this problem was provided in the same volume of Nature by Lord Rayleigh.

The random walk on a line is much simpler. The positions are spaced regularly along a line. The walker has two possibilities: either one step to the right ($+1$) with probability p or one step to the left (-1) with probability $q = 1 - p$. The symmetric case (pure diffusion) means $p = q = 1/2$.

6 One-Dimensional Diffusion

Table 6.1 Random walk experiment with $m = 10$ stochastic realizations ($j = 1, 2, \ldots, 10$) each consisting of $n = 10$ steps ($i = 1, 2, \ldots, 10$). The tabulated numbers are the series of displacements $\ell_i = \pm 1$, and the position $s = \sum_i \ell_i$ after $n = 10$ steps and the squared deviation for each realization of the random walk.

j \ i	1	2	3	4	5	6	7	8	9	10	s	$(s - \langle s \rangle)^2$
1	1	−1	1	−1	1	1	−1	1	1	1	4	9
2	−1	−1	1	1	−1	1	1	−1	−1	1	0	1
3	1	1	1	−1	1	1	1	1	−1	−1	4	9
4	−1	1	−1	−1	1	1	−1	−1	−1	−1	−4	25
5	1	1	−1	−1	−1	−1	−1	−1	1	−1	−4	25
6	−1	1	−1	1	1	1	−1	1	1	−1	2	1
7	−1	1	1	−1	−1	−1	1	1	1	1	2	1
8	1	1	−1	−1	1	−1	−1	1	−1	1	0	1
9	1	1	−1	−1	1	1	1	1	−1	1	4	9
10	−1	1	1	1	−1	1	−1	1	−1	1	2	1

After n steps, the position of the random walker is given by

$$s(n) = \sum_{i=1}^{n} \ell_i \quad \text{with} \quad \ell_i = \pm 1. \tag{6.1}$$

A series of fluctuating values $s(n)$ is obtained when the random walk is repeated many times. Interesting quantities are the averages over an ensemble of m different realizations. As an example, the results of $m = 10$ such random walk realizations, each consisting of $n = 10$ steps, are collected in Table 6.1.

For an infinitely large ensemble $m \to \infty$, the probability theory allows us to calculate the probability distribution for $s(n)$, as well as the mean values. For the symmetric case $p = q = 1/2$ we have zero mean value

$$\langle s(n) \rangle = 0, \tag{6.2}$$

since $s(n)$ and $s(-n)$ are equally likely. The root-mean-square

$$\sigma(n) = \sqrt{\langle s(n)^2 \rangle - \langle s(n) \rangle^2} = \sqrt{n} \tag{6.3}$$

characterizes the average deviation amplitude from the mean value.

In the experiment reflected in Table 6.1 we have $\langle s(10) \rangle = 1$ and $\sigma(10) = \sqrt{8.2} \approx 2.86$. Taking into account that our ensemble consists only of $m = 10$ realizations, the agreement with the theoretical values $\langle s(10) \rangle = 0$ and $\sigma(10) = \sqrt{10} \approx 3.16$ is good. In particular, the expected error for the $\langle s(n) \rangle$ value is about $\sigma(n)/\sqrt{m} = 1$, as in our experiment. However, ten realizations are still insufficient to evaluate the probability distribution $P(s, n)$ reasonably well, where $P(s, n)$ is the probability of the deviation s after n steps. For this we need more realizations. For illustration, the probability distributions estimated from 10 and 100 realizations (crosses and

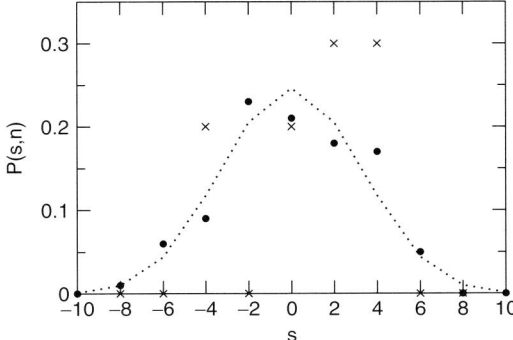

Figure 6.1 The probability distribution $P(s, n)$ for the position s after $n = 10$ steps of a symmetric random walk (dotted curve) in comparison with the empirical results for 10 (crosses) and 100 (circles) realizations.

circles, respectively) are shown in Figure 6.1 and compared with the theoretical result (dotted curve).

The well known demonstration experiment for the random walk on a line is provided by the Galton board. Here balls are dropped through a triangular array of nails. Every time a ball hits a nail it has a probability q of falling to the left of the nail and a probability p of falling to the right of the nail. For the usual symmetrical Galton board we have $p = q = 1/2$.

The theoretical results given by (6.2)–(6.3), as well as the probability distribution over the positions of the random walker, follow easily from the properties of the Markov chain describing the process. Namely, the probability $P(m, n+1)$ that the walker is at position m after $n + 1$ steps is given by the set of probabilities $P(m, n)$ after n steps in accordance with the master equation

$$P(m, n + 1) = p\, P(m - 1, n) + q\, P(m + 1, n). \tag{6.4}$$

The solution of (6.4) is the binomial distribution

$$P(m, n) = \frac{n!}{[(n+m)/2]!\,[(n-m)/2]!}\, p^{(n+m)/2}\, q^{(n-m)/2}. \tag{6.5}$$

The first moment of this probability distribution is

$$\langle m \rangle(n) = \sum_{m=-n}^{n} m P(m, n) = 2n\left(p - \frac{1}{2}\right) \tag{6.6}$$

and the second moment is

$$\langle m^2 \rangle(n) = \sum_{m=-n}^{n} m^2 P(m, n) = 4npq + 4n^2 \left(p - \frac{1}{2}\right)^2 \tag{6.7}$$

Hence, the root-mean-square is given by

$$\sigma(n) = \sqrt{\langle (m - \langle m \rangle)^2 \rangle} = \sqrt{\langle m^2 \rangle - \langle m \rangle^2} = \sqrt{4npq}, \tag{6.8}$$

and the relative width (error)

$$\frac{\sigma}{\langle m \rangle} = \frac{\sqrt{4np(1-p)}}{2n(p-1/2)} = \sqrt{\frac{p(1-p)}{(p-1/2)^2}} \frac{1}{\sqrt{n}} \simeq n^{-1/2} \qquad (6.9)$$

tends to zero when n goes to infinity.

The particular symmetric case considered before is recovered at $p = q = 1/2$, taking into account that $s(n) \equiv m$.

Up to this point we have considered the discrete random walk on an infinite line. In other words, the case with natural boundary conditions. In general, one can consider the random walk in a finite interval with different boundary conditions on the left and right. These can be reflecting or absorbing. In the first case it is not allowed for the random walker to cross the boundary, whereas in the second case the walker is absorbed when reaching it (i. e. the random walk is then terminated).

Random walks on a multidimensional hypercubic lattice with periodic boundary conditions are treated in [180] when studying the Polya problem of recurrence. This is the problem of finding the probability that a random walker will ever return to its starting point. In 1919 George Polya showed that walks occuring in one or two dimensions return to their starting point with absolute certainty (recurrent walks), but in higher dimensions the walker has a nonzero probability of never revisiting its starting point (transient walks) [195].

The discrete random walk serves as a toy model for diffusive motion. In a certain continuum limit of infinitely small steps, the probability distribution over coordinates of the random walker is described by the diffusion (or in a more general case drift–diffusion) equation with corresponding boundary conditions. In the following section we will show how this scheme works in the simplest case of the symmetric random walk with natural boundary conditions.

6.2
A Drunken Sailor as Random Walker

To provide an illustrative example of a simple stochastic process with the application of the continuum approximation we refer to the problem of the well known one-dimensional random walk with equal probability of making a step to the right (forward) or to the left (backward). The experimental equipment used to get the probabilities $P(m, n)$ of finding the particle at position m after n steps, when starting at $m = 0$, is the Galton board. After a series of n steps of equal length the particle could be found at any of the following points

$$m = \{-n, -n+1, \ldots, -1, 0, +1, \ldots, n-1, n\}. \qquad (6.10)$$

Position m consists of k steps in one direction (success) and $n - k$ in the opposite direction (failure)

$$m = k - (n - k) = 2k - n. \qquad (6.11)$$

For the k successes we get

$$k = \tfrac{1}{2}(n+m). \tag{6.12}$$

Starting with the well known binomial distribution for discrete probabilities

$$P(m,n) \equiv B(k,n) = \binom{n}{k} p^k (1-p)^{n-k} \tag{6.13}$$

we reduce this to the symmetric case ($p = 1/2$)

$$P(m,n) = \frac{n!}{k!(n-k)!}\left(\frac{1}{2}\right)^n = \frac{n!}{[(n+m)/2]!\,[(n-m)/2]!}\left(\frac{1}{2}\right)^n. \tag{6.14}$$

Further on we introduce the (still discrete) coordinate $x_m = a\,m$ and time $t_n = \tau\,n$, where a is the hopping distance (a length unit) and τ is the time step (a time unit) and rewrite the binomial distribution (6.14) as

$$P(x_m, t_n) = \binom{t_n/\tau}{t_n/(2\tau) + x_m/(2a)}\left(\frac{1}{2}\right)^{t_n/\tau}$$

$$= \frac{(t_n/\tau)!}{[t_n/(2\tau) + x_m/(2a)]!\,[t_n/(2\tau) - x_m/(2a)]!}\left(\frac{1}{2}\right)^{t_n/\tau}. \tag{6.15}$$

We are interested in the limit where the total number of steps n as well as the number of steps in one direction k is large. In this case we can apply the well known Stirling formula

$$n! \simeq e^{-n} n^n \sqrt{2\pi n} \tag{6.16}$$

or

$$\ln n! \simeq n \ln n - n + \tfrac{1}{2}\ln(2\pi n) \tag{6.17}$$

to evaluate the factorials in (6.14). This yields

$$P(m,n) \simeq \exp\left(n \ln n - k \ln k - (n-k)\ln(n-k) - n \ln 2\right)$$
$$\times n^{1/2}[2\pi k(n-k)]^{-1/2}. \tag{6.18}$$

According to our substitutions $x_m = a\,m$ and $t_n = \tau\,n$ made in (6.15) we have

$$k = \frac{t_n}{2\tau}(1+\delta), \tag{6.19}$$

$$n - k = \frac{t_n}{2\tau}(1-\delta), \tag{6.20}$$

where

$$\delta = \frac{m}{n} = \frac{x_m}{a}\frac{\tau}{t_n}. \tag{6.21}$$

This is a property of the symmetrical random walk, following on from (6.14), that δ is a small quantity for relevant stochastic realizations at $n \to \infty$, since typically the deviations m from the origin $m = 0$ are of order $\sqrt{n}$. According to this, we can insert (6.19)–(6.21) into (6.18) and make an expansion in the Taylor series of δ. Retaining only the terms up to the second order in δ we obtain

$$P(m, n) \equiv P(x_m, t_n) \simeq 2 (2\pi t_n)^{-1/2} \tau^{1/2} \left(1 - \delta^2\right)^{-1/2} \exp\left(-\frac{x_m^2 \tau}{2a^2 t_n}\right). \quad (6.22)$$

The term δ^2 in the prefactor is vanishingly small and can be omitted.

Now we introduce a new parameter

$$D = \frac{1}{2} \frac{a^2}{\tau}, \quad (6.23)$$

called the diffusion coefficient, by considering the continuum limit where length unit a and time unit τ both tend to zero in such a way that D remains constant. In this case a physically interesting quantity is the probability density $p(x, t)$, that is, the probability of finding a particle within $[x, x + dx]$ divided by the interval length dx. Taking into account that the positions x_m which can be reached after n steps form a discrete grid with step size $2a$ [cf. (6.11)], any interval dx includes $dx/(2a)$ points of this grid at $a \to 0$. Hence, for any small enough dx the probability of finding a particle in this interval is $P(x_m, t_n) \, dx/(2a) = p(x, t) \, dx$ with $x = x_m$ and $t = t_n$, where the probability density is given by

$$p(x, t) = \frac{P(x_m, t_n)}{2a} \simeq (4\pi t)^{-1/2} \left(\frac{2\tau}{a^2}\right)^{1/2} \exp\left(-\frac{x^2 \tau}{2a^2 t}\right). \quad (6.24)$$

Taking into account the definition (6.23), we finally obtain the Gaussian distribution

$$p(x, t) = \frac{1}{\sqrt{4\pi D t}} \exp\left(-\frac{x^2}{4Dt}\right) \quad (6.25)$$

in the considered limit where $a \to 0$ at a constant diffusion coefficient D.

6.3
Diffusion with Natural Boundaries

The dynamics of the probability density $p(x, t)$ for a one-dimensional random walk (Wiener process, Brownian motion) is given by the one-dimensional diffusion equation

$$\frac{\partial p(x, t)}{\partial t} = D \frac{\partial^2 p(x, t)}{\partial x^2}. \quad (6.26)$$

This deals with a nonlocal problem in coordinate space. To obtain a certain solution, the diffusion equation (6.26) has to be completed by initial and boundary conditions. We consider the initial condition $p(x, t = 0) = \delta(x - x_0)$ given by the delta function (a sharp peak at $x = x_0$), which physically means that the random walk starts at $x = x_0$, as well as the natural boundary conditions $\lim_{x \to \pm\infty} p(x, t) = 0$. Since we have a closed system, the probability density function $p(x, t)$ obeys the normalization condition (or global probability conservation law)

$$\int_{-\infty}^{\infty} p(x, t) \, dx = 1, \tag{6.27}$$

as well as the continuity equation (local conservation)

$$\frac{\partial p(x, t)}{\partial t} + \frac{\partial}{\partial x} j_{\text{diff}}(x, t) = 0. \tag{6.28}$$

The diffusion equation originates from the continuity equation via Fick's laws:
Fick's 1st law (continuity equation)

$$\frac{\partial p(x, t)}{\partial t} + \frac{\partial}{\partial x} j_{\text{diff}}(x, t) = 0 \tag{6.29}$$

Fick's 2nd law (diffusion flow)

$$j_{\text{diff}}(x, t) = -D \frac{\partial p(x, t)}{\partial x}. \tag{6.30}$$

By inserting (6.30) into (6.29) we arrive at the diffusion equation (6.26).

The solution of the diffusion equation with the initial condition as a delta-peak at $x = x_0$ and natural boundary conditions, is the Gaussian (normal) distribution

$$p(x, t) = \frac{1}{\sqrt{4\pi D t}} \exp\left(-\frac{(x - x_0)^2}{4 D t}\right), \tag{6.31}$$

which is an extension of (6.25) to an arbitrary starting point x_0 of the random walk. It is symmetric with respect to x_0. Figure 6.2 illustrates the behavior of $p(x, t)$ for different observation times.

A generalization for an arbitrary initial distribution $p(x, t = 0) = p_0(x)$ is possible due to the superposition of probability densities created by different sources (initial distributions), which is a property of our diffusion equation due to its linearity. Hence, for $p_0(x) \equiv \int_{-\infty}^{\infty} p_0(y) \delta(x - y) \, dy$ the solution is an integral over the normal distributions corresponding to $p(x, t = 0) = \delta(x - y)$ weighted by the source intensity $p_0(y)$, so

$$p(x, t) = \frac{1}{\sqrt{4\pi D t}} \int_{-\infty}^{+\infty} p_0(y) \exp\left(-\frac{(x - y)^2}{4 D t}\right) dy. \tag{6.32}$$

For $p_0(x) = \delta(x - x_0)$ we recover the known Gaussian distribution (6.31).

Now we show how the solution of the diffusion equation can be obtained by a one-dimensional Fourier transformation to $\widetilde{p}(k, t)$ (transformation to the inverse space by a generating function) which is defined by

$$p(x, t) = \frac{1}{\sqrt{2\pi}} \int_{-\infty}^{+\infty} e^{ikx} \widetilde{p}(k, t) \, dk, \tag{6.33}$$

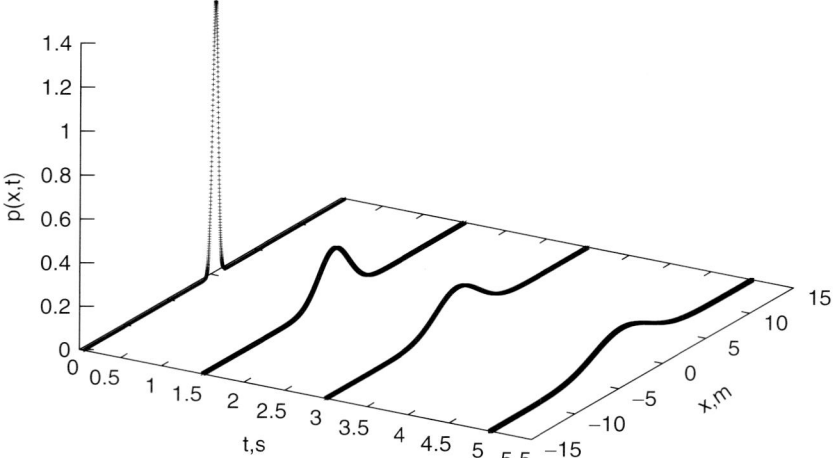

Figure 6.2 The solution of the diffusion equation (6.26) showing times $t = 0.05$ s, 1.5 s, 3 s, 5 s (from left to right). Three-dimensional view of Gaussian distribution $p(x, t)$ (6.31) in m^{-1} with initial condition $x_0 = 0$ and diffusion constant $D = 1$ m^{2}s^{-1}.

$$\widetilde{p}(k, t) = \frac{1}{\sqrt{2\pi}} \int_{-\infty}^{+\infty} e^{-ikx} p(x, t) \, dx, \quad (6.34)$$

where k is the wave number.

The left-hand side of the diffusion equation (6.26) is transformed as

$$\frac{\partial}{\partial t} p(x, t) = \frac{\partial}{\partial t} \frac{1}{\sqrt{2\pi}} \int_{-\infty}^{\infty} dk \, e^{ikx} \widetilde{p}(k, t) = \frac{1}{\sqrt{2\pi}} \int_{-\infty}^{\infty} dk \, e^{ikx} \frac{\partial \widetilde{p}(k, t)}{\partial t}, \quad (6.35)$$

whereas the right-hand side, starting with

$$\frac{\partial}{\partial x} p(x, t) = \frac{\partial}{\partial x} \frac{1}{\sqrt{2\pi}} \int_{-\infty}^{\infty} dk \, e^{ikx} \widetilde{p}(k, t) = \frac{1}{\sqrt{2\pi}} \int_{-\infty}^{\infty} (ik) e^{ikx} \widetilde{p}(k, t) \, dk, \quad (6.36)$$

becomes

$$\frac{\partial^2}{\partial x^2} p(x, t) = \frac{\partial}{\partial x} \frac{1}{\sqrt{2\pi}} \int_{-\infty}^{\infty} (ik) e^{ikx} \widetilde{p}(k, t) \, dk = \frac{1}{\sqrt{2\pi}} \int_{-\infty}^{\infty} (ik)^2 e^{ikx} \widetilde{p}(k, t) \, dk$$

$$= -\frac{1}{\sqrt{2\pi}} \int_{-\infty}^{\infty} k^2 e^{ikx} \widetilde{p}(k, t) \, dk. \quad (6.37)$$

Hence the transformed equation reads

$$\frac{1}{\sqrt{2\pi}} \int_{-\infty}^{\infty} dk\, e^{ikx} \frac{\partial \widetilde{p}(k,t)}{\partial t} = -\frac{D}{\sqrt{2\pi}} \int_{-\infty}^{\infty} dk\, k^2 e^{ikx} \widetilde{p}(k,t) \tag{6.38}$$

and the integrands must be equal

$$\frac{\partial \widetilde{p}(k,t)}{\partial t} = -k^2 D\, \widetilde{p}(k,t), \tag{6.39}$$

which leads to a local problem in the k-space. An elementary integration yields the solution

$$\widetilde{p}(k,t) = \widetilde{p}(k,t=0) e^{-k^2 Dt} \tag{6.40}$$

in the form of an exponentially decaying kth Fourier mode. The initial condition transforms as

$$\widetilde{p}(k,t=0) = \frac{1}{\sqrt{2\pi}} \int_{-\infty}^{\infty} e^{ikx} p(x,t=0)\, dx = \frac{1}{\sqrt{2\pi}} \int_{-\infty}^{\infty} e^{-ikx} \delta(x-x_0)\, dx$$

$$= \frac{1}{\sqrt{2\pi}} e^{ikx_0}, \tag{6.41}$$

so that the solution in the Fourier space (spectral representation) is

$$\widetilde{p}(k,t) = \frac{1}{\sqrt{2\pi}} e^{-ikx_0} e^{-k^2 Dt}. \tag{6.42}$$

Now we make the inverse transformation to the coordinate space

$$p(x,t) = \frac{1}{\sqrt{2\pi}} \int_{-\infty}^{\infty} dk\, e^{ikx} \underbrace{\frac{1}{\sqrt{2\pi}} e^{-ikx_0} e^{-k^2 Dt}}_{\widetilde{p}(k,t)}$$

$$= \frac{1}{2\pi} \int_{-\infty}^{\infty} dk\, e^{ik(x-x_0)-k^2 Dt}$$

$$= \frac{1}{2\pi} \int_{-\infty}^{\infty} dk\, e^{-Dt\left[k^2 - \frac{ik(x-x_0)}{Dt}\right]}$$

$$= \frac{1}{2\pi} \int_{-\infty}^{\infty} dk\, e^{-Dt\left[\left(k - \frac{i(x-x_0)}{2Dt}\right)^2 + \left(\frac{x-x_0}{2Dt}\right)^2\right]}. \tag{6.43}$$

Further on we change the variable $z = k - i(x-x_0)/(2Dt)$; $dz = dk$. According to this we have to make the integration in the complex plane along the line which is parallel to the real axis although shifted by $c = -i(x-x_0)/(2Dt)$. The integral does not depend on c and therefore we can shift the integration path to the real axis. This can be proven by considering a closed integration contour including the real axis, the line $\operatorname{Im} z = c$, and additional lines closing the contour at $\operatorname{Re} z = \pm\infty$. Since the integrand function is analytical inside the enclosed part of the complex plane, the integral over the whole contour is zero; the integrals over the connecting parts at infinity are also zero, and hence the integral over $\operatorname{Im} z = c$ from $\operatorname{Re} z = +\infty$ to $\operatorname{Re} z = -\infty$ is compensated by the integral over the real axis from $-\infty$ to ∞. The integral changes its sign when the integration direction is reversed and, therefore, both integrals are equal if taken from $-\infty$ to $+\infty$.

Thus we have

$$p(x,t) = \frac{1}{2\pi} \int_{-\infty}^{\infty} dz \, e^{-Dtz^2} \cdot e^{-Dt\frac{(x-x_0)^2}{4D^2 t^2}} = \frac{1}{2\pi} e^{\frac{(x-x_0)^2}{4Dt}} \int_{-\infty}^{\infty} dz \, e^{-Dtz^2}. \quad (6.44)$$

By using the known formula

$$I = \int_{-\infty}^{\infty} e^{-\alpha x^2} \, dx = \sqrt{\frac{\pi}{\alpha}} \quad (6.45)$$

with $\alpha = Dt$, to calculate the last integral in (6.44), we again obtain the solution as a Gaussian distribution

$$p(x,t) = \frac{1}{\sqrt{4\pi Dt}} \exp\left(-\frac{(x-x_0)^2}{4Dt}\right). \quad (6.46)$$

The Gaussian integral (6.45) (often also called the Poisson integral) can be calculated as follows. Consider the square of this integral

$$I^2 = \left(\int_{-\infty}^{\infty} e^{-\alpha x^2} \, dx\right)^2 = \int_{-\infty}^{\infty} dx \int_{-\infty}^{\infty} dy \, e^{-\alpha(x^2+y^2)} \quad (6.47)$$

and then make a transformation to polar coordinates by

$$\begin{aligned} x &= r\cos\alpha, & r &= \sqrt{x^2 + y^2} \\ y &= r\sin\alpha, & \phi &= \arctan(y/x) \end{aligned} \quad (6.48)$$

$$dx \, dy = r \, dr \, d\varphi. \quad (6.49)$$

The factor r in the latter equality represents the determinant of the transformation Jacobian

$$J = \begin{vmatrix} \frac{\partial x}{\partial r} & \frac{\partial y}{\partial r} \\ \frac{\partial x}{\partial \alpha} & \frac{\partial y}{\partial \alpha} \end{vmatrix} = \begin{vmatrix} \cos\alpha & \sin\alpha \\ r(-\sin\alpha) & r(\cos\alpha) \end{vmatrix} = r. \quad (6.50)$$

Thus we have

$$I^2 = \int_0^{2\pi} d\varphi \int_0^\infty dr\, re^{-\alpha r^2} = 2\pi \int_0^\infty re^{-\alpha r^2}\, dr = \pi \int_0^\infty e^{-\alpha z}\, dz. \quad (6.51)$$

By substituting

$$\alpha z = u; \quad du = \alpha\, dz \quad (6.52)$$

the integral becomes

$$I^2 = \frac{\pi}{\alpha} \int_0^\infty e^{-u}\, du = \frac{\pi}{\alpha} \left[-e^{-u}\right]\bigg|_0^\infty = \frac{\pi}{\alpha}, \quad (6.53)$$

and hence (cf. (6.45))

$$I = \sqrt{\frac{\pi}{\alpha}}. \quad (6.54)$$

Now let us calculate the moments of the density distribution function $p(x,t)$ (6.31). The moment of nth order is given by

$$\langle x^n \rangle (t) = \int_{-\infty}^\infty dx\, x^n p(x,t). \quad (6.55)$$

The moment of zeroth order has to be one, as it represents the normalization integral

$$\langle x^0 \rangle \equiv 1 = \int_{-\infty}^\infty p(x,t)\, dx. \quad (6.56)$$

To check it we insert the solution (6.31)

$$\langle x^0 \rangle = \int_{-\infty}^\infty \frac{1}{\sqrt{4\pi Dt}} e^{-\frac{(x-x_0)^2}{4Dt}}\, dx. \quad (6.57)$$

This integral is further transformed by substituting

$$x - x_0 = z; \quad dx = dz \quad (6.58)$$

and denoting $a^2 = 1/(4Dt)$. It yields

$$\langle x^0 \rangle = \frac{a}{\sqrt{\pi}} 2 \int_0^\infty e^{-a^2 z^2}\, dz = \frac{a}{\sqrt{\pi}} 2 \frac{\sqrt{\pi}}{2a} = 1. \quad (6.59)$$

The first moment of the probability distribution function represents the mean value

$$\langle x^1 \rangle = \int_{-\infty}^{\infty} x\, p(x,t)\, dx. \tag{6.60}$$

In our case it has to be equal to x_0, since the distribution is symmetric with respect to this point. We check it by calculation

$$\langle x^1 \rangle = \frac{1}{\sqrt{4\pi Dt}} \int_{-\infty}^{\infty} x e^{-\frac{(x-x_0)^2}{4Dt}}\, dx = \frac{1}{\sqrt{4\pi Dt}} \int_{-\infty}^{\infty} (x_0 + z) e^{-\frac{z^2}{4Dt}}\, dz$$

$$= x_0 \underbrace{\frac{1}{\sqrt{4\pi Dt}} \int_{-\infty}^{\infty} e^{-\frac{z^2}{4Dt}}\, dz}_{\langle x^0 \rangle = 1} + \underbrace{\frac{1}{\sqrt{4\pi Dt}} \int_{-\infty}^{\infty} z e^{-\frac{z^2}{4Dt}}\, dz}_{=0,\ \text{symmetry around } z = x - x_0 = 0} \tag{6.61}$$

$$= x_0 + 0 = x_0. \tag{6.62}$$

The mean value $\langle x \rangle = x_0$ does not change in time, it keeps the initial value.

The second moment

$$\langle x^2 \rangle = \int_{-\infty}^{\infty} x^2 p(x,t)\, dx \tag{6.63}$$

is related to the standard deviation σ via

$$\langle (x - \langle x \rangle)^2 \rangle = \langle (x - x_0)^2 \rangle \equiv \sigma^2. \tag{6.64}$$

To calculate the second moment we use the identity

$$\int_{-\infty}^{\infty} x^2 e^{-\alpha x^2}\, dx = -\frac{d}{d\alpha} \int_{-\infty}^{\infty} e^{-\alpha x^2}\, dx. \tag{6.65}$$

Denoting again $\alpha = 1/(4Dt)$, we have

$$\langle x^2 \rangle = \int_{-\infty}^{\infty} x^2 \frac{1}{\sqrt{4\pi Dt}} e^{-\frac{(x-x_0)^2}{4Dt}}\, dx \tag{6.66}$$

$$= \frac{1}{\sqrt{4\pi Dt}} \int_{-\infty}^{\infty} (y + x_0)^2 e^{-\alpha y}\, dy \tag{6.67}$$

$$= \frac{1}{\sqrt{4\pi Dt}} \int_{-\infty}^{\infty} (y^2 + 2x_0 y + x_0^2) e^{-\alpha y}\, dy \tag{6.68}$$

$$= x_0^2 + 0 + \frac{1}{\sqrt{4\pi Dt}} \int_{-\infty}^{\infty} y^2 e^{-\alpha y}\, dy \tag{6.69}$$

$$= x_0^2 + \frac{1}{\sqrt{4\pi Dt}} \left(-\frac{d}{d\alpha} \int_{-\infty}^{\infty} e^{-\alpha y^2} dy \right). \tag{6.70}$$

In the second line we have substituted $x - x_0 = y$, $dx = dy$. Using (6.45) further calculation yields

$$\langle x^2 \rangle = x_0^2 - \frac{1}{\sqrt{4\pi Dt}} \frac{d}{d\alpha} \left(\frac{\sqrt{\pi}}{\sqrt{\alpha}} \right) = x_0^2 - \frac{\sqrt{\pi}}{\sqrt{4\pi Dt}} \left(-\frac{1}{2} \right) \alpha^{-\frac{3}{2}} \tag{6.71}$$

$$= x_0^2 + \frac{\sqrt{\pi}}{2\sqrt{4\pi Dt}} \left(\frac{1}{4Dt} \right)^{-\frac{3}{2}} = x_0^2 + 2Dt \tag{6.72}$$

which finally gives

$$\sigma^2 \equiv \langle (x - \langle x \rangle)^2 \rangle = 2Dt \tag{6.73}$$

$$\sigma = \sqrt{2Dt} \sim \sqrt{t}. \tag{6.74}$$

The latter equality means that the width of the probability distribution increases with time as the square root of time t.

6.4
Diffusion in a Finite Interval with Mixed Boundaries

Here we consider an example of the initial and boundary value diffusion problem in a finite interval with one reflecting and one absorbing boundary. We calculate the breakdown probability density $\mathcal{P}(t, b)$ which is defined as the probability per time unit of reaching the absorbing boundary [82].

The problem is described by the following set of equations:

1. The equation of motion (dynamics)

$$\frac{\partial p(x, t)}{\partial t} = D \frac{\partial^2 p(x, t)}{\partial x^2}. \tag{6.75}$$

2. The initial condition (delta function)

$$p(x, t = 0) = \delta(x - x_0). \tag{6.76}$$

3. The reflecting boundary condition at $x = a$ (left border)

$$\left. \frac{\partial p(x, t)}{\partial x} \right|_{x=a} = 0 \tag{6.77}$$

4. The absorbing boundary condition at $x = b$ (right border)

$$p(x = b, t) = 0. \tag{6.78}$$

For convenience we make a transformation to a new variable $y = x - a$. The transformed equations read as follows:

1. The equation of motion (dynamics)

$$\frac{\partial p(y,t)}{\partial t} = D \frac{\partial^2 p(y,t)}{\partial y^2}. \tag{6.79}$$

2. The initial condition (delta function)

$$p(y, t=0) = \delta(y - y_0). \tag{6.80}$$

3. The reflecting boundary condition at $y = 0$ (left border)

$$\left.\frac{\partial p(y,t)}{\partial y}\right|_{y=0} = 0. \tag{6.81}$$

4. The absorbing boundary condition at $y = b - a$ (right border)

$$p(y = b - a, t) = 0. \tag{6.82}$$

To solve the problem, first we make a separation ansatz $p(y,t) = \chi(t) f(y)$, which yields

$$\frac{1}{\chi(t)} \frac{d\chi(t)}{dt} = D \frac{1}{f(y)} \frac{d^2 f(y)}{dy^2}. \tag{6.83}$$

Both sides should be equal to a constant, called $-\lambda$. Integration of the left-hand side gives an exponential decay function

$$\chi(t) = \chi_0 \exp(-\lambda t) \tag{6.84}$$

with $\chi(t=0) = \chi_0 = 1$.

Introducing the notion of wave number k given by

$$k^2 = \frac{\lambda}{D} \tag{6.85}$$

and integrating the right-hand side of (6.83) we obtain the wave equation

$$\frac{d^2 f(y)}{d^2 y} + k^2 f(y) = 0. \tag{6.86}$$

Its general solution is

$$f(y) = A \sin(ky) + B \cos(ky). \tag{6.87}$$

This solution (6.87) contains three unknown parameters k (or λ), A, and B. The two (left and right) boundary conditions thus allow us to determine particular solutions of (6.86) up to unknown prefactors, which further can be uniquely determined by constructing a time-dependent solution which fulfills the initial condition.

6.4 Diffusion in a Finite Interval with Mixed Boundaries

The left boundary condition

$$\left.\frac{df(y)}{dy}\right|_{y=0} = 0 \tag{6.88}$$

gives

$$A k \cos(k \cdot 0) - B k \sin(k \cdot 0) = 0 \tag{6.89}$$

from which we get $A = 0$.

The right boundary condition

$$f(y = b - a) = 0 \tag{6.90}$$

gives us

$$A \sin(k(b-a)) - B \cos(k(b-a)) = 0. \tag{6.91}$$

Taking into account that $A = 0$ and looking for a nontrivial solution with $B \neq 0$ we obtain the conditional equation

$$\cos(k(b-a)) = 0. \tag{6.92}$$

This yields discrete solutions for the wave numbers k and eigenvalues $\lambda = Dk^2$:

$$k_m = \frac{\pi}{b-a}\left(\frac{1}{2} + m\right), \tag{6.93}$$

$$\lambda_m = \frac{D\pi^2}{(b-a)^2}\left(\frac{1}{2} + m\right)^2 \tag{6.94}$$

with non-negative integer numbers $m = 0, 1, 2, \ldots$ (see Table 6.2).

A particular time-dependent solution, which fulfills the boundary conditions and corresponds to the eigenvalue λ_m, is thus the eigenfunction $p_m(y, t)$ given by

$$p_m(y, t) = B_m e^{-\lambda_m t} \cos(k_m y). \tag{6.95}$$

Table 6.2 Wave numbers k_m (in m^{-1}) and eigenvalues λ_m (in s^{-1}) for $m = 0, 1, \ldots, 5$ using the parameter set $a = 0, b = 1$ m, $D = 1$ m^2 s^{-1}.

m	k_m	λ_m
0	1.5708	2.4674
1	4.7124	22.2066
2	7.8539	61.6850
3	10.9956	120.9026
4	14.1372	199.8595
5	17.2787	298.5555

The complete solution of the problem is found as a superposition of these eigenfunctions

$$p(y,t) = \sum_{m=0}^{\infty} p_m(y,t) = \sum_{m=0}^{\infty} B_m e^{-\lambda_m t} \cos(k_m y), \qquad (6.96)$$

where the weight parameters B_m are obtained from the initial condition

$$p(y, t=0) = \sum_{m=0}^{\infty} B_m \cos(k_m y) = \delta(y - y_0). \qquad (6.97)$$

To get them we first multiply (6.97) by $\cos(k_n y)$, and integrate over y from 0 to $b-a$:

$$\sum_{m=0}^{\infty} B_m \int_0^{b-a} dy \cos(k_m y) \cos(k_n y) = \int_0^{b-a} dy \, \delta(y - y_0) \cos(k_n y). \qquad (6.98)$$

The integral on the left-hand side can be easily calculated using the orthogonality of the eigenfunctions, which can also be verified by direct calculation

$$\int_0^{b-a} dy \cos(k_n y) \cos(k_m y)$$

$$= \frac{1}{2} \int_0^{b-a} dy \left\{ \cos\left[(k_n + k_m)y\right] + \cos\left[(k_n - k_m)y\right] \right\} \qquad (6.99)$$

$$= \frac{1}{2}(b-a)\delta_{mn}, \qquad (6.100)$$

where the relations

$$k_n + k_m = \frac{\pi}{b-a}(1 + n + m), \qquad (6.101)$$

$$k_n - k_m = \frac{\pi}{b-a}(n - m) \qquad (6.102)$$

following from (6.94), as well as the well known limit

$$\lim_{x \to 0} \frac{\sin x}{x} = 1 \qquad (6.103)$$

have been used. Hence, (6.98) reduces to

$$\sum_{m=0}^{\infty} B_m \frac{b-a}{2} \delta_{mn} = \int_0^{b-a} \delta(y - y_0) \cos(k_n y) \qquad (6.104)$$

which yields

$$B_m = \frac{2}{b-a} \cos(k_m y_0). \qquad (6.105)$$

6.4 Diffusion in a Finite Interval with Mixed Boundaries

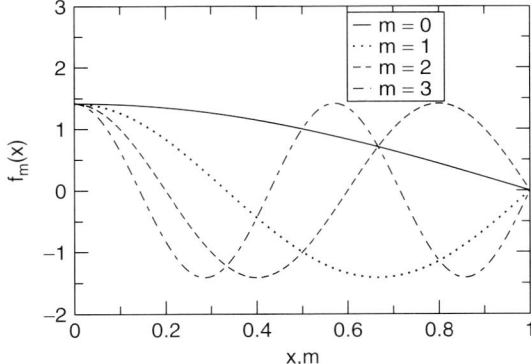

Figure 6.3 Eigenfunctions $f_m(x)$ (6.106) in $m^{-1/2}$ for $m = 0, 1, 2, 3$ with $a = 0$, $b = 1$ m and diffusion constant $D = 1$ m^2 s^{-1}.

Figure 6.3 shows the normalized eigenfunctions

$$f_m(x) = \sqrt{\frac{2}{b-a}} \cos(k_m(x-a)) \tag{6.106}$$

for the first four values of m.

According to this, the solution reads

$$p(y, t) = \frac{2}{b-a} \sum_{m=0}^{\infty} e^{-\lambda_m t} \cos(k_m y_0) \cos(k_m y). \tag{6.107}$$

The inverse transformation from y to x gives us the final probability distribution

$$p(x, t) = \frac{2}{b-a} \sum_{m=0}^{\infty} e^{-\lambda_m t} \cos(k_m(x_0 - a)) \cos(k_m(x - a)). \tag{6.108}$$

Figure 6.4 shows the probability density distribution $p(x, t)$ (6.108) which tends to zero with increasing time.

The first-passage time distribution (breakdown probability density) follows from the balance condition

$$\mathcal{P}(t, x = b) = -\frac{d}{dt} \int_a^b p(x, t)\, dx. \tag{6.109}$$

By inserting the solution (6.108) into the right-hand side of this equation, we obtain

$$\mathcal{P}(t, b) = \frac{2}{b-a} \sum_{m=0}^{\infty} \lambda_m e^{-\lambda_m t} \cos(k_m(x_0 - a)) \int_a^b \cos(k_m(x-a))$$

$$= \frac{2}{b-a} \sum_{m=0}^{\infty} \frac{\lambda_m}{k_m} e^{-\lambda_m t} \cos(k_m(x_0 - a)) \sin(k_m(b-a))$$

6 One-Dimensional Diffusion

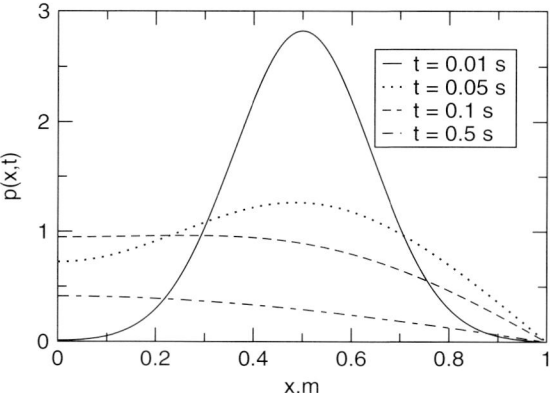

Figure 6.4 Time-dependent solution $p(x,t)$ (6.108) in m^{-1} of the diffusion equation in the finite interval $x \in [0, 1]$ with reflecting boundary at $a = 0$ and absorbing boundary at $b = 1$ m. Parameters are: initial condition $x_0 = 0.5$ m and diffusion constant $D = 1$ m^2 s^{-1}.

$$= \frac{2\pi D}{(b-a)^2} \sum_{m=0}^{\infty} (-1)^m \left(\frac{1}{2} + m\right) e^{-Dk_m^2 t} \cos\left(k_m(x_0 - a)\right). \tag{6.110}$$

The result fulfills the normalization condition

$$\int_0^\infty \mathcal{P}(t, b) \, dt = 1. \tag{6.111}$$

The first-passage time or breakdown probability density distribution $\mathcal{P}(t, x = b)$ indicating how long it takes for the system to reach the absorbing boundary at $x = b$ for the first time, is presented in Figure 6.5 for the same set of parameters as

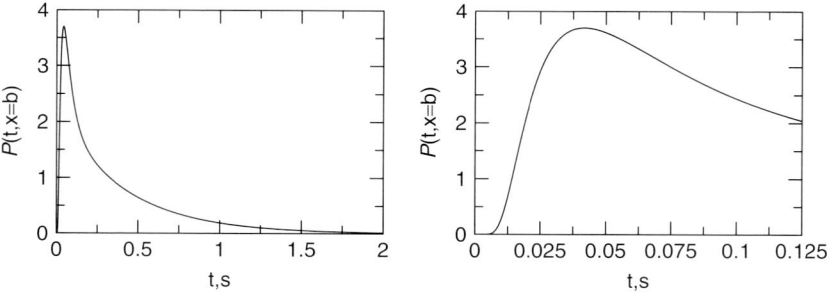

Figure 6.5 The first-passage time probability density distribution $\mathcal{P}(t, x = b)$ in s^{-1} with $b = 1$ m, initial condition $x_0 = 0.5$ m and diffusion constant $D = 1$ m^2 s^{-1}. The right curve shows the time lag in detail.

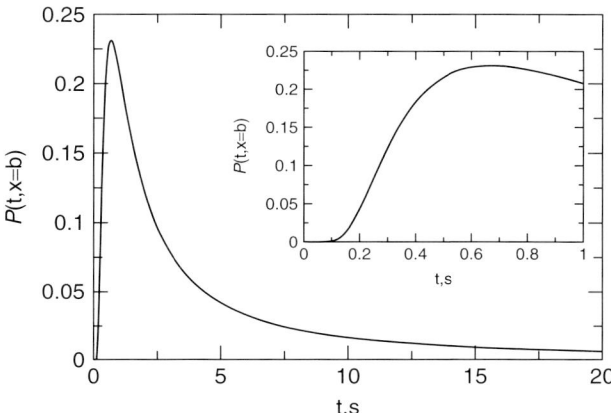

Figure 6.6 The first-passage time probability density distribution $\mathcal{P}(t, x = b)$ in s^{-1} when $a \to -\infty$ (6.112) with $b = 2$ m, initial condition $x_0 = 0$ and diffusion constant $D = 1$ m^2 s^{-1}. Internal plot is a scaled picture of the time lag.

in Figure 6.4. The right plot of Figure 6.5 is a scaled picture showing the time lag which will be explained later.

Finally, we present a well known result in the mathematical literature. If we move the left boundary very far away (limiting case: $a \to -\infty$) we receive from the infinite sum (6.110) the well known formula (see Figure 6.6)

$$\mathcal{P}(t, b) = \frac{b - x_0}{\sqrt{4\pi D t^3}} \exp\left(-\frac{(b - x_0)^2}{4Dt}\right). \tag{6.112}$$

The following suggests a way to get this result. The first-passage time distribution $\mathcal{P}(t, b)$ for the diffusion problem in a finite interval with reflecting boundary at $x = a$, absorbing boundary at $x = b$, and the initial probability distribution $\delta(x - x_0)$ is given by (6.110) together with wave numbers (6.94) written as

$$k_m(b - a) = \pi\left(m + \tfrac{1}{2}\right). \tag{6.113}$$

Taking into account (6.113), we make the following replacements

$$\cos\left(k_m[x_0 - a]\right) = \cos\left(k_m[b - a] - k_m[b - x_0]\right)$$
$$= \cos\left(\pi\left(m + \tfrac{1}{2}\right) - k_m s\right), \tag{6.114}$$

where $s = b - x_0$. By virtue of the well known trigonometric relationship $\cos(\alpha \pm \beta) = \cos\alpha \cos\beta \mp \sin\alpha \sin\beta$, it further transforms to

$$\cos\left(k_m[x_0 - a]\right) = \cos\left(\pi\left(m + \tfrac{1}{2}\right)\right) \cos\left(k_m s\right)$$
$$+ \sin\left(\pi\left(m + \tfrac{1}{2}\right)\right) \sin\left(k_m s\right)$$
$$= (-1)^m \sin\left(k_m s\right). \tag{6.115}$$

By inserting this into (6.110) we obtain

$$P(t,b) = 2D \sum_{m=0}^{\infty} \frac{\pi(m+1/2)}{(b-a)^2} e^{-Dk_m^2 t} \sin(k_m s)$$

$$= 2D \sum_{m=0}^{\infty} \frac{k_m}{b-a} e^{-Dk_m^2 t} \sin(k_m s). \tag{6.116}$$

Now we use the identity

$$\frac{\partial}{\partial s} \cos(k_m s) = -k_m \sin(k_m s), \tag{6.117}$$

and the relation for the wave number increment

$$\Delta k_m = k_{m+1} - k_m = \frac{\pi}{b-a} \tag{6.118}$$

to transform (6.116) into

$$P(t,b) = -\frac{2D}{\pi} \frac{\partial}{\partial s} \sum_{m=0}^{\infty} e^{-Dk_m^2 t} \cos(k_m s)\, \Delta k_m. \tag{6.119}$$

In the limit where the left boundary tends to minus infinity, which corresponds to $a \to -\infty$ or $b - a \to \infty$, the sum is replaced with the integral, which yields

$$P(t,b) = -\frac{2D}{\pi} \frac{\partial}{\partial s} \int_0^{\infty} e^{-Dk^2 t} \cos(ks)\, dk. \tag{6.120}$$

This integral is calculated by applying the known formula

$$\int_0^{\infty} e^{-u^2 x^2} \cos(sx)\, dx = \frac{\sqrt{\pi}}{2u} e^{-s^2/(4u^2)}, \quad u > 0 \tag{6.121}$$

which in our case ($u = \sqrt{Dt}$) yields

$$P(b,t) = -\frac{D}{\sqrt{\pi Dt}} \frac{\partial}{\partial s} e^{-\frac{s^2}{4Dt}} = \frac{s}{\sqrt{4\pi Dt^3}} e^{-\frac{s^2}{4Dt}}. \tag{6.122}$$

Returning to the original notations, we finally recover the known expression (6.112) for the first-passage time distribution in the semi–infinite interval.

6.5
The Mirror Method and Time Lag

Alternatively, the solution of the diffusion equation in shifted coordinates $y = x - a$, together with the initial condition $p(y, t = 0) = \delta(y - y_0)$ and different boundary conditions can be obtained as a superposition of solutions

$$p_{y_0}(y,t) = \frac{1}{\sqrt{4\pi Dt}} \exp\left(-\frac{(y-y_0)^2}{4Dt}\right) \tag{6.123}$$

for natural boundary conditions with different positions y_0 of the delta-peaks at the initial time moment $t = 0$. Any such superposition satisfies the diffusion equation (6.79) due to its linearity. The only problem is to fulfill the initial and boundary conditions.

For simplicity, let us consider first the case where there is a reflecting boundary at $y = 0$ and we are looking for the solution within $y \in [0, \infty)$, so the other boundary is located at $+\infty$. Formally we can extend the y interval from $-\infty$ to ∞ and make a mirror construction. Obviously, the superposition as the sum of $p_{y_0}(y, t)$ and $p_{-y_0}(y, t)$ fulfills the initial condition for the interval $y \in [0, \infty)$, as well as the reflecting boundary condition $\partial p(y, t)/\partial y = 0$ at $y = 0$ due to the mirror symmetry around this point (see Figure 6.7(a)). Hence

$$p(y, t) = p_{y_0}(y, t) + p_{-y_0}(y, t) \tag{6.124}$$

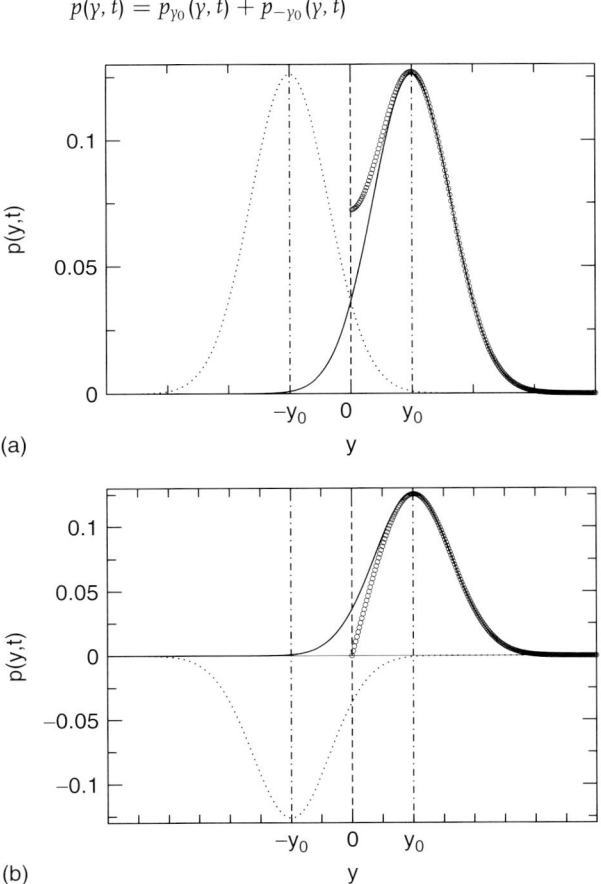

Figure 6.7 Construction of the solution for $y \in [0, +\infty)$ with reflecting boundary at $y = 0$ (a) and absorbing boundary at $y = 0$ (b). Circles (full lines) are the final results corresponding to (6.124) and (6.125), respectively.

is the solution of the problem with one reflecting boundary at $y = 0$. Similarly (see Figure 6.7(b)), we can construct a solution which is antisymmetric with respect to $y = b - a$, That is,

$$p(y, t) = p_{y_0}(y, t) - p_{2(b-a)-y_0}(y, t). \tag{6.125}$$

This is the solution of the problem with single absorbing boundary at $y = b - a$ (the other boundary located infinitely far away), since it obeys the condition $p(y, t) = 0$ at $y = b - a$. The problem with two boundaries located at a finite distance $b - a$ from each other is more complicated, since an infinite number of periodically placed peaks are necessary to satisfy the specific symmetry and/or antisymmetry conditions. In this case (Figure 6.8) we have

$$p(y, t) = \sum_{m=-\infty}^{\infty} (-1)^m \left[p_{y_0+2m(b-a)} + p_{-y_0+2m(b-a)} \right] \tag{6.126}$$

for our problem with the reflecting boundary at $y = 0$ and the absorbing one at $y = b - a$. For two reflecting boundaries (Figure 6.9) the solution is

$$p(y, t) = \sum_{m=-\infty}^{\infty} \left[p_{y_0+2m(b-a)} + p_{-y_0+2m(b-a)} \right] \tag{6.127}$$

and for two absorbing boundaries (Figure 6.10) it reads

$$p(y, t) = \sum_{m=-\infty}^{\infty} \left[p_{y_0+2m(b-a)} - p_{-y_0+2m(b-a)} \right]. \tag{6.128}$$

The first-passage time distribution (cf. (6.109))

$$\mathcal{P}(t, y = b - a) = -\frac{d}{dt} \int_0^{b-a} p(y, t) \, dy \tag{6.129}$$

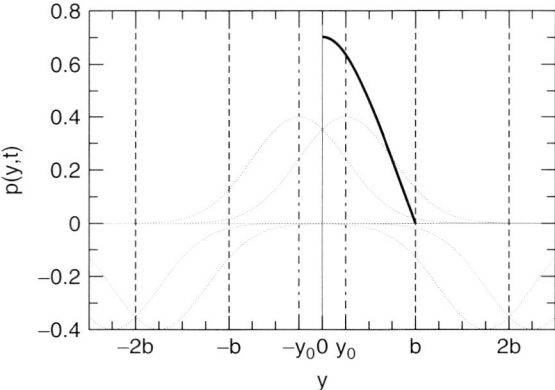

Figure 6.8 Construction of solution for finite interval $y \in [0, b - a)$ with $a = 0$ for the mixed boundary problem, that is, reflecting boundary at $y = 0$ and absorbing boundary at $y = b$. The bold curve is the final solution (6.126).

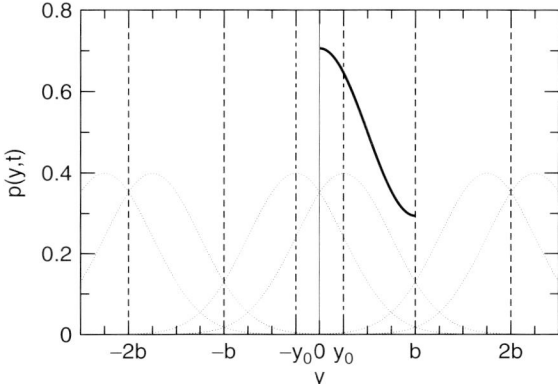

Figure 6.9 Schematic picture of the mirror method for constructing the solution of the Fokker–Planck equation with two reflecting boundary conditions at $y = 0$ and $y = b - a$ ($a = 0$). The bold curve is the final solution (6.127).

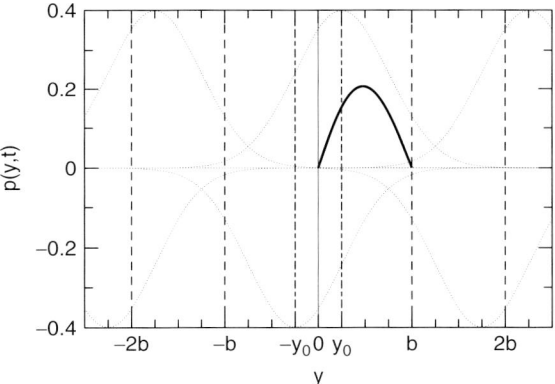

Figure 6.10 Schematic picture of the mirror method for constructing the solution of the Fokker–Planck equation with two absorbing boundary conditions at $y = 0$ and $y = b - a$ ($a = 0$). The bold curve is the final solution (6.128).

for our problem with the absorbing boundary at $x = b$ or $y = b - a$ in the limit where the other boundary is moved infinitely far away ($b - a \to \infty$ at a finite $b - x_0$ or $b - a - y_0$) can be calculated easily by using the solution (6.125). In this case it is useful to rewrite the probability outflow (6.129) as

$$\mathcal{P}(t, y = b - a) = -D \left. \frac{\partial p(y, t)}{\partial y} \right|_{y = b - a} \tag{6.130}$$

obtained by inserting the time derivative of $p(y, t)$ from (6.79) into (6.129). Substituting the solution (6.108) and (6.123) into (6.130) we obtain

$$\mathcal{P}(t, y = b - a) = \frac{b - a - y_0}{\sqrt{4\pi D t^3}} \exp\left(-\frac{(b - a - y_0)^2}{4Dt}\right). \tag{6.131}$$

The transformation to the original coordinate $x = y + a$ leads finally to

$$\mathcal{P}(t, x = b) = \frac{b - x_0}{\sqrt{4\pi D t^3}} \exp\left(-\frac{(b - x_0)^2}{4Dt}\right), \tag{6.132}$$

already stated as the result (6.112).

A remarkable property of the breakdown probability distribution (6.132) is its diverging mean value, which is related to the fact that stochastic trajectories are allowed to move arbitrarily far away from the origin in the negative direction, requiring as unlimited long time to return and reach the absorbing boundary. Another interesting property is the presence of a certain time lag t_{lag}, which means that the distribution function $\mathcal{P}(t, b)$ is very small and tends to zero extremely fast for $t < t_{\text{lag}}$. This implies that the breakdown is very rarely (or practically never) observed before $t = t_{\text{lag}}$, as typically one needs some minimal time to cover the distance from the origin $x = x_0$ to the absorbing boundary $x = b$. The value of t_{lag} can be defined by constructing a tangent at the inflection point of the $\mathcal{P}(t, b)$ plot and looking for the point $t = t_{\text{lag}}$ where this linear approximation gives zero. Figure 6.11 shows the graphical interpretation of this time lag construction.

The function $\mathcal{P}(t, b)$ has a maximum at

$$t = t_{\text{max}} = \frac{2}{3} \frac{(b - x_0)^2}{4D} \tag{6.133}$$

and two inflection points

$$t_{1,2} = t_{\text{max}} \left(1 \mp \sqrt{\frac{2}{5}}\right). \tag{6.134}$$

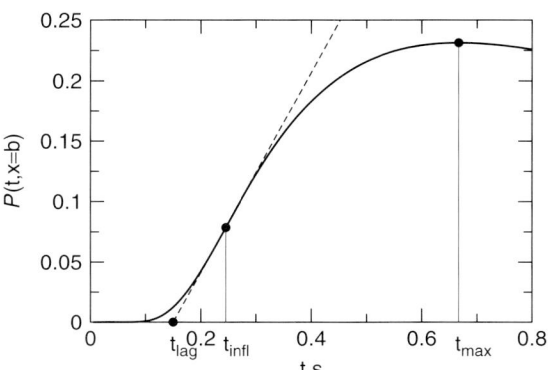

Figure 6.11 Construction of time lag t_{lag} of the first-passage time probability density $\mathcal{P}(t, x = b)$ in s^{-1} for limit case of $a \to -\infty$ with $x_0 = 0$ and $b = 2$ m (see inside plot in Figure 6.6). Time lag t_{lag} (6.136) is calculated as the intersection point of the t-axis and the tangent (dotted line) to $\mathcal{P}(t, x = b)$ at the inflection point $(t_{\text{infl}}, \mathcal{P}(t_{\text{infl}}, x = b))$ (6.134).

The smallest value $t_1 = t_{infl}$ is meaningful in our tangent construction. Thus, the time lag is given by the equation

$$(t_{infl} - t_{lag}) \left.\frac{d\mathcal{P}}{dt}\right|_{t=t_{infl}} = \mathcal{P}(t_{infl}, b) \tag{6.135}$$

from which we find

$$t_{lag} = \frac{1}{3}\left(7 - 2\sqrt{10}\right) t_{max} \approx 3.75 \times 10^{-2} \frac{(b-x_0)^2}{D}. \tag{6.136}$$

Finally, this time lag is calculated for the example which has been presented in Figure 6.11. In this case we get a numerical value of $t_{lag} = 0.15$ s for parameters $x_0 = 0$, $b = 2$ m and $D = 1$ m² s^{-1}. The maximum of $\mathcal{P}(t, b)$ corresponds to the point (0.667 s; 0.231 s^{-1}). There are two inflection points of the function $\mathcal{P}(t, b)$ but we confine our consideration to the inflection point located on the left-hand side of t_{max}, that is (0.245 s; 0.078 s^{-1}).

6.6
Maximum Value Distribution

Consider a one-dimensional random walk (called a diffusion process) with natural boundary conditions starting at point x_0. We require the distribution of maximum values of coordinate x for an ensemble of stochastic trajectories running within a given time interval from zero to t. Let $\mathcal{P}_{max}(x|x_0, t)$ be the probability density in x space of the maximum distribution. Note that $x \geq x_0$ always holds, and any trajectory wandering only in the region $x \leq x_0$ has the maximum at $x = x_0$. Let us denote by $P_{max}(x = x_0|x_0, t)$ the probability of the latter scenario. The probability $P_{max}(x \leq b|x_0, t)$ that the maximum of a trajectory does not exceed a certain value b reads

$$P_{max}(x \leq b|x_0, t) = P_{max}(x = x_0|x_0, t) + \int_{x_0}^{b} \mathcal{P}_{max}(x|x_0, t)\, dx. \tag{6.137}$$

In the following it will be shown that the probability $P_{max}(x = x_0|x_0, t)$ vanishes in our case of stochastic trajectories. In fact, $P_{max}(x \leq b|x_0, t)$ is the probability that x belongs to the interval $(-\infty, b]$ at the time moment t. Thus, we have

$$\int_{-\infty}^{b} p(x, t|x_0, b)\, dx = \int_{x_0}^{b} \mathcal{P}_{max}(x|x_0, t)\, dx, \tag{6.138}$$

where $p(x, t|x_0, b)$ is the probability distribution density over x at a time t provided that the initial distribution is $\delta(x - x_0)$ and the absorbing boundary is located at $x = b$. It can be related to the first-passage problem in a semi-infinite interval $x \in (-\infty, b]$ with absorbing boundary at $x = b$. According to (6.138), we have

$$\int_{x_0}^{b} \mathcal{P}_{max}(x|x_0, t)\, dx = 1 - \int_{0}^{t} \mathcal{P}(t'|x_0, b)\, dt', \tag{6.139}$$

where $\mathcal{P}(t'|x_0, b)$ is the first-passage time (t') distribution density to reach the absorbing boundary.

Taking the derivative with respect to b in (6.138), we obtain

$$\mathcal{P}_{\max}(x = b|x_0, t) = \frac{\partial}{\partial b} \int_{-\infty}^{b} p(x, t|x_0, b) \, dx. \tag{6.140}$$

Using (6.139), the derivation yields

$$\mathcal{P}_{\max}(x = b|x_0, t) = -\frac{\partial}{\partial b} \int_{0}^{t} \mathcal{P}(t'|x_0, b) \, dt' = -\int_{0}^{t} \frac{\partial}{\partial b} \mathcal{P}(t'|x_0, b) \, dt'. \tag{6.141}$$

For pure diffusion with the above given initial and boundary conditions we recall the solution

$$p(x, t|x_0, b) = \frac{1}{\sqrt{4\pi Dt}} \left\{ \exp\left[-\frac{(x - x_0)^2}{4Dt}\right] - \exp\left[-\frac{(x - 2b + x_0)^2}{4Dt}\right] \right\} \tag{6.142}$$

obtained by the mirror construction method. The maximum distribution can be in principle, calculated from the first-passage time distribution (6.112)

$$\mathcal{P}(t|x_0, b) = \frac{b - x_0}{\sqrt{4\pi Dt^3}} \exp\left(-\frac{(x - x_0)^2}{4Dt}\right) \tag{6.143}$$

according to (6.141), although in our case it is more easily obtained from (6.140):

$$\mathcal{P}_{\max}(x = b|x_0, t) = \frac{\partial}{\partial b} \int_{-\infty}^{b} p(x, t|x_0, b) \, dx$$

$$= p(b, t|x_0, b) + \int_{-\infty}^{b} \frac{\partial}{\partial b} p(x, t|x_0, b) \, dx. \tag{6.144}$$

The first term on the right-hand side of (6.144) is zero according to the absorbing boundary condition at $x = b$. By inserting (6.142) into the integral, we obtain

$$\mathcal{P}_{\max}(x = b|x_0, t) = -\frac{1}{\sqrt{4\pi(Dt)^3}} \int_{-\infty}^{b} (x - 2b + x_0)$$

$$\times \exp\left[-\frac{(x - 2b + x_0)^2}{4Dt}\right] dx. \tag{6.145}$$

This integral can be easily calculated using the substitution $z = (x - 2b + x_0)^2$, which yields the result in the form of the Gaussian distribution

$$\mathcal{P}_{\max}(x|x_0, t) = \frac{1}{\sqrt{\pi Dt}} \exp\left[-\frac{(x - x_0)^2}{4Dt}\right]. \tag{6.146}$$

This fulfills the normalization condition

$$\int_{x_0}^{\infty} \mathcal{P}_{\max}(x|x_0, t)\, \mathrm{d}x = 1. \tag{6.147}$$

Since the total probability is one, it means that there is no special additional contribution due to the trajectories located in the region $x \leq x_0$ in the whole time interval, so $\mathcal{P}_{\max}(x = x_0|x_0, t) = 0$. The latter probability can be calculated directly as

$$\mathcal{P}_{\max}(x = x_0|x_0, t) = \lim_{b \searrow x_0} \int_{-\infty}^{b} p(x, t|x_0, b)\, \mathrm{d}x = 0 \tag{6.148}$$

in accordance with the solution (6.142). It can be understood in such a way that any trajectory of the diffusion process, which starts at $x = x_0$, with probability one, reaches a value $x > x_0$ which is infinitely close to the origin. However, there is a continuum of trajectories which wander in the region $x < x_0$ after some small fluctuations around the origin $x = x_0$. For all these trajectories, the maximum is located near x_0, which explains the maximum of the probability distribution (6.146) at $x = x_0$.

In Figure 6.12 the theoretical formula (6.146) is compared to a numerical simulation on a lattice.

The formula (6.146) is well known in the mathematical literature, see e.g. [91], p. 95, where it is obtained from the probability density distribution over the variables a and b (where $a \leq b$ and $b \geq 0$)

$$P^0(a, b) = \frac{2(2b - a)}{\sqrt{2\pi(2Dt)^3}} \exp\left\{-\frac{(2b - a)^2}{4Dt}\right\} \tag{6.149}$$

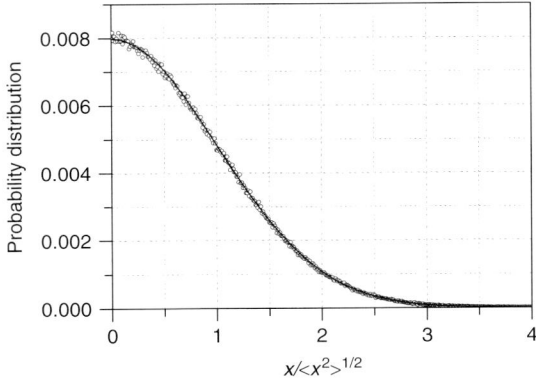

Figure 6.12 The probability distribution of maximal x values of the random walk depending on the normalized coordinate $x/\langle x^2 \rangle^{1/2}$ with $\langle x^2 \rangle = N$, where $N = 10^4$ is the number of discrete steps of unit length made by the random walker. The solid line shows the theoretical curve (6.146) for $2Dt = N$, whereas the simulation results obtained from a sample of 10^6 trajectories are represented by circles.

to have the maximum at $x = b$ and the end-coordinate at $x = a$ for stochastic trajectories starting at $x_0 = 0$ and running within the time interval from zero to t. The maximum distribution (6.146) is obtained by integrating the distribution (6.149) over all the possible a values from $-\infty$ to b.

6.7
Summary of Results for Diffusion in a Finite Interval

Summarizing the results of the probability density $p(x, t)$ for one-dimensional diffusion in a finite interval $a \leq x \leq b$ we distinguish between three different cases related to the properties of the boundaries.

6.7.1
Reflected Diffusion

Diffusion within two reflecting barriers (RR) at $x = a$ and $x = b$:

Mirror method:

$$p_{RR}(x, t) = \sum_{m=-\infty}^{\infty} \left[p_{x_0-a+2m(b-a)}(x, t) + p_{-(x_0-a)+2m(b-a)}(x, t) \right] \qquad (6.150)$$

with

$$p_{x_0}(x, t) = \frac{1}{\sqrt{4\pi Dt}} \exp\left(-\frac{(x - x_0)^2}{4Dt}\right). \qquad (6.151)$$

Eigenfunction expansion:

$$p_{RR}(x, t) = \frac{1}{b-a} \sum_{m=-\infty}^{\infty} e^{-Dk_m^2 t} \cos\left(k_m(x_0 - a)\right) \cos\left(k_m(x - a)\right) \qquad (6.152)$$

or

$$p_{RR}(x, t) = \frac{1}{b-a} + \frac{2}{b-a} \sum_{m=1}^{\infty} e^{-Dk_m^2 t} \cos\left(k_m(x_0 - a)\right) \cos\left(k_m(x - a)\right)$$

$$(6.153)$$

with

$$k_m = \frac{\pi}{b-a} m, \quad \lambda_m = Dk_m^2 = \frac{D\pi^2}{(b-a)^2} m^2. \qquad (6.154)$$

Reflected diffusion takes place in a closed system with a normalized survival function

$$G_{RR}(t, x_0) = \int_a^b p_{RR}(x, t) \, dx = 1 \quad \text{or} \quad \frac{d}{dt} G(t, x_0) = 0 \qquad (6.155)$$

without outflow at any border $\mathcal{P}(t, x = a) = 0$ and $\mathcal{P}(t, x = b) = 0$.

6.7.2
Diffusion in a Semi-Open System

Reflecting barrier (left at $x = a$) and absorbing border (right at $x = b$) (RA):

Mirror method:

$$p_{RA}(x,t) = \sum_{m=-\infty}^{\infty} (-1)^m \left[p_{x_0 - a + 2m(b-a)}(x,t) + p_{-(x_0 - a) + 2m(b-a)}(x,t) \right] \quad (6.156)$$

with

$$p_{x_0}(x,t) = \frac{1}{\sqrt{4\pi Dt}} \exp\left(-\frac{(x-x_0)^2}{4Dt}\right). \quad (6.157)$$

Eigenfunction expansion:

$$p_{RA}(x,t) = \frac{1}{b-a} \sum_{m=-\infty}^{\infty} e^{-Dk_m^2 t} \cos\left(k_m(x_0 - a)\right) \cos\left(k_m(x - a)\right) \quad (6.158)$$

or

$$p_{RA}(x,t) = \frac{2}{b-a} \sum_{m=0}^{\infty} e^{-Dk_m^2 t} \cos\left(k_m(x_0 - a)\right) \cos\left(k_m(x - a)\right) \quad (6.159)$$

with

$$k_m = \frac{\pi}{b-a}\left(\frac{1}{2} + m\right), \quad \lambda_m = Dk_m^2 = \frac{D\pi^2}{(b-a)^2}\left(\frac{1}{2} + m\right)^2. \quad (6.160)$$

Survival function in a semi-open system:

$$G_{RA}(t, x_0) = \int_a^b p_{RA}(x,t)\,dx$$

$$= \frac{2}{b-a} \sum_{m=0}^{\infty} e^{-Dk_m^2 t} \cos\left(k_m(x_0 - a)\right) \frac{(-1)^m}{k_m} \quad (6.161)$$

Outflow density at the right border:

$$\mathcal{P}(t, x=b) = \frac{d}{dt} G(t, x_0) = j(x=b, t) = -D \left.\frac{\partial p(x,t)}{\partial x}\right|_{x=b}$$

$$= \frac{2}{b-a} \sum_{m=0}^{\infty} \frac{\lambda_m}{k_m} e^{-\lambda_m t} \cos\left(k_m(x_0 - a)\right) \sin\left(k_m(b-a)\right)$$

$$= \frac{2\pi D}{(b-a)^2} \sum_{m=0}^{\infty} (-1)^m \left(\frac{1}{2} + m\right) e^{-Dk_m^2 t} \cos\left(k_m(x_0 - a)\right). \quad (6.162)$$

Mean first passage time:

$$\langle t_1(x_0 \to b)\rangle = \frac{1}{2}\frac{b-x_0}{D}(b-a+x_0-a) \tag{6.163}$$

including the special case

$$\langle t_1(x_0=a \to b)\rangle = \frac{1}{2}\frac{(b-a)^2}{D}. \tag{6.164}$$

Second moment:

$$\langle t_2(x_0 \to b)\rangle = \frac{5}{12}\frac{(b-a)^4}{D^2} + \frac{1}{12}\frac{(x_0-a)^2}{D^2}\left((x_0-a)^2 - 6(b-a)^2\right) \tag{6.165}$$

with variance

$$\sqrt{\langle t_2(x_0=a \to b)\rangle - \langle t_1(x_0=a \to b)\rangle^2} = \frac{1}{\sqrt{6}}\frac{(b-a)^2}{D}. \tag{6.166}$$

6.7.3
Diffusion in an Open System

Diffusion within two absorbing boundaries (AA):

Mirror method:

$$p_{AA}(x,t) = \sum_{m=-\infty}^{\infty}\left[p_{x_0-a+2m(b-a)}(x,t) - p_{-(x_0-a)+2m(b-a)}(x,t)\right] \tag{6.167}$$

with

$$p_{x_0}(x,t) = \frac{1}{\sqrt{4\pi Dt}}\exp\left(-\frac{(x-x_0)^2}{4Dt}\right). \tag{6.168}$$

Eigenfunction expansion:

$$p_{AA}(x,t) = \frac{1}{b-a}\sum_{m=-\infty}^{\infty} e^{-Dk_m^2 t}\sin\left(k_m(x_0-a)\right)\sin\left(k_m(x-a)\right) \tag{6.169}$$

or

$$p_{AA}(x,t) = \frac{2}{b-a}\sum_{m=1}^{\infty} e^{-Dk_m^2 t}\sin\left(k_m(x_0-a)\right)\sin\left(k_m(x-a)\right) \tag{6.170}$$

with

$$k_m = \frac{\pi}{b-a}m, \quad \lambda_m = Dk_m^2 = \frac{D\pi^2}{(b-a)^2}m^2. \tag{6.171}$$

Survival function in an open system:

$$G_{AA}(t, x_0) = \int_a^b p_{AA}(x, t)\, dx$$

$$= \frac{2}{b-a} \sum_{m=0}^{\infty} e^{-Dk_m^2 t} \cos\left(k_m(x_0 - a)\right) \frac{(-1)^{m+1}}{k_m}. \tag{6.172}$$

Outflow density at left border:

$$\mathcal{P}(t, x = a) = -j(x = a, t) = +D \left.\frac{\partial p(x, t)}{\partial x}\right|_{x=a}$$

$$= \frac{2\pi D}{(b-a)^2} \sum_{m=1}^{\infty} e^{-Dk_m^2 t}\, m \sin\left(k_m(x_0 - a)\right). \tag{6.173}$$

Outflow density at right border:

$$\mathcal{P}(t, x = b) = j(x = b, t) = -D \left.\frac{\partial p(x, t)}{\partial x}\right|_{x=b}$$

$$= \frac{2\pi D}{(b-a)^2} \sum_{m=0}^{\infty} e^{-Dk_m^2 t} (-1)^{m+1}\, m \sin\left(k_m(x_0 - a)\right). \tag{6.174}$$

6.8 Exercises

E 6.1 Random walk with right border

Consider a discrete random walk on a line starting at the origin $m = 0$. The walker can move with equal probability $p = q = 1/2$ one unit to the left or to the right at each time step. The motion is unbounded from the left (natural boundary condition at $m \to -\infty$), and there is an absorbing boundary at $m = m_b$. Find the probability $p(m_b, n)$ that the walker will be absorbed at $m = m_b$ after n time steps. Compare the obtained result with the simple model having natural boundary conditions.

E 6.2 Galton board

Sir Francis Galton studied stochastic motion by discrete probabilistic jumps on an experimental setup which is now called the Galton board. As for a drunken sailor, he mimics the outcome of a large number of independent experiments with balls on triangular arrangements of nails as scatterers.

The general task is to analyze the stochastic dynamics of an asymmetrical Galton board including the continuum limit.

Divide the task into three parts. Start with the historical background and learn from Galton's life about the motivation of Sir Francis Galton to study human heredity and his interest on qualitative statistics caused by Charles Darwin's 'Origin of Species'.

The Galton board is a discrete version of Brownian motion. Study the time evolution of the probability of finding the ball (as a point particle) after the nth collision in a certain bin m. Having in mind the Markov property, solve this Markov chain dynamics using elementary left (p) and right ($q = 1 - p$) jump probabilities. Consider the general case $p \neq q$ and discuss two special geometries: pure symmetry $p = q = 1/2$, and total asymmetry $p = 0$.

Francis Galton had already made the point that, for large n or a large number of independent experiments, the probability of the outcome approaches the normal distribution. Derive this Gaussian normal distribution as a continuum limit out of the binomial distribution. Introduce continuous spatial and time variables as well as microscopic parameters for drift and diffusion in order to get the corresponding Fokker–Planck equation.

E 6.3 Brownian motion

This motion named after Robert Brown shows the stochastic displacement of a particle. Brown was a botanist and he did not realize that the motion he saw in 1827 was associated with collisions on a molecular scale. It took over 75 years before Albert Einstein recognized the connection between Brownian motion and the physical process called diffusion.

Study Einstein's concept of Brownian motion to derive the well known diffusion equation (6.26) by reading the orginal paper 'Über die von der molekularkinetischen Theorie der Wärme geforderte Bewegung von in ruhenden Flüssigkeiten suspendierten Teilchen' [40] in Annalen der Physik 1905, pp. 549–60.

The solution of the diffusion equation is known as a Gaussian distribution. Show its profile and discuss the properties, especially the second moment called the mean-square displacement.

E 6.4 Levy walks

The mathematician Paul Levy examined random walks with self-similar dynamics and power-law scaling. Prepare an overview talk with handout about these so-called Levy walks (or Levy flights) and show the relationship with the traditional random walk or ordinary diffusion studied perviously.

7
Bounded Drift–Diffusion Motion

7.1
Drift–Diffusion Equation with Natural Boundaries

Let us consider the following Fokker–Planck equation

$$\frac{\partial p(x,t)}{\partial t} = -v \frac{\partial p(x,t)}{\partial x} + D \frac{\partial^2 p(x,t)}{\partial x^2} \tag{7.1}$$

with constant (positive or negative) drift v and diffusion $D > 0$ in the interval $-\infty < x < +\infty$ called natural boundaries, together with the initial condition as a delta function $p(x, t = 0) = \delta(x - x_0)$.

We would like to find an analytical solution $p(x,t)$ of (7.1) by the linear transformation

$$P(y, T) \, dy = p(x, t) \, dx \tag{7.2}$$

with

$$y = y(x, t) = x - vt, \tag{7.3}$$

$$T = T(x, t) = Dt. \tag{7.4}$$

According to (7.2) we have

$$p(x,t) = P(y,t) \frac{dy}{dx} = P(y, T) \cdot 1 = P(y, T). \tag{7.5}$$

The time derivative on the left-hand side of (7.1) transforms to

$$\frac{\partial p(x,t)}{\partial t} = \frac{\partial P(y,T)}{\partial t} = \frac{\partial P(y,T)}{\partial T} \frac{\partial T}{\partial t} + \frac{\partial P(y,T)}{\partial y} \frac{\partial y}{\partial t}, \tag{7.6}$$

whereas the coordinate derivatives on the right-hand side are

$$\frac{\partial p(x,t)}{\partial x} = \frac{\partial P(y,T)}{\partial x} = \frac{\partial P(y,T)}{\partial y} \frac{\partial y}{\partial x} \tag{7.7}$$

Physics of Stochastic Processes: How Randomness Acts in Time
Reinhard Mahnke, Jevgenijs Kaupužs and Ihor Lubashevsky
Copyright © 2009 WILEY-VCH Verlag GmbH & Co. KGaA, Weinheim
ISBN: 978-3-527-40840-5

and

$$\frac{\partial^2 p(x,t)}{\partial x^2} = \frac{\partial^2 P(y,T)}{\partial x^2} = \frac{\partial}{\partial y}\left(\frac{\partial P(y,T)}{\partial y}\frac{\partial y}{\partial x}\right)\frac{\partial y}{\partial x}. \tag{7.8}$$

According to (7.3) and (7.4) we have $\partial y/\partial x = 1$, $\partial y/\partial t = -v$, and $\partial T/\partial t = D$. By inserting these relations into (7.6)–(7.8) and then all into (7.1) we obtain the well known partial differential equation called the diffusion law (cf. 6.26)

$$\frac{\partial P(y,T)}{\partial T} = \frac{\partial^2 P(y,T)}{\partial y^2} \tag{7.9}$$

with vanishing distribution $P(y,T)$ at infinite boundaries and initial condition $P(y, T=0) = \delta(y - y_0)$.

The Gaussian solution of (7.9) reads

$$P(y,T) = \frac{1}{\sqrt{4\pi T}} e^{-\frac{(y-y_0)^2}{4T}}, \tag{7.10}$$

and after inverse transformation to original variables we get

$$p(x,t) = \frac{1}{\sqrt{4\pi Dt}} e^{-\frac{(x-x_0-vt)^2}{4Dt}}. \tag{7.11}$$

Figure 7.1 illustrates the dynamics given by the Gaussian moving and widening profile (7.11).

The mean value (first moment) changes linearly in time

$$\langle x \rangle(t) = x_0 + vt, \tag{7.12}$$

as does the variance (second moment minus the squared first moment)

$$\langle x^2 \rangle(t) - \langle x \rangle^2(t) = 2Dt. \tag{7.13}$$

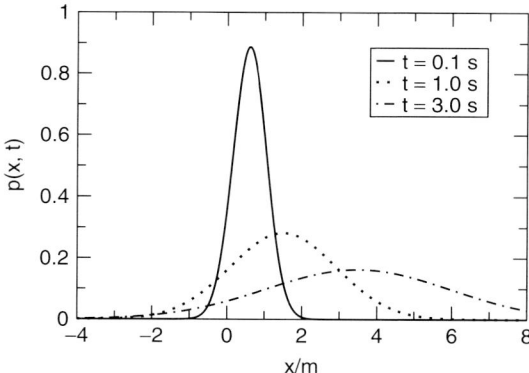

Figure 7.1 Evolution of the probability density $p(x,t)$ of drift–diffusion dynamics for different time moments. Parameters: $v = 1\,\mathrm{ms}^{-1}$; $D = 1\,\mathrm{m^2 s^{-1}}$; $x_0 = 0.5\,\mathrm{m}$.

7.2
Drift–Diffusion Problem with Absorbing and Reflecting Boundaries

Let us consider the initial boundary-value problem (shown schematically in Figure 7.2) with constant diffusion coefficient D and constant drift coefficient v. Our task is to calculate analytically the probability density $p(x, t)$ of finding the system in state x (exactly in the small interval $[x; x + dx]$) at time t. The state space is defined as closed on the left-hand side and open on the right-hand side. Due to these properties we introduce boundary conditions which determine the behavior of the solution. Another important quantity is the outflow (or breakdown) probability density at the right border which is found from the solution of the Fokker–Planck equation using the balance equation [22, 82, 83, 185]. Applications for the calculated outflow probability are the many-car systems used to define and describe traffic breakdown on roads, depending stochastically on the vehicular density [111, 112, 151].

The dynamics of $p(x, t)$ is given by the drift–diffusion equation, as well as the initial and boundary conditions:

1. The drift–diffusion equation (Fokker–Planck dynamics, see (7.1))

$$\frac{\partial p(x, t)}{\partial t} = -v\frac{\partial p(x, t)}{\partial x} + D\frac{\partial^2 p(x, t)}{\partial x^2}. \tag{7.14}$$

2. The initial condition (delta function)

$$p(x, t = 0) = \delta(x - x_0). \tag{7.15}$$

3. The reflecting boundary at $x = a$ (flux j vanishes at left border)

$$j(x = a, t) = vp(x = a, t) - D\left.\frac{\partial p(x, t)}{\partial x}\right|_{x=a} = 0. \tag{7.16}$$

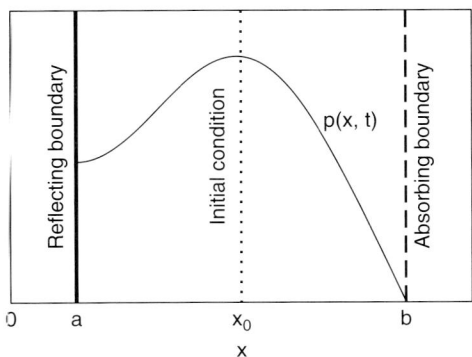

Figure 7.2 Schematic picture of the boundary-value problem showing the probability density $p(x, t)$ in the finite interval $a \leq x \leq b$.

4. The absorbing boundary at $x = b$ (probability density p vanishes at the right border)

$$p(x = b, t) = 0. \tag{7.17}$$

The balance equation in our open system, see Figure 7.2,

$$\frac{\partial}{\partial t} \int_a^b p(x, t)\, dx + \mathcal{P}(t, x = b) = 0 \tag{7.18}$$

relates the probability that the system is still in a state $x \in [a, b]$ with the probability flux $\mathcal{P}(t, x = b)$ out of this interval at the right absorbing boundary $x = b$ at time t (exactly in the interval $[t; t + dt]$).

7.3
Dimensionless Drift–Diffusion Equation

It is convenient to formulate the drift–diffusion problem in dimensionless variables. For this purpose we define a new variable $0 \leq y \leq 1$ instead of $a \leq x \leq b$ by

$$y = \frac{x - a}{b - a}, \tag{7.19}$$

a new time T by

$$T = \frac{D}{(b - a)^2} t, \tag{7.20}$$

one dimensionless control parameter (scaled drift v which may have positive, zero, or negative values)

$$\Omega = \frac{v}{D}(b - a), \tag{7.21}$$

a new probability density $P(y, T)$ by

$$P(y, T)\, dy = p(x, t)\, dx \tag{7.22}$$

and therefore

$$P(y, T) = (b - a)p(x, t). \tag{7.23}$$

As a result, (7.14)–(7.17) can be rewritten as follows:

1. The drift–diffusion equation (Fokker–Planck dynamics)

$$\frac{\partial P(y, T)}{\partial T} = -\Omega \frac{\partial P(y, T)}{\partial y} + \frac{\partial^2 P(y, T)}{\partial y^2}. \tag{7.24}$$

2. The initial condition (delta–function)

$$P(y, T = 0) = \delta(y - y_0). \tag{7.25}$$

3. The reflecting boundary at $y = 0$ (flux J vanishes at left border)

$$J(y=0, T) = \Omega P(y=0, T) - \left.\frac{\partial P(y, T)}{\partial y}\right|_{y=0} = 0. \tag{7.26}$$

4. The absorbing boundary at $y = 1$ (probability density P vanishes at right border)

$$P(y = 1, T) = 0. \tag{7.27}$$

Our aim is to calculate the first-passage time distribution function $\overline{\mathcal{P}}(t)$ as the probability flux out of the system, given in dimensionless variables by

$$\overline{\mathcal{P}}(T, y = 1) = -\frac{\partial}{\partial T} G(T) \quad \text{with} \quad G(T) = \int_0^1 P(y, T) \, dy. \tag{7.28}$$

To switch over to cumulative probability functions we define

$$W(T \leq T_{\text{obs}}) = \int_0^{T_{\text{obs}}} \overline{\mathcal{P}}(T, y = 1) \, dT \tag{7.29}$$

and using (7.28)

$$W(T \leq T_{\text{obs}}) = -\int_0^{T_{\text{obs}}} \frac{\partial G(T)}{\partial T} \, dT = -G(T_{\text{obs}}) + G(0) \tag{7.30}$$

calculate the following relationship

$$W(T \leq T_{\text{obs}}) = 1 - G(T_{\text{obs}}). \tag{7.31}$$

Both (7.28) (differential form) and (7.31) (integral notation) consider the balance between the probability which, at time T, is still inside the system $G(T)$ and the flux which is passing the surface at that moment $\overline{\mathcal{P}}(T, y = 1)$ or the amount of probability which has already left the system during the time interval $0 \leq T \leq T_{\text{obs}}$ over the surface to the outside $W(t \leq T_{\text{obs}})$.

By analogy with the statistics of the durability of technical components we may call the probability distribution $G(T_{\text{obs}})$ the survival or life-time function indicating that the equipment or particle is still alive at observation time T_{obs}. The breakdown function $W(T \leq T_{\text{obs}})$ gives the probability distribution for the components which have already been broken from the beginning up to the observation time T_{obs}. In this context, the cumulative life-time distribution $G(T_{\text{obs}})$ is often estimated on the basis of experiments of limited duration [26].

7.4
Solution in Terms of Orthogonal Eigenfunctions

To find the solution of the well defined drift–diffusion problem, first we take the dimensionless form (7.24)–(7.27) and use a transformation to a new function Q by

$$Q(y, T) = e^{-\frac{\Omega}{2} y} P(y, T). \tag{7.32}$$

This results in a dynamics without a first derivative, called the reduced Fokker–Planck equation

$$\frac{\partial Q(y, T)}{\partial T} = -\frac{\Omega^2}{4} Q(y, T) + \frac{\partial^2 Q(y, T)}{\partial y^2}. \tag{7.33}$$

According to (7.32) the initial condition is transformed to

$$Q(y, T = 0) = e^{-\frac{\Omega}{2} y_0} P(y, T = 0), \tag{7.34}$$

whereas the reflecting boundary condition at $y = 0$ becomes

$$\frac{\Omega}{2} Q(y = 0, T) - \left. \frac{\partial Q(y, T)}{\partial y} \right|_{y=0} = 0, \tag{7.35}$$

and the absorbing boundary condition at $y = 1$ now reads

$$Q(y = 1, T) = 0. \tag{7.36}$$

The solution of the reduced equation (7.33) can be found by the method of separation of variables. Making a separation ansatz $Q(y, T) = \chi(T)\psi(y)$, we obtain

$$\frac{1}{\chi(T)} \frac{d\chi(T)}{dT} = -\frac{\Omega^2}{4} + \frac{1}{\psi(y)} \frac{d^2 \psi(y)}{dy^2}. \tag{7.37}$$

Both sides should be equal to a constant. This constant is denoted by $-\lambda$, where λ has the meaning of an eigenvalue. The eigenvalue λ should be real and non-negative.

Integration of the left-hand side gives the exponential decay

$$\chi(T) = \chi_0 \exp\{-\lambda T\} \tag{7.38}$$

with $\chi(T = 0) = \chi_0$ where the constant χ_0 can be set equal to 1.

Let us now define the dimensionless wave number k as $k^2 = \lambda$. The right-hand side of (7.37) then transforms into the following wave equation

$$\frac{d^2 \psi(y)}{dy^2} + \left(k^2 - \frac{\Omega^2}{4}\right) \psi(y) = 0. \tag{7.39}$$

Further on, we introduce a modified wave number $\tilde{k}^2 = k^2 - \Omega^2/4$. Note that $\tilde{k} = +\sqrt{k^2 - \Omega^2/4}$ may be complex (either pure real or pure imaginary).

First we consider the case where $\tilde{k}$ is real. A suitable complex ansatz for the solution of the wave equation (7.39) reads

$$\psi(y) = C^* \exp\{+i\tilde{k}y\} + C \exp\{-i\tilde{k}y\} \tag{7.40}$$

with complex coefficients $C = A/2 + iB/2$ and $C^* = A/2 - iB/2$ chosen in such a way as to ensure a real solution

$$\psi(y) = A \cos(\tilde{k}y) + B \sin(\tilde{k}y). \tag{7.41}$$

The two boundary conditions (7.35) and (7.36) can be used to determine the modified wave number $\tilde{k}$ and the ratio A/B. The particular solutions are eigenfunctions $\psi_m(y)$, which form a complete set of orthogonal functions. As the third condition, we require that these eigenfunctions are normalized

$$\int_0^1 \psi_m^2(y)\,dy = 1. \tag{7.42}$$

In this case all three parameters $\tilde{k}$, A, and B are defined.

The condition for the left boundary (7.35) reads

$$\frac{\Omega}{2}\psi(y=0) - \frac{d\psi(y)}{dy}\bigg|_{y=0} = 0. \tag{7.43}$$

After substitution by (7.40) this reduces to

$$\frac{\Omega}{2}(C^* + C) = i\tilde{k}(C^* - C) \tag{7.44}$$

or

$$\frac{\Omega}{2}A = \tilde{k}B. \tag{7.45}$$

The condition for the right boundary (7.36)

$$\psi(y=1) = 0 \tag{7.46}$$

gives us

$$C^* \exp\{+i\tilde{k}\} + C \exp\{-i\tilde{k}\} = 0 \tag{7.47}$$

or

$$A \cos\left(\tilde{k}\right) + B \sin\left(\tilde{k}\right) = 0. \tag{7.48}$$

By putting both equalities (7.45) and (7.48) together and looking for a nontrivial solution, we arrive at a transcendental equation

$$i\frac{\Omega}{2}\left(\exp\{+i\tilde{k}\} - \exp\{-i\tilde{k}\}\right) = \tilde{k}\left(\exp\{+i\tilde{k}\} + \exp\{-i\tilde{k}\}\right) \tag{7.49}$$

or

$$\frac{\Omega}{2}\sin\left(\tilde{k}\right) + \tilde{k}\cos\left(\tilde{k}\right) = 0, \tag{7.50}$$

and, respectively,

$$\tan\left(\tilde{k}\right) = -\frac{2}{\Omega}\tilde{k}, \tag{7.51}$$

which gives the spectrum of values $\tilde{k}_m$ with $m = 0, 1, 2, \ldots$ (numbered in such a way that $0 < \tilde{k}_0 < \tilde{k}_1 < \tilde{k}_2 < \ldots$) and the discrete eigenvalues $\lambda_m > 0$.

Due to (7.41) and (7.48), the eigenfunctions can be written as

$$\psi_m(y) = R_m\left[\cos\left(\tilde{k}_m y\right)\sin\left(\tilde{k}_m\right) - \cos\left(\tilde{k}_m\right)\sin\left(\tilde{k}_m y\right)\right], \tag{7.52}$$

where $R_m = A_m/\sin(\tilde{k}_m) = -B_m/\cos(\tilde{k}_m)$. Taking into account the identity $\sin(\alpha - \beta) = \sin\alpha\cos\beta - \cos\alpha\sin\beta$, (7.52) reduces to

$$\psi_m(y) = R_m \sin\left[\tilde{k}_m(1-y)\right]. \tag{7.53}$$

The normalization constant R_m is found by inserting (7.53) into (7.42). Calculation of the normalization integral by using the transcendental equation (7.50) gives us

$$R_m^2 \int_0^1 \sin^2\left[\tilde{k}_m(1-y)\right] dy = R_m^2 \left[\frac{1}{2} - \frac{1}{4\tilde{k}_m}\sin\left(2\tilde{k}_m\right)\right]$$

$$= \frac{R_m^2}{2}\left(1 + \frac{\Omega}{2}\frac{1}{\tilde{k}_m^2 + \Omega^2/4}\right) = 1, \tag{7.54}$$

and hence (7.53) becomes

$$\psi_m(y) = \sqrt{\frac{2}{1 + \frac{\Omega}{2}\frac{1}{\tilde{k}_m^2+\Omega^2/4}}} \sin\left[\tilde{k}_m(1-y)\right] \tag{7.55}$$

or

$$\psi_m(y) = \sqrt{\frac{2}{1 + \frac{\Omega}{2}\frac{1}{k_m^2}}} \sin\left[\sqrt{k_m^2 - \Omega^2/4}\,(1-y)\right]. \tag{7.56}$$

This calculation refers to the case $\Omega > -2$ where all wave numbers k_m or $\tilde{k}_m = \sqrt{k_m^2 - \Omega^2/4}$ are real and positive.

However, the smallest or ground-state wave vector $\tilde{k}_0$ vanishes when Ω tends to -2 from above, and no continuation of this solution exists on the real axis for $\Omega < -2$. A purely imaginary solution $\tilde{k}_0 = i\kappa_0$ appears instead, where κ_0 is real, see Figure 7.3. In this case (for $\Omega < -2$) a real ground state eigenfunction $\psi_0(y)$ can be found in the form (7.40) where $C = A/2 + B/2$ and $C^* = A/2 - B/2$, so,

$$\psi_0(y) = A\cosh(\kappa_0 y) + B\sinh(\kappa_0 y). \tag{7.57}$$

The transcendental equation for the wave number $\tilde{k}_0 = i\kappa_0$ can be written as the following equation for κ_0

$$\frac{\Omega}{2}\sinh(\kappa_0) + \kappa_0\cosh(\kappa_0) = 0. \tag{7.58}$$

As compared to the previous case $\Omega > -2$, trigonometric functions are replaced by the corresponding hyperbolic ones. Similar calculations as before yield

$$\psi_0(y) = \sqrt{-\frac{2}{1 + \frac{\Omega}{2}\frac{1}{-\kappa_0^2+\Omega^2/4}}} \sinh\left[\kappa_0(1-y)\right]. \tag{7.59}$$

Note that $\kappa_0 = -i\tilde{k}_0$ is the imaginary part of $\tilde{k}_0$ and $\kappa_0^2 = -\tilde{k}_0^2$. As regards other solutions of (7.50), called excited states, which are those for $\tilde{k}_m$ with $m > 0$, nothing special happens at $\Omega = -2$, so that these wave numbers are always real. The

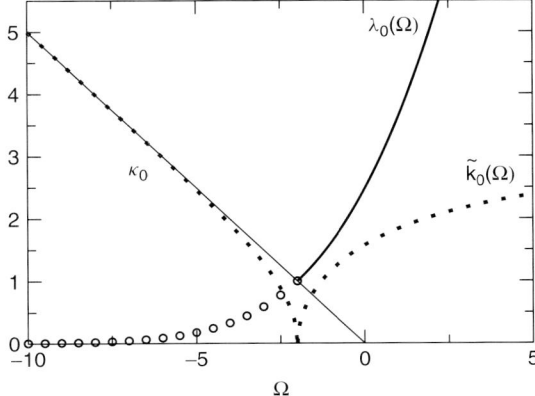

Figure 7.3 The wave number $\tilde{k}_0$ ($\Omega \geq -2$) respectively κ_0 ($\Omega \leq -2$) and eigenvalue λ_0 for the ground state $m = 0$. The thin straight line shows the approximation $\kappa_0 \approx -\Omega/2$ valid for large negative $\Omega < -5$.

situation for the ground state $m = 0$ at different values of the dimensionless drift parameter Ω is summarized in Table 7.1, which presents the solutions of the transcendental equation (7.58) κ_0 together with $\lambda_0 = -\kappa_0^2 + \Omega^2/4$ and those of (7.50) $\tilde{k}_0$ together with eigenvalues $\lambda_0 = \tilde{k}_0^2 + \Omega^2/4$. Table 7.2 shows the behavior of the lowest wave numbers $\tilde{k}_m$ with $m = 0, 1, \ldots, 5$. The results are plotted in Figure 7.4.

Table 7.1 The ground-state wave number κ_0 (for $\Omega \leq -2$) and $\tilde{k}_0$ (for $\Omega \geq -2$) and eigenvalue λ_0 depending on the dimensionless drift parameter Ω.

Ω	κ_0	λ_0	Ω	$\tilde{k}_0$	λ_0
−9.00	4.499	0.010	−2.00	0.000	1.000
−8.50	4.248	0.015	−1.50	0.845	1.276
−8.00	3.997	0.021	−1.00	1.165	1.608
−7.50	3.745	0.031	−0.50	1.393	2.004
−7.00	3.493	0.045	0.00	1.571	2.468
−6.50	3.240	0.064	0.50	1.715	3.005
−6.00	2.984	0.091	1.00	1.836	3.623
−5.50	2.726	0.128	1.50	1.939	4.325
−5.00	2.464	0.178	2.00	2.028	5.116
−4.50	2.195	0.245	2.50	2.106	5.999
−4.00	1.915	0.333	3.00	2.174	6.979
−3.50	1.617	0.446	3.50	2.235	8.058
−3.00	1.288	0.591	4.00	2.288	9.239
−2.50	0.888	0.774	4.50	2.337	10.525
−2.00	0.000	1.000	5.00	2.381	11.917

Table 7.2 The wave numbers $\tilde{k}_m$ ($m = 0, 1, \ldots, 5$) depending on the dimensionless drift parameter Ω.

Ω	−10.0	−5.0	−2.0	−1.0	0.0	1.0	2.0	5.0	10.0
$m = 0$		4.999	2.464	0.000	1.165	1.571	1.836	2.028	2.381	2.653
$m = 1$	3.790	4.172	4.493	4.604	4.712	4.816	4.913	5.163	5.454
$m = 2$	7.250	7.533	7.725	7.789	7.854	7.917	7.979	8.151	8.391
$m = 3$	10.553	10.767	10.904	10.949	10.995	11.040	11.085	11.214	11.408
$m = 4$	13.789	13.959	14.066	14.101	14.137	14.172	14.207	14.310	14.469
$m = 5$	16.992	17.133	17.220	17.249	17.279	17.308	17.336	17.421	17.556

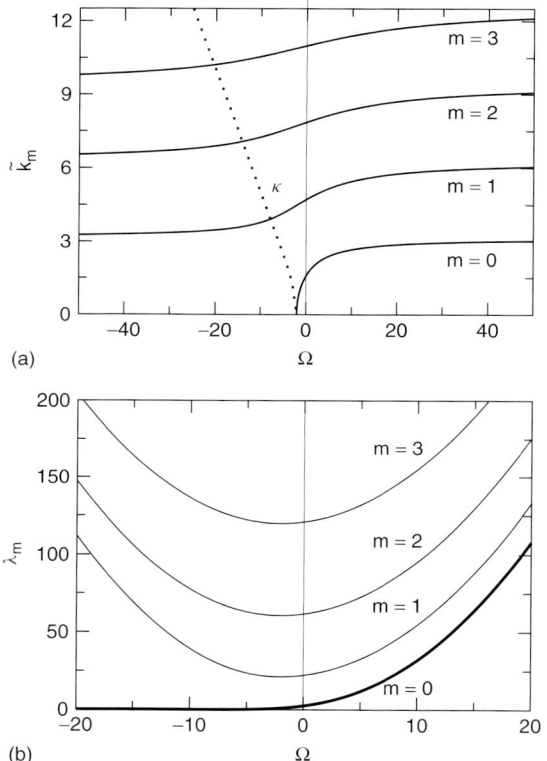

Figure 7.4 The parameter dependence of wave numbers $\tilde{k}_m(\Omega)$ (a) and eigenvalues $\lambda_m(\Omega)$ (b) for ground state $m = 0$ and excited states $m = 1, 2, 3$.

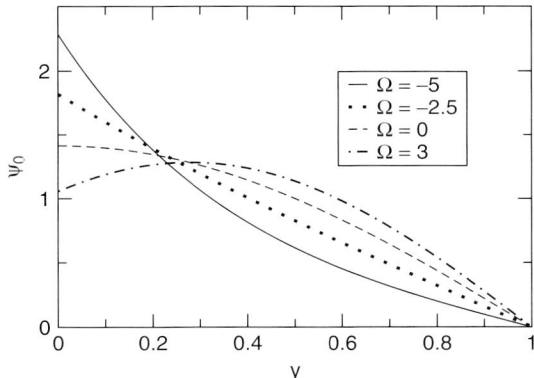

Figure 7.5 The eigenfunction $\psi_0(y)$ for different values of control parameter Ω.

In general (for arbitrary Ω), the eigenfunctions are orthogonal and normalized:

$$\int_0^1 \psi_l(y)\psi_m(y)\,dy = 0, \qquad m \neq l \tag{7.60}$$

$$\int_0^1 \psi_m(y)\psi_m(y)\,dy = 1, \qquad m = l. \tag{7.61}$$

Figure 7.5 shows the ground eigenstate ($m = 0$) for different parameter values Ω, whereas Figure 7.6 gives a collection of eigenstate functions ($m = 0, 1, \ldots, 5$) for $\Omega = -5.0$ and $\Omega = 3.0$.

In the following, explicit formulas (where $\psi_m(y)$ is specified) are written for the case $\Omega > -2$.

In order to construct the time-dependent solution for $Q(y, t)$, which fulfills the initial condition, we consider the superposition of all particular solutions with different eigenvalues λ_m

$$Q(y, T) = \sum_{m=0}^{\infty} C_m e^{-\lambda_m T} \psi_m(y). \tag{7.62}$$

By inserting the initial condition

$$P(y, T = 0) = e^{\frac{\Omega}{2}y} Q(y, T = 0) = \delta(y - y_0) \tag{7.63}$$

into (7.62) we obtain

$$\sum_{m=0}^{\infty} C_m \psi_m(y) = e^{-\frac{\Omega}{2}y} \delta(y - y_0). \tag{7.64}$$

Now we expand the right-hand side of this equation by using the basis of orthonormal eigenfunctions (7.55) and identify C_m with the corresponding coefficient at ψ_m, so

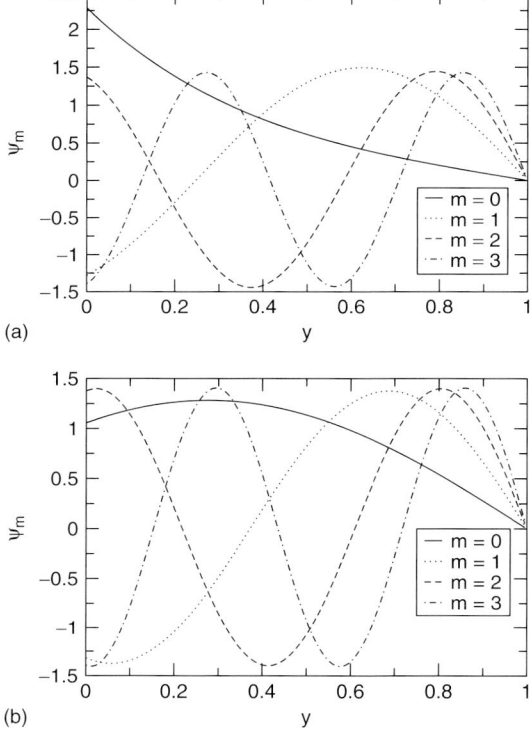

Figure 7.6 The eigenfunctions $\psi_m(y)$ for $m = 0, 1, 2, 3$ and for $\Omega = -5.0$ (a) and $\Omega = 3.0$ (b).

$$C_m = \int e^{\frac{\Omega}{2} y} \delta(y - y_0) \psi_m \, dy = e^{-\frac{\Omega}{2} y_0} \psi_m(y_0). \tag{7.65}$$

This allows us to write the solution for $P(y, T)$ as

$$P(y, T) = e^{\frac{\Omega}{2}(y - y_0)} \sum_{m=0}^{\infty} e^{-\lambda_m T} \psi_m(y_0) \psi_m(y), \tag{7.66}$$

or, more specifically,

$$P(y, T) = 2 e^{\frac{\Omega}{2}(y - y_0)} \sum_{m=0}^{\infty} \frac{e^{-\left(\tilde{k}_m^2 + \Omega^2/4\right) T}}{1 + \frac{\Omega}{2} \frac{1}{\tilde{k}_m^2 + \Omega^2/4}} \sin\left[\tilde{k}_m (1 - y_0)\right] \sin\left[\tilde{k}_m (1 - y)\right]. \tag{7.67}$$

The set of Figures 7.7 illustrates the time evolution of the probability density (7.67) choosing different parameter values Ω.

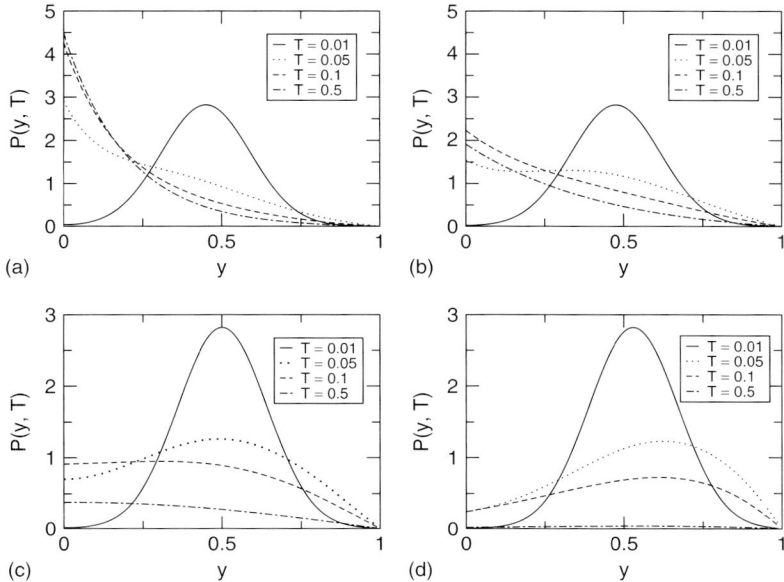

Figure 7.7 The solution of the drift–diffusion Fokker–Planck equation with initial condition $y_0 = 0.5$ for different values of the control parameter Ω, so: $\Omega = -5.0$ (a); $\Omega = -2.5$ (b); $\Omega = 0.1$ (c); $\Omega = 3.0$ (d).

Finally, we make the inverse transformation to original variables as follows:

$$p(x, t) = \frac{1}{b-a} P(y, T), \qquad (7.68)$$

$$\frac{\Omega}{2}(y - y_0) = \frac{v}{2D}(b-a)\left[\frac{x-a}{b-a} - \frac{x_0-a}{b-a}\right] = \frac{v}{2D}(x - x_0), \qquad (7.69)$$

$$\tilde{k}_m(1-y) = \tilde{k}_m \left(\frac{b-a}{b-a} - \frac{x-a}{b-a}\right) = \frac{\tilde{k}_m}{b-a}(b-x), \qquad (7.70)$$

$$\left(\tilde{k}_m^2 + \frac{\Omega^2}{4}\right) T = \left(\tilde{k}_m^2 + \frac{1}{4}\frac{v^2}{D^2}(b-a)^2\right) \frac{Dt}{(b-a)^2}$$

$$= \left[\left(\frac{\tilde{k}_m}{b-a}\right)^2 + \left(\frac{v}{2D}\right)^2\right] Dt, \qquad (7.71)$$

$$1 + \frac{\Omega}{2}\frac{1}{\tilde{k}_m^2 + \Omega^2/4} = 1 + \frac{v}{2D}(b-a) \frac{1}{\frac{\tilde{k}_m^2(b-a)^2}{(b-a)^2} + \left(\frac{v}{2D}(b-a)\right)^2}$$

$$= 1 + \frac{v}{2D(b-a)} \frac{1}{\left(\frac{\tilde{k}_m}{b-a}\right)^2 + \left(\frac{v}{2D}\right)^2}. \qquad (7.72)$$

Hence, the solution in original variables reads

$$p(x,t) = \frac{2}{a-b} e^{\frac{v}{2D}(x-x_0)} \tag{7.73}$$

$$\times \sum_{m=0}^{\infty} \frac{e^{-\left[\left(\frac{\tilde{k}_m}{b-a}\right)^2 + \left(\frac{v}{2D}\right)^2\right]Dt}}{1 + \frac{v}{2D(b-a)} \frac{1}{\left[\left(\frac{\tilde{k}_m}{b-a}\right)^2 + \left(\frac{v}{2D}\right)^2\right]}} \sin\left[\frac{\tilde{k}_m}{b-a}(b-x_0)\right] \sin\left[\frac{\tilde{k}_m}{b-a}(b-x)\right]$$

where the values $\tilde{k}_m$ are solutions of the transcendental equation (7.50)

$$\frac{1}{2}\frac{v}{D}(b-a)\sin\left(\tilde{k}_m\right) + \tilde{k}_m \cos\left(\tilde{k}_m\right) = 0. \tag{7.74}$$

7.5
First-Passage Time Probability Density

Let us recall the balance equation (7.18) for open systems. Using dimensionless variables, the quantity $\overline{\mathcal{P}}(T, y = 1) \, dT$, given by (7.28), is the probability that the absorbing boundary at $y = 1$ is reached for the first time within the time interval $[T, T + dT]$ (since it is forbiddenr to return and then reach it once again). Hence, $\overline{\mathcal{P}}(T, y = 1)$ is the first-passage time probability density sometimes called the breakdown probability density.

First we calculate the integral

$$G(T) = \int_0^1 P(y, T) \, dy \tag{7.75}$$

$$= 2 \int_0^1 dy \, e^{\frac{\Omega}{2}(y-y_0)} \sum_{m=0}^{\infty} \frac{e^{-\left(\tilde{k}_m^2 + \Omega^2/4\right)T}}{1 + \frac{\Omega}{2}\frac{1}{\tilde{k}_m^2 + \Omega^2/4}} \sin\left[\tilde{k}_m(1 - y_0)\right] \sin\left[\tilde{k}_m(1 - y)\right]$$

$$= 2 \sum_{m=0}^{\infty} \frac{e^{-\frac{\Omega}{2}y_0} e^{-\left(\tilde{k}_m^2 + \Omega^2/4\right)T}}{1 + \frac{\Omega}{2}\frac{1}{\tilde{k}_m^2 + \Omega^2/4}} \sin\left[\tilde{k}_m(1 - y_0)\right] \int_0^1 e^{\frac{\Omega}{2}y} \sin\left[\tilde{k}_m(1 - y)\right].$$

The latter integral is transformed as follows

$$I_n = \int_0^1 e^{\frac{\Omega}{2}y} \sin\left[\tilde{k}_m(1-y)\right] dy = e^{\frac{\Omega}{2}} \int_0^1 e^{-\frac{\Omega}{2}z} \sin\left[\tilde{k}_m z\right] dz. \tag{7.76}$$

By using the known integration formula

$$\int e^{ax} \sin(bx) \, dx = \frac{e^{ax}}{a^2 + b^2} \left[a \sin(bx) - b \cos(bx)\right], \tag{7.77}$$

7.5 First-Passage Time Probability Density | 227

where in our case $a = -\Omega/2$ and $b = \tilde{k}_m$, we obtain

$$I_n = e^{\frac{\Omega}{2}} \frac{\tilde{k}_m}{\tilde{k}_m^2 + \Omega^2/4}. \tag{7.78}$$

Hence, the integral (7.75) is

$$G(T) = 2 \sum_{m=0}^{\infty} \frac{e^{\frac{\Omega}{2}(1-y_0)} e^{-(\tilde{k}_m^2 + \Omega^2/4)T}}{1 + \frac{\Omega}{2}\frac{1}{\tilde{k}_m^2 + \Omega^2/4}} \frac{\tilde{k}_m}{\tilde{k}_m^2 + \Omega^2/4} \sin\left[\tilde{k}_m(1 - y_0)\right]. \tag{7.79}$$

By means of (7.79) the first-passage time probability density (7.28) can be calculated easily:

$$\overline{\mathcal{P}}(T, y = 1) = -\frac{\partial}{\partial T} \int_0^1 P(y, T) \, dy$$

$$= -2 \sum_{m=0}^{\infty} \frac{e^{\frac{\Omega}{2}(1-y_0)} \tilde{k}_m \sin\left[\tilde{k}_m(1-y_0)\right]}{1 + \frac{\Omega}{2}\frac{1}{\tilde{k}_m^2 + \Omega^2/4}} \left(\tilde{k}_m^2 + \Omega^2/4\right) \frac{\partial}{\partial T}\left[e^{-(\tilde{k}_m^2 + \Omega^2/4)T}\right]$$

$$= 2 e^{\frac{\Omega}{2}(1-y_0)} \sum_{m=0}^{\infty} \frac{e^{-(\tilde{k}_m^2 + \Omega^2/4)T}}{1 + \frac{\Omega}{2}\frac{1}{\tilde{k}_m^2 + \Omega^2/4}} \tilde{k}_m \sin\left[\tilde{k}_m(1 - y_0)\right]. \tag{7.80}$$

The outflow distribution $\overline{\mathcal{P}}(T, y = 1)$ is shown in Figure 7.8 (with different values of the dimensionless drift Ω) as well as in Figure 7.9 (with different values of the initial condition y_0).

Making the inverse transformation to original variables, we obtain

$$\overline{\mathcal{P}}(T, y = 1) \, dT = \mathcal{P}(t, x = b) \, dt, \tag{7.81}$$

$$\mathcal{P}(t, x = b) = \overline{\mathcal{P}}(T, y = 1) \frac{dT}{dt} = \overline{\mathcal{P}}(T, y = 1) \frac{D}{(b - a)^2}, \tag{7.82}$$

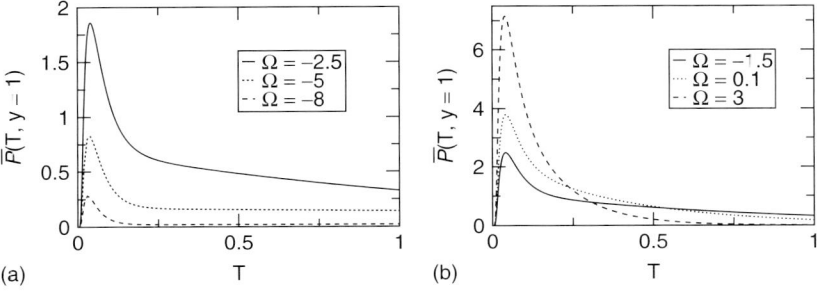

Figure 7.8 The first-passage time probability density distribution $\overline{\mathcal{P}}(T, y = 1)$ for $\Omega < -2$ (a) and $\Omega > -2$ (b).

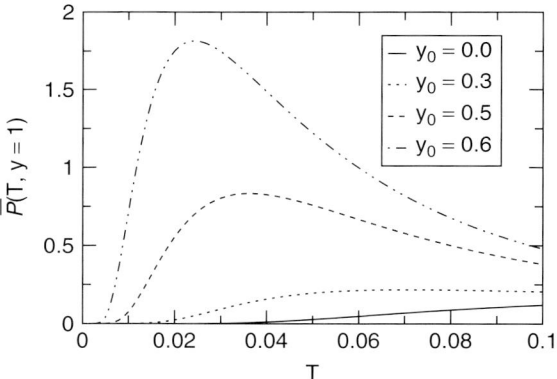

Figure 7.9 Short-time behavior of first-passage time probability density distribution $\overline{\mathcal{P}}(T, y=1)$ for different initial conditions $0 \leq y_0 \leq 1$ showing time lag.

and

$$\mathcal{P}(t, x = b) = \frac{2D}{b-a} e^{\frac{v}{2D}(b-x_0)} \tag{7.83}$$

$$\times \sum_{m=0}^{\infty} \frac{e^{-\left[\left(\frac{\tilde{k}_m}{b-a}\right)^2 + \left(\frac{v}{2D}\right)^2\right]Dt}}{1 + \frac{v}{2D} \frac{1}{(b-a)\left[\left(\frac{\tilde{k}_m}{b-a}\right)^2 + \left(\frac{v}{2D}\right)^2\right]}} \frac{\tilde{k}_m}{b-a} \sin\left[\frac{\tilde{k}_m}{b-a}(b-x_0)\right].$$

7.6
Cumulative Breakdown Probability

The probability that the absorbing boundary $y = 1$ is reached within a certain observation time interval $0 \leq T \leq T_{\text{obs}}$ is given by the cumulative (breakdown) probability

$$W(\Omega, T \leq T_{\text{obs}}) = \int_0^{T_{\text{obs}}} \overline{\mathcal{P}}(T, y = 1)\, dT \tag{7.84}$$

with (7.80)

$$\overline{\mathcal{P}}(T, y = 1) = 2e^{\frac{\Omega}{2}(1-y_0)} \sum_{m=0}^{\infty} \frac{e^{-\left(\tilde{k}_m^2 + \Omega^2/4\right)T}}{1 + \frac{\Omega}{2} \frac{1}{\tilde{k}_m^2 + \Omega^2/4}} \tilde{k}_m \sin\left[\tilde{k}_m(1-y_0)\right]. \tag{7.85}$$

For $T_{\text{obs}} \to \infty$ we have $W \to 1$. Exchanging the order of summation and integration we can integrate (7.84) term by term. In particular, we calculate

$$I = \int_0^{T_{\text{obs}}} e^{-\left(\tilde{k}_m^2 + \Omega^2/4\right)T} \, dT = \left. \frac{-1}{\tilde{k}_m^2 + \Omega^2/4} e^{-\left(\tilde{k}_m^2 + \Omega^2/4\right)T} \right|_0^{T_{\text{obs}}}$$

$$= \frac{1}{\tilde{k}_m^2 + \Omega^2/4} \left(1 - e^{-\left(\tilde{k}_m^2 + \Omega^2/4\right)T_{\text{obs}}}\right). \tag{7.86}$$

Hence we obtain

$$W(\Omega, T_{\text{obs}}) = 2e^{\frac{\Omega}{2}(1-y_0)}$$

$$\times \sum_{m=0}^{\infty} \frac{1 - e^{-\left(\tilde{k}_m^2 + \Omega^2/4\right)T_{\text{obs}}}}{\left(\tilde{k}_m^2 + \Omega^2/4\right)\left(1 + \frac{\Omega}{2}\frac{1}{\tilde{k}_m^2 + \Omega^2/4}\right)} \tilde{k}_m \sin\left[\tilde{k}_m(1-y_0)\right]$$

$$= 2e^{\frac{\Omega}{2}(1-y_0)} \sum_{m=0}^{\infty} \frac{1 - e^{-\left(\tilde{k}_m^2 + \Omega^2/4\right)T_{\text{obs}}}}{\tilde{k}_m^2 + \Omega^2/4 + \Omega/2} \tilde{k}_m \sin\left[\tilde{k}_m(1-y_0)\right], \tag{7.87}$$

displayed in Figure 7.10 as a function of the observation time T_{obs} (a) and the parameter dependence Ω (b).

Carrying out an inverse transformation on the original variables, we get the same (breakdown) probability w as a function of all the parameters v, D, a, b, and the initial condition x_0, so

$$w(T_{\text{obs}}; v, D, a, b, x_0) = 2e^{\frac{v}{2D}(b-x_0)} \tag{7.88}$$

$$\times \sum_{m=0}^{\infty} \frac{1 - e^{-\left[\left(\frac{\tilde{k}_m}{b-a}\right)^2 + \left(\frac{v}{2D}\right)^2\right]DT_{\text{obs}}}}{\tilde{k}_m^2 + \left(\frac{v}{2D}\right)^2 (b-a)^2 + \left(\frac{v}{2D}\right)(b-a)} \tilde{k}_m \sin\left[\frac{\tilde{k}_m}{b-a}(b-x_0)\right].$$

In the following we want to consider an approximation for large control parameter values Ω.

7.7
The Limiting Case for Large Positive Values of the Control Parameter

Consider the parameter limit $\Omega \to +\infty$ which corresponds either to large positive drift v and/or large interval $b - a$, or to a small diffusion coefficient D. In this case, for a given m, the solution of the transcendental equation can be found in the form $\tilde{k}_m = \pi(m+1) - \varepsilon_m$, where ε_m is small and positive. From the periodicity property we obtain

$$\cos \tilde{k}_m = \cos(\pi(m+1) - \varepsilon_m) = -(-1)^m \cos(\varepsilon_m) = -(-1)^m + \mathcal{O}(\varepsilon_m^2),$$

$$\sin \tilde{k}_m = \sin(\pi(m+1) - \varepsilon_m) = (-1)^m \sin(\varepsilon_m) = (-1)^m \varepsilon_m + \mathcal{O}(\varepsilon_m^3).$$

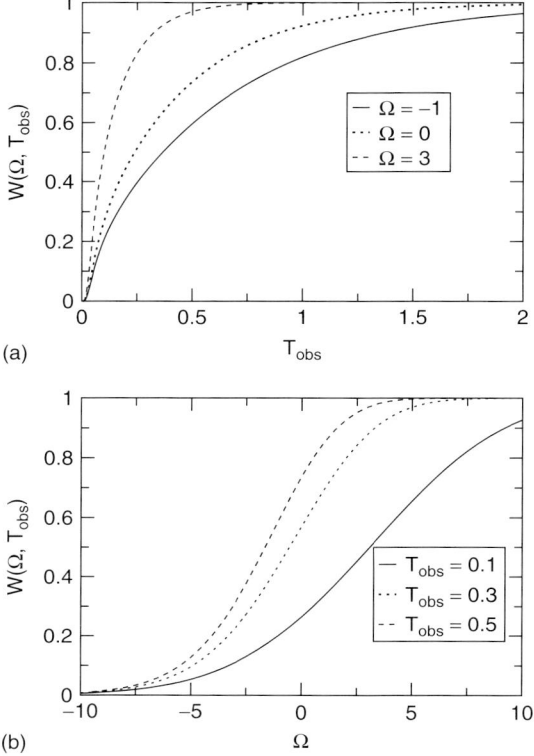

Figure 7.10 The probability $W(\Omega, T_{\text{obs}})$ (7.84), (7.87) as a function of the observation time T_{obs} with fixed Ω (a) and vice versa (b).

By inserting this into the transcendental equation (7.50), we obtain

$$\varepsilon_m = \frac{2}{\Omega}\pi(m+1) + \mathcal{O}\left(\Omega^{-2}\right), \tag{7.89}$$

$$\sin(\tilde{k}_m) = \frac{2}{\Omega}(-1)^m \pi(m+1) + \mathcal{O}\left(\Omega^{-2}\right). \tag{7.90}$$

In this approximation the normalization integral for large Ω and the initial condition $y_0 \to 0$ can be written as

$$I = \int_0^\infty \overline{\mathcal{P}}(T, y=1)\,dT = 2e^{\Omega/2}\sum_{m=0}^\infty \frac{\tilde{k}_m \sin\left(\tilde{k}_m\right)}{\lambda_m + \Omega/2}$$

$$\simeq e^{\Omega/2}\sum_{m=1}^\infty \frac{-4}{\Omega}\frac{(-1)^m(\pi m)^2}{\pi^2 m^2 + \Omega^2/4} = e^{\Omega/2}\sum_{m=-\infty}^\infty \frac{-2}{\Omega}\frac{(-1)^m(\pi m)^2}{\pi^2 m^2 + \Omega^2/4}. \tag{7.91}$$

7.7 The Limiting Case for Large Positive Values of the Control Parameter

Further on we set $(-1)^m = e^{i\pi m}$ and, in a continuum approximation, replace the sum by the integral

$$I \simeq e^{\Omega/2} \int_{-\infty}^{\infty} \frac{-2}{\Omega} \frac{e^{i\pi m}(\pi m)^2}{\pi^2 m^2 + \Omega^2/4} \, dm. \tag{7.92}$$

Now we make an integration contour in the complex plane, closing it in the upper plane ($\mathrm{Im}\, m > 0$) at infinity where $|e^{i\pi m}|$ is exponentially small. According to the residue theorem, this yields

$$I = 2\pi i \sum_i \mathrm{Res}(m_i) = 2\pi i \, \mathrm{Res}(m_0), \tag{7.93}$$

where $m_0 = i\Omega/(2\pi)$ is the location of the pole in the upper plane, found as a root of the equation $\pi^2 m^2 + \Omega^2/4 = 0$. According to the well known rule, the residue is calculated by setting $m = m_0$ in the enumerator of (7.92) and replacing the denominator with its derivative at $m = m_0$. This gives the desired result $I = 1$, that is, the considered approximation gives the correct normalization of the outflow probability density $\overline{\mathcal{P}}(T, y = 1)$ at the right boundary.

The probability distribution function $P(y, T)$ given by (7.67) can also be calculated in such a continuum approximation. In this case the increment in the wave numbers is

$$\Delta \tilde{k}_m = \tilde{k}_{m+1} - \tilde{k}_m = \pi + \varepsilon_m - \varepsilon_{m+1} \simeq \pi\left(1 - \frac{2}{\Omega}\right) \simeq \frac{\pi}{1 + 2/\Omega}. \tag{7.94}$$

Note that, in this approximation, for $\Omega \to \infty$ the normalization constant R_m in (7.54) is related to the increment $\Delta \tilde{k}$ via

$$R_m^2 = \frac{2}{1 + \frac{\Omega}{2}\frac{1}{\tilde{k}_m^2 + \Omega^2/4}} \simeq \frac{2}{1 + 2/\Omega} \simeq \frac{2}{\pi}\Delta \tilde{k}_m. \tag{7.95}$$

Hence, (7.67) for the probability density can be written as

$$P(y, T) = 2e^{\frac{\Omega}{2}(y-y_0)} \sum_{m=0}^{\infty} R_m^2 e^{-\lambda_m T} \sin\left[\tilde{k}_m(1 - y_0)\right] \sin\left[\tilde{k}_m(1 - y)\right]$$

$$\simeq \frac{2}{\pi} e^{\frac{\Omega}{2}(y-y_0)} \sum_{m=0}^{\infty} e^{-\left(\tilde{k}_m^2 + \Omega^2/4\right)T} \sin\left[\tilde{k}_m(1 - y_0)\right] \sin\left[\tilde{k}_m(1 - y)\right] \Delta \tilde{k}_m. \tag{7.96}$$

In the continuum approximation we replace the sum by the integral

$$P(y, T) \simeq \frac{2}{\pi} e^{\frac{\Omega}{2}(y-y_0)} \int_0^{\infty} e^{-\left(\tilde{k}^2 + \Omega^2/4\right)T} \sin\left[\tilde{k}(1 - y_0)\right] \sin\left[\tilde{k}(1 - y)\right] d\tilde{k}$$

$$= \frac{1}{\pi} e^{\frac{\Omega}{2}(y-y_0)} \int_0^{\infty} e^{-\left(\tilde{k}^2 + \Omega^2/4\right)T} \left(\cos\left[\tilde{k}(y - y_0)\right] - \cos\left[\tilde{k}(2 - y - y_0)\right]\right) d\tilde{k}. \tag{7.97}$$

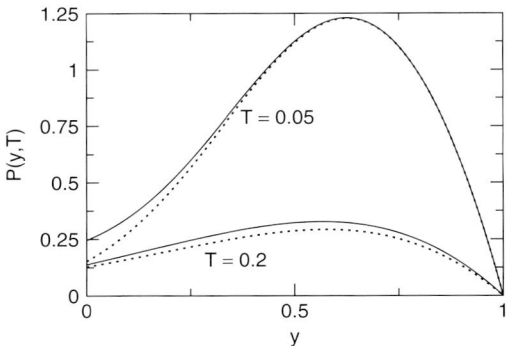

Figure 7.11 Comparison of the probability density $P(y, T)$ in drift–diffusion dynamics with finite boundaries for two times. The parameter value is $\Omega = 3.0$; initial condition is $y_0 = 0.5$. The solid lines represent the exact result (7.67); dotted lines show the approximation (7.98).

In the latter transformation we have used the trigonometric identity $\sin\alpha \sin\beta = \frac{1}{2}\left(\cos(\alpha - \beta) - \cos(\alpha + \beta)\right)$. The resulting known integrals yield

$$P(y, T) \simeq \frac{1}{\sqrt{4\pi T}} e^{\frac{\Omega}{2}\left(y - y_0 - \frac{\Omega}{2}T\right)} \left[e^{-\frac{(y-y_0)^2}{4T}} - e^{-\frac{(2-y-y_0)^2}{4T}} \right]. \tag{7.98}$$

The comparison between the exact distribution (7.67) and the approximation (7.98) is shown in Figure 7.11. For short enough times $4T \ll (2 - y - y_0)^2$ the second term is very small. Neglecting this term, (7.98) reduces to the known exact solution for natural boundary conditions.

Based on (7.98), it is easy to calculate the probability flux

$$J(y, T) = \Omega P(y, T) - \frac{\partial}{\partial y} P(y, T) \tag{7.99}$$

and the first-passage time distribution $\overline{P}(T) = J(y = 1, T)$ which takes a particularly simple form

$$\overline{P}(T) = \frac{1 - y_0}{\sqrt{4\pi T^3}} e^{-\frac{(1-y_0-\Omega T)^2}{4T}}. \tag{7.100}$$

The cumulative breakdown probability (7.84) is then

$$W(\Omega, T \leq T_{\mathrm{obs}}) = \int_0^{T_{\mathrm{obs}}} \frac{1 - y_0}{\sqrt{4\pi T^3}} e^{-\frac{(1-y_0-\Omega T)^2}{4T}} \, dT. \tag{7.101}$$

7.8
A Brief Survey of the Exact Solution

Finally, we want to summarize the exact results written in dimensionless form.

7.8.1
Probability Density

The general solution of the initial-boundary-value Fokker–Planck equation (7.24) reads (7.67)

$$P(y, T) = e^{\frac{\Omega}{2}(y-y_0)} \sum_{m=0}^{\infty} e^{-\lambda_m T} \psi_m(y_0) \psi_m(y) \qquad (7.102)$$

with eigenfunctions (7.55) and (7.59) of the ground state ($m = 0$)

$$\psi_0(y) = \begin{cases} \sqrt{\dfrac{2}{1 + \frac{\Omega}{2} \frac{1}{\tilde{k}_0^2 + \Omega^2/4}}} \sin\left[\tilde{k}_0(1-y)\right], & \Omega > -2 \\ \sqrt{3}\,(1-y), & \Omega = -2 \\ \sqrt{-\dfrac{2}{1 + \frac{\Omega}{2} \frac{1}{-\kappa_0^2 + \Omega^2/4}}} \sinh\left[\kappa_0(1-y)\right], & \Omega < -2 \end{cases} \qquad (7.103)$$

and all other eigenfunctions (7.55)

$$\psi_m(y) = \sqrt{\dfrac{2}{1 + \frac{\Omega}{2} \frac{1}{\tilde{k}_m^2 + \Omega^2/4}}} \sin\left[\tilde{k}_m(1-y)\right] \quad m = 1, 2, \ldots. \qquad (7.104)$$

The eigenvalue of the ground state ($m = 0$) is given by

$$\lambda_0 = \begin{cases} \tilde{k}_0^2 + \Omega^2/4, & \Omega > -2 \\ 1, & \Omega = -2 \\ -\kappa_0^2 + \Omega^2/4, & \Omega < -2 \end{cases} \qquad (7.105)$$

and all others are

$$\lambda_m = \tilde{k}_m^2 + \Omega^2/4 \quad m = 1, 2, \ldots, \qquad (7.106)$$

where the wave numbers are calculated from the transcendental equation (7.51)

$$\tilde{k}_0: \quad \tan \tilde{k}_0 = -\frac{2}{\Omega} \tilde{k}_0, \quad \Omega > -2 \qquad (7.107)$$

$$\kappa_0: \quad \tanh \kappa_0 = -\frac{2}{\Omega} \kappa_0, \quad \Omega < -2 \qquad (7.108)$$

$$\tilde{k}_m: \quad \tan \tilde{k}_m = -\frac{2}{\Omega} \tilde{k}_m, \quad m = 1, 2, \ldots. \qquad (7.109)$$

7.8.2
Outflow Probability Density

The first-passage time probability density distribution $\overline{\mathcal{P}}$ depending on Ω reads as follows

1. $\Omega > -2$ (see (7.80))

$$\overline{\mathcal{P}}(T, y=1) = 2e^{\frac{\Omega}{2}(1-y_0)} \sum_{m=0}^{\infty} \frac{e^{-\left(\tilde{k}_m^2 + \Omega^2/4\right)T}}{1 + \frac{\Omega}{2} \frac{1}{\tilde{k}_m^2 + \Omega^2/4}} \tilde{k}_m \sin\left[\tilde{k}_m(1-y_0)\right]. \qquad (7.110)$$

2. $\Omega = -2$

$$\overline{\mathcal{P}}(T, y=1) = e^{-(1-y_0)} \Bigg[3(1-y_0)e^{-T}$$

$$+ 2\sum_{m=1}^{\infty} \frac{e^{-\left(\tilde{k}_m^2+1\right)T}}{1 - \frac{1}{\tilde{k}_m^2+1}} \tilde{k}_m \sin\left[\tilde{k}_m(1-y_0)\right] \Bigg]. \qquad (7.111)$$

3. $\Omega < -2$

$$\overline{\mathcal{P}}(T, y=1) = 2e^{\frac{\Omega}{2}(1-y_0)}$$

$$\times \Bigg[-\frac{e^{-\left(-\kappa_0^2 + \Omega^2/4\right)T}}{1 + \frac{\Omega}{2}\frac{1}{-\kappa_0^2 + \Omega^2/4}} \kappa_0 \sinh\left[\kappa_0(1-y_0)\right]$$

$$+ \sum_{m=1}^{\infty} \frac{e^{-\left(\tilde{k}_m^2+\Omega^2/4\right)T}}{1 + \frac{\Omega}{2}\frac{1}{\tilde{k}_m^2+\Omega^2/4}} \tilde{k}_m \sin\left[\tilde{k}_m(1-y_0)\right] \Bigg]. \qquad (7.112)$$

7.8.3
First Moment of the Outflow Probability Density

The mean first-passage time (or average time of breakdown) T_1 corresponding to the breakdown probability density distribution $\overline{\mathcal{P}}$ is

$$T_1 = \int_0^{\infty} T \overline{\mathcal{P}}(T, y=1) \, dT = -\int_0^{\infty} T \frac{\partial}{\partial T} G(T) \, dT \qquad (7.113)$$

or

$$T_1 = \int_0^{\infty} G(T) \, dT \quad \text{with} \quad G(T) = \int_0^1 P(y, T) \, dy. \qquad (7.114)$$

The mean value T_1 depending on the control parameter Ω reads:

1. $\Omega > -2$

$$T_1 = 2e^{\frac{\Omega}{2}(1-y_0)} \sum_{m=0}^{\infty} \frac{\tilde{k}_m}{\left(\tilde{k}_m^2 + \Omega^2/4 + \frac{\Omega}{2}\right)\left(\tilde{k}_m^2 + \Omega^2/4\right)} \sin\left[\tilde{k}_m(1-y_0)\right]. \tag{7.115}$$

2. $\Omega = -2$

$$T_1 = e^{-(1-y_0)}\left[3(1-y_0) + 2\sum_{m=1}^{\infty} \frac{\sin\left[\tilde{k}_m(1-y_0)\right]}{\tilde{k}_m^3}\right]. \tag{7.116}$$

3. $\Omega < -2$

$$T_1 = 2e^{\frac{\Omega}{2}(1-y_0)} \tag{7.117}$$

$$\times \left[-\frac{\kappa_0}{\left(-\kappa_0^2 + \Omega^2/4 + \frac{\Omega}{2}\right)\left(-\kappa_0^2 + \Omega^2/4\right)} \sinh\left[\kappa_0(1-y_0)\right] \right.$$

$$\left. + \sum_{m=1}^{\infty} \frac{\tilde{k}_m}{\left(\tilde{k}_m^2 + \Omega^2/4 + \frac{\Omega}{2}\right)\left(\tilde{k}_m^2 + \Omega^2/4\right)} \sin\left[\tilde{k}_m(1-y_0)\right] \right].$$

The analytical expressions for the mean breakdown time T_1 depending on the control parameter Ω and fixed initial value y_0 are summarized in Figure 7.12. The straight line $T_1 = 0.5$ shows the result for $\Omega = 0$ (driftless diffusion).

7.8.4
Second Moment of the Outflow Probability Density

By definition, the second moment is

$$T_2 = \int_0^{\infty} T^2 \overline{\mathcal{P}}(T, y = 1)\, dT \tag{7.118}$$

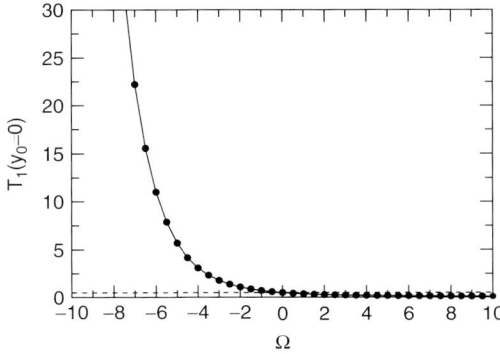

Figure 7.12 Mean first-passage time $T_1(y_0 = 0)$ as a function of the drift parameter Ω.

or

$$T_2 = 2\int_0^\infty TG(T)\,dT \quad \text{with} \quad G(T) = \int_0^1 P(y,T)\,dy. \tag{7.119}$$

The second moment T_2 depending on the control parameter Ω reads

1. $\Omega > -2$

$$T_2 = 4e^{\frac{\Omega}{2}(1-y_0)} \sum_{m=0}^{\infty} \frac{\tilde{k}_m}{\left(\tilde{k}_m^2 + \Omega^2/4 + \frac{\Omega}{2}\right)\left(\tilde{k}_m^2 + \Omega^2/4\right)^2} \sin\left[\tilde{k}_m(1-y_0)\right]. \tag{7.120}$$

2. $\Omega = -2$

$$T_2 = 2e^{-(1-y_0)}\left[3(1-y_0) + 2\sum_{m=1}^{\infty} \frac{\sin\left[\tilde{k}_m(1-y_0)\right]}{\tilde{k}_m^4}\right]. \tag{7.121}$$

3. $\Omega < -2$

$$T_2 = 4e^{\frac{\Omega}{2}(1-y_0)} \tag{7.122}$$

$$\times \left[-\frac{\kappa_0}{\left(-\kappa_0^2 + \Omega^2/4 + \frac{\Omega}{2}\right)\left(-\kappa_0^2 + \Omega^2/4\right)^2} \sinh\left[\kappa_0(1-y_0)\right] \right.$$

$$\left. + \sum_{m=1}^{\infty} \frac{\tilde{k}_m}{\left(\tilde{k}_m^2 + \Omega^2/4 + \frac{\Omega}{2}\right)\left(\tilde{k}_m^2 + \Omega^2/4\right)^2} \sin\left[\tilde{k}_m(1-y_0)\right] \right].$$

7.8.5
Outflow Probability

The cumulative probability W depending on Ω is given by

1. $\Omega > -2$ (see (7.87))

$$W(\Omega, T_{\text{obs}}) = 2e^{\frac{\Omega}{2}(1-y_0)} \sum_{m=0}^{\infty} \frac{1 - e^{-\left(\tilde{k}_m^2 + \Omega^2/4\right)T_{\text{obs}}}}{\tilde{k}_m^2 + \Omega^2/4 + \Omega/2} \tilde{k}_m \sin\left[\tilde{k}_m(1-y_0)\right]. \tag{7.123}$$

2. $\Omega = -2$

$$W(\Omega, T_{\text{obs}}) = e^{-(1-y_0)}\left[3\left(1 - e^{-T_{\text{obs}}}\right)(1-y_0)\right.$$

$$\left. + 2\sum_{m=1}^{\infty} \frac{1 - e^{-\left(\tilde{k}_m^2+1\right)T_{\text{obs}}}}{\tilde{k}_m} \sin\left[\tilde{k}_m(1-y_0)\right]\right]. \tag{7.124}$$

3. $\Omega < -2$

$$W(\Omega, T_{obs}) = 2e^{\frac{\Omega}{2}(1-y_0)}$$
$$\times \left[-\frac{1 - e^{-(-\kappa_0^2 + \Omega^2/4)T_{obs}}}{-\kappa_0^2 + \Omega^2/4 + \Omega/2} \kappa_0 \sinh\left[\kappa_0(1 - y_0)\right] \right.$$
$$\left. + \sum_{m=1}^{\infty} \frac{1 - e^{-(\tilde{k}_m^2 + \Omega^2/4)T_{obs}}}{\tilde{k}_m^2 + \Omega^2/4 + \Omega/2} \tilde{k}_m \sin\left[\tilde{k}_m(1 - y_0)\right] \right]. \quad (7.125)$$

Going back to the original variables $t = t_{obs}$ including parameters v, D, a, b, x_0 and defining the modified wave numbers (in units of m^{-1}) as $\hat{k}_m = \tilde{k}_m/(b-a)$ and $\kappa z_0 = z_0/(b-a)$ the cumulative (breakdown) probability $w(t = t_{obs}; v, D, a, b, x_0)$ has the form

1. $\frac{v(b-a)}{D} > -2$ (see (7.88))

$$w = \frac{2}{b-a} e^{\frac{v}{2D}(b-x_0)} \sum_{m=0}^{\infty} \frac{1 - e^{-D\left(\hat{k}_m^2 + \left(\frac{v}{2D}\right)^2\right)t_{obs}}}{\hat{k}_m^2 + \left(\frac{v}{2D}\right)^2 + \frac{v}{2D(b-a)}} \hat{k}_m \sin\left[\hat{k}_m(b - x_0)\right]. \quad (7.126)$$

2. $\frac{v(b-a)}{D} = -2$

$$w = e^{-\frac{b-x_0}{b-a}} \left[\frac{3}{b-a} \left(1 - e^{-\frac{Dt_{obs}}{(b-a)^2}}\right)(b - x_0) \right. \quad (7.127)$$
$$\left. + \frac{2}{b-a} \sum_{m=1}^{\infty} \frac{1 - e^{-D\left(\hat{k}_m^2 + \frac{1}{(b-a)^2}\right)t_{obs}}}{\hat{k}_m} \sin\left[\hat{k}_m(b - x_0)\right] \right].$$

3. $\frac{v(b-a)}{D} < -2$

$$w = \frac{2}{b-a} e^{\frac{v}{2D}(b-x_0)}$$
$$\times \left[-\frac{1 - e^{-D\left(-\hat{\kappa}_0^2 + \left(\frac{v}{2D}\right)^2\right)t_{obs}}}{-\hat{\kappa}_0^2 + \left(\frac{v}{2D}\right)^2 + \frac{v}{2D(b-a)}} \hat{\kappa}_0 \sinh\left[\hat{\kappa}_0(b - x_0)\right] \right. \quad (7.128)$$
$$\left. + \sum_{m=1}^{\infty} \frac{1 - e^{-D\left(\hat{k}_m^2 + \left(\frac{v}{2D}\right)^2\right)t_{obs}}}{\hat{k}_m^2 + \left(\frac{v}{2D}\right)^2 + \frac{v}{2D(b-a)}} \hat{k}_m \sin\left[\hat{k}_m(b - x_0)\right] \right]$$

with, compare (7.110)–(7.112),

$$\hat{k}_0: \quad \tan\left(\hat{k}_0 (b-a)\right) = -\frac{2D}{v}\hat{k}_0, \qquad \frac{v(b-a)}{D} > -2 \qquad (7.129)$$

$$\hat{\kappa}_0: \quad \tanh\left(\hat{\kappa}_0 (b-a)\right) = -\frac{2D}{v}\hat{\kappa}_0, \qquad \frac{v(b-a)}{D} < -2 \qquad (7.130)$$

$$\hat{k}_m: \quad \tan\left(\hat{k}_m (b-a)\right) = -\frac{2D}{v}\hat{k}_m, \qquad m = 1, 2, \ldots. \qquad (7.131)$$

7.9
Relationship to the Sturm–Liouville Theory

The particular drift–diffusion problem over a finite interval with reflecting (left) and absorbing (right) boundaries belongs to the following general mathematical theory named after Jacques Charles Francois Sturm (1803–1855) and Joseph Liouville (1809–1882).

The classical Sturm–Liouville theory [249] considers a real second-order linear differential equation of the form

$$-\frac{d}{dx}\left[p(x)\frac{d\psi}{dx}\right] + q(x)\psi = \lambda w(x)\psi \qquad (7.132)$$

together with boundary conditions at the ends of interval $[a, b]$ given by

$$\alpha_1 \psi(x = a) + \alpha_2 \left.\frac{d\psi}{dx}\right|_{x=a} = 0, \qquad (7.133)$$

$$\beta_1 \psi(x = b) + \beta_2 \left.\frac{d\psi}{dx}\right|_{x=b} = 0. \qquad (7.134)$$

The particular functions $p(x), q(x), w(x)$ are real and continuous on the finite interval $[a, b]$ together with specified values at the boundaries. The aim of the Sturm–Liouville problem is to find the values of λ (called eigenvalues λ_n) for which there are nontrivial solutions of the differential equation (7.132) satisfying the boundary conditions (7.133) and (7.134). The corresponding solutions (for such λ_n) are called eigenfunctions $\psi_n(x)$ of the problem.

Defining the Sturm–Liouville differential operator over the unit interval $[0, 1]$ by

$$\mathcal{L}\psi = -\frac{d}{dx}\left[p(x)\frac{d\psi}{dx}\right] + q(x)\psi \qquad (7.135)$$

and making the weight $w(x)$ equal to unity ($w = 1$) the general equation (7.132) can be represented as an eigenvalue problem

$$\mathcal{L}\psi = \lambda \psi \qquad (7.136)$$

with boundary conditions (7.133) ($a = 0$) and (7.134) ($b = 1$) written as

$$\mathcal{B}_0 \psi = 0, \qquad \mathcal{B}_1 \psi = 0. \qquad (7.137)$$

Assuming a differentiable positive function $p(x) > 0$, the Sturm–Liouville operator is called regular and it is self-adjoint, to fulfill

$$\int_0^1 \mathcal{L}\psi_1 \cdot \psi_2 = \int_0^1 \psi_1 \cdot \mathcal{L}\psi_2. \tag{7.138}$$

Any self-adjoint operator has real non-negative eigenvalues $\lambda_0 < \lambda_1 < \ldots < \lambda_n < \ldots \to \infty$. The corresponding eigenfunctions $\psi_n(x)$ have exactly n zeros in $(0, 1)$ and form an orthogonal set

$$\int_0^1 \psi_n(x)\psi_m(x)\,dx = \delta_{mn}. \tag{7.139}$$

The eigenvalues λ_n of the classical Sturm–Liouville problem (7.132) with positive function $p(x) > 0$ as well as positive weight function $w(x) > 0$, together with separated boundary conditions (7.133) and (7.134) can be calculated by the following expression

$$\lambda_n \int_a^b \psi_n(x)^2 w(x)\,dx = \int_a^b \left[p(x)\left(d\psi_n(x)/dx\right)^2 + q(x)\psi_n(x)^2 \right] dx$$
$$- \left| p(x)\psi_n(x)\left(d\psi_n(x)/dx\right) \right|_a^b. \tag{7.140}$$

The eigenfunctions are mutually orthogonal ($m \neq n$) and usually normalized ($m = n$) in accordance with the equation

$$\int_a^b \psi_n(x)\psi_m(x)w(x)\,dx = \delta_{mn} \tag{7.141}$$

known as orthogonality relation (similar to (7.139)).

Coming back to the original drift–diffusion problem written in dimensionless variables over unit interval $0 \leq y \leq 1$ and recalling (7.37), the separation constant λ appears in the following differential equation

$$-\frac{d^2\psi(y)}{dy^2} + \frac{\Omega^2}{4}\psi(y) = \lambda\psi(y) \tag{7.142}$$

which can be related to the regular Sturm–Liouville eigenvalue problem via $p(y) = 1 > 0$; $w(y) = 1 > 0$ and $q(y) = \Omega^2/4$.

The boundary conditions given by (7.43) and (7.46) can be expressed as

$$\frac{\Omega}{2} \cdot \psi(y = 0) + (-1) \cdot \left.\frac{d\psi}{dy}\right|_{y=0} = 0, \tag{7.143}$$

$$1 \cdot \psi(y = 1) + 0 \cdot \left.\frac{d\psi}{dy}\right|_{y=1} = 0 \tag{7.144}$$

in agreement with (7.133) and (7.134).

The previously unknown separation constant λ has a spectrum of real positive eigenvalues which can be calculated using (7.140) from

$$\lambda_n = \int_0^1 \left[\left(\frac{d\psi_n(y)}{dy}\right)^2 + \frac{\Omega^2}{4} \psi_n(y)^2 \right] dx - \left. \psi_n(y) \frac{d\psi_n(y)}{dy} \right|_0^1 \quad (7.145)$$

taking into account the orthogonality relation (7.141)

$$\int_0^1 \psi_n(y) \psi_m(y) \, dy = \delta_{mn} \quad \text{for} \quad w(y) = 1. \quad (7.146)$$

In this way, the first-passage time problem can be solved on the basis of the Sturm–Liouville theory. An alternative approach will be considered in the following section.

7.10
Alternative Method by the Backward Fokker–Planck Equation

Here we treat the first-passage time problem based on the backward Fokker–Planck equation. Random walks inside the one-dimensional domain $(0, 1)$ are considered assuming its boundaries $x = 0$ and $x = 1$ to be reflecting and absorbing, respectively. In addition, the random walks are assumed to undergo drift with constant rate Ω; as previously the positive values of Ω denoted the drift directed from left to right. The backward Fokker–Planck equation for the distribution function $P(y, y_0, T)$ is of the form

$$\frac{\partial P(y, y_0, T)}{\partial T} = \frac{\partial^2 P(y, y_0, T)}{\partial y_0^2} + \Omega \frac{\partial P(y, y_0, T)}{\partial y_0} \quad (7.147)$$

subject to the boundary conditions

$$\left. \frac{\partial P(y, y_0, T)}{\partial y_0} \right|_{y_0 = 0} = 0, \quad (7.148)$$

$$P(y, y_0 = 1, T) = 0, \quad (7.149)$$

and the initial condition

$$P(y, y_0, T = 0) = \delta(y - y_0). \quad (7.150)$$

Here, as before, y_0 is the starting position of a random walker.

The probability that the walker will leave the interval $(0, 1)$ some time is equal to unity. So the probability that the walker is located inside this interval up to the current moment T is equal to the probability of getting to the boundary $y = 1$ for the first time in the future. Therefore, if $\overline{\mathcal{P}}(T, y_0)$ is the probability density of getting to the boundary $y = 1$ for the first time at the moment T then the following relation

$$\int_T^\infty \overline{\mathcal{P}}(T', y_0) \, dT' = \int_0^1 P(y, y_0, T) \, dy \quad (7.151\text{a})$$

or, which is actually the same as (7.28),

$$\overline{\mathcal{P}}(T, y_0) = -\frac{d}{dT} \int_0^1 P(y, y_0, T) \, dy \qquad (7.151b)$$

has to hold. It should be noted that the symbol $\overline{\mathcal{P}}(T, y = 1)$ was used previously to denote the first-passage time probability and the argument y_0 was omitted because it plays the role of constant parameter. In this description the variable y_0 is shown explicitly and the rather formal symbol $y = 1$, denoting the position of the absorbing boundary, is omitted for brevity.

The variable y plays the role of the parameter in (7.147) and the operators ∂_T, ∂_{y_0} are commutative. Therefore, first integrating (7.147) with respect to y over the interval $(0, 1)$ and then acting by operator ∂_T on both sides, we immediately obtain the governing equation for the function $\overline{\mathcal{P}}(T, y_0)$

$$\frac{\partial \overline{\mathcal{P}}}{\partial T} = \frac{\partial^2 \overline{\mathcal{P}}}{\partial y_0^2} + \Omega \frac{\partial \overline{\mathcal{P}}}{\partial y_0}. \qquad (7.152)$$

Since the boundary $y = 0$ is reflecting, the probability distribution $P(y, y_0, T)$ does not exhibit any anomalous behavior as $y_0 \to 0$. Therefore, applying again the same method of conversion from the probability distribution to the first-passage probability $\overline{\mathcal{P}}(T, y_0)$ we conclude that the boundary condition (7.148) also holds for the function $\overline{\mathcal{P}}(T, y_0)$, so

$$\left. \frac{\partial \overline{\mathcal{P}}(T, y_0)}{\partial y_0} \right|_{y_0 = 0} = 0. \qquad (7.153)$$

At the initial time $T = 0$ the probability of getting the boundary $y = 1$ from any internal point $y_0 \in [0, 1)$ is equal to zero, so

$$\overline{\mathcal{P}}(T = 0, y_0) = 0 \quad \text{for} \quad y_0 < 1. \qquad (7.154)$$

is the initial condition for $\overline{\mathcal{P}}(T, y_0)$.

At the absorbing boundary $y_0 = 1$ the probability density $\overline{\mathcal{P}}(T, y_0)$ exhibits a singularity. Indeed, on one hand, if the walker was born exactly at the point $y_0 = 1$, then it will be absorbed immediately and, thus,

$$\overline{\mathcal{P}}(T, y_0 = 1) = 0 \quad \text{for} \quad T > 0. \qquad (7.155)$$

On the other hand, for a point $y_0 < 1$ located arbitrarily near this boundary, equality (7.154) holds, so

$$\lim_{y_0 \to 1-0} \overline{\mathcal{P}}(T, y_0,) = 0. \qquad (7.156)$$

In addition, the characteristic time interval during which such a walker will be absorbed at the boundary $y_0 = 1$ tends to zero as $y_0 \to 1$. The latter feature is equivalent to the equality

$$\lim_{\tau \to +0} \lim_{y_0 \to 1-0} \int_0^\tau \overline{\mathcal{P}}(T, y_0) \, dT = 1. \qquad (7.157)$$

In the general case we can only state that the walker definitely leaves the interval $(0, 1)$ some time so, for an arbitrary point y_0, we have

$$\int_0^\infty \overline{\mathcal{P}}(T, y_0) \, dT = 1. \tag{7.158}$$

To tackle this singularity let us convert from the function $\overline{\mathcal{P}}(y_0, T)$ to its Laplace transform

$$\mathcal{L}(s, y_0) = \int_0^\infty e^{-sT} \overline{\mathcal{P}}(T, y_0) \, dT. \tag{7.159}$$

Performing the Laplace transformation with respect to both sides of (7.152) and integrating by parts, the left-hand side reduces to

$$s\mathcal{L} = \frac{\partial^2 \mathcal{L}}{\partial y_0^2} + \Omega \frac{\partial \mathcal{L}}{\partial y_0} + \overline{\mathcal{P}}(T = 0, y_0). \tag{7.160}$$

So, by virtue of (7.154) the governing equation for the Laplace transform $\mathcal{L}(s, y_0)$ has the form

$$s\mathcal{L} = \frac{\partial^2 \mathcal{L}}{\partial y_0^2} + \Omega \frac{\partial \mathcal{L}}{\partial y_0}. \tag{7.161}$$

Using the operator ∂_{y_0} on expression (7.159) we see that the boundary condition (7.153) holds also for the given Laplace transform. As far the other boundary is concerned, only an infinitely narrow neighborhood of $T = 0$ contributes to the integral (7.159) in this case and, thus, the exponential cofactor $\exp(-sT)$ can be omitted when $y_0 \to 1$, reducing (7.159) to (7.158). It turns out that (7.161) must be accompanied by the following boundary conditions

$$\left. \frac{\partial \mathcal{L}(s, y_0)}{\partial y_0} \right|_{y_0 = 0} = 0 \tag{7.162}$$

and

$$\mathcal{L}(s, y_0 = 1) = 1. \tag{7.163}$$

Let us analyze in detail the first-passage time properties in the situation under consideration. As a first step, the general solution of (7.161) is written in the form

$$\mathcal{L}(s, y_0) = A_+ e^{y_0 \kappa_+} + A_- e^{y_0 \kappa_-}, \tag{7.164}$$

where $A_\pm$ are constants and $\kappa_\pm$ are the values that should be found substituting (7.164) into (7.161). In this way we conclude that the values $\kappa_\pm$ are the roots of the equation

$$\kappa^2 + \Omega \kappa - s = 0, \tag{7.165}$$

so

$$\kappa_\pm = -\frac{1}{2} \Omega \pm \sqrt{\frac{1}{4} \Omega^2 + s}. \tag{7.166}$$

7.10 Alternative Method by the Backward Fokker–Planck Equation

The boundary conditions (7.162)–(7.163) in this case become

$$\left.\frac{\partial \mathcal{L}(s, y_0)}{\partial y_0}\right|_{y_0=0} = \kappa_+ A_+ + \kappa_- A_- = 0, \tag{7.167}$$

$$\mathcal{L}(s, y_0 = 1) = A_+ e^{\kappa_+} + A_- e^{\kappa_-} = 1, \tag{7.168}$$

specifying the constants $A_\pm$, namely,

$$A_+ = -\frac{\kappa_-}{\kappa_+ e^{\kappa_-} - \kappa_- e^{\kappa_+}} \tag{7.169a}$$

$$A_- = \frac{\kappa_+}{\kappa_+ e^{\kappa_-} - \kappa_- e^{\kappa_+}}. \tag{7.169b}$$

The substitution of (7.166) and (7.169) into (7.164) yields the desired expression

$$\mathcal{L}(s, y_0) = e^{-\frac{1}{2}\Omega(y_0-1)} \frac{k \cosh(y_0 k) + \frac{1}{2}\Omega \sinh(y_0 k)}{k \cosh(k) + \frac{1}{2}\Omega \sinh(k)}. \tag{7.170}$$

where the symbol k denotes the expression

$$k = k(s) := \sqrt{\frac{1}{4}\Omega^2 + s}. \tag{7.171}$$

In order to find the inverse image $\overline{\mathcal{P}}(T, y_0)$ of (7.170) we make use of the inverse Laplace transform (γ is any positive number)

$$\overline{\mathcal{P}}(T, y_0) = \frac{1}{2\pi i} \int_{\gamma-i\infty}^{\gamma+i\infty} e^{sT} \mathcal{L}(s, y_0) \, ds. \tag{7.172}$$

Function (7.170) is analytic everywhere on the complex plane $s = \operatorname{Re} s + i \operatorname{Im} s$ except for the roots of its denominator because it contains only the even orders of the variable $\sqrt{\frac{1}{4}\Omega^2 + s}$. The denominator set of zeros $\{s_n\}$, as will be clear later, is made up of isolated points of the real axis ($\operatorname{Re} s$) on the left-hand side from the origin $s = 0$ shown in Figure 7.13.

Therefore, to calculate integral (7.172) the residue theory is used. Let us consider a loop $\mathcal{C}_s$ of radius R on the complex plane that contains the fragment of the line $(\gamma - i\infty, \gamma + i\infty)$ (see Figure 7.13) and analyze the integral

$$I(T, y_0) := \frac{1}{2\pi i} \int_{\mathcal{C}_s} e^{sT} \mathcal{L}(s, y_0) \, ds. \tag{7.173}$$

along it. According to the residue technique developed for functions of a complex variable its value is equal to the sum running over all the residues of the integrand which are located in the domain $\mathcal{Q}_s$ bounded by the loop $\mathcal{C}_s$:

$$I(T, y_0) = \sum_{s_n \in \mathcal{Q}_s} \operatorname{Res}\left[e^{s_n T} \mathcal{L}(s_n, y_0)\right]. \tag{7.174}$$

We recall that, by definition, the residue $\operatorname{Res}[f(s_p)]$ of a function $f(s)$ of complex variable s is the coefficient c_{-1} of its expansion around a singularity point s_p (a pole of the function $f(s)$):

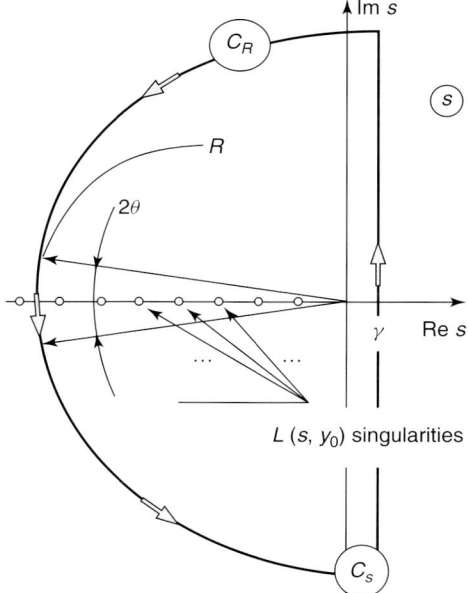

Figure 7.13 Technique for calculating integral (7.172).

$$f(s) = \ldots + \frac{c_{-2}}{(s-s_p)^2} + \frac{c_{-1}}{(s-s_p)} + c_0 + c_1(s-s_p) + c_2(s-s_p)^2 + \ldots$$

In particular, let a function $f(s)$ be of the form

$$f(s) = \frac{P(s)}{Q(s)}, \tag{7.175}$$

and exhibit a singularity of the first order at the point s_p, so $c_1 \neq 0$ whereas all the other coefficients $c_{-2}, c_{-3}, \ldots = 0$, and $P(s_p) \neq 0$. Then

$$\mathrm{Res}\left[f(s_p)\right] = P(s_p) \left(\left. \frac{dQ(s)}{ds} \right|_{s=s_p} \right)^{-1}. \tag{7.176}$$

If we are able to find the integral along the circular fragment $\mathcal{C}_R$ of $\mathcal{C}_s$ then expression (7.174) immediately gives us the inverse Laplace transform (7.172).

The value of this integral part can be estimated in the following way. We note that for $y_0 < 1$ the function $\mathcal{L}(s, y_0)$ tends to zero exponentially as the argument s goes to infinity along the half-lines $s = -re^{\pm i\theta}$ where $0 < \theta < \pi$, namely

$$\mathcal{L}(s, y_0) \propto \exp\left[-(1-y_0)\sqrt{r} \left| \sin\left(\frac{1}{2}\phi\right) \right| \right]. \tag{7.177}$$

Therefore the integral along $\mathcal{C}_R$, except for an arbitrary small-angle neighborhood of the real axis (Re s), tends to zero as $R \to \infty$. The set of zeros of the denominator is practically equidistant far from the origin. We can choose an arbitrary large circular fragment passing between these singularities and, so, the Laplace transform at $\mathcal{C}_R$ is bounded in the given neighborhood, where the exponential cofactor $e^{sT} \sim e^{-RT} \to 0$ decreases exponentially. Summarizing the aforesaid argument we conclude that the integral vanishes

$$\int_{\mathcal{C}_R} e^{sT} \mathcal{L}(s, y_0) \, ds \to 0 \quad \text{as} \quad R \to \infty. \tag{7.178}$$

Whence it follows that the desired inverse Laplace transform can be represented as the series

$$\overline{\mathcal{P}}(T, y_0) = \sum_n \mathrm{Res}\left[e^{s_n T} \mathcal{L}(s, y_0), s_n\right]. \tag{7.179}$$

As follows from (7.170) the residue set $\{s_n\}$ is determined by the equation

$$k \cosh k + \frac{\Omega}{2} \sinh k = 0, \tag{7.180}$$

where $k = k(s)$ is given by (7.171). Equation (7.180) will be analyzed in detail in Section 7.11. Applying the results of this analysis, it is possible to assert that the collection of poles are located at $\mathrm{Im}\, s = 0$ and $\mathrm{Re}\, s < 0$, containing the points

$$s_n = -\frac{1}{4}\Omega^2 - \widetilde{k}_n^2, \quad n = 1, 2, 3, \ldots \tag{7.181}$$

and an additional point either

$$s_* = -\frac{1}{4}\Omega^2 - \widetilde{k}_*^2 \quad \text{for } -2 < \Omega < 0 \tag{7.181a}$$

or

$$s_* = -\frac{1}{4}\Omega^2 + \widetilde{k}_*^2 \quad \text{for } \Omega < -2 \tag{7.181b}$$

depending on the value of the drift rate Ω. Here the set $\{\widetilde{k}_n\}$ of real numbers as well as the value $\widetilde{k}_*$ are the roots of equation

$$k \cos k + \frac{\Omega}{2} \sin k = 0, \tag{7.182}$$

with the latter value only meeting it for $-2 < \Omega < 0$, whereas for $\Omega < -2$ the value $\widetilde{k}_*$ is a root of the equation

$$k \cosh k + \frac{\Omega}{2} \sinh k = 0. \tag{7.183}$$

Figure 7.14 shows the solutions $\widetilde{k}$ from (7.182) or (7.183) geometrically constructed as the intersection between the trigonometric function $\tan k$ or hyperbolic tangent $\tanh k$ with the linear function $-(\Omega/2)k$ for different values of the control parameter Ω. It should be noted that the zero-value roots of (7.181a) and (7.181b) do not contribute to the set of poles because the numerator of the ratio (7.180) is also equal to zero in this case.

In order to calculate the residues of function (7.170) using formula (7.176) we will first calculate the derivative of its denominator

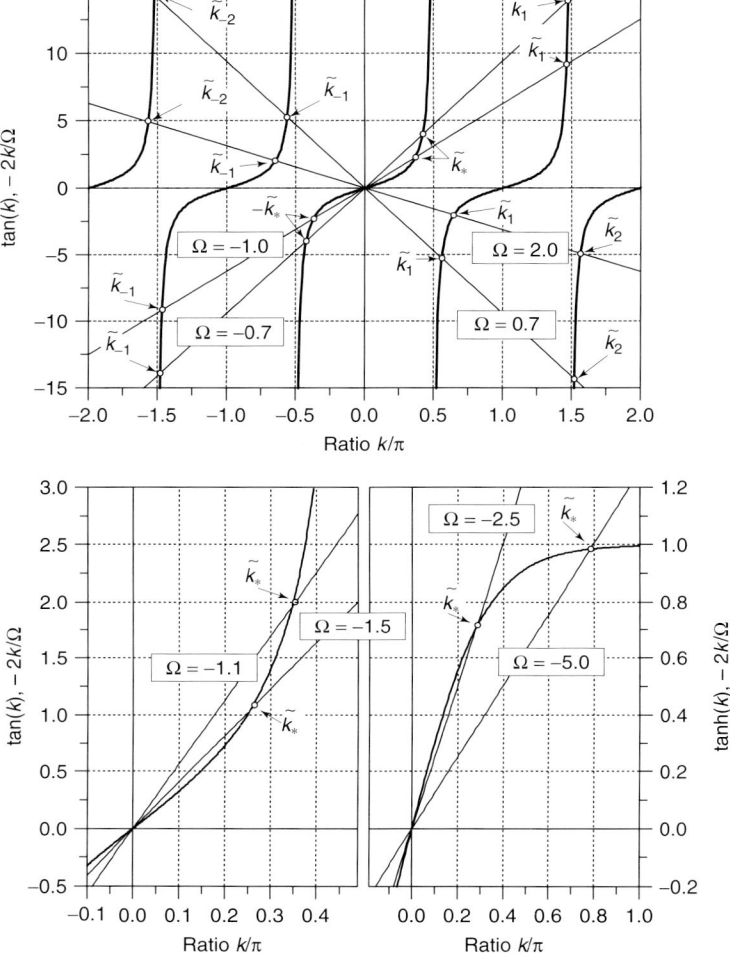

Figure 7.14 Set of solutions $\tilde{k}$ obtained from (7.182) as the intersection between $\tan k$ and the straight line $-(\Omega/2)k$ and those of (7.183) as the intersection between $\tanh k$ and the same straight line (bottom right figure).

$$D(s) := \sqrt{\frac{1}{4}\Omega^2 + s} \cosh\left(\sqrt{\frac{1}{4}\Omega^2 + s}\right) + \frac{1}{2}\Omega \sinh\left(\sqrt{\frac{1}{4}\Omega^2 + s}\right)$$

at the given points. Taking into account equality (7.180) we have

$$\left.\frac{dD(s)}{ds}\right|_{s_n} = \frac{\left(\frac{\Omega}{2} - s_n\right)}{\Omega\sqrt{\frac{1}{4}\Omega^2 + s_n}} \cosh\left(\sqrt{\frac{1}{4}\Omega^2 + s_n}\right).$$

The numerator of ratio (7.180) at the poles is equal to

$$N(s_n) := \sqrt{\frac{1}{4}\Omega^2 + s_n} \frac{\sinh\left[(1-y_0)\sqrt{\frac{1}{4}\Omega^2 + s_n}\right]}{\sinh\left(\sqrt{\frac{1}{4}\Omega^2 + s_n}\right)}.$$

In these terms the desired probability density $\overline{\mathcal{P}}(T, y_0)$ is specified by the sum

$$\overline{\mathcal{P}}(T, y_0) = e^{[-\frac{1}{2}\Omega(y_0-1)+s_n t]} \sum_n \frac{N(s_n)}{dD(s)/ds|_{s_n}}. \tag{7.184}$$

Using the expressions obtained and again the equality (7.180) this sum can be rewritten as

$$\overline{\mathcal{P}}(T, y_0) = 2\Omega e^{-\frac{1}{2}\Omega(y_0-1)} \sum_n \frac{\left(\frac{1}{4}\Omega^2 + s_n\right)e^{s_n T}}{\frac{\Omega}{2} - s_n} \cdot \frac{\sinh\left[(1-y_0)\sqrt{\frac{1}{4}\Omega^2 + s_n}\right]}{\sinh\left[2\sqrt{\frac{1}{4}\Omega^2 + s_n}\right]}$$

$$= 2e^{\frac{1}{2}\Omega(1-y_0)} \left\{ \sum_{n=1}^{\infty} \frac{e^{-\left(\widetilde{k}_n^2 + \Omega^2/4\right)T}}{1 + \frac{\Omega}{2}\frac{1}{\widetilde{k}_n^2 + \Omega^2/4}} \widetilde{k}_n \sin\left[(1-y_0)\widetilde{k}_n\right] \right.$$

$$\left. \begin{array}{l} -2<\Omega<0 \\ + \end{array} \frac{e^{-\left(\widetilde{k}_*^2 + \Omega^2/4\right)T}}{1 + \frac{\Omega}{2}\frac{1}{\widetilde{k}_*^2 + \Omega^2/4}} \widetilde{k}_* \sin\left[(1-y_0)\widetilde{k}_*\right] \right.$$

$$\left. \begin{array}{l} \Omega<-2 \\ - \end{array} \frac{e^{-\left(-\widetilde{k}_*^2 + \Omega^2/4\right)T}}{1 + \frac{\Omega}{2}\frac{1}{-\widetilde{k}_*^2 + \Omega^2/4}} \widetilde{k}_* \sinh\left[(1-y_0)\widetilde{k}_*\right] \right\}. \tag{7.185}$$

The latter formula stems from the former one after several steps of arithmetical manipulation also taking into account (7.182) and (7.183). Expression (7.185) is the desired formula for the probability density of getting to the absorbing boundary for the first time at moment T. It exactly coincides with expressions (7.110), (7.112) and contains (7.111) as the limiting case $\Omega \to -2$ where the summation index is related to $n = m + 1$.

Let us consider the case corresponding to a significant potential barrier separating the terminal points of the diffusion interval, that is the limit of large negative values $\Omega < 0$ and $|\Omega| \gg 1$. For such as Ω the value $\widetilde{k}_*$ is given by the root of (7.183) and its magnitude is much larger than unity. Approximating we get $\widetilde{k}_* \approx -\Omega/2 = |\Omega|/2$ because, in this case, $\tanh \widetilde{k}_* \approx 1$ (see Figure 7.14). The straight line in Figure 7.3 shows the same root $\kappa_0 \equiv \widetilde{k}_* = \Omega/2$ in its zeroth approximation.

In order to obtain the deviation of $\widetilde{k}_*$ from the zeroth estimate we have found, (7.183) can be solved by iteration. So, let us rewrite (7.183) as

$$k = -\frac{\Omega}{2} \tanh k = -\frac{\Omega}{2}\left(\frac{1-e^{-2k}}{1+e^{-2k}}\right)$$

$$= -\frac{\Omega}{2}\left(1 - 2e^{-2k} + 2e^{-4k} - 2e^{-6k} + \dots\right). \tag{7.186}$$

Obviously the zeroth approximation with respect to the small parameter $\exp(-2k)$ yields the same estimate and the first iteration gives the desired value

$$\widetilde{k}_* \approx -\frac{\Omega}{2}\left(1 - 2e^{\Omega}\right). \tag{7.187}$$

All the terms in expression (7.185) containing the time-dependent cofactors $\exp\left[-(\widetilde{k}_n^2 + \Omega^2/4)T\right]$ become ignorable on scales $T \gg 4/\Omega^2$. The last one is the exception because by virtue of (7.187)

$$\Omega^2/4 - \widetilde{k}_*^2 = \Omega^2 e^{\Omega} - \Omega^2 e^{2\Omega} \approx \Omega^2 e^{\Omega} \tag{7.188}$$

its cofactor is of the form

$$\exp\left[-(-\widetilde{k}_*^2 + \Omega^2/4)T\right] \approx \exp\left[-\Omega^2 e^{\Omega} T\right] \tag{7.189}$$

and on time scales $T < e^{|\Omega|}/\Omega^2$ the corresponding term keeps its initial value at $T = 0$. So as time T goes beyond a rather narrow initial interval $T \ll 4/\Omega^2$ only the last term contributes to the function $\overline{\mathcal{P}}(T, y_0)$. In this case, taking into account estimate (7.187) expression (7.185) can be calculated as

$$\overline{\mathcal{P}}(T, y_0) = -2e^{\frac{1}{2}\Omega(1-y_0)} \frac{e^{-\left(-\widetilde{k}_*^2 + \Omega^2/4\right)T}}{1 + \frac{\Omega}{2}\frac{1}{-\widetilde{k}_*^2 + \Omega^2/4}} \widetilde{k}_* \sinh\left[(1 - y_0)\widetilde{k}_*\right]. \tag{7.190}$$

In this expression the value $\Omega^2/4 - \widetilde{k}_*^2$ should be estimated using the second term in (7.187) which gives (7.188) because otherwise the result is equal to zero. In the cases where the value $\widetilde{k}_*$ enters individually it is possible to use the leading term only, that is set $\widetilde{k}_* = -\Omega/2$. In this way also using the definition of the hyperbolic sine, $\sinh z = \left(e^z - e^{-z}\right)/2$, (7.190) can be represented as

$$\overline{\mathcal{P}}(T, y_0) = \left[1 - e^{(1-y_0)\Omega}\right]\Omega^2 e^{\Omega} \exp\left[-\Omega^2 e^{\Omega} T\right]. \tag{7.191}$$

It should be noted that (7.191) matches the standard formulas for the escape rate from a potential well. In fact, for $1 - y_0 \gg 1/|\Omega|$, (7.191) considered as a function of time T, has the form

$$\overline{\mathcal{P}}(T, y_0) = \frac{1}{\langle T \rangle} \exp\left[-\frac{T}{\langle T \rangle}\right] \tag{7.192}$$

with the time scale

$$\langle T \rangle = \frac{1}{\Omega^2} \exp\left(-\Omega\right). \tag{7.193}$$

It is exactly the time distribution for escaping from a potential well and (7.193) is the characteristic dependence of the mean life-time in the potential well, on its height.

Finally, we display the different approximations together with the exact solution shown in Figure 7.15. In the long-time behavior, without considering the influence of the initial delta peak at y_0, all curves coincide up to a certain extent very well.

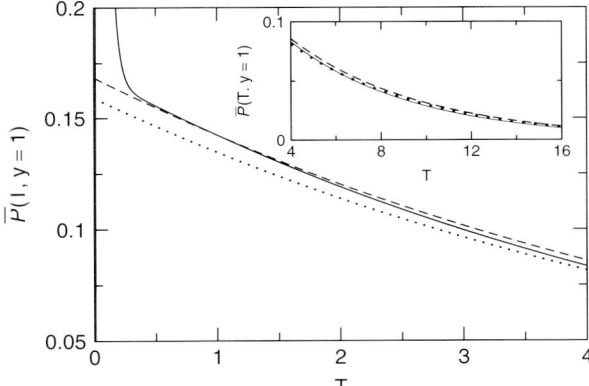

Figure 7.15 Behavior of first-passage time probability density distribution $\overline{\mathcal{P}}(T, y = 1) \equiv \overline{\mathcal{P}}(T, y_0)$ for different approximation levels. Parameters are $\Omega = -5.0$ (scaled negative drift coefficient) and $y_0 = 0.5$ (initial condition). The exact solution (7.185) is given by the full line, the approximation (7.192) by the dotted and (7.191) by the dashed line.

7.11 Roots of the Transcendental Equation

For the complex number $z = \zeta + i\xi$ the following identities

$$\sinh z = \sinh \zeta \cos \xi + i \cosh \zeta \sin \xi$$
$$\cosh z = \cosh \zeta \cos \xi + i \sinh \zeta \sin \xi$$

hold, therefore, the equation $z \cosh z = \beta \sinh z$ becomes

$$(\zeta + i\xi)\left(\cosh \zeta \cos \xi + i \sinh \zeta \sin \xi\right) = \beta \left(\sinh \zeta \cos \xi + i \cosh \zeta \sin \xi\right).$$

Then, separating the real and imaginary parts, we get

$$\zeta \cosh \zeta \cos \xi - \xi \sinh \zeta \sin \xi = \beta \sinh \zeta \cos \xi, \qquad (7.194a)$$
$$\zeta \sinh \zeta \sin \xi + \xi \cosh \zeta \cos \xi = \beta \cosh \zeta \sin \xi. \qquad (7.194b)$$

If $\zeta = 0$, then the system of equations (7.194) is reduced to the equality imposed on ξ:

$$\xi \cos \xi = \beta \sin \xi. \qquad (7.195)$$

In its turn, for $\xi = 0$ the coupled equation (7.194) converts into the equality imposed on ζ:

$$\zeta \cosh \zeta = \beta \sinh \zeta. \qquad (7.196)$$

We will demonstrate that system (7.194) does not admit other solutions. Let $\zeta \neq 0$ and $\xi \neq 0$. In this case we also have $\cos \xi \neq 0$ and $\sin \xi \neq 0$ because, otherwise, either (7.194a) or (7.194b) cannot hold. This enables us to divide (7.194a) and (7.194b) by $\sinh \zeta \cos \xi$ and $\cosh \zeta \sin \xi$, respectively, reducing (7.194) to

$$\zeta \frac{\cosh \zeta}{\sinh \zeta} - \xi \frac{\sin \xi}{\cos \xi} = \beta, \tag{7.197}$$

$$\zeta \frac{\sinh \zeta}{\cosh \zeta} + \xi \frac{\cos \xi}{\sin \xi} = \beta, \tag{7.198}$$

thus,

$$\zeta \frac{\cosh \zeta}{\sinh \zeta} - \xi \frac{\sin \xi}{\cos \xi} = \zeta \frac{\sinh \zeta}{\cosh \zeta} + \xi \frac{\cos \xi}{\sin \xi} \tag{7.199}$$

or

$$\frac{\sinh 2\zeta}{2\zeta} = \frac{\sin 2\xi}{2\xi}. \tag{7.200}$$

Since $|\sin \xi / \xi| < 1$ and $|\sinh \zeta / \zeta| > 1$ for $\zeta, \xi \neq 0$ there are no solutions with $\zeta, \xi \neq 0$ simultaneously.

Summarizing the aforesaid we get the conclusion that the equation

$$z \cosh z = \beta \sinh z$$

admits only solutions meeting one of the following conditions

$$\zeta \cosh \zeta = \beta \sinh \zeta, \qquad \xi = 0, \tag{7.201a}$$

or

$$\xi \cos \xi = \beta \sin \xi, \qquad \zeta = 0. \tag{7.201b}$$

Whence it follows that, first, the collection of solutions contains the series

$$\zeta_0 = 0, \quad \xi_0 = 0,$$

for $n = 1, 2, 3, \ldots$

$$\zeta_n = 0, \quad \xi_n \in \left(-\frac{\pi}{2} + n\pi, \frac{\pi}{2} + n\pi\right)$$

and for $n = -1, -2, -3, \ldots$

$$\zeta_n = 0, \quad \xi_n = -\xi_{-n}. \tag{7.202}$$

Second, for $\beta > 0$ and $\beta \neq 1$ there is an additional pair of roots $\{\zeta_*, \xi_*\}$ and $\{-\zeta_*, -\xi_*\}$ such that

$$\zeta_* = 0, \qquad \xi_* \in \left(0, \frac{\pi}{2}\right) \qquad \text{for } \beta < 1 \tag{7.203a}$$

$$\zeta_* \in (0, \beta), \qquad \xi_* = 0 \qquad \text{for } \beta > 1. \tag{7.203b}$$

In particular, when $\beta \to 1$ the value $\xi_* \to 0$ or $\zeta_* \to 0$.

In the limiting case $\beta \gg 1$, the first terms of the root series for n greater and similar to unity are

$$\xi_n \approx n\pi \tag{7.204}$$

and the root of the second type is

$$\zeta_* \approx \beta\left[1 - 2\exp(-2\beta)\right]. \tag{7.205}$$

7.12 Exercises

E 7.1 Fokker–Planck dynamics with linear potential

Remember the general one-dimensional Fokker–Planck equation (5.53)

$$\frac{\partial}{\partial t}p(x,t) = -\frac{\partial}{\partial x}\left[f(x)p(x,t)\right] + \frac{\sigma^2}{2}\frac{\partial^2 p(x,t)}{\partial x^2},$$

which corresponds to the stochastic differential equation of Langevin type with white noise (5.46)

$$dx(t) = f(x(t))\,dt + \sigma\,dW(t).$$

Introduce the diffusion coefficient $D = \sigma^2/2$ and the potential function $V(x)$ via $V(x) = -\int f(x)\,dx$ to get, as a starting point, the following Fokker–Planck partial differential equation

$$\frac{\partial p(x,t)}{\partial t} = \frac{\partial}{\partial x}\left(\frac{dV(x)}{dx}p(x,t) + D\frac{\partial p(x,t)}{\partial x}\right) \tag{7.206}$$

together with the delta-like initial condition $p(x, t=0) = \delta(x - x_0)$.

Investigate the dynamics (7.206) with linear potential $V(x) = -\alpha x$, where the constant drift coefficient α may have value ranges from minus to plus infinity, shown in Figure 7.16.

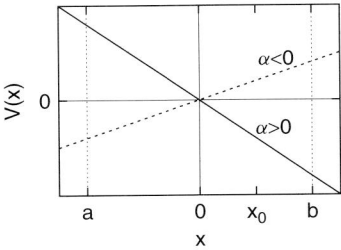

Figure 7.16 Sketch of the linear potential $V(x) = -\alpha x$ within the finite interval $a \leq x \leq b$ for two different parameter values $\alpha > 0$ (solid line) and $\alpha < 0$ (broken line). The initial condition is marked by x_0.

Find the solution of the drift–diffusion problem in a finite interval $a \leq x \leq b$ with well defined boundary conditions. Take mixed boundaries, meaning closed on the left-hand side $x = a$ and open on the right-hand side $x = b$.

Discuss the first-passage time problem (7.28). Follow the analytical method outlined in Section 7.5 and compare it with results presented by Fox and Choi [29, 48] and Linetsky [125, 126]. Is the Péclet number, known from Redner [185], related to the dimensionless drift parameter Ω (7.21)?

E 7.2 Fokker–Planck dynamics with V-shaped potential

Find the solution of the drift–diffusion problem for a particle moving in a V-shaped potential given by $V(x) = \gamma |x|$ ($\gamma \geq 0$). The dynamics is described by the Fokker–Planck equation (7.206) with natural boundary conditions $-\infty \leq x \leq +\infty$ and initial value x_0.

Hint: use the analogy with the solution of (7.1).

E 7.3 First-passage time problem with absorbing boundary

Find the solution of the drift–diffusion problem given by the Fokker–Planck equation (7.206) with the, already known, potential $V(x) = \gamma |x|$ in the case of a natural boundary at $x \to -\infty$ and an absorbing one at $x = b$.

E 7.4 First-passage time problem with mixed boundaries

Find the solution of the drift–diffusion problem given by the Fokker–Planck equation (7.206) with the potential $V(x) = \gamma |x|$ in the case of a reflecting boundary at $x = a$ and an absorbing one at $x = b$.

Compare the results with the previous task.

8
The Ornstein–Uhlenbeck Process

8.1
Definitions and Properties

In Chapter 5 we discussed the Brownian motion in the velocity space based on the Langevin equation. The corresponding stochastic process is known as the Ornstein–Uhlenbeck process. The mathematical model developed with some modifications also has other applications, e.g. in finance [119]. Here we consider a more general case where the mean value of the stochastic variable relaxes to a nonzero value. In the following we will study the process in the space of two variables corresponding to the velocity and the coordinate of the Brownian particle. However, at first we restrict our analysis to motion in one dimension. As distinct from Chapter 5, here we will consider the probability density functions in detail by solving the Fokker–Planck rather than the Langevin equation.

From a mathematical point of view [7, 25, 201] the Ornstein–Uhlenbeck process is defined as a mean-reverting process given by the following stochastic differential equation

$$dx(t) = (a - c\,x(t))\,dt + b\,dW(t) \tag{8.1}$$

together with initial conditions $x(t=0) = x_0$ and $W(t=0) = 0$.

All control parameters a, b and c are non-negative. The special case with $a = 0$ is a process for which the mean value tends to zero. In general, it tends to a nonzero value a/c in the long-time limit $t \to \infty$. If we set $c = 0$, the arithmetic Brownian motion (cf. Section 5.8) remains.

By using the identity

$$d\left(\phi x(t)\right) = \phi(t)\left(c\,x(t)\,dt + dx(t)\right), \tag{8.2}$$

where $\phi(t) = \exp\{c\,t\}$ acts as an integrating factor, then (8.1) can be written as

$$d\left(\phi x(t)\right) = \phi(t)\left(a\,dt + b\,dW(t)\right). \tag{8.3}$$

This is a stochastic differential equation with respect to $\phi\,x(t)$. The formal solution, obtained by integrating both sides of (8.3), reads

Physics of Stochastic Processes: How Randomness Acts in Time
Reinhard Mahnke, Jevgenijs Kaupužs and Ihor Lubashevsky
Copyright © 2009 WILEY-VCH Verlag GmbH & Co. KGaA, Weinheim
ISBN: 978-3-527-40840-5

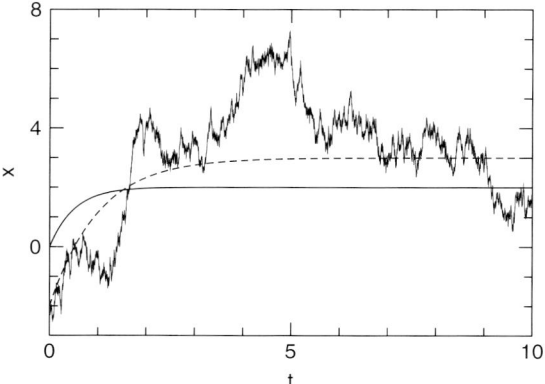

Figure 8.1 A stochastic trajectory of the Ornstein–Uhlenbeck process (8.1) with the initial condition $x_0 = -2$ for parameter values $a = 3$, $b = 2$ and $c = 1$. The theoretical mean value and the variance are shown by dashed and solid curves, respectively.

$$x(t) = \frac{a}{c}[1 - \exp\{-ct\}] + x_0 \exp\{-ct\} + b \int_0^t \exp\{-c(t-s)\}\,dW(s). \qquad (8.4)$$

The first and the second moments are

$$\langle x(t) \rangle = \frac{a}{c}[1 - \exp\{-ct\}] + x_0 \exp\{-ct\}, \qquad (8.5)$$

$$\langle x(t)^2 \rangle = \langle x(t) \rangle^2 + \frac{b^2}{2c}\left(1 - \exp\{-2ct\}\right). \qquad (8.6)$$

As an illustrative example, one stochastic trajectory of the mean-reverting process is shown in Figure 8.1, comparing it with the theoretical mean value. The variance $\langle x(t)^2 \rangle - \langle x(t) \rangle^2$, calculated from (8.6) and (8.5), indicates the amplitude of the stochastic fluctuations.

The random real-valued function $x(t)$ (Figure 8.1 shows one realization) is defined on the set of outcomes of a random experiment as a solution of the stochastic differential equation (8.1). Details of stochastic modeling including numerical solution techniques can be found in [4] and [104].

Over the years, the Ornstein–Uhlenbeck process [231] has become to play an important role in several branches of natural sciences related, in some form, to Brownian motion and diffusion [13, 103, 154, 186].

8.2
The Ornstein–Uhlenbeck Process and its Solution

We consider the Ornstein–Uhlenbeck process in the space of coordinate x and velocity v [231]. In the Langevin formalism it is defined by the following system of stochastic differential equations [180, 193]

$$dx = v\,dt, \tag{8.7}$$
$$dv = -\gamma v\,dt + \sqrt{2B}\,dW(t), \tag{8.8}$$

where $W(t)$ is the Wiener process. The ensemble of stochastic trajectories given by these equations is characterized by the probability distribution function $p(x,v,t)$ which obeys the Fokker–Planck equation

$$\frac{\partial}{\partial t}p = -\frac{\partial}{\partial x}[v\,p] + \frac{\partial}{\partial v}[\gamma\,v\,p] + \frac{\partial^2}{\partial v^2}[B\,p]. \tag{8.9}$$

This is a particular case of (5.4) with a two-dimensional state vector (x,v). We set

$$p(x,v,t=0) = \delta(x-x_0)\,\delta(v-v_0) \tag{8.10}$$

as the initial condition, which means that the process starts at the position $x = x_0$ with given velocity v_0.

Our aim is to solve the above Fokker–Planck equation (8.9) in agreement with (8.10) to get the probability density $p = p(x,v,t)$ analytically. Here we will give a sketch of the approach worked out by Ralf Remer [189].

Since the equation contains only the first derivative with respect to the coordinate x, it is helpful to make the Fourier transformation relative to this variable. It is defined as

$$p(x,v,t) = \frac{1}{\sqrt{2\pi}}\int_{-\infty}^{+\infty} dk_x\,\exp[i\,k_x\,x]\,\overline{p}(k_x,v,t). \tag{8.11}$$

In this way the Fokker–Planck equation (8.9) is transformed to

$$\frac{\partial}{\partial t}\overline{p} = -i\,k_x\,v\,\overline{p} + \frac{\partial}{\partial v}[\gamma\,v\,\overline{p}] + \frac{\partial^2}{\partial v^2}[B\,\overline{p}]. \tag{8.12}$$

The resulting equation (8.12) for $\overline{p} = \overline{p}(k_x,v,t)$ contains the second derivative with respect to v, whereas the velocity v itself is contained only in the first power. Therefore we can easily make the Fourier transformation with respect also to this variable, so

$$\overline{p}(k_x,v,t) = \frac{1}{\sqrt{2\pi}}\int_{-\infty}^{+\infty} dk_v\,\exp[i\,k_v\,v]\,\tilde{p}(k_x,k_v,t), \tag{8.13}$$

$$v\,\overline{p}(k_x,v,t) = -\frac{1}{\sqrt{2\pi}}\int_{-\infty}^{+\infty} dk_v\,i\,\frac{\partial}{\partial k_v}\exp[i\,k_v\,v]\,\tilde{p}(k_x,k_v,t)$$
$$= \frac{1}{\sqrt{2\pi}}\int_{-\infty}^{+\infty} dk_v\,i\,\exp[i\,k_v\,v]\,\frac{\partial}{\partial k_v}\tilde{p}(k_x,k_v,t). \tag{8.14}$$

Hence we arrive at the following equation for $\tilde{p} = \tilde{p}(k_x,k_v,t)$ in the Fourier space of x and v

$$\frac{\partial}{\partial t}\tilde{p} = [k_x - \gamma\,k_v]\frac{\partial}{\partial k_v}\tilde{p} - B\,k_v^2\,\tilde{p} \tag{8.15}$$

or

$$\frac{\partial}{\partial t}\tilde{p} + [\gamma k_v - k_x]\frac{\partial}{\partial k_v}\tilde{p} = -B k_v^2 \tilde{p}. \tag{8.16}$$

This equation represents the Cauchy problem which can be solved by the method of characteristics. For this purpose we introduce a time-dependent quantity $\bar{k}_v = \bar{k}_v(\tau)$ which depends on the intrinsic time $\tau \in [0, t]$ and obeys the boundary condition

$$\bar{k}_v(t) = k_v. \tag{8.17}$$

Then we consider the equation

$$\frac{\partial}{\partial t}\tilde{p} + [\gamma \bar{k}_v - k_x]\frac{\partial}{\partial \bar{k}_v}\tilde{p} = -B \bar{k}_v^2 \tilde{p} \tag{8.18}$$

which is obtained from (8.16) by replacing k_v with $\bar{k}_v$. Due to the boundary condition (8.17), the solution of (8.18) agrees with that of (8.16) at $\tau = t$. The function $\bar{k}_v(\tau)$ can be chosen such that it satisfies the equation

$$\frac{d}{d\tau}\bar{k}_v = \gamma \bar{k}_v - k_x. \tag{8.19}$$

In this case we have

$$[\gamma \bar{k}_v - k_x]\frac{\partial}{\partial \bar{k}_v}\tilde{p} = \frac{\partial \tilde{p}}{\partial \bar{k}_v}\frac{d\bar{k}_v}{d\tau} \tag{8.20}$$

and hence we obtain an important relation

$$\frac{d\tilde{p}}{d\tau} = \frac{\partial \tilde{p}}{\partial \tau} + \frac{\partial \tilde{p}}{\partial \bar{k}_v}\frac{d\bar{k}_v}{d\tau} = \frac{\partial \tilde{p}}{\partial \tau} + [\gamma \bar{k}_v - k_x]\frac{\partial}{\partial \bar{k}_v}\tilde{p} = -B \bar{k}_v^2 \tilde{p} \tag{8.21}$$

for the total derivative of $\tilde{p}(\tau)$. Thus we have an ordinary differential equation

$$\frac{d}{d\tau}\tilde{p} = -B \bar{k}_v^2 \tilde{p} \tag{8.22}$$

the solution of which gives us $\tilde{p}(k_x, k_v, t)$ at $\tau = t$. Equation (8.19) can be solved with respect to the variable $\bar{k}_v$ by the method of variation. It yields

$$\bar{k}_v(\tau) = \left[\bar{k}_0 - \frac{k_x}{\gamma}\right]\exp[\gamma \tau] + \frac{k_x}{\gamma}. \tag{8.23}$$

Taking into account the boundary condition (8.17), we obtain the complete solution

$$\bar{k}_v(\tau) = \left[k_v - \frac{k_x}{\gamma}\right]\exp[\gamma (\tau - t)] + \frac{k_x}{\gamma}. \tag{8.24}$$

This equation is used to solve the differential equation (8.22) as follows

$$\tilde{p} = \tilde{p}_0 \exp[-\Phi(t)],$$

$$\Phi(t) = B \int_0^t \bar{k}_v^2(\tau)\,d\tau, \tag{8.25}$$

$$\Phi(t) = B\left[\frac{A^2}{2\gamma}(\exp[2\gamma t] - 1) + C^2 t + 2\frac{AC}{\gamma}(\exp[\gamma t] - 1)\right], \tag{8.26}$$

$$A = \left[k_v - \frac{k_x}{\gamma}\right]\exp[-\gamma t], \quad C = \frac{k_x}{\gamma}.$$

8.2 The Ornstein–Uhlenbeck Process and its Solution

The initial condition for the probability density distribution in the Fourier space reads

$$\tilde{p}(k_x, k_v, t=0) = \frac{1}{2\pi} \exp[-i k_x x_0] \exp[-i k_v v_0]. \tag{8.27}$$

By inserting the initial condition in the solution for p we replace k_v with the equivalent term $\bar{k}_v(0)$. Then the solution for the probability distribution in the Fourier space becomes

$$\tilde{p}(k_x, k_v, t) = \frac{1}{2\pi} \exp[-i k_x x_0] \exp[-i v_0 (A+C)] \exp[-\Phi(t)], \tag{8.28}$$

$$\Phi(t) = B \left[\frac{A^2}{2\gamma} (\exp[2\gamma t] - 1) + C^2 t + 2 \frac{AC}{\gamma} (\exp[\gamma t] - 1) \right].$$

$$A = \left[k_v - \frac{k_x}{\gamma} \right] \exp[-\gamma t], \quad C = \frac{k_x}{\gamma}. \tag{8.29}$$

In order to obtain the one-dimensional probability density distribution in the space of velocities v, we use the following relation

$$p_v(v,t) = \int dx\, p(x,v,t) = \sqrt{2\pi}\, \bar{p}(0,v,t) = \int_{-\infty}^{+\infty} dk_v \, \exp[i k_v v] \tilde{p}(0, k_v, t)$$

$$= \sqrt{2\pi}\, F[\tilde{p}(0, k_v, t)](v,t). \tag{8.30}$$

The probability distribution for the velocity v at time t is thus given by the inverse transformation of the solution in the Fourier space at $k_x = 0$. According to this, we calculate

$$p_v(v,t) = \sqrt{2\pi}\, F[\tilde{p}(0, k_v, t)](v,t),$$

$$\tilde{p}(0, k_v, t) = \frac{1}{2\pi} \exp[-i k_v v_0 \exp[-\gamma t]] \exp[-\phi(t) k_v^2],$$

$$\phi(t) = \frac{B}{2\gamma} (1 - \exp[-2\gamma t]), \tag{8.31}$$

$$F\left[\exp[-\phi(t) k_v^2]\right](v,t) = \frac{1}{\sqrt{2\phi(t)}} \exp\left[-\frac{1}{2} \frac{v^2}{2\phi(t)}\right], \tag{8.32}$$

$$F\left[\bar{f}(k_v, t)\right](v,t) = f(v,t),$$

$$F\left[\exp[i k_v b] \bar{f}(k_v, t)\right](v,t) = f(v+b, t). \tag{8.33}$$

It leads to the complete solution in v which reads

$$p_v(v,t) = \frac{1}{\sqrt{2\pi \sigma_v^2(t)}} \exp\left[-\frac{1}{2} \frac{(v - v_0 \exp[-\gamma t])^2}{\sigma_v^2(t)}\right], \tag{8.34}$$

$$\sigma_v^2(t) = \frac{B}{\gamma} (1 - \exp[-2\gamma t]).$$

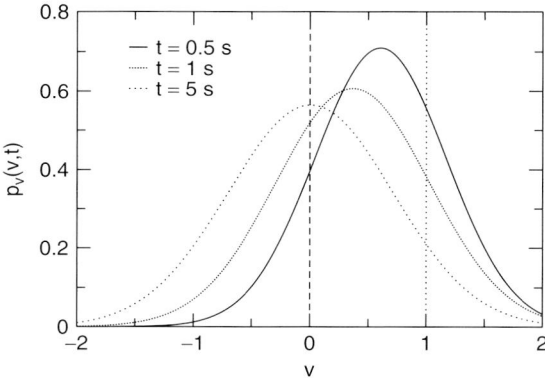

Figure 8.2 The probability density distribution $p_v(v, t)$ given by (8.34) at three different times for the parameter values $B = 0.5$ m^2 s^{-3} and $\gamma = 1$ s^{-1}. The initial condition is $v_0 = 1$ m s^{-1}.

The probability density distribution for velocity v at different times t is shown in Figure 8.2. The dotted line indicates our initial condition $p_v(v, t = 0) = \delta(v - v_0)$ with $v_0 = 1$ m s^{-1} (the initial velocity). The dashed line corresponds to the mean value of the velocity in the limit $t \to \infty$. The relaxation of the distribution to the stationary (equilibrium) one can be clearly seen.

Considering the long-time limit $t \to \infty$ in (8.34), we obtain the well known Maxwell distribution [85, 193]

$$p_v(v, t) = \frac{m}{\sqrt{2\pi k_B T}} \exp\left[-\frac{1}{2} \frac{m v^2}{k_B T}\right] \tag{8.35}$$

$$\frac{B}{\gamma} = \frac{k_B T}{m} \tag{8.36}$$

with temperature T, mass of particles m, and the Boltzmann constant k_B. The diffusion coefficient in the velocity space B characterizes the fluctuation strength, whereas γ is the friction coefficient, which is related to the energy dissipation. As we can see, the ratio of these two quantities is proportional to the temperature and, respectively, the thermal energy $k_B T$. Therefore, this relation (8.36), which has already appeared in Chapter 5 (cf. (5.103)) and is known as Einstein's formula representing some form of the fluctuation-dissipation theorem.

In the same way we can calculate the probability distribution over the coordinate x

$$p_x(x, t) = \sqrt{2\pi} \, F[\tilde{p}(k_x, 0, t)], \tag{8.37}$$

$$\tilde{p}(k_x, 0, t) = \frac{1}{2\pi} \exp[-i k_x x_0]$$
$$\times \exp\left[-i k_x v_0 (1 - \exp[-\gamma t])/\gamma\right] \exp[-\omega(t) k_x^2],$$

$$\omega(t) = -\frac{3}{2}\frac{B}{\gamma^3} - \frac{B}{2\gamma^3}\exp[-2\gamma t] + \frac{B}{\gamma^2}t + 2\frac{B}{\gamma^3}\exp[-\gamma t].$$

It gives us the solution

$$p_x(x,t) = \frac{1}{\sqrt{2\pi\sigma_x^2(t)}}\exp\left[-\frac{1}{2}\frac{(x-\mu_x(t))^2}{\sigma_x^2(t)}\right], \quad (8.38)$$

$$\mu_x(t) = x_0 + \frac{v_0}{\gamma}(1-\exp[-\gamma t]),$$

$$\sigma_x^2(t) = \frac{2B}{\gamma^2}t - 3\frac{B}{\gamma^3} + 4\frac{B}{\gamma^3}\exp[-\gamma t] - \frac{B}{\gamma^3}\exp[-2\gamma t].$$

The probability density distribution over the coordinate x at different times is presented in Figure 8.3. The dotted line again shows our initial condition $p_x(x, t=0) = \delta(x-x_0)$ with $x_0 = 0$. The dashed line indicates the mean value of the coordinate in the long-time limit $t \to \infty$. The broadening of the distribution with time can be clearly seen.

Considering the variance in more detail, we find the following relation

$$\sigma_x^2 \sim \frac{2B}{\gamma^2}t = 2Dt \quad (8.39)$$

for long times $t \to \infty$. This linear growth of the variance, shown in Figure 8.4, had already been discovered by Albert Einstein [40].

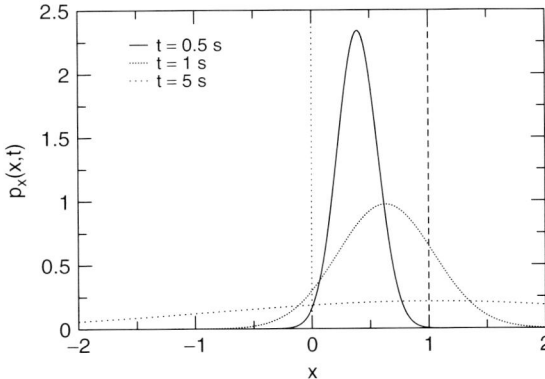

Figure 8.3 The probability density distribution $p_x(x,t)$ given by (8.38) at three different times for parameter values $B = 0.5$ m^2 s^{-3} and $\gamma = 1$ s^{-1}. The initial condition is $x_0 = 0$ m and $v_0 = 1$ m s^{-1}.

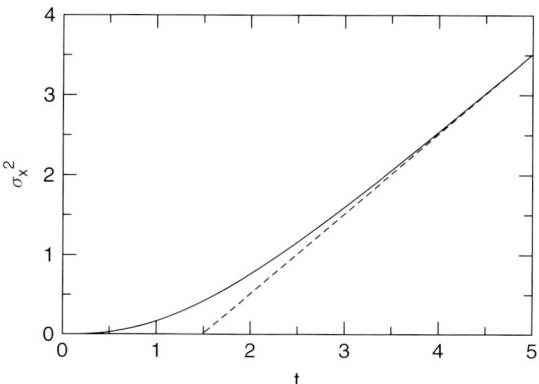

Figure 8.4 Time dependence of the variance σ_x^2 given by (8.38). The dashed line shows the asymptotic slope $D = B/\gamma^2$ for large times (see (8.39)). The values of the parameters are $B = 0.5$ m^2 s^{-3} and $\gamma = 1$ s^{-1}.

In order to make the inverse transformation of the solution in Fourier space, we modify (8.28) as follows

$$\tilde{p}(k_x, k_v, t) = \frac{1}{2\pi} \exp[-i k_x \mu_x] \exp[-i k_v \mu_v]$$
$$\times \exp[-\sigma_x^2 k_x^2 - \sigma_v^2 k_v^2 - \sigma_{xv}^2 k_v k_x] \tag{8.40}$$

$$\mu_x = x_0 + \frac{v_0}{\gamma}(1 - \exp[-\gamma t]),$$

$$\mu_v = v_0 \exp[-\gamma t],$$

$$\sigma_x^2 = \frac{B}{2\gamma^3}(1 - \exp[-2\gamma t]) + \frac{B}{\gamma^2} t - 2\frac{B}{\gamma^3}(1 - \exp[-\gamma t]),$$

$$\sigma_v^2 = \frac{B}{2\gamma}(1 - \exp[-2\gamma t]),$$

$$\sigma_{xv}^2 = -\frac{B}{\gamma^2}(1 - \exp[-2\gamma t]) + 2\frac{B}{\gamma^2}(1 - \exp[-\gamma t]).$$

In the following we make the inverse transformation with respect to the variable k_v. For this purpose we rewrite (8.40) in a slightly modified form:

$$\tilde{p}(k_x, k_v, t) = \frac{1}{2\pi} \exp[-i k_x \mu_x] \exp[-i k_v \overline{\mu}_v] \exp[-\sigma_x^2 k_x^2 - \sigma_v^2 k_v^2], \tag{8.41}$$

$$\overline{\mu}_v = \mu_v - i\sigma_{xv}^2 k_x.$$

By using the inverse Fourier transformation given in (8.33), we obtain the probability distribution $\bar{p}(k_x, v, t)$ which reads

$$\bar{p} = \frac{1}{2\pi\sqrt{2\sigma_v^2}} \exp[-ik_x\mu_x] \exp[-\sigma_x^2 k_x^2] \exp\left[-\frac{1}{2}\frac{(v-(\mu_v - i\sigma_{xv}^2 k_x))^2}{2\sigma_v^2}\right]$$

$$= \frac{1}{2\pi\sqrt{2\sigma_v^2}} \exp[-ik_x\bar{\mu}_x] \exp[-\bar{\sigma}_x^2 k_x^2] \exp\left[-\frac{1}{2}\frac{(v-\mu_v)^2}{2\sigma_v^2}\right], \quad (8.42)$$

$$\bar{\mu}_x = \mu_x + (v - \mu_v)\sigma_{xv}^2/(2\sigma_v^2),$$
$$\bar{\sigma}_x^2 = \sigma_x^2 - \sigma_{xv}^4/(4\sigma_v^2).$$

The resulting probability density distribution $\bar{p}$ can be easily transformed back to the original distribution $p(x, v, t)$ which we wanted to calculate, so

$$p(x, v, t) = \frac{1}{\sqrt{2\pi 2\sigma_v^2}} \exp\left[-\frac{1}{2}\frac{(v-\mu_v)^2}{2\sigma_v^2}\right] \frac{1}{\sqrt{2\pi 2\bar{\sigma}_x^2}} \exp\left[-\frac{1}{2}\frac{(x-\bar{\mu}_x)^2}{2\bar{\sigma}_x^2}\right].$$
(8.43)

The probability density $p(x, v, t)$ is given in units of s m^{-2}. We have thus calculated the probability distribution from which we can determine the probability of finding a particle within any small coordinate interval $[x, x + dx]$ and velocity interval $[v, v + dv]$ at moment t. The probability density distribution obtained is presented in Figure 8.5. The broadening over the coordinate axis with increasing time, as well as the stationary profile over the velocity axis can be easily recognized. Furthermore, it is easy to see that a simple multiplication of one-dimensional distributions for the coordinate x and velocity v does not reproduce mutual dependence.

8.3
The Ornstein–Uhlenbeck Process with Linear Potential

In the previous section we have treated the well known Ornstein–Uhlenbeck process and have obtained the time-dependent solution for the distribution of the probability density. The process considered there takes place in a spatial region with constant potential. Now we would like to consider the same process in the presence of a potential which is linear in x. It is defined by

$$U(x) = -m\gamma\theta x. \quad (8.44)$$

By means of the known relation between force F and potential U

$$F(x, t) = -\frac{\partial}{\partial x} U(x, t) \quad (8.45)$$

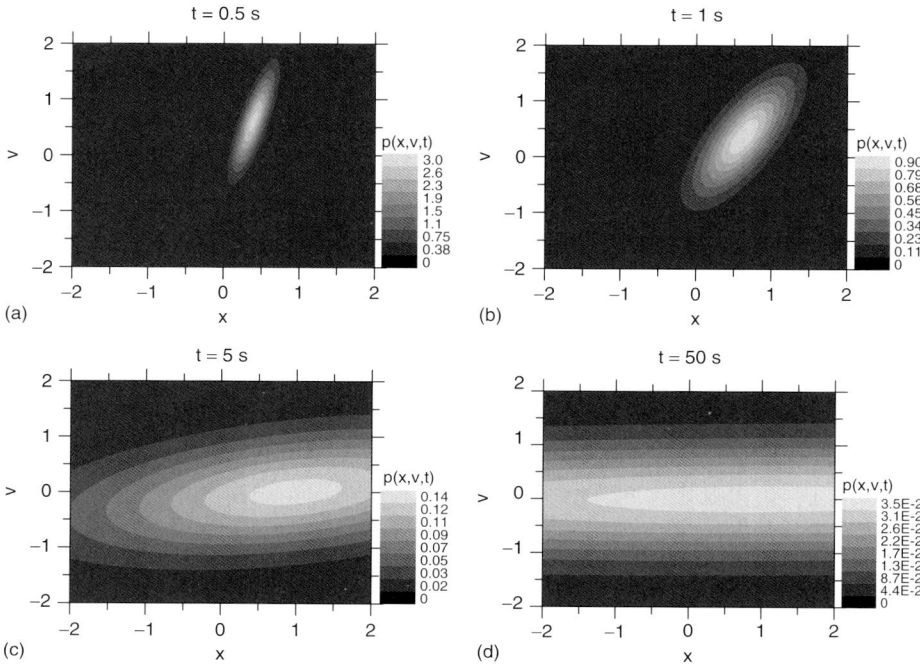

Figure 8.5 The probability density distribution $p(x, v, t)$ given by (8.43) at four different times $t = 0.5$ s (a), $t = 1$ s (b), $t = 5$ s c), and $t = 50$ s (d). The values of the parameters are $B = 0.5$ m^2 s^{-3} and $\gamma = 1$ s^{-1}. The initial conditions are $x_0 = 0$ m and $v_0 = 1$ m s^{-1}.

the equations of motion in the Langevin formalism can be written as

$$dx = v\, dt, \tag{8.46}$$

$$dv = \gamma(\theta - v)\, dt + \sqrt{2B}\, dW(t). \tag{8.47}$$

As distinct from the standard case we now have an additional term $\gamma \theta\, dt$. The corresponding Fokker–Planck equation for the probability density $p(x, v, t)$ reads

$$\frac{\partial}{\partial t} p = -\frac{\partial}{\partial x}(v\, p) - \frac{\partial}{\partial v}(\gamma \theta\, p) + \frac{\partial}{\partial v}(\gamma v\, p) + \frac{\partial^2}{\partial v^2}(B\, p). \tag{8.48}$$

Applying the Fourier transformation twice as in the previous section, with $\bar{p} = \bar{p}(k_x, v, t)$ and $\tilde{p} = \tilde{p}(k_x, k_v, t)$, we obtain the following equation

$$\frac{\partial}{\partial t}\tilde{p} + [\gamma k_v - k_x]\frac{\partial}{\partial k_v}\tilde{p} = -[B k_v^2 + i\gamma\theta k_v]\tilde{p} \tag{8.49}$$

which has to be solved. It is easy to see that the system of characteristic equations changes only with respect to $\tilde{p}$ and not with respect to $\bar{k}_v$. Hence, the solution for $\bar{k}_v(\tau)$ can be taken from (8.24). The solution for $\tilde{p}$ is calculated as follows

8.3 The Ornstein–Uhlenbeck Process with Linear Potential

$$\frac{d}{d\tau}\tilde{p} = -[B\overline{k}_v^2 + i\gamma\theta\overline{k}_v]\tilde{p} \quad \text{gives} \quad \tilde{p} = \tilde{p}_0 \exp[-\Phi(t)], \tag{8.50}$$

$$\Phi(t) = \int_0^t d\tau \left[B\overline{k}_v^2(\tau) + i\gamma\theta\overline{k}_v(\tau) \right],$$

$$\Phi(t) = \frac{BA^2}{2\gamma} (\exp[2\gamma t] - 1) + BC^2 t + 2\frac{BAC}{\gamma} (\exp[\gamma t] - 1)$$
$$+ i\theta A (\exp[\gamma t] - 1) + i\gamma\theta C t,$$

$$A = \left[k_v - \frac{k_x}{\gamma} \right] \exp[-\gamma t], \quad C = \frac{k_x}{\gamma}.$$

The initial conditions are the same as in the previous section. Therefore we also have the same constant $\tilde{p}_0$.

In the following, we transform the above solution for the probability density distribution in the Fourier space into a form similar to (8.40), that is,

$$\tilde{p}(k_x, k_v, t) = \frac{1}{2\pi} \exp[-ik_x\mu_x] \exp[-ik_v\mu_v]$$
$$\times \exp[-\sigma_x^2 k_x^2 - \sigma_v^2 k_v^2 - \sigma_{xv}^2 k_v k_x] \tag{8.51}$$

$$\mu_x = x_0 + \left(\frac{v_0}{\gamma} - \frac{\theta}{\gamma} \right) (1 - \exp[-\gamma t]) + \theta t,$$

$$\mu_v = v_0 \exp[-\gamma t] + \theta (1 - \exp[-\gamma t]),$$

$$\sigma_x^2 = \frac{B}{2\gamma^3} (1 - \exp[-2\gamma t]) + \frac{B}{\gamma^2} t - 2\frac{B}{\gamma^3} (1 - \exp[-\gamma t]),$$

$$\sigma_v^2 = \frac{B}{2\gamma} (1 - \exp[-2\gamma t]),$$

$$\sigma_{xv}^2 = -\frac{B}{\gamma^2} (1 - \exp[-2\gamma t]) + 2\frac{B}{\gamma^2} (1 - \exp[-\gamma t]).$$

In this way, for the Ornstein–Uhlenbeck process with linear potential we obtain a similar solution as previously (for constant or zero potential) with only slightly modified parameters

$$p(x, v, t) = \frac{1}{\sqrt{2\pi 2\sigma_v^2}} \exp\left[-\frac{1}{2} \frac{(v - \mu_v)^2}{2\sigma_v^2}\right] \frac{1}{\sqrt{2\pi 2\overline{\sigma}_x^2}} \exp\left[-\frac{1}{2} \frac{(x - \overline{\mu}_x)^2}{2\overline{\sigma}_x^2}\right], \tag{8.52}$$

$$\overline{\mu}_x = \mu_x + (v - \mu_v) \sigma_{xv}^2 / (2\sigma_v^2),$$
$$\overline{\sigma}_x^2 = \sigma_x^2 - \sigma_{xv}^4 / (4\sigma_v^2).$$

The resulting probability density distribution $p(x, v, t)$ is presented in Figure 8.6. The drift and broadening over the coordinate x with increasing of time and also the stationary profile over the velocity v axis (now with an other than zero mean value) can be easily recognized.

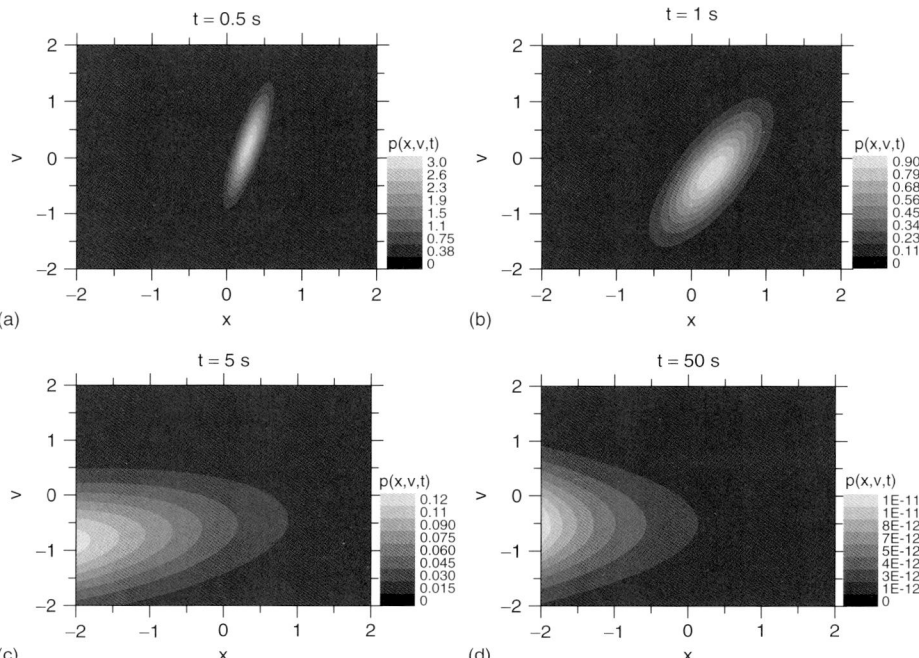

Figure 8.6 The probability density distribution $p(x, v, t)$ given by (8.52) at four different times $t = 0.5$ s (a), $t = 1$ s (b), $t = 5$ s (c), and $t = 50$ s (d). The values of the parameters are $\theta = -1$ m s^{-1}, $B = 0.5$ m^2 s^{-3}, and $\gamma = 1$ s^{-1}. The initial conditions are $x_0 = 0$ m and $v_0 = 1$ m s^{-1}.

Comparing the solutions for the Ornstein–Uhlenbeck process with constant potential (previous section) and with the linear potential (see (8.52)) we state the same analytical form of a double Gaussian distribution. The variance (and the correlation between the coordinate x and the velocity v) in both cases is the same. However, the mean values μ_v and $\overline{\mu}_x$ have been changed by the potential. The velocity v now relaxes to the mean value θ for large times $t \to \infty$. In this case we also have a deterministic drift $\theta\, t$ in the coordinate x in addition to the influence of the initial condition, via the difference term $1/\gamma\, (v_0 - \theta)$.

By means of the general solution (8.52) one can again calculate the one-dimensional probability density distributions for any one of the variables. For the velocity v we obtain

$$p_v(v, t) = \frac{1}{\sqrt{2\pi 2\sigma_v^2}} \exp\left[-\frac{1}{2}\frac{(v - \mu_v)^2}{2\sigma_v^2}\right], \tag{8.53}$$

$$\mu_v = v_0 \exp[-\gamma t] + \theta(1 - \exp[-\gamma t]),$$

$$\sigma_v^2 = \frac{B}{2\gamma}(1 - \exp[-2\gamma t]).$$

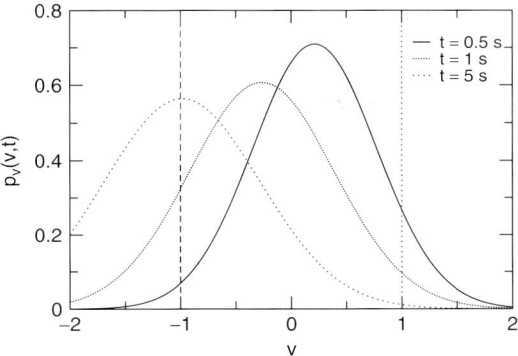

Figure 8.7 Probability density distribution $p_v(v, t)$ given by (8.54) for three different times. The values of the parameters are $\theta = -1$ m s^{-1}, $B = 0.5$ m^2 s^{-3}, and $\gamma = 1$ s^{-1}. The initial condition is $v_0 = 1$ m s^{-1}.

The time-dependent solution for the velocity v is presented in Figure 8.7. We again find that the probability density distribution corresponds to the Gaussian distribution. Due to the linear potential in coordinate space, the mean value of the stationary probability distribution is shifted by θ. The variance (the mean squared fluctuation of the velocity), however, is not influenced by the linear potential.

Now we calculate the one-dimensional distribution over the coordinate x by means of the probability density $p(x, v, t)$. The time-dependent solution reads

$$p_x(x, t) = \frac{1}{\sqrt{2\pi 2\sigma_x^2}} \exp\left[-\frac{1}{2} \frac{(x - \mu_x)^2}{2\sigma_x^2}\right], \tag{8.54}$$

$$\mu_x = x_0 + \left(\frac{v_0}{\gamma} - \frac{\theta}{\gamma}\right)(1 - \exp[-\gamma t]) + \theta t,$$

$$\sigma_x^2 = \frac{B}{2\gamma^3}(1 - \exp[-2\gamma t]) + \frac{B}{\gamma^2} t - 2\frac{B}{\gamma^3}(1 - \exp[-\gamma t]).$$

The distribution is shown in Figure 8.8. This function also corresponds to the Gaussian distribution. The variance (the mean squared fluctuation in the coordinate space) is not influenced by the linear potential. While the velocity tends to a certain value θ, the time dependence of the mean value of the coordinate x is dominated by the linear term θt in the long-time limit $t \to \infty$.

The Ornstein–Uhlenbeck process with the linear potential differs from the standard one only in the mean value of the double Gaussian distribution. Therefore, we have shown in Figure 8.9 the mean value of the velocity μ_v depending on the mean value of the coordinate μ_x for times between 0 and 50 s. This relation is always linear for the standard Ornstein–Uhlenbeck process and has a fixed starting point (corresponding to the given initial condition) $\mu_x = x_0$, $\mu_v = v_0$ and a fixed end-point $\mu_x = x_0 + v_0/\gamma$, $\mu_v = 0$ for $t \to \infty$. To the contrary, the relation between both mean values is nonlinear for $\theta \neq 0$ in the case of the Ornstein–Uhlenbeck

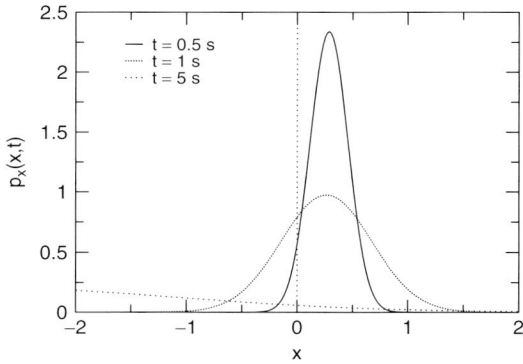

Figure 8.8 The probability density distribution $p_x(x,t)$ given by (8.55) at three different times. The values of the parameters are $\theta = -1$ m s^{-1}, $B = 0.5$ m^2 s^{-3}, and $\gamma = 1$ s^{-1}. The initial condition is $x_0 = 0$ m and $v_0 = 1$ m s^{-1}.

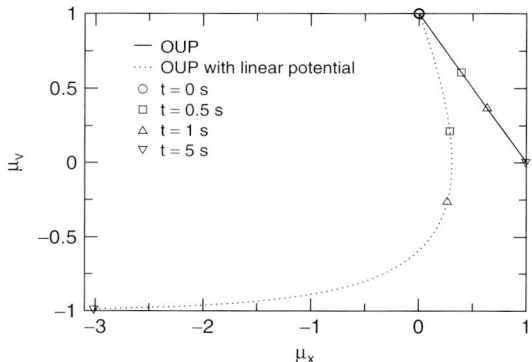

Figure 8.9 The mean value of the velocity μ_v versus the mean value of the coordinate μ_x for times t from 0 s to 5 s. OUP is the abbreviation for the Ornstein–Uhlenbeck process. The values of parameters are $\theta = -1$ m s^{-1}, $B = 0.5$ m^2 s^{-3}, and $\gamma = 1$ s^{-1}. The initial condition is given by $x_0 = 0$ m and $v_0 = 1$ m s^{-1}.

process with linear potential. The starting point in both cases is the same because of the same initial condition. The end point at $t \to \infty$ is $\mu_x = \infty$ and $\mu_v = \theta$ for the linear potential. This results in the nonlinear relation between both mean values.

8.4
The Exponential Ornstein–Uhlenbeck Process

Here we consider the Ornstein–Uhlenbeck process with a potential of exponential form and derive the equation of motion in the space of Laplace and Fourier transformed variables. We consider the following system of equations in the Langevin formalism

$$dx = \theta \exp[a\,v]\,dt, \tag{8.55}$$

$$dv = -\gamma\,v\,dt + \sqrt{2\,B}\,dW(t). \tag{8.56}$$

Since the dependence on the velocity v is exponential, the coordinate x can take only positive values if we start with a positive one. Therefore the coordinate can only increase with time. The Fokker–Planck equation for the probability density $p = p(x, v, t)$ reads

$$\frac{\partial}{\partial t}p = -\frac{\partial}{\partial x}\left(\theta \exp[a\,v]\,p\right) + \frac{\partial}{\partial v}(v\,p) + \frac{\partial^2}{\partial v^2}(B\,p). \tag{8.57}$$

We set

$$p(x, v, t = 0) = \delta(x - x_0)\,\delta(v - v_0), \tag{8.58}$$

$$p(0, v, t) = 0 \quad \text{for} \quad x_0 > 0 \tag{8.59}$$

as the initial condition. In this case it is reasonable to consider the Laplace rather than the Fourier transformation with respect to the coordinate x, since the latter is positively defined. The Laplace transformation is defined by

$$\bar{p}(k_x, v, t) = \int_0^{+\infty} dx\,\exp[-k_x\,x]\,p(x, v, t). \tag{8.60}$$

Hence we obtain for $\bar{p} = \bar{p}(k_x, v, t)$

$$\frac{\partial}{\partial t}\bar{p} = -\theta \exp[a\,v]\,k_x\,\bar{p} + \frac{\partial}{\partial v}(v\,\bar{p}) + \frac{\partial^2}{\partial v^2}(B\,\bar{p}). \tag{8.61}$$

We also make the Fourier transformation $\tilde{p} = \tilde{p}(k_x, k_v, t)$ with respect to the velocity v

$$\bar{p}(k_x, v, t) = \frac{1}{\sqrt{2\pi}}\int_{-\infty}^{+\infty} dk_v\,\exp[i\,k_v\,v]\,\tilde{p}(k_x, k_v, t), \tag{8.62}$$

$$\exp[a\,v]\,\bar{p}(k_x, v, t) = \frac{1}{\sqrt{2\pi}}\int_{-\infty-ia}^{+\infty-ia} dk_v\,\exp[i\,k_v\,v]\,\tilde{p}(k_x, k_v + i\,a, t)$$

$$= \frac{1}{\sqrt{2\pi}}\int_{-\infty}^{+\infty} dk_v\,\exp[i\,k_v\,v]\,\tilde{p}(k_x, k_v + i\,a, t). \tag{8.63}$$

Here we have assumed that $\tilde{p}(k_x, k_v, t)$ is vanishing at $\operatorname{Re} k_v \to \pm\infty$ and analytical within $-a \leq \operatorname{Im} k_v \leq 0$ in the complex plane of k_v, which allows us to shift the integration path in (8.63) to the real axis. This leads to the following equation of motion

$$\frac{\partial}{\partial t}\tilde{p}(k_x, k_v, t) = -\theta k_x \tilde{p}(k_x, k_v + ia, t) - \gamma k_v \frac{\partial}{\partial k_v}\tilde{p}(k_x, k_v, t) - B k_v^2 \tilde{p}(k_x, k_v, t). \tag{8.64}$$

An essential difference from the equations we obtained earlier for the cases with constant and linear potentials is that (8.64) contains $\tilde{p}$ with shifted argument $k_v + i\,a$. This equation therefore cannot be solved analytically in the same way as previously.

8.5
Outlook on Econophysics

Over the last few years, research activities in the field of quantitative finance have greatly increased [119, 180]. The main topics of interest, especially for practitioners, are risk management and derivatives. But in order to handle risk optimally it is necessary to know some basic facts about stock price dynamics, at least the basic characteristics of the dynamics (chaotic or stochastic, Gaussian-like or not). This is why, we want to consider stock price dynamics as a subject itself [189].

The stock price development is described as a stochastic process following geometric Brownian motion (cf. Section 5.9). The Langevin equation in Ito notation (5.166) is well known as

$$dk = \mu\, k\, dt + \sigma\, k\, dW(t), \tag{8.65}$$

with the stock price k at time t, where μ is the time constant drift (or growth rate), σ is the time constant fluctuation and $dW(t)$ is the time-dependent increment of a Wiener process $W(t)$. As a result of this description, the logarithmic stock price $x(t) = \ln[k(t)/k_0]$ follows a Gaussian distribution, with a time-dependent mean of $(\mu - \sigma^2/2)\,t + x_0$ and a time-dependent variance of $\sigma^2\, t$, if the initial value at time $t = 0$ is fixed at x_0.

The geometric Brownian motion was the basis for further investigations in econophysics, for instance, of the Black–Scholes formula [180]. But in recent years it was found that the description did not agree with the empirical facts of the stock market [189–192]. Mainly, the changes $dx(t)$ in the logarithmic stock price $x(t)$ are not Gaussian distributed. The empirical distribution is a leptokurtic distribution. It has, compared to that of a Gaussian, fat tails (higher probability density of (absolute) high values) and a higher concentration of probability density around the mean, see Figure 8.10.

In models with stochastic volatility, the variance $v = \sigma^2$ is no longer a parameter, but is itself a variable. In addition to this, the variance is, like the stock price, a stochastic variable. This coincides with empirical observations, that exhibit random behavior of the variance. The corresponding stochastic differential equations with the Langevin formalism in the Ito notation are as follows:

$$dk = \mu\, k\, dt + \sqrt{v(t)}\, k\, dW_k(t)$$
$$dv = a_v(v, k, t)\, dt + b_v(v, k, t)\, dZ_v(t)$$
$$dZ_v(t) = \rho\, dW_k(t) + \sqrt{1 - \rho^2}\, dW_v(t). \tag{8.66}$$

The noise terms $dW_i(t)$ are again the increments of a Wiener process and they are independent and identically distributed.

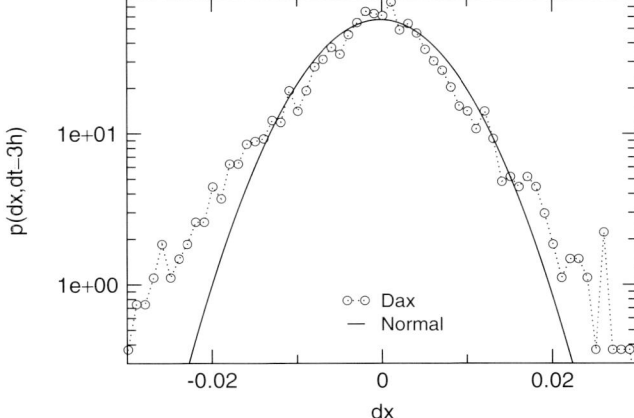

Figure 8.10 The empirical probability density distribution of the DAX (German stock index) for an increment dx in the logarithmic stock price $x(t)$ for $dt = 3$ h is plotted against the normal distribution with estimated moments.

The analysis of empirical stock prices shows that the variance exhibits mean-reversion, which means that it fluctuates randomly around a long-run average. Therefore, for the drift term $a_v(v, k, t)$ we use the following relation (Ito notation):

$$a_v(v, k, t) = \gamma (\theta - v). \tag{8.67}$$

The parameter θ is the value of v (usually taken as the long-run average), that the variance reaches within the relaxation time $1/\gamma$.

Both approaches that we want to analyze, the Hull–White and the Heston model, belong to the models with stochastic volatility and describe the variance by a mean reverting process. Only the diffusion terms are different. The fluctuation term of the Hull–White model $b_v^{HW}(v, k, t)$ and of the Heston model $b_v^{H}(v, k, t)$ are known as [192]

$$b_v^{HW}(v, k, t) = \tilde{\kappa} v \tag{8.68}$$

$$b_v^{H}(v, k, t) = \kappa \sqrt{v}, \tag{8.69}$$

where the parameters κ and $\tilde{\kappa}$ represent the strength of the stochastic fluctuations of the variance v.

In fitting the models against empirical data, the correlation coefficient ρ is often set to zero. The empirical analysis shows that $\rho \ll 1$ holds, and we assume it in our further analysis. In the Heston model without correlation, the dynamics of the variance $v(t)$ is described by a stochastic differential equation in the Langevin formalism with Ito notation as follows

$$dv = \gamma (\theta - v) dt + \kappa \sqrt{v} dW_v(t). \tag{8.70}$$

The corresponding equation for the Hull–White model is

$$dv = \gamma(\theta - v)\,dt + \tilde{\kappa}\,v\,dW(t). \tag{8.71}$$

For a further analysis of the variance we transform the Langevin equation into a Fokker–Planck equation

$$\frac{\partial}{\partial t} p(v,t) + \frac{\partial}{\partial v} S(v,t) = 0 \tag{8.72}$$

with probability $p(v,t)\,dv$ that the variance is in the interval $[v, v+dv]$ at time t, where $S(v,t)$ is the probability flux. It is

$$S(v,t) = \gamma\theta p(v,t) - \gamma v p(v,t) - \frac{\kappa^2}{2} \frac{\partial}{\partial v}\left[v\,p(v,t)\right] \tag{8.73}$$

for the Heston model and

$$S(v,t) = \gamma\theta p(v,t) - \gamma v p(v,t) - \frac{\tilde{\kappa}^2}{2} \frac{\partial}{\partial v}\left[v^2\,p(v,t)\right] \tag{8.74}$$

for the Hull–White model.

We are now interested in the stationary solution of the probability density distribution of the variance v. In this case the probability current $S(v,t)$ in (8.73) is equal to zero. We obtain the stationary probability density distribution $p_{st}^H(v)$ for the variance $v(t)$ of the Heston model without correlation as

$$p_{st}^H(v) = \frac{a^a}{\Gamma(a)\,\theta^a}\,v^{a-1}\,\exp\left[-\frac{av}{\theta}\right]; \qquad a = \frac{2\gamma\theta}{\kappa^2}. \tag{8.75}$$

For the Hull–White model this reads as follows

$$p_{st}^{HW}(v) = \frac{\tilde{a}^{\tilde{a}}\,\theta^{\tilde{a}+1}}{\Gamma(\tilde{a})}\,v^{-(\tilde{a}+2)}\,\exp\left[-\frac{\tilde{a}\theta}{v}\right]; \qquad \tilde{a} = \frac{2\gamma}{\tilde{\kappa}^2}. \tag{8.76}$$

The empirical analysis of stock price data reveals that the relaxation time $1/\gamma$ of the variance v is around 22 days (especially for indices). Compared to the time scale of the logarithmic returns of about 1 h, that we are interested in, this is very long and leads to the conclusion that we can regard the variance v as constant in the short time window we are considering. Therefore, we consider the probability density distribution of the logarithmic changes y of the stock prices for short time windows τ.

We apply the relation for the conditional probabilities

$$p(k,t) = \int p(k,t\,|\,v,t)\,p(v,t)\,dv. \tag{8.77}$$

In accordance with the above discussion, we can use the stationary distribution $p_{st}(v)$ instead of the time-dependent probability density distribution $p(v,t)$. Furthermore, we do not focus on the stock price k itself, but on the changes $y = dx$ of the logarithmic stock price $x = \ln[k(t)/k_0]$ in the time interval $\tau = dt$. Therefore, according to (8.77) we derive the relation

$$p_s(y, \tau) = \int_{v=0}^{\infty} p(y, \tau \mid v) \, p_{st}(v) \, dv, \qquad (8.78)$$

where $p_s(y, \tau)$ indicates that the variance has to be in the stationary regime. The stationary distribution $p_{st}(v)$ we have already calculated. But we need an ansatz for the conditional probability density distribution $p(y, \tau \mid v)$.

We know that, in the case of constant variance $v' = \sigma^2$, the Heston model and the Hull–White model transform into geometric Brownian motion (for the parameters $\kappa = 0 \, d^{-1}$, $\tilde{\kappa} = 0 \, d^{-1/2}$, $v(t=0) = v'$, $\theta = v'$), where the logarithmic changes y are normally distributed with mean $\mu \tau + \frac{1}{2} v' \tau$ and variance $v' \tau$. Because the integral (8.77) is also valid for the geometric Brownian motion (with $p(v, t) = \delta_{v-v'}$) we obtain the ansatz for the conditional probability density distribution

$$p(y, \tau \mid v) = \frac{1}{\sqrt{2 \pi v \tau}} \exp\left[-\frac{1}{2} \frac{(y - \mu \tau + \frac{1}{2} v \tau)^2}{v \tau}\right], \qquad (8.79)$$

which also coincides with an analytically obtained result for the Heston model.

With the help of the conditional probability density distribution (8.79) and the specific stationary distributions of the variance we calculate the probability density distributions of the logarithmic returns for short times τ. We have fitted the obtained solutions against the returns of 1 hour ($\tau = 1 \, h$) for the DAX index presented in Figures 8.11 and 8.12. More examples and an extended analysis can be found in [189–192].

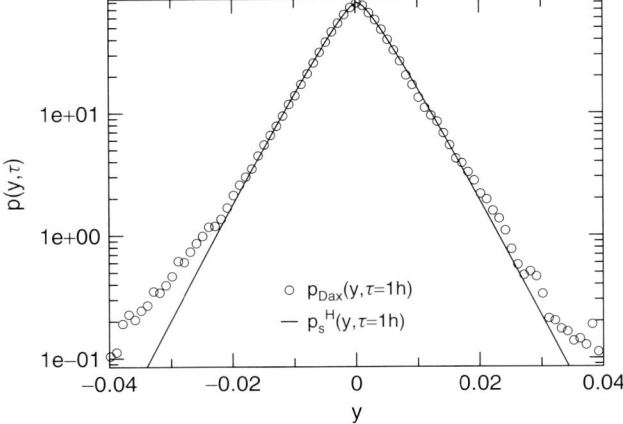

Figure 8.11 The probability density distribution of the logarithmic return y for the Heston model $p_s^H(y, \tau)$ is plotted against the calculated probability density distribution $p_{Dax}(y, \tau)$ of the DAX index (02.05.1996–28.12.2001) for time $\tau = 1 \, h$ with $\chi^2 = 24.2$, $a = 1.36$, $\theta = 5.15 \times 10^{-5} \, h^{-1}$ and $\mu = 3.03 \times 10^{-4} \, h^{-1}$.

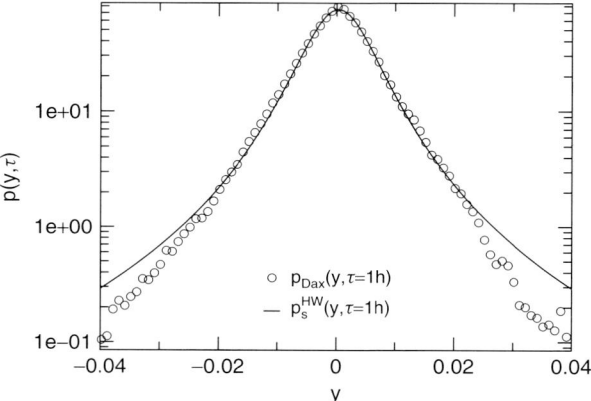

Figure 8.12 The probability density distribution of the logarithmic return y for the Hull–White model $p_s^{HW}(y, \tau)$ is plotted against the calculated probability density distribution $p_{Dax}(y, \tau)$ of the DAX index (02.05.1996–28.12.2001) for time $\tau = 1$ h with $\chi^2 = 54.1$, $a = 0.08$, $\theta = 3.21 \times 10^{-4}$ h^{-1} and $\mu = 2.97 \times 10^{-4}$ h^{-1}.

8.6 Exercises

E 8.1 Ornstein–Uhlenbeck's paper from 1930

Study the historical paper by George Eugene Uhlenbeck (1880–1941) and Leonard Salomon Ornstein (1900–1988) 'On the Theory of Brownian motion' (Physical Review, 1930, vol. 36, pp. 823–841) in detail. Consider an extension to the situation without additional force (cf. Section 8.2) the case with the external force as the Brownian motion of a harmonically bonded particle.

Hint: Take [231] as the starting point for your analysis.

E 8.2 The Ornstein–Uhlenbeck process

Consider the one-dimensional stochastic process (8.1) with $a = 0$, $c = \gamma$ (friction coefficient) and $b = \sqrt{2B}$ (fluctuation strength). Based on the given Langevin equation (8.8) with initial value $v(t = 0) = v_0$ write down the corresponding Fokker–Planck equation.

Find the time-dependent solution for the stochastic process given by $v(t)$ (Langevin approach) as well as $p(v, t)$ (probability density) and $p(v_1, t_1; v_2, t_2)$ (joint probability density, JPD, (3.1)) or $p(v_2, t_2 \mid v_1, t_1)$ (conditional probability density, CPD, (3.2)) from the Fokker–Planck approach.

Remember the average or mean value $\langle v(t) \rangle$

$$\mu(t) = \int v\, p(v, t)\, dv \qquad (8.80)$$

and the covariance or correlation function $\langle (v(t) - \langle v(t) \rangle)(w(s) - \langle w(s) \rangle) \rangle$

$$\sigma^2(t,s) = \iint (v - \mu(t))(w - \mu(s)) \, p(v,t;w,s) \, dw \, dv \qquad (8.81)$$

and show that the solution is a Gaussian normal distribution $N(\mu(t); \sigma^2(t,t))$ with mean $\mu(t)$ and variance $\sigma(t,t)$, which should be calculated.

E 8.3 *The Heston model*

Verify that (8.75) is the stationary solution of the Fokker–Planck equation (8.72) with the probability flux (8.73) describing the variance in the Heston model. Perform simulations of stochastic trajectories for the corresponding Langevin equation and evaluate the stationary probability distribution in the long-time limit. Compare the result with the analytical solution (8.75).

9
Nucleation in Supersaturated Vapors

9.1
Dynamics of First-Order Phase Transitions in Finite Systems

Stochastic processes have many classical applications in physics such as diffusion or Brownian motion as discussed in Chapters 6–8. Here we consider a further important application to the dynamics of first-order phase transitions, considering formation of vapor droplets in a supersaturated vapor as a particular example.

A complete theory of the dynamics of first-order phase transitions would have to account for a variety of physical processes on different time and length scales including nucleation, spinodal decomposition, growth of clusters of new phase and late-stage coarsening process like Ostwald ripening and coagulation, see e. g. [237, 244].

If the system is infinitely large, then the depletion of the medium can be neglected in the nucleation process. This allows a theoretical description of first-order phase transition ignoring any interaction between the growing clusters as presumed in the classical nucleation theory [16, 45, 220, 236, 248]. In this case we obtain a simultaneous nucleation and independent growth of already formed supercritical clusters. The basic kinetic model, underlying classical nucleation theory, was proposed by Leo Szilard. It yields a steady-state nucleation rate provided that the system is continuously supplied by monomers while large clusters are removed. For a finite system the formation and growth of the clusters result in depletion of the surrounding medium. Thus we come to another scenario of first-order phase transitions in finite closed systems. This general scenario is characterized as follows. First, a process of nucleation occurs in the initial homogeneous supersaturated state requiring a very short time. The critical energy for the formation of a stable droplet is determined through a competition between a volume term (which favors creation of the droplet), and a surface term (which favors its dissolution). On average, the droplets with $n > n_{cr}$ (critical cluster size) grow, while those with $n < n_{cr}$ shrink. Therefore, the formation of critical nuclei in the initially homogeneous state is possible due only to the stochastic fluctuations. Further stable growth of droplets over the critical size ($n > n_{cr}$) can be described in a deterministic manner.

In a second stage, the growth of the already formed supercritical clusters predominates and the nucleation rate decreases. This stage is succeeded by a third stage of competitive growth of the clusters, the so-called Ostwald ripening period. It is characterized by a decrease in the number of clusters and an increase in their mean size or radius during a large interval of time. The theory of Ostwald ripening goes back to the pioneering work by Lifshitz and Slyozov [124] and, independently, that of Wagner [238]. The classical Lifshitz–Slyozov–Wagner (LSW) theory describes a homogeneous or heterogeneous system (binary mixture) in the two-phase region within the droplet model using a cluster size distribution which changes due to monomer condensation and/or evaporation. Ostwald ripening is the process of phase separation in a supersaturated system or binary mixture by diffusional growth of spherical nuclei of the minority phase. At low initial supersaturation, spherical droplets will typically be nucleated at large mutual separation so that the diffusional droplet growth can be described in a single-droplet picture. In the LSW theory of Ostwald ripening the droplet growth is coupled to the concentration field to obey global mass conservation. The LSW theory provides evidence for the existence of a universal cluster distribution function if appropriately scaled variables are used and for universal power laws in time for the physical quantities such as cluster density and mean cluster size. Universality here means that the asymptotic long-time behavior is independent of the details of the initial nucleation process. One finds, in particular, that the critical cluster size increases as $t^{1/3}$, the supersaturation follows the inverse $t^{-1/3}$ law, but the number of clusters decreases as $1/t$ at large times t. Ostwald ripening, the late-stage process of droplet growth by evaporation and condensation, is well understood. To reduce the interfacial free energy of the system, material diffuses away from small, high-curvature droplets (which dissolve), and condenses onto large, low-curvature droplets (which grow). In this way, the large droplets swallow the small ones. A single large droplet survives in the final stage of this competition. The classical LSW theory considers this process in the noninteracting limit with the vanishing volume fraction ϕ of the condensating minority phase.

Because of the depletion of the medium, the three stages are not independent of each other. In particular the nucleation rate depends on the growth of the already formed clusters. The outlined scenario of first-order phase transitions is valid, if the initial supersaturation is not too high. In this case the first stage of the transition can be described by nucleation as a formation of fluctuations in small regions of space with large differences in the density compared with the initial state. If the initial supersaturation increases, nucleation is replaced continuously by spinodal decomposition.

Although the limiting cases are understood, much less is known about the complete evolution of the system from the early nucleation to the late Ostwald ripening stage. This problem studied in the work by Schmelzer [202] and others [113, 145, 155] has experimental evidence which points to the importance of the interparticle diffusional interactions and of the spatial locations of particles in nucleation and growth. The experiments have confirmed the prediction of self-similar coarsening behavior at long times. However, the measured distributions over cluster sizes generally are broader and more symmetric than the LSW theory predicts.

9.2 Condensation of Supersaturated Vapor

A simple example of a system of many particles, is a gas consisting of identical molecules called monomers. If the gas is dilute (the density, as the number of molecules per unit volume, is small), the average separation length between the monomers is large and, correspondingly, their interaction is negligible. The gas is said to be ideal if the average separation length is much larger than the de Broglie wavelength. We treat the monomers as indistinguishable particles moving in a closed volume and making reactive collisions to form aggregates called molecular clusters [139, 229].

If we consider a vapor at equilibrium then a certain change in the thermodynamic parameters enables us to move the system into a nonequilibrium state. The vapor becomes supersaturated. The basic quantity describing the situation is the cluster distribution function **N** at time t

$$\mathbf{N}(t) = (N_0, N_1, N_2, \ldots, N_n, \ldots, N_N) \tag{9.1}$$

which gives the number of clusters N_n of size n. The free particles (molecules) are called monomers of size $n = 0$. It is supposed that N_1 molecules are excited. These molecules may be named a precluster of size $n = 1$. The bound states are clusters of size $n \geq 2$. Investigating a finite system the overall number of particles N_{total} as well as the volume V and the temperature T are fixed. The particle conservation law

$$N_{\text{total}} = N_0 + N_1 + \sum_{n=2}^{N} n N_n = \text{constant} \tag{9.2}$$

takes into account that particles are either free, excited, or bounded in clusters. There is always some difficulty in describing the initial stage of formation of a cluster. We have introduced the precluster as an intermediate state between free and bounded states to provide an easy and unified description of the aggregation process.

Further on, we consider a simplified case where only one single cluster of size n (i. e. $N_n = 1$) coexists with $N_0 = N_{\text{total}} - n$ unbounded (free) particles, which means that the cluster distribution (9.1) reduces to

$$\mathbf{N}(t) = (N_0, 0, \ldots, 0, N_n = 1, 0, \ldots, 0) \tag{9.3}$$

and the overall particle conservation (9.2) to

$$N_{\text{total}} = N_0 + \sum_{n=1}^{N} n N_n = N_0 + n \cdot 1 = \text{constant}, \tag{9.4}$$

where the stochastic variable $n = n(t)$ is the number of particles bounded in the cluster at time t. The nucleation box (volume V) embedded in a heat bath (temperature T) displaying the situation schematically is shown in Figure 9.1.

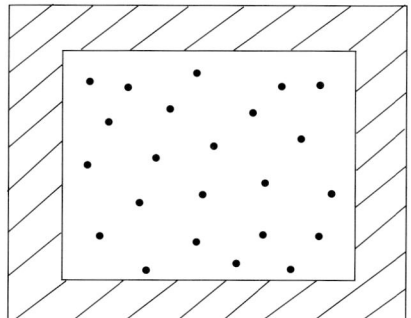 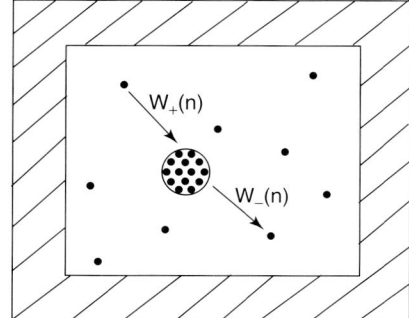

Figure 9.1 Isothermal–isochoric nucleation in supersaturated vapor: free molecules (black dots) called monomers in the initial stage (left) of the aggregation process form a cluster (spherical droplet) of certain size n, coexisting with a gas of free molecules, afterwards (right).

The starting point is the one-dimensional one-step master equation (3.38) describing the condensation and evaporation of a single particle on or from a molecular cluster, where $w_+(n)$ and $w_-(n)$ are the transition rates of condensation and evaporation, respectively, which have now to be formulated.

The attachment probability per time of a monomer to a (spherical) cluster of size n is proportional to the cluster surface $A(n)$ and to the density of free monomers N_0/V_{free}, so

$$w_+(n) = \alpha A(n)\,(N_{\text{total}} - n)/V_{\text{free}}, \quad 1 \leq n \leq N_{\text{total}}, \tag{9.5}$$

where

$$V_{\text{free}} = V - n\,c_{\text{clust}}^{-1} \tag{9.6}$$

is the free volume not occupied by the cluster. Here we have assumed that each particle has his own eigenvolume c_{liquid}^{-1} related to the density of liquid c_{liquid} equivalent to the density of particles in a cluster, so $c_{\text{liquid}} \equiv c_{\text{clust}}$. The coefficient α has not yet been specified. We can interpret it as the velocity of sticking. The surface of a spherical droplet is given by

$$A(n) = 4\pi r^2 = 4\pi\,(c_{\text{clust}} 4\pi/3)^{-2/3}\,n^{2/3} \sim n^{2/3} \tag{9.7}$$

with known incompressible particle density inside the cluster $c_{\text{clust}} = \text{constant}$ (liquid density as given experimental value). A special case is the formation of a precluster $n = 1$ out of an elementary particle ($n = 0$). The precluster can be understood as an excited monomer which is able to react with some other monomer to form a dimer ($n = 2$). In the free-particle state $n = 0$ any of the N_{total} monomers can become excited, so that we can write

$$w_+(0) = \frac{p}{\tau}\,N_{\text{total}} \tag{9.8}$$

where the parameter p in this case means the excitation probability per time multiplied by the time constant τ.

By using the detailed balance relation (3.49), the evaporation rate $w_-(n)$ of a monomer from a cluster of size n is calculated from the known attachment rate (9.5). For the equilibrium cluster distribution, the most probable value of the cluster size n corresponds to the minimum of the thermodynamic potential, in this case free energy F, as it is evident from (3.50). The latter equation (with $\Omega \equiv F$) combined with the detailed balance condition (3.49) allows to find the relation between transition rates $w(\mathbf{N}' \mid \mathbf{N})$ and $w(\mathbf{N} \mid \mathbf{N}')$ of opposite stochastic events (transition from state $\mathbf{N}$ to state $\mathbf{N}'$ and vice versa), so

$$\frac{w(\mathbf{N}' \mid \mathbf{N})}{w(\mathbf{N} \mid \mathbf{N}')} = \exp\left(\frac{F(T, V, \mathbf{N}) - F(T, V, \mathbf{N}')}{k_B T}\right). \tag{9.9}$$

In this way, to find the relation between transition probabilities we need the knowledge of the free energy F. Following the basic principles of the statistical mechanics [167], we define the Hamiltonian of our many-particle system and calculate the statistical sum (or integral) Z which is related to free energy via $F = -k_B T \ln Z$. Our system is described by the cluster distribution $\mathbf{N}$ defined by (9.1). The total Hamiltonian $H(\mathbf{N})$ reads

$$H = \sum_{n=0}^{N} H_n \tag{9.10}$$

with the contribution H_n for the N_n clusters of size n at coordinates $\mathbf{r}_i^{(n)}$ and momenta $\mathbf{p}_i^{(n)}$ written as kinetic energy and interaction potential

$$H_n\left(\mathbf{p}^{(n)}, \mathbf{r}^{(n)}\right) = \sum_i \frac{\left(\mathbf{p}_i^{(n)}\right)^2}{2m_i^{(n)}} + \sum_{i<j} U_{ij}^{(n,n)}\left(\mid \mathbf{r}_i - \mathbf{r}_j \mid\right). \tag{9.11}$$

The mass $m_i^{(n)}$ of a cluster containing n monomers ($n \geq 1$) is given by

$$m_i^{(n)} \equiv m_n = nm \tag{9.12}$$

where $m \equiv m_0$ is the mass of one monomer. The canonical partition function, that is, the statistical integral, is an integral over all space and momentum coordinates. In the semi-classical approximation it reads [167]

$$Z(T, V, \mathbf{N}) = \prod_{n=0}^{N} \frac{1}{N_n! h^{3N_n}} \int d^{3N_n}p \, d^{3N_n}q \, \exp(-\beta H_n), \tag{9.13}$$

where $\beta = 1/(k_B T)$. This partition function can be divided in two factors, one of which represents an ideal part due to the kinetic energy

$$Z_{\text{ideal}}(T, V, \mathbf{N}) = \prod_{n=0}^{N} \frac{V_{\text{eff}}^{N_n}}{N_n! h^{3N_n}} \left(\sqrt{2\pi m_n k_B T}\right)^{3N_n} \tag{9.14}$$

and the second part $Z_{\text{binding}}(T, V, \mathbf{N})$ is responsible for the energy stored in clusters. Here $V_{\text{eff}} < V$ is the effective volume. By introducing this quantity in (9.14) instead of the total volume V we take into account the fact that particles are not point-like, so the integration over the spatial coordinates in (9.13) effectively takes place in a reduced volume V_{eff}. In a certain approximation, both terms together read

$$Z(T, V, \mathbf{N}) = \prod_{n=0}^{N} \frac{V_{\text{eff}}^{N_n}}{N_n!} \left[\left(\frac{\sqrt{2\pi m_n k_B T}}{h} \right)^3 \exp\left(-\frac{f_n}{k_B T}\right) \right]^{N_n} \quad (9.15)$$

where the binding energy $f_n(T)$ is the minimum value of the potential energy sought over all spatial arrangements of the n bounded monomers

$$f_n(T) = \min_{\mathbf{r}} \sum_{i<j} U_{ij}^{(n,n)} \left(|\mathbf{r}_i - \mathbf{r}_j| \right). \quad (9.16)$$

From the canonical partition function Z we can calculate the thermodynamic quantities using the relation between Z and the state function free energy F via

$$F(T, V, \mathbf{N}) = -k_B T \ln Z(T, V, \mathbf{N}). \quad (9.17)$$

According to (9.15) and (9.17), in the isothermal–isochoric situation we obtain

$$F(T, V, \mathbf{N}) = k_B T \sum_{n=0}^{N} \left[N_n \ln \left(\lambda_n(T)^3 / V_{\text{eff}} \right) + \ln N_n! \right] + \sum_{n=0}^{N} N_n f_n(T) \quad (9.18)$$

with de Broglie wavelength $\lambda_n(T) = n^{-1/2} \lambda_0(T) = n^{-1/2} \left(h^2/(2\pi m k_B T) \right)^{1/2}$ (at $n \geq 1$). Here $\lambda_0(T)$ is the de Broglie wavelength of a monomer (particle of mass m) given by

$$\lambda_0 = h/(2\pi m k_B T)^{1/2} \approx 10^{-10} \text{ m}. \quad (9.19)$$

This is the wavelength of a quantum-mechanical free particle with energy $E = p^2/2m = \hbar^2 k^2/2m$, where k is the wave number relating to the wavelength λ_0.

Using the Stirling formula $\ln N_n! \simeq N_n \ln N_n - N_n$ in (9.18) we obtain an approximation for large cluster numbers N_n

$$F(T, V, \mathbf{N}) = k_B T \sum_{n=0}^{N} N_n \left[\ln \left(\lambda_n(T)^3 N_n / V_{\text{eff}} \right) - 1 \right] + \sum_{n=0}^{N} N_n f_n(T). \quad (9.20)$$

In the case of one cluster of size n only (see (9.3)) and at $N_0 = N_{\text{total}} - n \to \infty$ (that is in the thermodynamic limit by expansion of $\ln N_0!$) the free energy (9.18) reads

$$F(T, V, n) = k_B T \left\{ (N_{\text{total}} - n) \left[\ln \left(\lambda_0(T)^3 (N_{\text{total}} - n)/V_{\text{eff}} \right) - 1 \right] \right.$$
$$\left. + (1 - \delta_{n,0}) \ln \left(\lambda_n(T)^3 / V_{\text{eff}} \right) \right\} + f_n(T). \quad (9.21)$$

In our special case, where only one cluster of size n is possible, (9.9) in the thermodynamic limit reduces to

$$\frac{w_-(n)}{w_+(n-1)} = \frac{V_{\text{eff}} (1 - 1/n)^{3/2}}{\lambda_0^3(T)(N_{\text{total}} - n)} \exp\left(\frac{f_n(T) - f_{n-1}(T)}{k_B T} \right) : n \geq 2 \quad (9.22)$$

$$\frac{w_-(1)}{w_+(0)} = \frac{1}{N_{\text{total}}} \exp\left(\frac{f_1(T)}{k_B T} \right). \quad (9.23)$$

The reduction of the effective integration volume in (9.13) is relevant at large densities when the cluster includes a relatively large part of all particles. According to this, the effective volume V_{eff} can be approximately replaced with the free volume V_{free} outside the cluster given by (9.6). Then from (9.22) we obtain an approximation

$$\frac{w_-(n)}{w_+(n)} = \frac{V_{\text{free}}}{\lambda_0^3(T)(N_{\text{total}} - n)} \exp\left(\frac{f_n(T) - f_{n-1}(T)}{k_B T}\right), \tag{9.24}$$

which is true for large enough n. In a rough approximation, we have assumed that (9.24) can be extrapolated up to $n = 1$. Taking into account (9.5), this yields

$$w_-(n) = \alpha A(n) \frac{1}{\lambda_0^3} \exp\left(\frac{f_n - f_{n-1}}{k_B T}\right), \tag{9.25}$$

where $f_n - f_{n-1}$ is the difference in the binding energies between clusters of size n and $n - 1$. Based on (9.23) the stochasticity parameter p in (9.8) can be written as

$$\frac{p}{\tau} = w_-(1) \exp\left(-\frac{f_1(T)}{k_B T}\right). \tag{9.26}$$

By (9.5), (9.25), and (9.26) all the transition rates are well defined in a way which is consistent with the basic principles of statistical mechanics.

Clusters as bound states of elementary particles (monomers) have negative potential energy, which is the so-called binding energy. The potential function $f_n(T)$ is well known from atomic and nuclear theory and also as the Bethe–Weizsäcker formula which, in a simple nonlinear approximation, reads [141–143, 203]

$$f_n(T) = \mu_\infty(T) n + \sigma A(n). \tag{9.27}$$

The binding energy consists of a negative volume term ($\mu_\infty < 0$) and a positive surface contribution. The quantity $\mu_\infty(T)$ is the chemical potential of one monomer or, in other words, the energy necessary for taking away one elementary particle (monomer) from a cluster with a flat surface. The parameter $\sigma > 0$ can be understood as the surface tension of a flat surface. Ansatz (9.27) is a good approximation for large enough sizes n and also provides the correct normalization condition $f_0 = 0$ for a free particle ($n = 0$). Substituting (9.27) into the detachment rate (9.25) we obtain the approximation

$$w_-(n) = \alpha A(n) \frac{1}{\lambda_0^3} \exp\left\{\frac{\mu_\infty(T) + \sigma[A(n) - A(n-1)]}{k_B T}\right\}$$

$$\approx \alpha A(n) \frac{1}{\lambda_0^3} \exp\left(\frac{\mu_\infty(T)}{k_B T}\right) \exp\left(\frac{2\sigma k(n)}{c_{\text{clust}} k_B T}\right). \tag{9.28}$$

This result is valid for large enough clusters (starting with $n \approx 10$) and contains the curvature $k(n)$ of a size-n droplet

$$k(n) = 1/r = (c_{\text{clust}} 4\pi/3)^{1/3} n^{-1/3}. \tag{9.29}$$

The difference in the surface areas $A(n) - A(n-1)$ in (9.28) can be evaluated by using a series expansion of $A(n-1)$ around n and retaining the leading term only, so

$$n^{2/3} - (n-1)^{2/3} \approx n^{2/3} - n^{2/3}\left(1 - \frac{2}{3n}\right) \sim n^{-1/3}. \tag{9.30}$$

Taking into account the ideal gas model, the chemical potential μ_∞ is related in a simple way to the equilibrium density (concentration) $c_{eq}(\infty)$ of monomers in the case of flat interface ($r \to \infty$) between the liquid phase (droplet) and the gaseous phase (free monomers). Thus, we have

$$\mu_\infty(T) = k_B T \ln\left[\lambda_0(T)^3 c_{eq}(\infty)\right]. \tag{9.31}$$

However, in reality, due to the spherical droplets, the interface is curved, therefore the concentration of free monomers in equilibrium is larger than $c_{eq}(\infty)$.

For large enough clusters, the detachment probability

$$w_-(n) = \alpha A(n) c_{eq}(\infty) \exp(\ell k(n)) \tag{9.32}$$

is obtained by inserting (9.31) into (9.28), where the length $\ell = \ell(T)$ defined as

$$\ell(T) = \frac{2\sigma}{c_{clust} k_B T}, \tag{9.33}$$

is explained and depicted in Figure 9.2.

Considering the cluster sizes $n(t)$ as a continuous variable which can be measured experimentally, the equation of motion dn/dt with a given velocity function $v(n)$, showing the time evolution of the cluster size, is well known. Putting forward phenomenological arguments such as Fick's law, the dynamical equation of interface reaction limited aggregation reads as:

$$\frac{dn}{dt} = v(n) \quad \text{with} \quad v(n) = \frac{D}{\ell} A(n) \left(c_{free} - c_{eq}(n)\right), \tag{9.34}$$

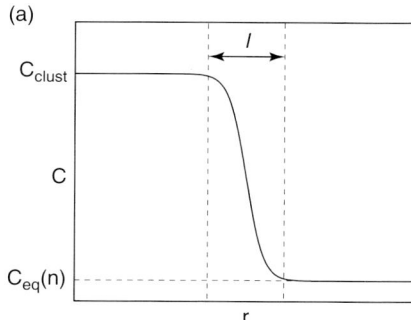 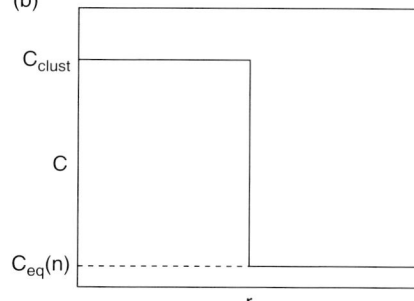

Figure 9.2 Particle concentration c depending on the distance r from the center of the cluster: (a) shows a real density profile, whereas (b) approximates the profile corresponding to the model of a cluster with sharp border. The length ℓ shows the width of the interface between the dense phase (c_{clust}) and the dilute surrounding ($c_{eq}(n)$) over a size-n droplet.

where $c_{\text{free}} = N_0/V_{\text{free}} = (N_{\text{total}} - n)/V_{\text{free}} = (c - n/V)/(1 - c_{\text{clust}}^{-1} n/V)$ is the density of free particles, and $c = N_{\text{total}}/V$ is the total density of particles. The constant D is called the diffusion coefficient; the coefficient ℓ, called the capillary length, is a small interface thickness defined by (9.33).

In the stochastic approach we also have an equation (10.70) for the average cluster size $\langle n \rangle$, which is similar to (9.34). This is a deterministic equation for the mean value

$$\frac{d}{dt}\langle n \rangle = \frac{d}{dt}\sum_n n\, p(n,t) = \langle w^+(n) \rangle - \langle w^-(n) \rangle, \tag{9.35}$$

which is obtained by an averaging of the master equation. In a certain approximation (9.35) may be written as follows

$$\frac{d}{dt}\langle n \rangle \approx w^+(\langle n \rangle) - w^-(\langle n \rangle), \tag{9.36}$$

describing the time evolution of the average cluster size $\langle n \rangle$.

According to the definitions of the transition frequencies (9.5) and (9.32), the time evolution of the mean cluster size (9.36) can be written in the same form as (9.34),

$$\frac{d}{dt}\langle n \rangle = \alpha A(\langle n \rangle) \left[\frac{N_{\text{total}} - \langle n \rangle}{V_{\text{free}}} - c_{eq}(\infty) e^{\ell k(\langle n \rangle)} \right]. \tag{9.37}$$

By comparing these two equations we find the previously unknown coefficient α and the equilibrium concentration $c_{eq}(n)$:

$$\alpha = D/\ell \tag{9.38}$$

and

$$c_{eq}(n) = c_{eq}(\infty)\, e^{\ell k(n)}. \tag{9.39}$$

The only difference between (9.34) and (9.37) is that, in the latter case, we always have the average value of the cluster size $\langle n \rangle$ instead of n. Rewriting (9.37) accounting for (9.6) and (9.38) we finally obtain

$$\frac{d\langle n \rangle}{dt} = \frac{D}{\ell} A(\langle n \rangle) \left(\frac{c - \langle n \rangle/V}{1 - c_{\text{clust}}^{-1} \langle n \rangle/V} - c_{eq}(\infty) e^{\ell k(\langle n \rangle)} \right). \tag{9.40}$$

In the stationary state $d\langle n \rangle/dt = 0$ holds. Equation (9.40) is valid at large enough n only, whereas at $\langle n \rangle \to 0$ it should be modified to ensure that the transition rates and, therefore, the expression in the brackets does not diverge. Then we obtain three stationary solutions. The homogeneous situation without any cluster corresponds to $A(\langle n \rangle_{\text{st}}) = 0$, or to zero value of the stationary cluster size $\langle n \rangle_{\text{st}} = 0$. The other two solutions, describing the heterogeneous situation, originate from the identity $c_{\text{free}}(\langle n \rangle_{\text{st}}) = c_{eq}(\langle n \rangle_{\text{st}})$ or in an extended version

$$\frac{c - \langle n \rangle_{\text{st}}/V}{1 - c_{\text{clust}}^{-1} \langle n \rangle_{\text{st}}/V} = c_{eq}(\infty) \exp\left[\ell\, (c_{\text{clust}} 4\pi/3)^{1/3}\, \langle n \rangle_{\text{st}}^{-1/3} \right]. \tag{9.41}$$

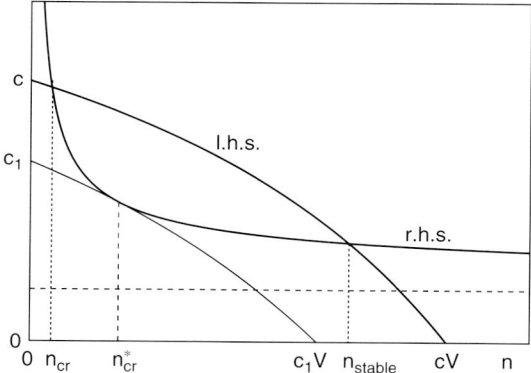

Figure 9.3 Terms on the l. h. s. and on the r. h. s. of (9.41) depending on the cluster size n. The two crossing points at $n = n_{\text{stable}}$ (stable cluster size) and $n = n_{\text{cr}}$ (unstable or critical cluster size) correspond to two different solutions of (9.41). The thin line related to the critical density c_1 has a single common point with the thick curved line at $n = n_{\text{cr}}^*$. The horizontal dashed line shows the value of the equilibrium concentration $c_{\text{eq}}(\infty)$.

From this equation we can find the stationary cluster size $\langle n \rangle_{\text{st}}$ as a function of the total density c. This is a nonlinear equation which cannot be solved analytically. However, it can be easily analyzed (solved) graphically. In Figure 9.3, terms on the left-hand side (l.h.s.) and on the right-hand side (r.h.s.) versus the mean cluster size $\langle n \rangle$ are shown. The two crossing points $\langle n \rangle = n_{\text{cr}}$ and $\langle n \rangle = n_{\text{stable}}$ correspond to two different solutions of (9.41). The quantity n_{cr} is known as the critical cluster size in nucleation theory, whereas n_{stable} represents the stable stationary cluster size. Their meaning will be clarified in further discussion. The crossing points exist only if the total concentration c exceeds some critical value $c > c_1$ which corresponds to a bifurcation point where both solutions merge into one, as shown by the thin line which has only one common point with the thick line at the marginal (largest possible) value of the critical cluster size $\langle n \rangle_{\text{st}} = n_{\text{cr}}^*$. At $c > c_1$ three different regions can be distinguished for the cluster size $\langle n \rangle_{\text{st}}$:

1. At $\langle n \rangle < n_{\text{cr}}$ we have $d\langle n \rangle/dt < 0$ which means that the cluster dissolves.
2. At $n_{\text{cr}} < \langle n \rangle < n_{\text{stable}}$ we have $d\langle n \rangle/dt > 0$ which means that the cluster grows until it reaches the stable stationary size n_{stable}.
3. At $\langle n \rangle > n_{\text{stable}}$ we have $d\langle n \rangle/dt < 0$ which means that the cluster reduces its size (dissolves) to the stationary value n_{stable}.

According to this, the solution $\langle n \rangle = n_{\text{stable}}(c)$ corresponds to a stable cluster size, whereas the solution $\langle n \rangle = n_{\text{cr}}(c)$ corresponds to an unstable stationary cluster size. In Figure 9.4 we have shown both solutions of (9.41) (branches $n_1(c)$ and $n_2(c)$), providing the stationary cluster size in a supersaturated vapor as a function of the total density c. Branch $n_1(c)$ depicted by a solid line corresponds to the stable cluster size, whereas branch $n_2(c)$ indicated by a dashed line corresponds to the

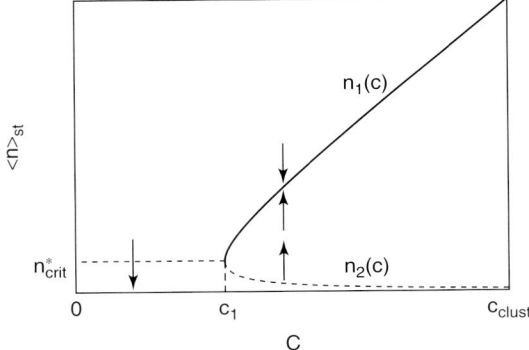

Figure 9.4 The stationary cluster size $\langle n \rangle_{st}$ depending on the total density c of particles in a supersaturated vapor. The stable cluster size (the horizontal line and the branch $n_1(c)$) is shown by solid lines, whereas the unstable cluster size (the branch $n_2(c)$) is shown by a dashed line. Arrows indicate the time evolution of $\langle n \rangle$.

unstable (critical) cluster size. Several trajectories showing the time evolution of $\langle n \rangle$ to one of the stable stationary values (solid lines) are indicated by arrows. The bifurcation diagram in Figure 9.4 around the critical density c_1 is similar to those in Figure 10.25 describing the nucleation on roads. An essential difference, however, is that no increase in the critical cluster size $n_2(c)$ is observed in a supersaturated vapor at large densities [142, 146, 149, 151].

Expansion of the exponent in (9.39), retaining the linear term only, yields

$$c_{eq}(\langle n \rangle) = c_{eq}(\infty)(1 + \ell k(\langle n \rangle)). \tag{9.42}$$

By using this linearization around the critical cluster size n_{cr}, (9.40) can be written in the well known form [203]

$$\frac{d\langle n \rangle}{dt} = D c_{eq}(\infty) A(\langle n \rangle) \left[k(n_{cr}) - k(\langle n \rangle) \right], \tag{9.43}$$

where

$$k(n_{cr}) = \frac{1}{\ell} \frac{c - n_{cr}/V - c_{eq}(\infty)}{c_{eq}(\infty) \left(1 - c_{clust}^{-1} n_{cr}/V\right)} \tag{9.44}$$

is the curvature of the critical cluster. From this equation the above discussed property that clusters with an overcritical size ($n > n_{cr}$, so $k(n) < k(n_{cr})$) grow, whereas those with an undercritical size ($n < n_{cr}$, so $k(n) > k(n_{cr})$) dissolve, is obvious. In the actual bistable situation the growth from an undercritical to an overcritical cluster size cannot be described by deterministic equations of motion like (9.43). This phenomenon of noise-induced transitions over the critical value of cluster size can be treated in the stochastic approach only [149, 156, 229, 230].

Using the Monte Carlo method we have simulated and presented in Figure 9.5 three different stochastic trajectories showing the time evolution, that is the cluster

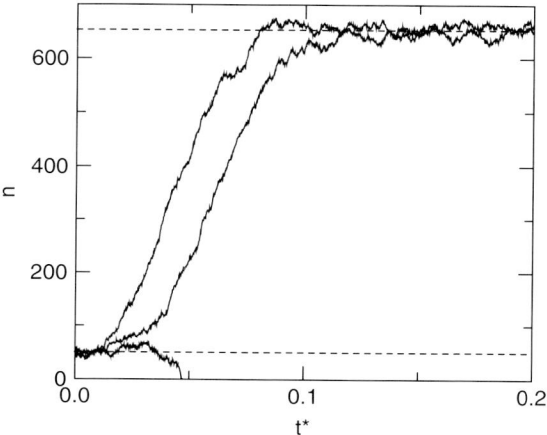

Figure 9.5 Three different stochastic trajectories showing the time evolution of the cluster size n vs the dimensionless time t^*. The lower dashed line indicates the critical cluster size $n_{cr} = 54$; the upper dashed line represents the stable cluster size $n_{stable} = 650$.

size n vs the dimensionless time $t^* = (\alpha A(1)/V) \cdot t$, of the system with $N_{total} = 1000$ point-like particles ($c_{clust}^{-1} \to 0$) starting with n values around the critical cluster size $n_{cr} \approx 54$ (the lower dashed line). The parameters of the system are chosen such that $\sigma A(1)/(k_B T) = 10$ and $Vc_{eq}(\infty) = 160$. In one of the cases the cluster dissolves, whereas in other two cases it grows over the critical size. Also, as distinct form the prediction of the deterministic equation (9.40), in one of the cases the growth up to the stable cluster size $n \approx 650$ (the upper dashed line) occurs, starting with $n = 45 < n_{cr}$. The probability distribution at three different times: $t^* = 0.003, 0.04$, and 0.3, have been calculated by averaging over a large number of stochastic trajectories starting with $n = n_{cr} = 54$. The results are shown in Figure 9.6. The probability maximum moves towards larger values of the cluster size n with increasing time, and at $t^* = 0.3$ the probability distribution agrees approximately with the equilibrium distribution $P^{eq}(n) \propto \exp(-F(n)/(k_B T))$ (smooth curve). The maximum of the equilibrium distribution corresponds to the minimum of the free energy, whereas the critical cluster size corresponds to the local maximum of the free energy, as shown in Figures 9.4 and 9.6.

9.3
The General Multi-Droplet Scenario

The basic equation describing the whole process of nucleation, cluster growth, and coarsening, is the stochastic master equation (3.19)

$$\frac{\partial}{\partial t} P(\mathbf{N}, t) = \sum_{\mathbf{N}'} W(\mathbf{N} \mid \mathbf{N}') P(\mathbf{N}', t) - W(\mathbf{N}' \mid \mathbf{N}) P(\mathbf{N}, t) \quad (9.45)$$

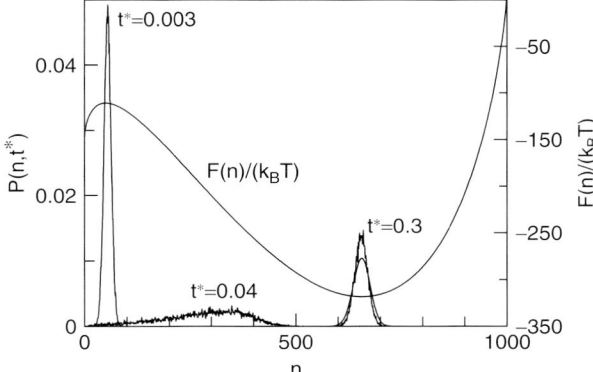

Figure 9.6 The probability distribution $P(n, t^*)$ at three different dimensionless times $t^* = 0.003$ (left), $t^* = 0.04$ (middle), and $t^* = 0.3$ (right) calculated by averaging over 20 000 stochastic trajectories simulated by the Monte Carlo method. The equilibrium distribution is shown by a smooth solid line. The maximum of the equilibrium distribution corresponds to the minimum of the free energy, $F(n)/(k_B T)$ (smooth curve).

which gives the probability $P(\mathbf{N}, t)$ at time t of the system having the cluster distribution $\mathbf{N}$ defined by (9.1). Here $W(\mathbf{N}' \mid \mathbf{N})$ is the transition frequency of the system from state $\mathbf{N}$ to state $\mathbf{N}'$.

The growth of liquid droplets in a supersaturated vapor, discussed in the previous section, follows the general scenario outlined here. For simplicity, our calculations have been made for the single-droplet case which corresponds to the final stage of the Ostwald ripening. A complete description of the process, however, has to include the multi-droplet picture, as consistent with the master-equation (9.45). The stationary cluster distribution $\mathbf{N}^{st}$ is determined by the stationary solution of this master-equation ($\partial P_{st}/\partial t = 0$). Following the master equation approach (see Chapter 3), the transition frequencies of opposite stochastic events are related by the detailed balance condition (9.9) which in a general case of thermodynamic system reads

$$W(\mathbf{N}' \mid \mathbf{N}) = W(\mathbf{N} \mid \mathbf{N}') \exp\left(-\frac{\Delta \Omega}{k_B T}\right) \tag{9.46}$$

where $\Delta \Omega = \Omega(\mathbf{N}') - \Omega(\mathbf{N})$ is the difference between the thermodynamic potentials of the states $\mathbf{N}$ and $\mathbf{N}'$. In this case the stationary solution is also the equilibrium solution $P^{eq}(\mathbf{N})$. For any thermodynamic constraints, reflected by the characteristic thermodynamic potential Ω, we may write

$$P^{eq}(\mathbf{N}) = P_{norm}^{-1} \exp\left(-\Omega(\mathbf{N})/k_B T\right) \tag{9.47}$$

in accordance with (3.50), where P_{norm} is a normalizing factor. The equilibrium states $\mathbf{N}^{eq}$ have the highest possible equilibrium (stationary) probabilities $P_{st}(\mathbf{N}^{eq}) \to \max$.

Similar to classical nucleation theory, we assume that the cluster size distribution $\mathbf{N}(t)$ is changed by monomer–cluster reactions only, so

$$N_1 + N_n \longleftrightarrow N_{n+1}. \tag{9.48}$$

In this case the master equation (9.45) can be written as

$$\frac{\partial}{\partial t} P(N_1, N_2, \ldots, N_n, \ldots, N_N, t) \quad (9.49)$$
$$= W_1^+(N_1 + 2) P(N_1 + 2, N_2 - 1, N_3, \ldots, N_N, t)$$
$$+ W_2^-(N_2 + 1) P(N_1 - 2, N_2 + 1, N_3, \ldots, N_N, t)$$
$$+ \sum_{n=3}^{N} W_{n-1}^+(N_{n-1} + 1) P(N_1 + 1, \ldots, N_{n-1} + 1, N_n - 1, \ldots, N_N, t)$$
$$+ \sum_{n=2}^{N-1} W_{n+1}^-(N_{n-1} + 1) P(N_1 - 1, \ldots, N_n - 1, N_{n+1} + 1, \ldots, N_N, t)$$
$$- \sum_{n=1}^{N} \left[W_n^+(N_n) + W_n^-(N_n) \right] P(N_1, N_2, \ldots, N_n, \ldots, N_N, t).$$

Here W_n^+ and W_{n+1}^- are the transition rates of the forward and backward monomer–cluster reaction (9.48). The master equation (9.49) also describes the formation of a dimer from two monomers and its decay as special cases. If we do not use the precluster–cluster concept introduced in Section 9.2, then we have to consider the probability that two monomers meet together and form a dimer, to estimate the transition frequency W_1^+. This depends on the density of monomers N_1 and, according to simple geometrical considerations, in our previous notation is given by

$$W_1^+(N_1) = \frac{D}{\ell} A(1) \frac{N_1(N_1 - 1)}{V}. \quad (9.50)$$

Other transition rates are also consistent with our consideration in Section 9.2. One has to take into account that the rate of the reaction (9.48) is proportional to the number of clusters N_n involved. Therefore, we have

$$W_n^+(N_n) = N_n \, w_+(n) \quad (9.51)$$
$$W_n^-(N_n) = N_n \, w_-(n), \quad (9.52)$$

where $w_+(n)$ is the attachment frequency to a given cluster of size n, defined by (9.5), and $w_-(n)$ is the detachment frequency (9.28). Coming back to the interface reaction limited aggregation law (9.34)

$$\frac{dn}{dt} = \frac{D}{\ell} A(n) \left(c_{\text{free}} - c_{\text{eq}}(n) \right) \quad (9.53)$$

with the concentration over a curved surface (9.39) given by

$$c_{\text{eq}}(n) = c_{\text{eq}}(\infty) \, e^{\ell k(n)} \approx c_{\text{eq}}(\infty) \left(1 + \ell k(n) \right) \quad (9.54)$$

we get the following linearized growth law

$$\frac{dn}{dt} = D \, c_{\text{eq}}(\infty) A(n) \left(k(n_{\text{cr}}) - k(n) \right) \quad (9.55)$$

where the curvature of the critical cluster is determined by the concentration of free particles c_{free} or supersaturation $y = (c_{\text{free}} - c_{\text{eq}}(\infty))/c_{\text{eq}}(\infty)$

$$k(n_{\text{cr}}) \equiv (c_{\text{clust}} 4\pi/3)^{1/3} n_{\text{cr}}^{-1/3} = \frac{1}{\ell} \frac{c_{\text{free}} - c_{\text{eq}}(\infty)}{c_{\text{eq}}(\infty)} = \frac{y}{\ell} \qquad (9.56)$$

and therefore the critical cluster size given by

$$n_{\text{cr}} = (c_{\text{clust}} 4\pi/3) \left(\frac{\ell}{y}\right)^3. \qquad (9.57)$$

Taking into account a multi–cluster distribution of different sizes $\{n_i \mid i = 1, \ldots m\}$ and the particle conservation law for a finite system (9.4)

$$c_{\text{free}} = c_{\text{total}} - \sum_{i=1}^{m} n_i N_i / V \qquad (9.58)$$

we get, for a temporal change in the monomer concentration

$$\frac{d}{dt} c_{\text{free}} = -\sum_i \frac{N_i}{V} \frac{dn_i}{dt}. \qquad (9.59)$$

Using the growth law for each cluster (9.55)

$$\frac{dn_i}{dt} = D c_{\text{eq}}(\infty) A(n_i) \left(\frac{1}{\ell} \frac{c_{\text{free}} - c_{\text{eq}}(\infty)}{c_{\text{eq}}(\infty)} - k(n_i)\right) \qquad (9.60)$$

we receive from (9.59)

$$\frac{d}{dt} c_{\text{free}} = -\frac{D}{\ell} \frac{c_{\text{free}} - c_{\text{eq}}(\infty)}{c_{\text{eq}}(\infty)} \sum_i \frac{N_i}{V} A(n_i) + D c_{\text{eq}}(\infty) \sum_i \frac{N_i}{V} A(n_i) k(n_i). \qquad (9.61)$$

Inserting the supersaturation, calculated from (9.61), into (9.60) we obtain the final set of ordinary nonlinear differential equations governing the constrained growth and Ostwald ripening of an ensemble of m droplets of sizes n_i

$$\frac{dn_i}{dt} = \frac{D}{\ell} A(n_i) \left[c_{\text{eq}}(\infty) \ell \left(<k> - k(n_i)\right) - \frac{\ell}{D} \frac{V}{A_{\text{clust}}} \frac{dc_{\text{free}}}{dt} \right] \qquad (9.62)$$

together with

$$\frac{dc_{\text{free}}}{dt} = -\frac{D}{\ell} \frac{A_{\text{clust}}}{V} \left[c_{\text{total}} - c_{\text{clust}} \frac{V_{\text{clust}}}{V} - c_{\text{eq}}(\infty) - c_{\text{eq}}(\infty) \ell <k> \right] \qquad (9.63)$$

where

$$A_{\text{clust}} = \sum_i N_i A(n_i) \quad \text{total surface of all droplets}$$

$$V_{\text{clust}} = c_{\text{clust}}^{-1} \sum_i N_i n_i \quad \text{total volume of all droplets} \qquad (9.64)$$

$$\langle k \rangle = \frac{\sum_i k(n_i) N_i A(n_i)}{\sum_i N_i A(n_i)} \quad \text{mean curvature}$$

with $c_{total} = N_{total}/V =$ constant being the total number of free and bounded monomers per volume. The surface of a spherical drop $A(n)$ is given by (9.7).

The kinetic equations (9.62, 9.63) describe the rapid growth of the droplet ensemble (second term dc_{free}/dt in (9.62)) as well as the slow selection between the droplets (first term in (9.62)) as a competition process. The numerical solution of the coupled nonlinear equations yields the time evolution of a given droplet ensemble with certain initial sizes. In a short time regime the size of all droplets increases and the two-phase system reaches so-called internal equilibrium ($dc_{free}/dt \approx 0$), a nonstationary situation where the liquid is in quasi-equilibrium with its surrounding vapor. Since the row material (monomers) is limited at the end of this growth stage, the total volume of the new phase V_{clust} is very close, but not equal, to its equilibrium value. In the final long-time regime a competitive ripening process takes place, which minimizes the total surface of all droplets A_{clust}, the total volume V_{clust} being almost unchanged. At the end of the coarsening process the dynamical system has reached a stable fixed point where only one droplet is present. The other smaller clusters having dissolved to give monomers to the winner; the largest drop.

9.4
Detailed Balance and Free Energy

In Section 9.2 we discussed how the detailed balance (9.9) can be used to determine the transition rates. For this purpose one needs to calculate the free energy of the system based on the first principles of statistical mechanics. In many applications the inverse problem can be of interest. Namely, if the transition rates are known due to some physical assumptions, then the free energy and chemical potentials can be derived, based on the principle of detailed balance. We will show how this works for the liquid–gas system with the idea of applying this scheme to other systems such as traffic in transportation and biology [32, 123, 150, 241].

For simplicity, we consider a situation where only one cluster of molecules coexists with the vapor phase. The number n of molecules called monomers bounded in the cluster is a stochastic variable, whereas their total number N in a given volume V is fixed. The stochastic events of adding or removing one monomer are characterized by the transition rates $w_+(n)$ and $w_-(n)$ depending on the actual cluster size n. According to (9.9), the detailed balance reads

$$\frac{w_+(n-1)}{w_-(n)} = \exp\left(-\frac{F(n) - F(n-1)}{k_B T}\right), \qquad (9.65)$$

where T is the temperature, k_B the Boltzmann constant, and $F(n)$ is the free energy of state (including all possible microscopic distributions of coordinates and momenta of free monomers) with cluster size n.

For large enough n, (9.65) can be approximated as

$$\frac{w_+(n)}{w_-(n)} \simeq \exp\left(-\frac{\partial F/\partial n}{k_B T}\right), \qquad (9.66)$$

which leads to the equation

$$\ln\left[\frac{w_+(n)}{w_-(n)}\right] = -\frac{1}{k_B T}\frac{\partial F}{\partial n}. \tag{9.67}$$

From this we get

$$F = F_0 - k_B T \int_0^n \ln\left[\frac{w_+(n')}{w_-(n')}\right] dn', \tag{9.68}$$

where $F_0 = F(n=0)$ does not depend on the cluster size n. It is the free energy of the system without cluster; in this case the free energy of an ideal gas. We insert here the physical ansatz for the transition rates for a large system having small fraction of total volume V occupied by the condensed (droplet) phase (cf. (9.24) with $V_{\text{free}} \to V$)

$$\frac{w_+(n)}{w_-(n)} = \frac{\lambda_0^3(T)(N-n)}{V}\exp\left(\frac{f_{n-1}(T) - f_n(T)}{k_B T}\right), \tag{9.69}$$

where N is the total number of particles previously written as N_{total}, and $f_n(T)$ is the binding energy of a cluster of size n. By using the approximation $f_{n-1}(T) - f_n(T) \simeq -\partial f_n(T)/\partial n$, we obtain

$$F = F_0 - k_B T \int_0^n \ln\left[\frac{\lambda_0^3(T)(N-n')}{V}\right] dn' + f_n(T). \tag{9.70}$$

The integration, using $\int \ln x\, dx = x\ln x - x$, yields

$$F = F_0 - k_B T N\left[\ln\left(\lambda_0^3(T)\frac{N}{V}\right) - 1\right] \tag{9.71}$$
$$+ k_B T(N-n)\left[\ln\left(\lambda_0^3(T)\frac{N-n}{V}\right) - 1\right] + f_n(T).$$

The free energy of an ideal system (gas) F_0 cannot be obtained from the detailed balance relation. It is given by $F_0 = -k_B T \ln Z_{\text{ideal}}$, where Z_{ideal} is the partition function of the ideal gas (9.14), which for the one-cluster system with volume $V = L^3$ reads

$$Z_{\text{ideal}} = \frac{1}{N!}\left(\frac{L}{\lambda_0(T)}\right)^{3N}. \tag{9.72}$$

Hence, applying the Stirling formula $\ln N! \simeq N\ln N - N$, we obtain

$$F_0 = k_B T N\left[\ln\left(\lambda_0^3(T)\frac{N}{V}\right) - 1\right]. \tag{9.73}$$

By inserting (9.73) into (9.71) we recover the known expression (cf. (9.21) for $n > 0$)

$$F = k_B T(N-n)\left[\ln\left(\lambda_0^3(T)\frac{N-n}{V}\right) - 1\right] + f_n(T) \tag{9.74}$$

for the free energy of a liquid–gas system under isothermal and isochoric conditions. Equation (9.71) can be written as

$$\frac{F - F_0}{V k_B T} = \rho \left\{ \left(1 - \frac{n}{N}\right) \left[\ln\left(1 - \frac{n}{N}\right) - 1\right] + 1 - \frac{n}{N} \ln\left(\lambda_0^3(T)\rho\right) \right. \quad (9.75)$$
$$\left. + \frac{\mu_\infty(T)}{k_B T} \frac{n}{N} + \frac{3}{2}\ell(T)\left(c_{\text{clust}} 4\pi/3\right)^{1/3} N^{-1/3} \left(\frac{n}{N}\right)^{2/3} \right\},$$

where $\rho = N/V$ is the overall density and $\ell(T)$ is the diffusion length (width) of the liquid–gas interface given by (9.33).

Later on we introduce the dimensionless density $\tilde{\rho} = \lambda_0^3(T)\rho$ and the dimensionless volume $\tilde{V} = V/\lambda_0^3(T)$. In this notation (9.69) transforms to

$$\frac{w_+(n)}{w_-(n)} = \tilde{\rho}\left(1 - \frac{n}{N}\right) \exp\left(-\frac{\mu_\infty(T)}{k_B T}\right) \quad (9.76)$$
$$\times \exp\left(-\ell(T)(c_{\text{clust}} 4\pi/3)^{1/3} \tilde{V}^{-1/3} \tilde{\rho}^{-1/3} \left(\frac{n}{N}\right)^{-1/3}\right),$$

whereas (9.75) becomes

$$\frac{F - F_0}{\tilde{V} k_B T} = \tilde{\rho}\left\{\left(1 - \frac{n}{N}\right)\left[\ln\left(1 - \frac{n}{N}\right) - 1\right] + 1 - \frac{n}{N}\ln(\tilde{\rho}) + \frac{\mu_\infty(T)}{k_B T}\frac{n}{N} \right.$$
$$\left. + \frac{3}{2}\ell(T)(c_{\text{clust}} 4\pi/3)^{1/3} \tilde{V}^{-1/3} \tilde{\rho}^{-1/3}\left(\frac{n}{N}\right)^{2/3}\right\}. \quad (9.77)$$

These equations allow us to calculate the ratio $w_+(n)/w_-(n)$, as well as the normalized (dimensionless) free energy difference $(F - F_0)/(\tilde{V} k_B T)$ depending on the fraction of condensed molecules n/N at a given overall density for fixed volume and temperature. The results of the calculation for three different dimensionless densities $\tilde{\rho} = 5 \times 10^{-7}, 10^{-5}, 1.2 \times 10^{-5}$ at the values of dimensionless control parameters $\mu_\infty/(k_B T) = -12$ and $\ell(T)(c_{\text{clust}} 4\pi/3)^{1/3} \tilde{V}^{-1/3} = 0.003$ are shown in Figures 9.7 and 9.8. Note that the extrema of $F - F_0$ in Figure 9.8 correspond to the crossing points with the horizontal line $w_+(n)/w_-(n) = 1$ in Figure 9.7. At the smallest density (dot-dashed line) there are no crossing points and the free energy is a monotonously increasing function of n/N, showing that the stable state of the liquid–gas system contains no liquid droplet. Stable droplet appears at larger densities (dashed and solid lines) by overcoming a nucleation barrier (local free energy maximum in Figure 9.8).

The parameters we have chosen are quite realistic, and are comparable with those of water at $T = 300$ K and $V = 5 \times 10^{-23}$ m^3 with about 37 250 molecules (mass $m = 2.99 \times 10^{-23}$ kg) at $\tilde{\rho} = 10^{-5}$. For water at $T = 300$ K we have $\lambda_0(T) = 2.377 \times 10^{-11}$ m and $c_{\text{clust}} = 3.346 \times 10^{28}$ m^{-3}. Hence, the dimensionless density in the cluster $\tilde{\rho}_{\text{clust}} = c_{\text{clust}} \lambda_0^3 = 4.491 \times 10^{-4}$ exceeds about 50 times the critical mean density $\tilde{\rho} = \tilde{\rho}_c \simeq 9.2 \times 10^{-6}$ at which the condensation (that is, a minimum of free energy at $n/N > 0$) appears in our calculation. Assuming the above parameters

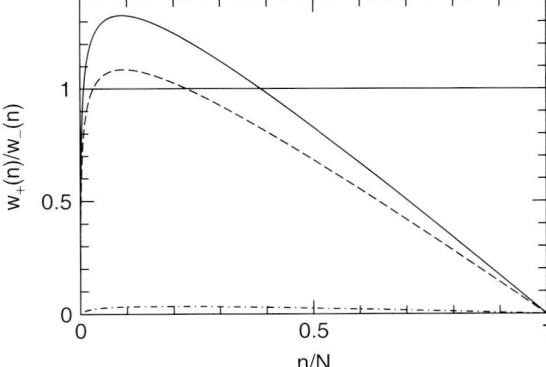

Figure 9.7 The ratio of transition rates $w_+(n)/w_-(n)$ depending on the fraction of condensed particles n/N for three dimensionless densities $\tilde{\rho} = 5 \times 10^{-7}$ (dot–dashed line), $\tilde{\rho} = 10^{-5}$ (dashed line), and $\tilde{\rho} = 1.2 \times 10^{-5}$ (solid line).

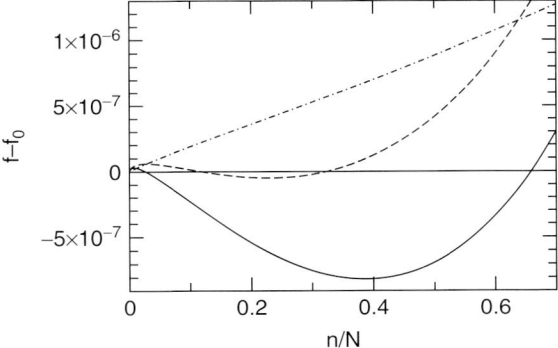

Figure 9.8 Normalized free energy difference $(F - F_0)/(\tilde{V} k_B T) = f - f_0$ depending on the fraction of condensed particles n/N for three dimensionless densities $\tilde{\rho} = 5 \times 10^{-7}$ (dot-dashed line), $\tilde{\rho} = 10^{-5}$ (dashed line), and $\tilde{\rho} = 1.2 \times 10^{-5}$ (solid line).

of water, we obtain $\ell(T) = 8.953 \times 10^{-10}$ m for the width of the liquid–gas interface and surface tension $\sigma = 6.20 \times 10^8$ N m^{-1}. It is about three times the characteristic intermolecular distance in the cluster, which is roughly $c_{\text{clust}}^{-1/3} \simeq 3.1 \times 10^{-10}$ m. The critical density $\tilde{\rho}_c$ increases with temperature and becomes equal to the cluster density $\tilde{\rho}_{\text{clust}}$ at the critical temperature $T = T_c$. In our description the physically meaningful densities are restricted by $\tilde{\rho} \leq \tilde{\rho}_{\text{clust}}$. This means that no condensation phase transition takes place for these physical densities at $T > T_c$. Assuming that μ_∞ and σ do not change with temperature and that the above given values of the dimensionless control parameters correspond to $T = 300$ K, we find $T_c \simeq 430$ K in our example.

9.5
Relaxation to the Free Energy Minimum

Now we consider the general behavior of a system in the vicinity of a local or global minimum of $F(n)$. In this case the argument of the exponent in (9.66) is small and we can make a Taylor expansion

$$\frac{w_+(n)}{w_-(n)} \simeq \exp\left(-\frac{\partial F/\partial n}{T^*}\right) \simeq 1 - \frac{1}{T^*}\frac{\partial F}{\partial n}, \qquad (9.78)$$

where $T^* = k_B T$ is the temperature measured in energy units. Such a notation is useful for generalization to other systems such as traffic flow discussed in the next chapter. Equation (9.78) can be rewritten as

$$w_+(n) - w_-(n) \simeq -\frac{w_-(n)}{T^*}\frac{\partial F}{\partial n}. \qquad (9.79)$$

On the other hand, we can write in a deterministic approximation

$$\frac{dn}{dt} = w_+(n) - w_-(n). \qquad (9.80)$$

Comparing (9.79) and (9.80), we obtain

$$\frac{dn}{dt} \simeq -\frac{w_-(n)}{T^*}\frac{\partial F}{\partial n}. \qquad (9.81)$$

As in the Landau theory of phase transitions, we can expand the free energy around the minimum point $n = n_0$ defined by

$$\left.\frac{\partial F}{\partial n}\right|_{n=n_0} = 0. \qquad (9.82)$$

In the first approximation, where we retain only the leading term, we have also $w_+(n) = w_-(n) = w_\pm(n_0)$, which leads to the kinetic equation [213]

$$\frac{dn}{dt} \simeq -\Gamma_0 (n - n_0), \qquad (9.83)$$

where

$$\Gamma_0 = \frac{w_\pm(n_0)}{T^*}\left.\frac{\partial^2 F}{\partial n^2}\right|_{n=n_0} \qquad (9.84)$$

is the relaxation rate. For $\Gamma_0 > 0$, which corresponds to the minimum of F, the solution is the exponential relaxation to $n = n_0$:

$$n(t) = n_0 + (n(0) - n_0) e^{-\Gamma_0 t}. \qquad (9.85)$$

This solution is valid also for $\Gamma_0 < 0$, in which case n_0 corresponds to a free energy maximum. In this case it describes the deviation from this maximum point.

9.6
Chemical Potentials

Our system can be considered as consisting of two phases: the cluster phase with n particles and free energy $F_{\text{clust}}(n)$, and the ideal gas phase with $N_{\text{ideal}} = N - n$ particles and free energy $F_{\text{ideal}}(N_{\text{ideal}})$. The total free energy then is $F = F_{\text{clust}} + F_{\text{ideal}}$. While the total number of particles N is fixed, the number of particles in any of the phases fluctuates. According to the definition, we can write $\mu_{\text{clust}} = \partial F_{\text{clust}}/\partial n$ and $\mu_{\text{ideal}} = \partial F_{\text{ideal}}/\partial N_{\text{ideal}} = -\partial F_{\text{ideal}}/\partial n$ for the chemical potentials of these phases. Hence

$$\frac{\partial F}{\partial n} = \frac{\partial F_{\text{clust}}}{\partial n} + \frac{\partial F_{\text{ideal}}}{\partial n} = \mu_{\text{clust}} - \mu_{\text{ideal}} \qquad (9.86)$$

and the kinetic equation (9.81) can be written as

$$\frac{dn}{dt} \simeq -\frac{w_{\pm}(n_0)}{T^*}(\mu_{\text{clust}} - \mu_{\text{ideal}}). \qquad (9.87)$$

The latter equation has a certain physical interpretation: the driving force pushing the system to the phase equilibrium is the difference in the chemical potentials in both phases. Equilibrium is reached when the chemical potentials of the coexisting phases are equal, so $\mu_{\text{clust}} = \mu_{\text{ideal}}$.

For the liquid–gas system $F_{\text{ideal}}(T, V, N, n)$ is given by (9.73), where N is replaced with $N_{\text{ideal}} = N - n$, so,

$$F_{\text{ideal}}(T, V, N, n) = k_B T (N - n) \left[\ln \left(\lambda_0^3(T) \frac{N-n}{V} \right) - 1 \right]. \qquad (9.88)$$

Hence the total free energy (9.74) can be written as

$$F = F_{\text{ideal}}(T, V, N, n) + f_n(T). \qquad (9.89)$$

The chemical potential of the liquid phase is thus given by the derivative of the binding energy $f_n(T) \equiv F_{\text{clust}}(T, V, N, n)$:

$$\mu_{\text{clust}} = \mu_\infty(T) + \sigma \frac{\partial A(n)}{\partial n} = \mu_\infty(T) + k_B T \ell(T) k(n), \qquad (9.90)$$

where $k(n) = 1/r$ is the curvature (9.29) of the liquid surface for a droplet of size n with radius r and surface area $A(n) = 4\pi r^2$. The chemical potential of the gaseous phase calculated from (9.88) is

$$\mu_{\text{ideal}} = -\frac{\partial F_{\text{ideal}}}{\partial n} = k_B T \ln \left(\lambda_0^3(T) \frac{N-n}{V} \right) = k_B T \ln (\tilde{\rho}_{\text{free}}), \qquad (9.91)$$

where $\tilde{\rho}_{\text{free}} = \lambda_0^3(T)(N - n)/V$ is the dimensionless density of molecules in the gaseous phase.

According to these expressions for the chemical potentials, the ansatz (9.69) can be written as

$$\frac{w_+(n)}{w_-(n)} = \exp\left(\frac{\mu_{\text{ideal}}}{k_B T}\right) \exp\left(\frac{f_{n-1}(T) - f_n(T)}{k_B T}\right) \simeq \exp\left(-\frac{\mu_{\text{clust}} - \mu_{\text{ideal}}}{k_B T}\right). \qquad (9.92)$$

The latter relation is consistent with (9.66) and (9.86).

This approach in the calculation of free energy and chemical potentials can be easily generalized to any system, where the principle of detailed balance is valid. In the following chapter we will apply this method to the description of traffic flow, which is a system of many interacting vehicles exhibiting similar features, such as phase transition and phase separation, as do many physical systems and supersaturated vapor, in particular.

9.7
Exercises

E 9.1 Szilard model of nucleation I

Consider the Szilard model of nucleation. The nucleation reactor consists of a reservoir of aggregates having size $n = a - 1$, a nucleation box with nuclea of sizes $n = a$, $a + 1, \ldots, b - 1, b$, and surrounding media, where large aggregates consisting of $n = b + 1$ monomers are collected, as shown in Figure 9.9.

Let $p(n, t)$ be the probability that a nucleus has size n at time t. The probability obeys the normalization condition

$$p(a - 1, t) + G(t) + p(b + 1, t) = 1, \qquad (9.93)$$

where $G(t) = \sum_{n=a}^{b} p(n, t)$ is the total probability of having the size $n \in [a, b]$. The aggregate of size n can add or lose a monomer with transition rates $w_+(n)$ and $w_-(n)$, respectively. However, there is no transition from the state $n = a$ to $n = a - 1$, as well as from $n = b + 1$ to $n = b$ (see Figure 9.9).

The task is to formulate the master equations describing the time evolution of $p(n, t)$ for this model. Another task is to do this for a modified model, where only the sizes $a \leq n \leq b$ are considered at a given probability inflow flux into the nucleation box, $\mathcal{P}_{\text{in}}$, and the outflow flux from it, $\mathcal{P}_{\text{out}}$, in the stationary situation where $\mathcal{P}_{\text{in}} = \mathcal{P}_{\text{out}}$ holds. In the latter case find the stationary solution for the probability distribution function over the aggregate sizes.

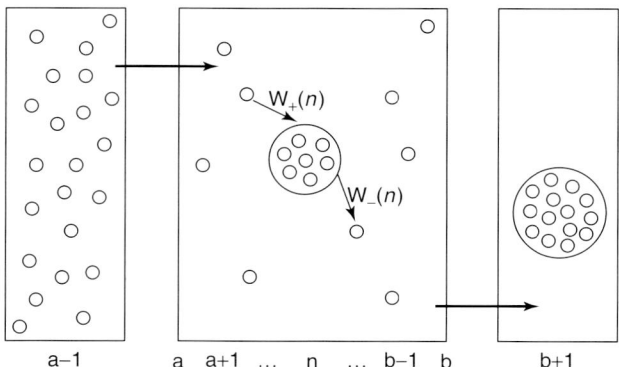

Figure 9.9 Schematic representation of the Szilard model.

E 9.2 Szilard model of nucleation II

Find the time-dependent solution $p(n,t)$ (where $a-1 \leq n \leq b$) for the Szilard model considered in the previous exercise in a special case of constant transition rates $w_+(n) = q$ and $w_-(n) = 0$ at the initial condition $p(n,0) = \delta_{n,a-1}$. Find the probability inflow and outflow fluxes $\mathcal{P}_{\text{in}}(t)$ and $\mathcal{P}_{\text{out}}(t)$ from the reservoir to the nucleation box and from the nucleation box to the surrounding media, respectively.

E 9.3 Nucleation in a supersaturated vapor

Construct the phase diagram of the supersaturated vapor–liquid system. Showing different states in the density–temperature plane. Explain the meaning of the critical point.
Hint: follow the analogy with the phase diagram for the many-car system showing free flow and congested traffic in Exercise E 10.3.

10
Vehicular Traffic

10.1
The Car-Following Theory

The stochastic processes and methods of their description are powerful tools in many applications. Apart from well known classical applications in physics such as diffusion or Brownian motion, stochastic processes have been successfully used also in many interdisciplinary fields related to physics.

Recently, the theoretical and empirical foundations of the *physics of traffic flow* have come into the focus of the physical community, see e.g. [76, 77, 100, 151]. Like every other field of physics the modeling of traffic flow described by generalized forces [78, 80, 227] should be based on the analysis of empirical data [105, 106] in order to understand the underlying stochastic process [240].

Different approaches such as deterministic and stochastic nonlinear dynamics as well as the statistical physics of many-particle systems have been very successful in understanding empirically observed structure formation on roads such as jam formation in freeway traffic [71–75, 168]. The motion of an individual vehicle has many peculiarities, since it is controlled by motivated driver behavior together with physical constraints. Nevertheless, on macroscopic scales the car ensemble displays phenomena such as phase formation (nucleation), widely encountered in different physical systems. This analogy can be clearly seen on a mesoscopic level of description where, instead of following the motion of each vehicle, a stochastic cluster of congested cars is considered [142, 147–149, 151, 153].

In a wide class of traffic models, the so-called car-following models, the behavior of a given vehicle is determined by the car ahead. One often assumes that each car tries to drive with a certain velocity, which is optimal for the actual headway distance. Car-following models with optimal velocity function, first proposed by Bando et al. [10–12], have been reviewed and analyzed in [44]. This optimal velocity models can be related to the traffic model with time lag [173, 243], in which the equation of motion of the ith car reads

$$\frac{dx_i(t+\tau)}{dt} = v_{\text{opt}}(\Delta x_i(t)), \qquad (10.1)$$

where $x_i(t)$ is the position of the actual vehicle at time t, $\Delta x_i(t) = x_{i+1}(t) - x_i(t)$ is the headway, and τ is the delay time. The meaning of the latter is the time lag that it takes the car velocity to reach the optimal velocity $v_{opt}(\Delta x_i(t))$. The optimal velocity function usually is expressed in terms of hyperbolic tangents [10, 44]

$$v_{opt}(\Delta x) = \frac{v_{max}}{2}\{\tanh(\Delta x - h_c) + \tanh(h_c)\} \tag{10.2}$$

or Hill's function [151, 153]

$$v_{opt}(\Delta x) = v_{max}\frac{(\Delta x)^2}{D^2 + (\Delta x)^2}. \tag{10.3}$$

Both these are sigmoidal functions. A common parameter is the maximal velocity v_{max}, whereas h_c and D are parameters with similar meaning which are called the safety distance and interaction distance, respectively.

By making the Taylor expansion with respect to the delay time τ in (10.1),

$$\frac{dx_i(t+\tau)}{dt} = \frac{dx_i(t)}{dt} + \tau\frac{d^2x_i(t)}{dt^2} + \ldots, \tag{10.4}$$

and taking two leading terms only, we obtain

$$\tau\frac{d^2x_i(t)}{dt^2} + \frac{dx_i(t)}{dt} = v_{opt}(\Delta x_i). \tag{10.5}$$

A rearrangement of terms yields the known optimal velocity model

$$\frac{d^2x_i(t)}{dt^2} = \frac{1}{\tau}\left(v_{opt}(\Delta x_i) - \frac{dx_i(t)}{dt}\right). \tag{10.6}$$

The parameter $1/\tau$ is called driver's sensitivity.

Car-following models with certain optimal distance function are also considered. In [250] a force model has been proposed, where the acceleration of a car consists of two components. One of them is only velocity (v) dependent and describes a tendency to keep the maximal speed

$$a_1(v) = a_0\left(1 - \frac{v}{v_{max}}\right), \tag{10.7}$$

where a_0 is the acceleration when the vehicle starts moving. The other component describes a tendency to keep some optimal distance Δx_{opt} from the vehicle in front, and has been defined as

$$\Delta x_{opt}(v) = d_0 + k_0 v^2, \tag{10.8}$$

where d_0 and k_0 are two parameters chosen as 9 m and 0.1 m^{-1}s^2 in [250]. The idea of keeping a certain distance is similar to that in molecule dynamics, therefore this component of acceleration $a_2(\Delta x, \Delta x_{opt})$ is related to the interaction potential of Lennard–Jones type

$$U(\Delta x, \Delta x_{opt}) \propto \frac{1}{4}\left(\frac{\Delta x_{opt}}{\Delta x}\right)^4 - \frac{1}{2}\left(\frac{\Delta x_{opt}}{\Delta x}\right)^2. \tag{10.9}$$

The difference from molecular dynamics is that the force

$$F(\Delta x_i) = m\, a_2(\Delta x_i, \Delta x_{\text{opt}}(v_i)) = -\frac{\partial U}{\partial x_i} = \frac{\partial U}{\partial \Delta x_i} \qquad (10.10)$$

is passed from the front vehicle to its follower but not vice versa, so Newton's third law does not apply here. In this notation, index i refers to the ith car and $\Delta x_i = x_{i+1} - x_i$ is the headway. Thus the force, normalized to the vehicle mass m, is given by

$$\frac{F(\Delta x)}{m} = a_2(\Delta x, \Delta x_{\text{opt}}(v)) = \frac{k}{\Delta x}\left[-\left(\frac{\Delta x_{\text{opt}}}{\Delta x}\right)^4 + \left(\frac{\Delta x_{\text{opt}}}{\Delta x}\right)^2\right], \qquad (10.11)$$

where k is a constant coefficient taken as $50\,\text{m}^2\text{s}^{-2}$ in [250]. For the actual control parameters, the normalized force depending on the headway distance Δx at different velocities v is shown in Figure 10.1.

The equation of motion is obtained by summing up both components of the acceleration $a_1(v)$ and $a_2(\Delta x, \Delta x_{\text{opt}}(v))$ to obtain the resulting acceleration $d^2 x_i/dt^2$ of the vehicle i, which depends on the headway $\Delta x_i = x_{i+1} - x_i$ and velocity $v_i = dx_i/dt$

$$\frac{d^2 x_i}{dt^2} = \frac{k}{\Delta x_i}\left[-\left(\frac{\Delta x_{\text{opt}}(v_i)}{\Delta x_i}\right)^4 + \left(\frac{\Delta x_{\text{opt}}(v_i)}{\Delta x_i}\right)^2\right] + a_0\left(1 - \frac{v_i}{v_{\max}}\right). \qquad (10.12)$$

The above are specific car-following models. In a more general formulation, the acceleration can depend on the headway distance, velocity, and also on the velocity difference $v_{i+1} - v_i$.

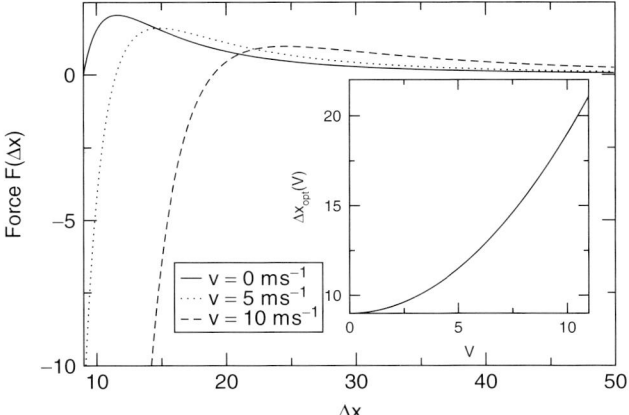

Figure 10.1 The force (10.11), normalized to the vehicle mass, in the car-following model (10.12) depending on the headway distance Δx at different car velocities v. The insertion shows the optimal distance Δx_{opt} as a function of velocity.

10.2
The Optimal Velocity Model and its Langevin Approach

Although the cooperative behavior of cars when treated as active particles seems to be more general, we concentrate on a well-investigated approximation known as the safety distance or optimal velocity (OV) model (10.6), first proposed by Bando et al. [10–12].

This model defines, on a *microscopic* level, the law of motion for circular traffic in terms of headway distances and velocity differences between neighboring cars. We start with the deterministic case, following the OV model, to include finally the stochastic fluctuations resulting in the Langevin equation.

The dynamics of the ensemble of N cars is given by the system of $2N$ differential equations of first order with respect to time. We consider a model of point-like cars moving on a circular road of length L. For each car, it is necessary to find its coordinate $x_i(t)$ and velocity $v_i(t)$, where $i = 1, \ldots, N$, at any time t. Within the framework of the optimal velocity model (OVM), the law of motion reads

$$\frac{dv_i}{dt} = \frac{1}{\tau}(v_{\text{opt}}(\Delta x_i) - v_i); \tag{10.13}$$

$$\frac{dx_i}{dt} = v_i \tag{10.14}$$

where $v_{\text{opt}}(\Delta x)$ is the optimal velocity function. As distinct from the original work [10] with (10.2), here we use a simpler expression (10.3)

$$v_{\text{opt}}(\Delta x) = v_{\text{max}} \frac{(\Delta x)^2}{D^2 + (\Delta x)^2} \tag{10.15}$$

for the optimal velocity function, where $\Delta x_i = x_{i+1} - x_i$ is the headway (bumper-to-bumper distance), v_{max} is the maximum speed allowed and D is a given positive control parameter called the interaction distance. We introduce new dimensionless variables for velocities, coordinates and time in the following way

$$u_i = v_i/v_{\text{max}}, \quad y_i = x_i/D, \quad T = t/\tau. \tag{10.16}$$

After transformation (10.16) the dynamical system (10.13)–(10.14) can be rewritten as

$$\frac{du_i}{dT} = u_{\text{opt}}(\Delta y_i) - u_i; \tag{10.17}$$

$$\frac{dy_i}{dT} = \frac{1}{b} u_i, \tag{10.18}$$

where the dimensionless optimal velocity u_{opt} and the dimensionless control parameter b are given as

$$u_{\text{opt}}(\Delta y) = \frac{(\Delta y)^2}{1 + (\Delta y)^2} \quad \text{and} \quad b = \frac{D}{\tau v_{\text{max}}}. \tag{10.19}$$

The steady-state or free-flow solution for all vehicles $n = 1\ldots N$ is

$$u_n^{\text{st}} = u_{\text{opt}}(\Delta y_{\text{hom}}) \tag{10.20}$$

$$y_n^{\text{st}} = y_0 + n\,\Delta y_{\text{hom}} + \frac{1}{b} u_{\text{opt}}(\Delta y_{\text{hom}})\,T \tag{10.21}$$

where y_0 is an arbitrary constant, $\Delta y_{\text{hom}} = \mathcal{L}/N$ is the headway distance in homogeneous flow, and $\mathcal{L} = L/D$ is the dimensionless length of the road.

The stability of the steady-state solution can be investigated by the method of small perturbations. Taking into account the periodic boundary conditions, we consider a perturbation of the form

$$y_n = y_n^{\text{st}} + \delta y\, e^{\lambda T + ik\Delta y_{\text{hom}}\, n}. \tag{10.22}$$

Here $\delta y \to 0$ is the amplitude of periodic perturbation with the wave vector $k = 2\pi m/\mathcal{L}$, where $m = 1, 2, \ldots, N-1$. In this case it is appropriate to reduce the system of $2N$ first-order differential equations (10.17)–(10.18) to N equations of the second order

$$\frac{\partial^2 y_n}{\partial T^2} + \frac{\partial y_n}{\partial T} - \frac{1}{b} u_{\text{opt}}(\Delta y_n) = 0. \tag{10.23}$$

By inserting (10.22) into (10.23) and expanding $u_{\text{opt}}(\Delta y_n)$ in the vicinity of $\Delta y = \Delta y_{\text{hom}}$, we obtain the following quadratic equation

$$\lambda^2 + \lambda + \frac{u'_{\text{opt}}}{b}\left(1 - \exp\left(\frac{2\pi i m}{N}\right)\right) = 0 \tag{10.24}$$

for the Lyapunov exponents λ, where $u'_{\text{opt}} = du_{\text{opt}}(\Delta y)/d(\Delta y)$ at $\Delta y = \Delta y_{\text{hom}}$. This can be easily solved by putting $\lambda = \text{Re}\,\lambda + i\,\text{Im}\,\lambda = \alpha + i\beta$. Hence, α and β are solutions of the system of two equations

$$\alpha^2 - \beta^2 + \alpha + \frac{u'_{\text{opt}}}{b}\left(1 - \cos\left(\frac{2\pi m}{N}\right)\right) = 0, \tag{10.25}$$

$$2\alpha\beta + \beta - \frac{u'_{\text{opt}}}{b}\sin\left(\frac{2\pi m}{N}\right) = 0. \tag{10.26}$$

The homogeneous solution loses stability at a certain value of u'_{opt}/b when α vanishes for some m. According to (10.25)–(10.26), this takes place at

$$\alpha = 0, \quad \beta = \frac{u'_{\text{opt}}}{b}\sin\left(\frac{2\pi m}{N}\right), \quad \frac{u'_{\text{opt}}}{b} = \frac{1}{1 + \cos\left(\frac{2\pi m}{N}\right)}. \tag{10.27}$$

The stability border corresponds to the value of u'_{opt}/b at which the real part of one of the Lyapunov exponents vanishes for the first time and then becomes positive if u'_{opt}/b is increased. This occurs first with the exponent indexed by $m = 1$.

Consider now the stability regions in the b-c plane, where $c = N/\mathcal{L}$ is the dimensionless density of cars. These regions for homogeneous and heterogeneous solutions depend on the initial conditions as well as the total number of cars N. The homogeneous flow described by the steady-state solution of (10.17)–(10.18), that is $\Delta y_i^{st} = \Delta y_{\text{hom}} = \mathcal{L}/N = 1/c$ and $u_i^{st} = u_{\text{opt}}(\Delta y_i^{st})$, becomes unstable when entering the region $b < b(c)$, where $b(c)$ is given by

$$b(c) = u'_{\text{opt}}\left(\frac{1}{c}\right)\left(1 + \cos\left(\frac{2\pi}{N}\right)\right) \tag{10.28}$$

with

$$u'_{\text{opt}}\left(\frac{1}{c}\right) = \frac{2c^3}{(1+c^2)^2}, \tag{10.29}$$

as consistent with the stability condition (10.27) at $m = 1$. The maximum of $b(c)$ corresponds to the critical concentration $c_{\text{cr}} = \sqrt{3}$. The finite size effect on the phase diagram, where the regions of stable and unstable homogeneous flow are separated by the $b(c)$ curve, is illustrated in Figure 10.2.

The homogeneous stationary solution (10.20)–(10.21) transforms into a heterogeneous limit-cycle solution when entering the region below the $b(c)$ curve. The limit cycle is formed in the phase space of headway distances Δy and velocities u, as shown in Figure 10.3 for $c = 1.5$ and $b = 1.1$. The trajectory of each car goes around this limit cycle in the long-time limit $T \to \infty$, whereas the homogeneous flow is represented by the unstable fixed point $(\Delta y_{\text{hom}}, u_{\text{opt}}(\Delta y_{\text{hom}}))$ (solid circle) on the optimal velocity curve (dashed line).

The numerical study of a system of $N = 60$ cars shows that the limit-cycle solution becomes unstable when exiting the region below the dashed curve shown in Figure 10.2(b). In other words, as in many physical systems (e.g. supersaturated vapor), we observe a hysteresis effect which is a property of the first-order phase transition. The stationary state of the system between the solid and dashed curves (at a given N) is of the same type as the initial conditions.

A phase transition from the heterogeneous to the homogeneous state takes place at a fixed density c when the parameter b is increased and reaches the dashed curve in Figure 10.2(b). The heterogeneous state is characterized by the minimum u_{min} and the maximum u_{max} values of the velocity u in the limit cycle. At the critical density $c = c_{\text{cr}} = \sqrt{3}$ both branches $u_{\text{min}} = u_{\text{min}}(b)$ and $u_{\text{max}} = u_{\text{max}}(b)$ continuously merge into one branch $u_{\text{min}}(b) = u_{\text{max}}(b) = u_{\text{hom}}$ at the critical point $b = b_{\text{cr}} = (3\sqrt{3}/8)(1 + \cos(2\pi/N))$, where $u_{\text{hom}} = 1/(1 + c^2)$ is the steady-state velocity of the homogeneous flow. It is the supercritical bifurcation diagram shown in Figure 10.4(b) for a finite number of cars $N = 60$. The transformation is discontinuous at $c \neq c_{\text{cr}}$, as consistent with the subcritical bifurcation diagram at $c = 1.5$ in Figure 10.4(a). The power-like singularities of u_{max} and u_{min} at $c = c_{\text{cr}}$ and $b \to b_{\text{cr}}$, so

$$u_{\text{max}}(b) - u_{\text{hom}} \propto (b_{\text{cr}} - b)^{\beta_1} \tag{10.30}$$

$$u_{\text{hom}} - u_{\text{min}}(b) \propto (b_{\text{cr}} - b)^{\beta_2} \tag{10.31}$$

are described by the critical exponents β_1 and β_2. It is useful to define the effective critical exponents $\beta_1^{\text{eff}}(h)$ and $\beta_2^{\text{eff}}(h)$ as the slope of the log-log plot evaluated from

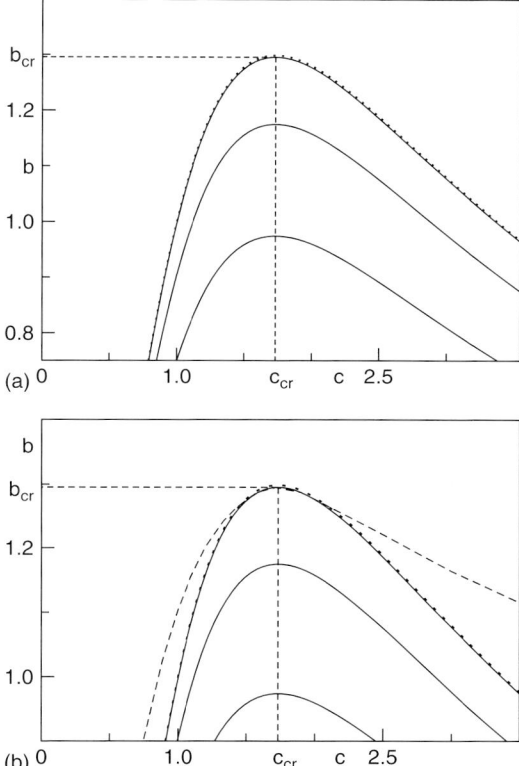

Figure 10.2 Phase diagram as b-c plane for a system with a fixed different number of cars N. From the bottom to the top, solid curves $b(c)$ show the stability border of the homogeneous traffic flow at $N = 6$, $N = 10$, and $N = 60$, respectively. The dotted curve is the function $b(c)$ when N tends to infinity. Each $b(c)$ plot has a maximum at the critical density $c_{cr} = \sqrt{3} \approx 1.732$. The maximum value at $N \to \infty$ is $b_{cr} = 3\sqrt{3}/4 \approx 1.299$. The dashed curve on figure (b) shows the stability border of the heterogeneous traffic flow at $N = 60$ (numerical simulation). It becomes unstable when exiting the region below this curve

the data at $b = b_{cr} - h$ and $b = b_{cr} - 2h$. The results are shown in Figure 10.5. The true (asymptotic) values of the critical exponents are obtained at $h \to 0$. Figure 10.5 represents the numerical evidence that both exponents β_1 and β_2 have the same universal mean-field value $1/2$. It also shows that an evaluation of the critical exponent by simply measuring the slope of the log-log plot at a quite small distance from the critical point, e.g. at $h \sim 0.006$, is rather misleading [94].

The dynamics of a car system is described by a trajectory in $2N$-dimensional space of all headway distances Δy_i and velocities u_i. For any given initial condition, one can consider N individual trajectories (one for each car) in the two-dimensional $(\Delta y, u)$ phase space. The dynamics represented by such a set of trajectories depends on the initial condition. However, for a certain set of initial conditions at a given

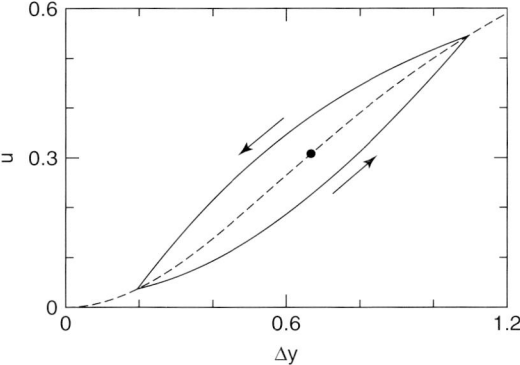

Figure 10.3 The limit cycle (solid curve) for the system of $N = 60$ cars at the dimensionless density $c = 1.5$ and parameter $b = 1.1$. The dashed line is the optimal velocity curve and the solid circle is the unstable fixed point. Arrows indicate the direction of motion along the limit cycle.

overall density of cars, the movement of the points in the phase space can be characterized by the vector field showing the average flow of these points. In particular, we consider an ensemble of initial conditions with random spatial distribution of cars with the only requirement that $\Delta y_i > \Delta y_{min}$ holds, where Δy_{min} is some minimum headway distance, the initial velocities being distributed uniformly within $u \in [\bar{u} - \Delta u, \bar{u} + \Delta u]$, where $\bar{u}$ is some mean velocity. Also the ensemble includes a set of $\bar{u}$ values to cover a certain region of the velocities of interest. At each time step of the numerical integration, we have N flow vectors $\mathbf{J}_i = (d\Delta y_i/dT, du_i/dT) = ([u_{i+1} - u_i]/b, du_i/dT)$ associated with N points $(\Delta y_i, u_i)$ in the phase space. The field of the averaged flow vectors $\langle \mathbf{J} \rangle$ is constructed by splitting the specifically considered region of the phase space in cells, and then calculating the mean flow vector for each cell by averaging over the ensemble of initial conditions and over all time steps from $T = 0$ to $T = T_{max}$. The mean flow vector is depicted by a suitably scaled arrow with the origin in the middle of the corresponding cell. It shows the mean direction in which the points belonging to this cell of the phase space move, and the length of the arrow is proportional to the average speed of this movement.

Such a vector field for the system of $N = 60$ cars at the values of the dimensionless control parameters $b = 1.3$ and $c = 0.75$, for which the homogeneous flow (that is, the fixed point) is stable, is depicted in Figure 10.6. The parameters $\Delta y_{min} = 0.2$, $\Delta u = 0.05$ with a grid of $\bar{u}$ values $\bar{u} = \Delta u, 2\Delta u, \ldots, 1 - \Delta u$ have been used for the set of initial conditions, including 100 different random initial distributions of coordinates and velocities for each value of $\bar{u}$. The integration has been performed up to $T_{max} = 100$. In Figure 10.6(b) a more detailed view of the region around the stable fixed point is shown, obtained by averaging over 1000 random realizations for each $\bar{u}$. This vector field or phase portrait of the car system represents a nontrivial

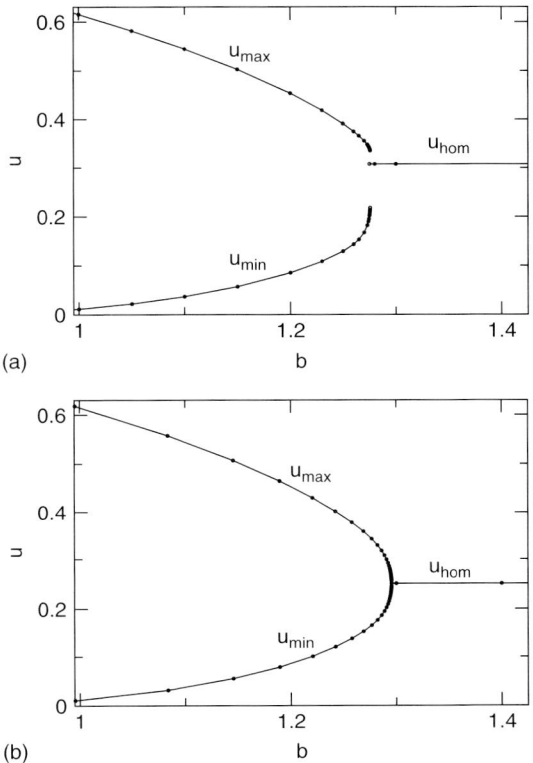

Figure 10.4 Subcritical bifurcation diagram at $c = 1.5$ (a) and supercritical bifurcation diagram at the critical density $c = c_{cr} = \sqrt{3}$ (b) for the minimum and maximum velocities depending on the control parameter b for a fixed number of cars $N = 60$.

(averaged) dynamics for reaching the fixed point. As we can see, the relaxation to the fixed point is not straightforward.

An important characteristic is the ratio $\langle |\mathbf{J}| \rangle / \sigma$, where $\langle |\mathbf{J}| \rangle$ is the averaged modulus of the flow vector and $\sigma = \sqrt{\langle (\mathbf{J} - \langle \mathbf{J} \rangle)^2 \rangle}$ is the square root of its variance, which characterizes the mean deviation or fluctuation of $\langle |\mathbf{J}| \rangle$. The region between the dotted lines in Figure 10.6 corresponds to $\langle |\mathbf{J}| \rangle < 3\sigma$, whereas $\langle |\mathbf{J}| \rangle > 3\sigma$ holds outside this region. In the latter case the motion in the phase plane of Δy and u is very predictable and is usually given with quite a good accuracy by the mean flow vector. The motion becomes more diffusive with decreasing $\langle |\mathbf{J}| \rangle / \sigma$. This is observed near the optimal velocity curve (the dashed line) and, particularly, in the vicinity of the fixed point (the solid circle), where $\langle |\mathbf{J}| \rangle \ll \sigma$ holds.

In Figure 10.7 the phase portrait is shown for a larger density $c = 1.5$ and $b = 1.28$. It is very close to the border $b = b(c) \simeq 1.2781$ given by (10.28), where the homogeneous stationary solution becomes unstable and transforms into the limit

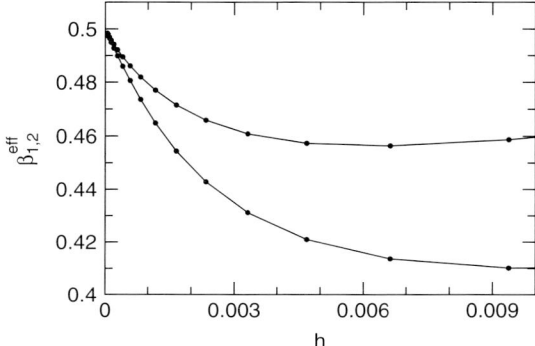

Figure 10.5 The effective critical exponents $\beta_1^{\text{eff}}(h)$ (solid circles) and $\beta_2^{\text{eff}}(h)$ (empty circles) estimated from the $u_{\max}(b)$ and $u_{\min}(b)$ data, respectively, at finite distances from the critical point: $b = b_{cr} - h$ and $b = b_{cr} - 2h$. The universal asymptotic value of the critical exponent $\beta = 0.5$ is obtained at $h \to 0$.

cycle. In fact, the flow vectors indicate an unstable limit cycle, and the convergence to the fixed-point stationary solution is very slow.

A disadvantage of the optimal velocity model (10.13)–(10.14) is that it is not collision free. In other words, it does not ensure the solution where all the headway distances are always positive. In particular, we never succeeded to obtain a limit-cycle solution without collisions at $b < 0.86$. An important feature of the original as well as our optimal velocity model (10.13)–(10.14) is the symmetry between acceleration and braking. In real traffic, however, the braking (deceleration) can be considerably larger than the acceleration, which is important to avoid collisions. Below, we propose a collision-free model which includes this asymmetry.

The dynamics of a car system can be formulated as Newton's law of motion. Returning to the dimensional quantities, we have the following set of $2N$ differential equations

$$m \frac{dv_i}{dt} = F_{\text{det}}(v_i, \Delta x_i); \tag{10.32}$$

$$\frac{dx_i}{dt} = v_i \tag{10.33}$$

where m is the mass of a point-like particle. Now we split the deterministic force $F_{\text{det}}(v_i, \Delta x_i)$ into two parts

$$F_{\text{det}}(v_i, \Delta x_i) = F_{\text{acc}}(v_i) + F_{\text{dec}}(v_i, \Delta x_i) \tag{10.34}$$

with acceleration and deceleration ansatz

$$F_{\text{acc}}(v_i) = \frac{m}{\tau}(v_{\max} - v_i) \geq 0 \tag{10.35}$$

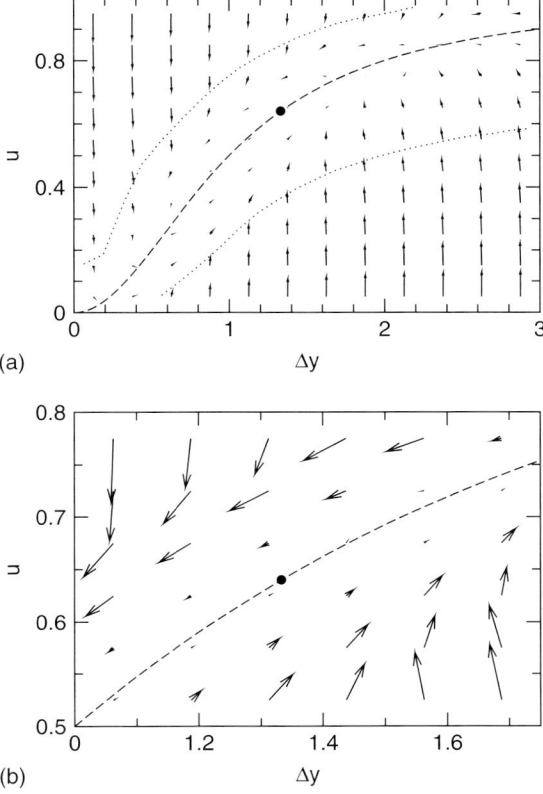

Figure 10.6 The vector field showing an averaged movement of points in the phase space of headway distances Δy and velocities u at the values of control parameters $c = 0.75$ and $b = 1.3$ for an ensemble of random initial distributions. The dashed line is the optimal velocity curve and the solid circle is the stable fixed point corresponding to the homogeneous flow. The dotted lines are isolines $\langle |J| \rangle = 3\sigma$.

$$F_{\text{dec}}(v_i, \Delta x_i) = \frac{m}{\tau}\left(v_{\text{opt}}(\Delta x_i) - v_{\max}\right)\left(1 + \left(p\frac{v_i}{\Delta x_i}\right)^2\right) \leq 0 \qquad (10.36)$$

taking into account the optimal velocity function (10.15). The deceleration force F_{dec} includes a symmetry-breaking term with new parameter p. A simple consideration shows that a deceleration $-dv_i/dt > v_0^2/(2\Delta x)$ is large enough for a car, moving with initial velocity v_0, to stop before a vehicle staying in front of it at an initial distance Δx. According to safety considerations, a car has to decelerate even faster at small distances Δx, therefore we have assumed in (10.36) that the additional term in F_{dec} is proportional to $(v_i/\Delta x_i)^2$. Under this condition a car can never reach with nonzero velocity $v' = v_i(t')$ (at a time $t = t'$) another car staying in front of it since the assumption $v' > 0$ leads to a contradiction: it implies that

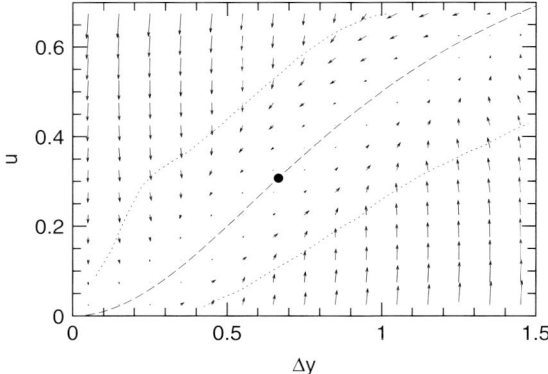

Figure 10.7 The vector field showing an averaged movement of points in the phase space of headway distances Δy and velocities u at the values of control parameters $c = 1.5$ and $b = 1.28$ for an ensemble of random initial distributions. The dashed line is the optimal velocity curve, and the solid circle is the stable fixed point corresponding to the homogeneous flow. The dotted lines are isolines $\langle |J| \rangle = 3\sigma$.

$-dv_i/dt > v_i^2(t_0)/(2\Delta x(t_0))$ holds within certain time interval $t_0 < t < t'$, which by itself rules out the collision. The original optimal velocity model (10.13)–(10.14) is recovered at $p = 0$. If p is a small parameter, then the new model behaves practically in the same way as the old one, except for those critical situations where some car produces an accident in the old model, and its motion is now corrected to avoid the collision. The deceleration force term due to interaction between cars is always negative

$$F_{\text{dec}} = -v_{\max} \frac{m}{\tau} \frac{1}{1 + (\Delta x_i/D)^2} \left(1 + \left(p\frac{v_i}{\Delta x_i}\right)^2\right). \tag{10.37}$$

Using the acceleration (10.35) and deceleration (10.37) ansatz we are able to write the deterministic force F_{det} as

$$F_{\text{det}}(v_i, \Delta x_i) = v_{\max} \frac{m}{\tau} \left(1 - \frac{v_i}{v_{\max}} - \frac{1}{1 + (\Delta x_i/D)^2}\right)\left(1 + \left(p\frac{v_i}{\Delta x_i}\right)^2\right). \tag{10.38}$$

As earlier, it is convenient to make our equations of motion dimensionless (10.16). Finally we obtain

$$\frac{du_i}{dT} = 1 - u_i - \frac{1}{1 + (\Delta y_i)^2}\left(1 + \left(\tilde{p}\frac{u_i}{\Delta y_i}\right)^2\right) \tag{10.39}$$

$$\frac{dy_i}{dT} = \frac{1}{b} u_i \tag{10.40}$$

where $\tilde{p} = p\, v_{\max}/D$.

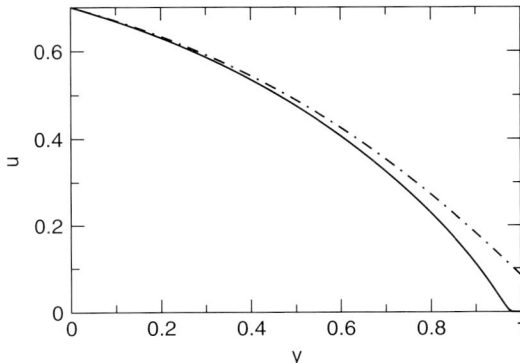

Figure 10.8 The dimensionless velocity u vs coordinate y for one car with the initial coordinate $y = 0$ and velocity $u = 0.7$ having a wall at the position $y = 1$ in front of it. The dashed line, showing a collision with the wall, corresponds to the optimal velocity model ($\tilde{p} = 0$), whereas the solid line corresponds to the advanced collision-free model with $\tilde{p} = 0.2$.

To illustrate the difference between the new and old models, we have made a numerical calculation at $b = 1$ for one car with the initial coordinate $y = 0$ and velocity $u = 0.7$ having a wall (or, equally, another staying car) in front of it at the position $y = 1$. In Figure 10.8 we have plotted the velocity u depending on the coordinate y for the OVM model with $\tilde{p} = 0$ (dot-dashed curve), as well as for the new advanced model with $\tilde{p} = 0.2$ (solid curve). As can be seen, the car reaches the wall with nonzero velocity in the original model. On the contrary, no accident occurs in the new model. In this case the car approaches the wall asymptotically at $T \to \infty$ with vanishing velocity $u \to 0$.

The homogeneous steady-state solution $du_i/dT = 0$ of system (10.39)–(10.40) reads

$$1 - u_i - \frac{1}{1 + (\Delta y_i)^2}\left(1 + \left(\tilde{p}\frac{u_i}{\Delta y_i}\right)^2\right) = 0. \tag{10.41}$$

This corresponds to equal headways $\Delta y_i = \mathcal{L}/N = 1/c$. The steady-state velocity can be found as the positive root of the quadratic equation

$$u_i^2 + \frac{(\Delta y_i)^2}{\tilde{p}^2}\left(1 + (\Delta y_i)^2\right)u_i - \frac{(\Delta y_i)^4}{\tilde{p}^2} = 0 \tag{10.42}$$

which follows from (10.41). It yields

$$u_i = \frac{(\Delta y_i)^2(1 + (\Delta y_i)^2)}{2\tilde{p}^2}\left(\sqrt{1 + \frac{4\tilde{p}^2}{(1 + (\Delta y_i)^2)^2}} - 1\right). \tag{10.43}$$

The result of the optimal velocity model $u_i = u_{\text{opt}}(\Delta y_i)$ is recovered at $\tilde{p} \to 0$.

In Figure 10.9 the steady-state velocity of the new model with $\tilde{p} = 1$ is compared to that of the optimal velocity model with $\tilde{p} = 0$. The difference is relatively small

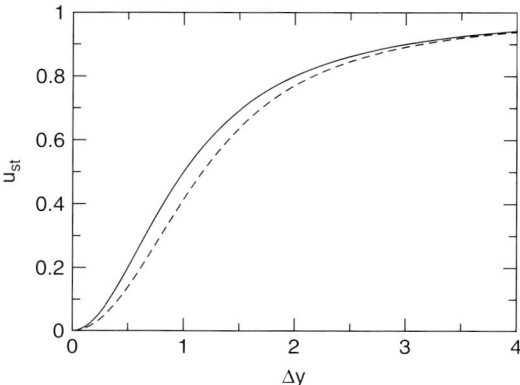

Figure 10.9 The steady state velocity of a homogeneous traffic flow depending on the headway distance Δy. The solid line corresponds to the optimal velocity model ($\tilde{p} = 0$), the dashed line indicates the result of the new model with $\tilde{p} = 1$.

at $\tilde{p} = 1$ and much smaller (about 100 times) at $\tilde{p} = 0.1$. The latter implies that the additional term with $\tilde{p}$ practically does not change the properties of the original optimal velocity model at $\tilde{p} \sim 0.1$, except that it makes the model collision free.

Up to now we have dealt with the deterministic car-following model. The stochastic optimal velocity (OV) model of a one-lane road with periodic boundary conditions is obtained by adding the noise term. Thus, within the Langevin approach [34, 157], we have the following set of acceleration equations

$$\frac{\mathrm{d}v_i}{\mathrm{d}t} = \frac{v_{\mathrm{opt}}(\Delta x_i) - v_i}{\tau} + \xi_i(t). \tag{10.44}$$

Neglecting the noise term ξ_i, the position $x_i(t)$ and the velocity $v_i(t)$ of each car $i = 1, \ldots, N$ at every time t can be calculated from the initial values by integrating the coupled equations of motion. The coupling is due to the interaction between two successive cars measured by the headway $\Delta x_i = x_{i+1} - x_i - \ell$ (note car length $\ell \to 0$ for point-like cars). The optimal or desired velocity $v_{\mathrm{opt}}(\Delta x)$ (10.15) is the steady-state velocity chosen by drivers as a function of the headway between cars. It increases monotonously with the distance and tends to a constant maximum value $v_{\max}$ for $\Delta x \to \infty$.

Our equations of motion can be written as a random dynamical system with multiplicative Gaussian white noise [152]

$$m\,\mathrm{d}v_i(t) = F_{\mathrm{det}}(v_i, \Delta x_i)\,\mathrm{d}t + \sigma v_i\,\mathrm{d}W_i(t) \tag{10.45}$$

$$\mathrm{d}x_i(t) = v_i\,\mathrm{d}t \tag{10.46}$$

with noise intensity $\sigma > 0$ for a point-like particle (vehicle i) of mass m with speed $v_i(t)$ at location $x_i(t)$. The choice of the multiplicative noise is motivated by the fact that, contrary to the additive noise, it ensures the positiveness of all velocities

v_i. Here $F_{\text{det}} = m\left(v_{\text{opt}}(\Delta x_i) - v_i\right)/\tau$ is the deterministic force. The fluctuations $dW_i = Z\sqrt{dt}$ are given by the increment of a Wiener process, where Z is a $\mathcal{N}(0,1)$ standard normal-distributed random number. By using the dimensionless variables introduced in (10.16), the system of equations (10.45)–(10.46) becomes

$$du_i(t) = (u_{\text{opt}}(\Delta y_i) - u_i)\,dT + a\,u_i\,dW_i(t) \tag{10.47}$$

$$dy_i(t) = \frac{1}{b}u_i\,dT, \tag{10.48}$$

where $a = \sigma\sqrt{\tau}/m$ is the dimensionless noise amplitude.

We have numerically solved, by simulation, the system of equations (10.47)–(10.48) for a finite system of 60 cars at nonvanishing noise intensity $a = 0.1$ by an algorithm called the explicit 1.5 order strong scheme [104]. We have fixed the dimensionless density of cars c and have varied the parameter b to estimate the value $b(c)$ at which the transition from the free-flow regime to congested traffic occurs. A jump-like increase in the variance of the velocity $var(u) = \langle u^2 \rangle - \langle u \rangle^2$ takes place around the stability curve $b = b(c)$, see Figure 10.2. The noise tends to wash out the phase transition in the finite system which we considered, therefore we could not observe a real jump and the value of $b(c)$ has been identified with the inflection point of the $var(u)$ vs b plot where this curve has maximum steepness. Moreover, the phase transition around the critical density $c_{\text{cr}} = \sqrt{3}$ becomes too diffuse to identify the location of the transition point. Due to the fluctuations at $a > 0$, our system of cars cannot stay unlimited for a long time in a metastable state. This means that a sufficiently long simulation will give us one line $b = b(c)$ irrespective of the initial conditions, rather than two branches of the phase diagram shown in Figure 10.2 (b) for the deterministic case $a = 0$. Nevertheless, even at positive $a = 0.1$ a hysteresis effect has been observed in finite-time simulations at certain densities, e.g. $c = 3$. The phase diagram in Figure 10.10 shows the parts of the $b = b(c)$ curve (dotted line) at $a = 0.1$ as well as the two branches with $a = 0$ for comparison.

The idea of using the variance of the velocity u to distinguish between the homogeneous and heterogeneous states of the system is similar to that applied in [169], where a closely related quantity; namely the variance of the density, has been considered in the framework of stochastic car-following models. In particular, we expect that the stochastic OVM model considered here would produce a similar variance plot depending on the density and noise intensity, as does the cellular automata model (CA) discussed in [169]. This variance plot is seen in Figure 10.11.

As distinct from the CA model considered in [169] (for the Nagel–Schreckenberg cellular automata model see also [168, 170, 198, 199, 205]), the noise can produce collisions in the model (10.45)–(10.46), which is highly likely at large noise intensities. A modification of the deterministic part, as proposed in (10.36), is helpful to solve this problem.

We have fixed the noise intensity $a = 0.1$ and the number of cars $N = 60$ and have studied the distribution of vehicular velocities and headway distances

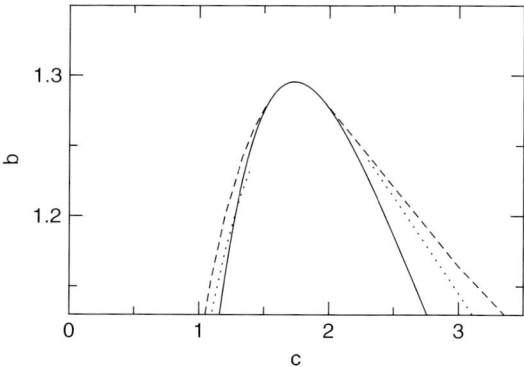

Figure 10.10 The phase diagram of a finite system of $N = 60$ cars, including the phase transition line (dotted curve) at a nonvanishing noise intensity $a = 0.1$. The solid and dashed lines refer to the case $a = 0$ and have the same meaning as in Figure 10.2.

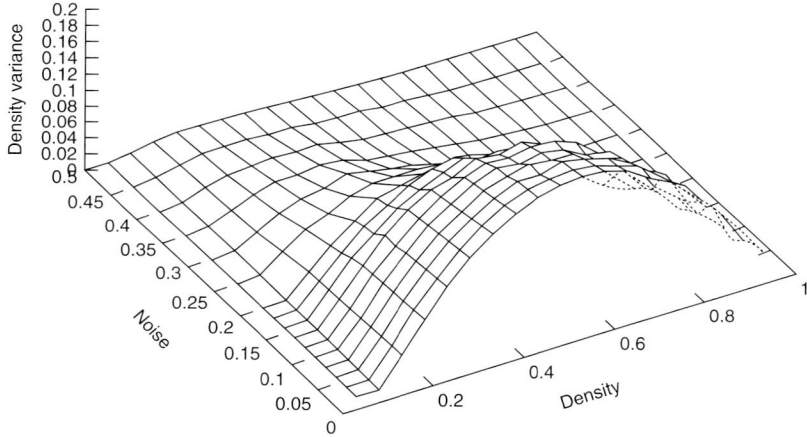

Figure 10.11 3D-plot of the density variance for a cellular automaton (CA) model with slow-to-start rule. The noise is given by a parameter $p_{>0}$ [169] of random deceleration. The figure is taken from [169].

depending on the dimensionless car density c and parameter b. In Figure 10.12(a) the velocity distribution function $f(u)$ is shown at $b = 1.1$ for a relatively small density $c = 0.5$ as well as for higher density $c = 2$. In the first case we have only one maximum located near $u = 0.8$ which is the steady-state velocity of the homogeneous free traffic flow without noise. The noise only smears out the delta-like distribution to yield the smooth maximum seen in Figure 10.12. The traffic flow is homogeneous with small fluctuations in velocities and also headway distances, as shown in Figure 10.12(b). The headway-distance distribution

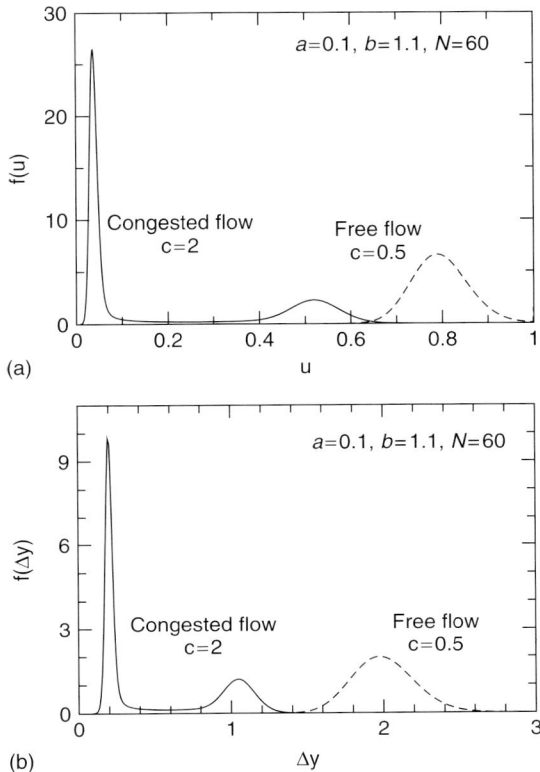

Figure 10.12 The probability density distribution of velocities (a) and headway distances (b) for two different car densities $c = 0.5$ and $c = 2$ at fixed dimensionless noise amplitude $a = 0.1$ and parameter $b = 1.1$.

function $f(\Delta y)$ has a maximum near $\Delta y = \Delta y_{\text{hom}} = 1/c = 2$ which is the average headway distance in a homogeneous flow. In the second case ($c = 2$) the velocity distribution function and the headway distribution function have two maxima. This indicates the coexistence of two phases – free flow with relatively large headways and velocities and jam with small headways and velocities.

The heavy traffic at $c = 3.5$, and the critical situation at $c = c_{\text{cr}}$ and $b = b_{\text{cr}}$ are illustrated in Figure 10.13 with the same notation as in Figure 10.12. In both cases the distributions have only one maximum, as is consistent with the existence of only one phase. A distinguishing feature of the critical point is that the distributions over the headway distances and velocities are relatively broad.

The simulation of the stochastic equations (10.45) and (10.46) in the coexistence region allows us also to find the probability $P(n,t)$ that just n cars are involved in the jam. In this case the jammed cars are defined as those vehicles which have the headway distance smaller than the homogeneous one $\Delta y_{\text{hom}} = 1/c$.

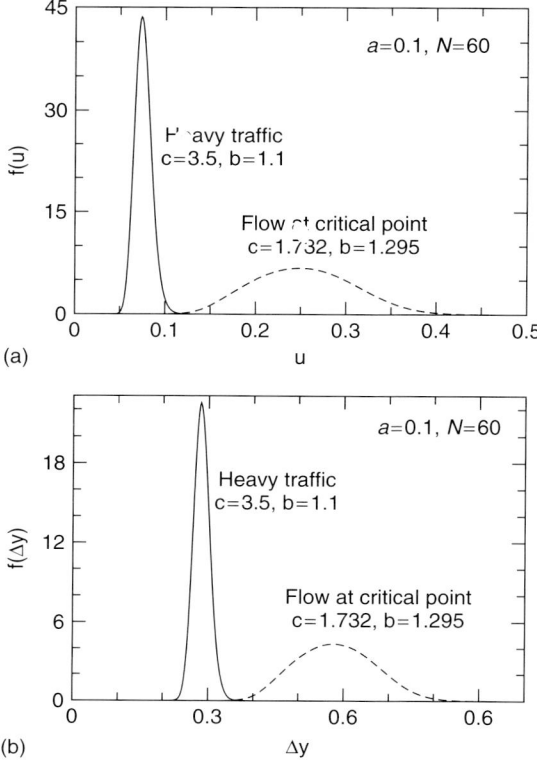

Figure 10.13 The probability density distribution of velocities (a) and headway distances (b) in the case of heavy traffic with large density of cars $c = 3.5$ (at $b = 1.1$) and at the critical point $c = c_{cr} = \sqrt{3} \simeq 1.73205$, $b = b_{cr} \simeq 1.29548$. In both cases the dimensionless noise intensity is $a = 0.1$.

The results of numerical simulation at $c = 2$ are shown in Figure 10.14. They coincide qualitatively with the investigations within the stochastic master equation approach [147–149, 152, 153] discussed in Sections 10.3–10.5.

10.3
Traffic Jam Formation on a Circular Road

In the previous section we have described the motion of a car ensemble on a *microscopic* level by considering each car individually. Many properties of traffic flow, including jam formation, can be more easily described on a less detailed *mesoscopic* level, where we are looking only for the number of jammed cars n. This type of model on a one-lane circular road is illustrated in Figure 10.15 where we have shown two different regimes of traffic flow: free traffic flow (a) and congested

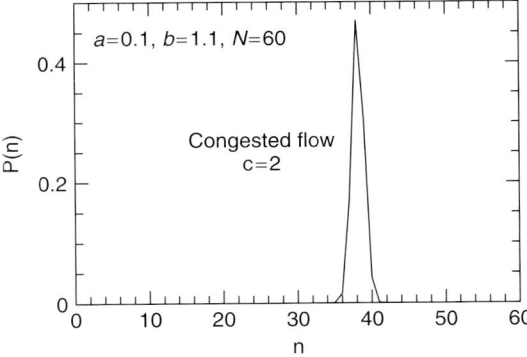

Figure 10.14 The cluster (jam) size distribution when the density of cars $c = 2$ for a system of $N = 60$ cars at the dimensionless noise intensity $a = 0.1$ and $b = 1.1$.

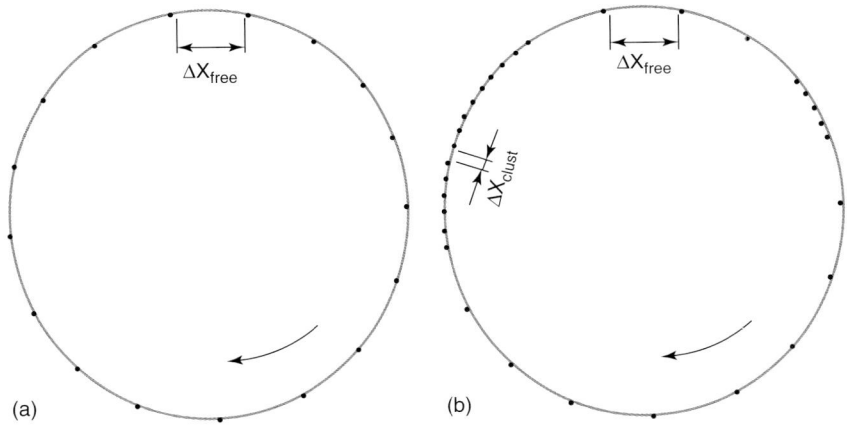

Figure 10.15 Free traffic flow (a) and congested traffic flow (b) on a one-lane circular road. In the case of congested traffic (b) there are two clusters of different length coexisting, with free flow shown as an example. The headway between the cars is Δx_{clust} inside a cluster and Δx_{free} in free flow. The direction of movement is indicated by an arrow.

traffic flow (b). In the congested traffic several jams can exist simultaneously, for example two clusters in Figure 10.15. Here we consider a simple model where only one car cluster (a queue of n cars) is allowed.

Installation of multiple loop detectors for queue detection in recent years has made it possible to measure the queue length or $n(t)$ as the number of congested cars directly [122].

In the following we consider the attachment of a vehicle to the car cluster and the detachment from it as elementary stochastic events. The traffic is thus treated as a one-step Markov process described by the general master equation (3.38)

$$\frac{\partial}{\partial t}p(n,t) = w_+(n-1)\,p(n-1,t) + w_-(n+1)p(n+1,t)$$
$$-[w_+(n) + w_-(n)]\,p(n,t). \tag{10.49}$$

Now the basic problem is to find an appropriate ansatz for both transition probabilities $w_+(n)$ and $w_-(n)$. Note that physical boundary conditions ($0 \leq n \leq N$) for the master equation (10.49) are ensured by formally setting $P(-1,t) = P(N+1,t) = 0$ and $w_+(N) = w_-(0) = 0$. The latter two transitions are impossible physically and they are not included in our further analysis. As before (3.76), we assume a constant value for the escape rate $w_-(n)$, so

$$w_-(n) = w_- = \frac{1}{\tau}. \tag{10.50}$$

The probability per time unit $w_+(n)$ that a vehicle is added to a car cluster of size n is estimated based on the following physical model. The total number of cars is N. They are moving along a circular one-lane road of length L. If a road is crowded with cars, each car requires some minimum space or length which, obviously, is larger than the real length of a car. We call this the effective length ℓ of a car. The distance between the front bumpers of two neighboring cars, in general, is $\ell + \Delta x$. The distance Δx can be understood as the headway between two 'effective' cars which, according to our definition, is always smaller than the real bumper-to-bumper distance. The maximum velocity of each car is v_{max}. The desired (optimum) velocity v_{opt}, depending on the distance between two cars Δx, is given by the formula

$$v_{opt}(\Delta x) = v_{max}\frac{(\Delta x)^2}{D^2 + (\Delta x)^2}, \tag{10.51}$$

where the parameter D, called the interaction distance, corresponds to the velocity value $v_{max}/2$. According to the ansatz (10.51) the optimum velocity, see Figure 10.16, is represented by a sigmoidal function with values ranging from 0, corresponding to zero distance between cars, to v_{max}, corresponding to an infinitely large distance or absence of interaction between cars. Our assumption is that a vehicle changes its velocity from $v_{opt}(\Delta x_{free})$ in free flow to $v_{opt}(\Delta x_{clust})$ in a jam and approaches the cluster as soon as the distance to the next car (the last car in the cluster) reduces from Δx_{free} to Δx_{clust}. This assumption allows one to calculate the average number of cars joining the cluster per time unit or the attachment frequency $w_+(n)$ to an existing car cluster. Thus, we have the ansatz valid for $1 \leq n < N$

$$w_+(n) = \frac{v_{opt}(\Delta x_{free}(n)) - v_{opt}(\Delta x_{clust})}{\Delta x_{free}(n) - \Delta x_{clust}}. \tag{10.52}$$

This equation (10.52) requires the knowledge of Δx_{free} and Δx_{clust} as a function of the cluster size n. Measurements on highways have shown that the density of

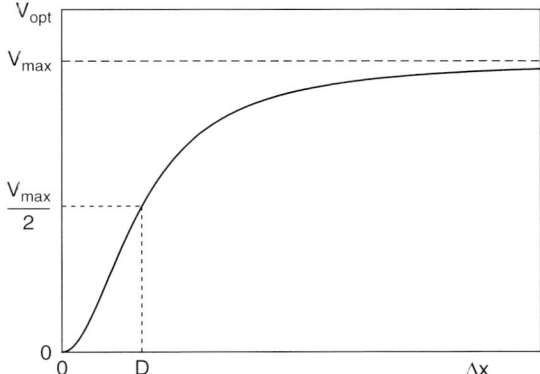

Figure 10.16 Analytical form of the optimal velocity function v_{opt} depending on headway Δx.

cars in congested traffic is independent of the size of the dense congested phase (jam). As a consequence, the distance between jammed cars, the spacing Δx_{clust}, has a constant value which has to be treated as a given measured quantity or known control parameter. We have defined the length of the car cluster or jam size depending on the number of congested cars n by

$$L_{\text{clust}} = \ell\, n + \Delta x_{\text{clust}}\, S(n), \tag{10.53}$$

where

$$S(n) = \begin{cases} 0 & : \quad n = 0 \\ n - 1 & : \quad n \geq 1 \end{cases} \tag{10.54}$$

is the number of spacings of size Δx_{clust}. In this way, we have for the total length of road

$$L = \underbrace{\ell\, n + \Delta x_{\text{clust}}\, S(n)}_{L_{\text{clust}}} + \underbrace{\ell(N - n) + \Delta x_{\text{free}}(N - S(n))}_{L_{\text{free}}}, \tag{10.55}$$

where

$$L_{\text{free}} = L - L_{\text{clust}} = L - \{\ell\, n + \Delta x_{\text{clust}}\, S(n)\} \tag{10.56}$$

denotes the length of the uncongested or free road. For L_{free} we can also write, according to (10.55)

$$L_{\text{free}} = \ell(N - n) + \Delta x_{\text{free}}(N - S(n)). \tag{10.57}$$

Comparing these two equations, we obtain the distance in free flow, depending on cluster size,

$$\Delta x_{\text{free}}(n) = \frac{L - \ell N - \Delta x_{\text{clust}}\, S(n)}{N - S(n)}. \tag{10.58}$$

By this, all the transition probabilities (10.52) are defined except the transition from the state without any cluster $n = 0$ to the smallest cluster size $n = 1$. This transition and the meaning of the state with a single congested car ($n = 1$) called a *precluster*, requires some explanation. Some stochastic event or perturbation of the free traffic flow, which is represented by $n = 0$, is necessary to initiate the formation of a cluster. These stochastic events are simulated assuming that one of the free cars can reduce its velocity to $v_{opt}(\Delta x_{clust})$ and thus become a single congested car or a cluster of size $n = 1$. This process is characterized by the transition frequency $w_+(0)$ which cannot be calculated from the ansatz (10.52), but has to be considered as one of the control parameters of the model. A cluster of size one appears also when a two-car cluster is reduced by one car. In this consideration, the vehicular cluster with size $n = 1$ is a car which still has not accelerated after this event. In any case, a precluster is defined as a single car moving with the velocity $v_{opt}(\Delta x_{clust})$. Since at $n = 0$ any of the N free cars has an opportunity to become a single congested car, an appropriate ansatz for the transition frequency $w_+(0)$ is

$$w_+(0) = \frac{p}{\tau} N, \tag{10.59}$$

where $p > 0$ is a dimensionless constant called the stochastic perturbation parameter or stochasticity.

In natural sciences, and especially in physics, it is usually accepted to write all the basic equations in dimensionless variables. It is appropriate to introduce the dimensionless time T via $T = t/\tau$ and the dimensionless distances normalized to ℓ, so, $\Delta y = \Delta x/\ell$, $d = D/\ell$, $\Delta y_{clust} = \Delta x_{clust}/\ell$ and $\Delta y_{free} = \Delta x_{free}/\ell$, as well as the dimensionless optimum velocity $w_{opt} = v_{opt}/v_{max}$.

Then the basic equations of this section can be rewritten as follows. The master equation for the scaled probability distribution $P(n, T)$ instead of $p(n, t)$:

$$\frac{1}{\tau}\frac{\partial}{\partial T}P(n, T) = w_+(n-1)\, P(n-1, T) + w_-(n+1)\, P(n+1, T)$$
$$- [w_+(n) + w_-(n)]\, P(n, T); \tag{10.60}$$

the optimal velocity definition:

$$w_{opt}(\Delta y) = \frac{(\Delta y)^2}{d^2 + (\Delta y)^2}; \tag{10.61}$$

the transition frequencies:

$$w_-(n) = w_- = \frac{1}{\tau}, \quad 1 \leq n \leq N, \tag{10.62}$$

$$w_+(0) = \frac{1}{\tau} p N, \tag{10.63}$$

$$w_+(n) = \frac{v_{max}}{\ell} \frac{\left[v_{opt}(\Delta x_{free}) - v_{opt}(\Delta x_{clust})\right]/v_{max}}{[\Delta x_{free} - \Delta x_{clust}]/\ell}$$
$$= \frac{1}{\tau} b \frac{w_{opt}(\Delta y_{free}(n)) - w_{opt}(\Delta y_{clust})}{\Delta y_{free}(n) - \Delta y_{clust}}, \quad 1 \leq n \leq N-1 \tag{10.64}$$

with dimensionless parameter

$$b = v_{\max}\tau/\ell; \tag{10.65}$$

and the ansatz for the cluster length and related quantities:

$$\frac{L_{\text{clust}}}{\ell} = n + \Delta y_{\text{clust}}\, S(n) = c_{\text{clust}}^{-1}\, n, \tag{10.66}$$

$$\frac{L_{\text{free}}}{\ell} = N - n + \Delta y_{\text{free}}(N - S(n)) = c_{\text{free}}^{-1}(N - n), \tag{10.67}$$

$$\Delta y_{\text{free}}(n) = \frac{L/\ell - N - \Delta y_{\text{clust}}\, S(n)}{N - S(n)}. \tag{10.68}$$

According to the definitions, $c = \ell N/L = \ell\varrho$ is the total density of cars, $c_{\text{clust}} = n\ell/L_{\text{clust}}$ and $c_{\text{free}} = (N-n)\ell/L_{\text{free}}$ are the densities in a jam and in the free flow, respectively.

In the stochastic approach an equation can be obtained for the average cluster size $\langle n \rangle$. Based on the master equation (3.38), we obtain a deterministic equation for the mean value

$$\frac{d}{dt}\langle n \rangle = \frac{d}{dt}\sum_n n p(n,t) = \langle w^+(n)\rangle - \langle w^-(n)\rangle, \tag{10.69}$$

which can be written in a certain approximation as follows

$$\frac{d}{dt}\langle n \rangle \approx w^+(\langle n \rangle) - w^-(\langle n \rangle), \tag{10.70}$$

describing the time evolution of the average cluster size $\langle n \rangle$. The stationary cluster size $\langle n \rangle_{\text{st}}$ can be calculated from the condition $d\langle n \rangle/dt = 0$ or

$$\frac{w_+(\langle n \rangle)}{w_-(\langle n \rangle)} = b\frac{w_{\text{opt}}(\Delta y_{\text{free}}(\langle n \rangle)) - w_{\text{opt}}(\Delta y_{\text{clust}})}{\Delta y_{\text{free}}(\langle n \rangle) - \Delta y_{\text{clust}}} = 1 \tag{10.71}$$

consistent with the ansatz for the transition probabilities (10.62) and (10.64). By using the definition (10.61) of the optimal velocity function $w_{\text{opt}}(\Delta y)$ we obtain the equation

$$b\left[\frac{(\Delta y_{\text{free}}(\langle n \rangle))^2}{d^2 + (\Delta y_{\text{free}}(\langle n \rangle))^2} - \frac{(\Delta y_{\text{clust}})^2}{d^2 + (\Delta y_{\text{clust}})^2}\right] = \Delta y_{\text{free}}(\langle n \rangle) - \Delta y_{\text{clust}} \tag{10.72}$$

which can be solved with respect to Δy_{free}. One solution of the third-order equation (10.72) is trivial $\Delta y_{\text{free}} = \Delta y_{\text{clust}}$. The other two solutions, which have a particular physical meaning, read

$$\Delta y_{\text{free}}^{(1,2)} = \frac{d}{2[d^2 + (\Delta y_{\text{clust}})^2]}$$
$$\times \left\{bd \pm \sqrt{b^2 d^2 + 4b\Delta y_{\text{clust}}[d^2 + (\Delta y_{\text{clust}})^2] - 4[d^2 + (\Delta y_{\text{clust}})^2]^2}\right\}. \tag{10.73}$$

According to (10.73), the headway Δy_{free} between cars in a free flow coexisting with a single cluster has a constant value depending merely on the control parameters of the model. Now, by means of (10.68), which states the relation between Δy_{free} and n, we are able to calculate the stationary cluster size $\langle n \rangle_{\text{st}}$. As already pointed out we set $S(n) = n - 1 + \delta_{n,0} \approx n$, assuming that the cluster (if it exists) contains a large number of cars. This leads to the equation

$$\frac{\langle n \rangle_{\text{st}}}{L/\ell} = \frac{c\left(1 + \Delta y_{\text{free}}^{(1,2)}\right) - 1}{\Delta y_{\text{free}}^{(1,2)} - \Delta y_{\text{clust}}}, \tag{10.74}$$

where $c = \ell N/L$ is the total density of cars and the term on the left-hand side of the equation is, in fact, the relative part of the road crowded by cars. In this case $\Delta y_{\text{free}}^{(1,2)}$ has a constant value given by (10.73). Result (10.74) makes sense at large enough densities where it provides a positive value of $\langle n \rangle_{\text{st}}$. The solution with the largest value $\Delta y_{\text{free}}^{(1)}$ (a positive sign in (10.73)) gives the average stationary size of a stable cluster depending on the total density c within the region $c > c_1$, where

$$c_1 = \frac{1}{1 + \Delta y_{\text{free}}^{(1)}} \tag{10.75}$$

is the critical density at which the spontaneous growth of a large car cluster starts. There is another critical density c_2, given by

$$c_2 = \frac{1}{1 + \Delta y_{\text{free}}^{(2)}}, \tag{10.76}$$

which defines the region $c > c_2$ where an unstable car cluster corresponding to the solution (10.74) with the smallest value $\Delta y_{\text{free}}^{(2)}$ of the headway can exist. In the special case of vanishing bumper-to-bumper distance in a jam $\Delta y_{\text{clust}} = 0$ our result (10.73) reduces to

$$\Delta y_{\text{free}}^{(1,2)} = \frac{b}{2} \pm \sqrt{\frac{b^2}{4} - d^2}, \tag{10.77}$$

and (10.74) to

$$\frac{\langle n \rangle_{\text{st}}}{L/\ell} = c + \frac{c - 1}{\frac{b}{2} \pm \sqrt{\frac{b^2}{4} - d^2}}. \tag{10.78}$$

The value of $\langle n \rangle_{\text{st}} \ell / L$ with sign $+$ in (10.78) corresponds to the stable, while that with sign $-$ corresponds to the unstable stationary cluster size. According to the deterministic equation (10.69), clusters of an undercritical (smaller than the unstable) size dissolve ($dn/dt < 0$) whereas those of the overcritical size relax to the stable stationary cluster size. The growth of the car cluster starting with the undercritical cluster size is possible, too, but only due to the stochastic fluctuations. Such a process can be described within a stochastic approach only.

The diagram shown in Figure 10.17 relates the stationary cluster size $\langle n \rangle_{\text{st}}$ to the total density of cars c in the case of the aggregation in traffic flow at vanishing

10.3 Traffic Jam Formation on a Circular Road

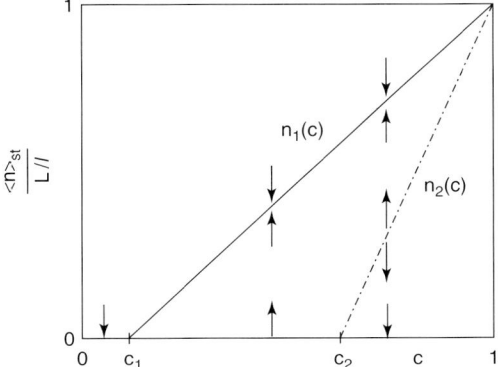

Figure 10.17 The stationary cluster size $\langle n \rangle_{st}$ normalized to L/ℓ (the relative part of the road crowded by cars) depending on the total density of cars c. The stable cluster size (branch $n_1(c)$ and horizontal lines) is shown by thick solid lines, whereas the unstable cluster size (branch $n_2(c)$ and a horizontal line) is shown by dot-dashed lines. Arrows indicate the time evolution of $\langle n \rangle$. Parameters: $b = 8.5$, $d = 13/6$, and $\Delta y_{clust} = 0$.

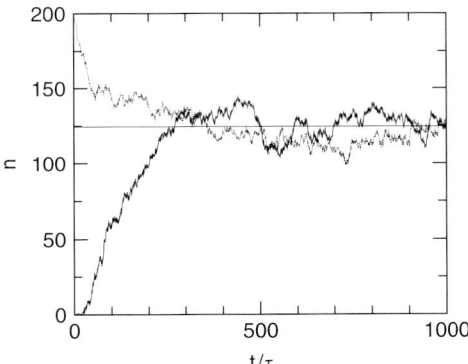

Figure 10.18 Two stochastic trajectories showing the time evolution of the cluster size $n(t)$ starting from a total congestion with $n(0) = N = 200$ (dotted line) and from a free flow with $n(0) = 0$ (solid line). The stationary mean value $\langle n \rangle_{st} = 125$ cars is indicated by a horizontal solid line. Parameters of the system are $L/\ell = 833.3$ ($L = 5000$ m, $\ell = 6$ m), $b = 8.5$, $d = 13/6$, $\Delta y_{clust} = 1/6$, and the stochasticity $p = 0.001$.

Δy_{clust}. The two branches given by (10.78) (with + and −, respectively) are denoted by $n_1(c)$ and $n_2(c)$. Several trajectories showing the time evolution of $\langle n \rangle$ to one of the stable stationary values (thick lines) are indicated by arrows.

The time evolution of the system to the stationary state, consistent with the master equation (10.60), is illustrated in Figure 10.18 by two typical stochastic trajectories. The stationary probability distribution (compare (3.24))

$$P(n) = \lim_{T \to \infty} P(n, T) \tag{10.79}$$

representing the long-time behavior of the master equation (10.60) can be found easily. This is exactly the general stationary solution for one-step processes in closed finite systems given by solution (3.46). In Figure 10.19 we have shown the stationary solution $P(n)$ depending on the total number of cars on a road of given length. The maximum of the probability distribution corresponds to the stable cluster size of congested cars.

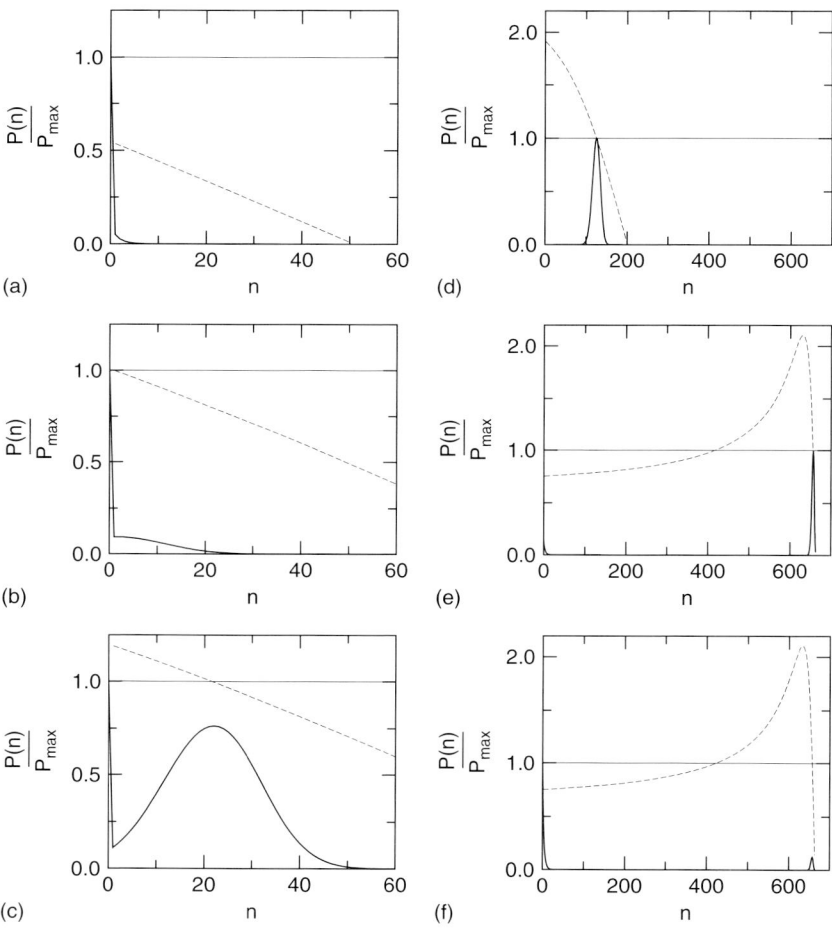

Figure 10.19 Series of different stationary probability distributions $P(n)$ (solid lines) and ratios of transition rates $w_+(n)/w_-(n)$ (dashed lines) showing the formation of a jam of size n depending on the total number of cars N on the road. The values of N and P_{max} (maximum of $P(n)$) are (a) $N = 50$, $P_{max} = 0.905$; (b) $N = 92$, $P_{max} = 0.451$; (c) $N = 110$, $P_{max} = 0.050$; (d) $N = 200$, $P_{max} = 0.042$; (e) $N = 663$, $P_{max} = 0.125$; (f) $N = 664$, $P_{max} = 0.220$. The parameters of the system are $L/\ell = 833.3$ ($L = 5000$ m, $\ell = 6$ m), $b = 8.5$, $d = 13/6$, $\Delta y_{clust} = 1/6$, and $p = 0.001$. The maximum of the probability distribution corresponds to the stable cluster size of the congested cars.

One of the most important characteristics of traffic flow is the *fundamental diagram* showing the flux J of vehicles on the road as a function of the total car density $\rho = N/L$ (or dimensionless total density $c = \ell\rho$). We define J as a local flux $\rho(x,t)v(x,t)$ averaged over an infinite time interval, where $\rho(x,t)$ is the local density and $v(x,t)$ is the local velocity of cars at a time t and space coordinate x, so

$$J = \frac{\Delta \mathcal{N}}{\Delta t} = \lim_{t \to \infty} \frac{1}{t} \int_0^t \rho(x,t')v(x,t')\,dt'. \tag{10.80}$$

In dimensionless variables this equation reduces to

$$j = \frac{\Delta \mathcal{N}}{\Delta T} = \tau \frac{\Delta \mathcal{N}}{\Delta t} = b \lim_{T \to \infty} \frac{1}{T} \int_0^T c(x, T')u(x, T')\,dT', \tag{10.81}$$

where $j = J\tau$ is the dimensionless flux, $u(x,T)$ is the velocity normalized to $v_{\max}$, and $b = v_{\max}\tau/\ell$ is the, already defined, dimensionless parameter (10.65).

In our model the local velocity and the density of cars are defined by the cluster size n and the distance $x - x'$ between the considered local coordinate x and the coordinate x' of the first car in the jam. Thus, we have $\varrho(x,t) = \varrho(x - x'(t), n(t))$ and $v(x,t) = v(x - x'(t), n(t))$, and after averaging over time we get

$$J = \sum_n \int dx'\, P(n, x')\, \varrho(x - x', n) v(x - x', n), \tag{10.82}$$

where $P(n, x')dx'$ denotes the part of the total time during which the size of the cluster is n and the coordinate of the first car of the jam is between x' and $x' + dx'$. The cluster can be found with equal probability at any coordinate x' along the circular road if an average over an infinite time interval t is considered. Thus we have $P(n, x') = P(n)/L$. According to our assumptions, the velocity of congested cars is $v_{\text{opt}}(\Delta x_{\text{clust}})$ and their density is $\varrho_{\text{clust}} = n/L_{\text{clust}} = c_{\text{clust}}/\ell$ inside the jam of length L_{clust}. Outside the cluster we have $v = v_{\text{opt}}(\Delta x_{\text{free}}(n))$ and $\varrho_{\text{free}} = (N - n)/L_{\text{free}} = c_{\text{free}}/\ell$. By these assumptions the integration (10.82) can be performed easily, and this yields for the dimensionless flux

$$j = b \sum_n P(n) \left\{ \frac{L_{\text{clust}}}{L} w_{\text{opt}}(\Delta y_{\text{clust}}) c_{\text{clust}} + \frac{L_{\text{free}}}{L} w_{\text{opt}}\left(\Delta y_{\text{free}}(n)\right) c_{\text{free}} \right\}. \tag{10.83}$$

After the substitution of (10.83) with the definitions of the densities c_{clust} (10.66) and c_{free} (10.67) we easily obtain

$$j = b \sum_n P(n) \left\{ w_{\text{opt}}(\Delta y_{\text{clust}}) \frac{n\ell}{L} + w_{\text{opt}}\left(\Delta y_{\text{free}}(n)\right)\left(c - \frac{n\ell}{L}\right) \right\}. \tag{10.84}$$

This equation is suitable for calculation of the flux-density fundamental diagram according to the known stationary probability distribution $P(n)$.

Now we consider the behavior of the system in the (thermodynamic) limit $N \to \infty$ under the condition

$$\sigma = (Rd)^2 + 4R\Delta y_{\text{clust}} - 4 > 0 \tag{10.85}$$

where $R = b/(d^2 + (\Delta y_{\text{clust}})^2)$. This is the condition at which the equation $w_+(n)/w_-(n) = 1$ has real physical solution(s) and a cluster with $n/N \neq 0$ emerges, that is, a phase transition takes place at some value of the car density c. In the opposite case there is no phase transition (cluster formation) at all. In the special situation $\Delta y_{\text{clust}} = 0$, condition (10.85) reduces to $b > 2d$. The analysis of solution (3.46) shows that $P(z) = N^{-1}\delta(z - z_0)$ holds in the thermodynamic limit $N \to \infty$, where z is defined as $z = n/N$ with the value z_0 corresponding to the absolute maximum of $P(z)$. $z_0 = 0$ holds, if $c \leq c_1$ or $c > c_2$. If $c_1 \leq c < c_2$, then $z_0 = z'_0$ where z'_0 is defined by $z'_0 = (1 + \Delta y_{\text{free}} - 1/c)/(\Delta y_{\text{free}} - \Delta y_{\text{clust}})$ and Δy_{free} has the constant value $(d/2)(Rd + \sqrt{\sigma})$, as follows from the equation $w_+(n)/w_-(n) = 1$. The critical densities c_1 and c_2 are defined by $z'_0 = 0$ or $c_1 = 1/(1 + (d/2)(Rd + \sqrt{\sigma}))$, and $\ln(P(z = 0)) = \ln(P(z = z'_0))$, respectively. Their physical meaning is the following. At $c = c_1$ the free traffic flow becomes unstable and a large cluster of cars emerges spontaneously in the thermodynamic limit $N \to \infty$. In a finite system, as seen in Figure 10.19, this situation corresponds to N about 92 (see (b)). At $c = c_2$ the probability distribution $P(n)$ has two competing maxima (with the same value of $\ln P(n)$ at $N \to \infty$). In Figure 10.19 this corresponds approximately to $N = 663$ and $N = 664$.

In the following an equation will be derived (under the assumption $\Delta y_{\text{clust}} = 0$) from which both critical densities c_1 and c_2 can be determined. Taking account of (3.44) the condition $\ln(P(z = 0)) = \ln(P(z = z'_0))$ in the thermodynamic limit reduces to

$$\int_0^{z'_0} \ln[Q(z)] \, dz = 0, \tag{10.86}$$

where $Q(z) = w_+(n = zN)/w_-(n = zN)$. This equation is satisfied both at $c = c_2$ and $c = c_1$ because in the latter case we have $z'_0 = 0$. Using partial integration (accounting for $Q(z'_0) = 1$) and changing the integration variable to $h = \Delta y_{\text{free}}$ (defined by (10.68)), we get

$$\int_{sd}^{Bd} \left(\frac{1}{h} - \frac{2h}{d^2 + h^2}\right)\left(1 - \frac{sd}{h}\right) dh = 0, \tag{10.87}$$

where $s = (1 - c)/(cd)$ and $B = b/(2d) + \sqrt{b^2/(4d^2) - 1}$. This integral can be calculated analytically, and this yields

$$\ln\left[\frac{B(1 + s^2)}{s(1 + B^2)}\right] + \frac{s}{B} - 1 + 2s(\arctan B - \arctan s) = 0. \tag{10.88}$$

One of the solutions is obviously $s_1 = B$, corresponding to the first critical value $c_1 = 1/(1 + Bd)$. A complete analytical solution is possible in some asymptotic cases. At $B = 1 + \epsilon$ where $\epsilon \to 0$ the solution can be found in the form $s = 1 + \delta$ where $\delta \to 0$. Neglecting terms of fourth and higher orders we get $(\delta - \epsilon)^2(\delta + 2\epsilon) = 0$. Thus, we have $s_1 = 1 + \epsilon = B$ and $s_2 \simeq 1 - 2\epsilon$. At the critical point $\epsilon = 0$ or $b = 2d$ (at $b > 2d$ the cluster emerges) we get $s_1 = s_2 = 1$ or $c_1 = c_2 = c_{\text{crit}}$ where $c_{\text{crit}} = 1/(d + 1)$ is the critical value of c. Another asymptotic case is $B \to \infty$ where we have a solution with $s_2 \to 0$. Retaining only the main terms in the equation we get $\ln(s_2 B) + 1 = 0$ or $s_2 = 1/(eB)$.

It should be noted that cases with $c \leq c_{\text{clust}}$ have a physical meaning only, because the total density c cannot exceed the density of cars in the cluster $c_{\text{clust}} = 1/(1 + \Delta y_{\text{clust}})$. In general (with $\Delta y_{\text{clust}} > 0$), a situation is possible where the equation for c_2 has no solution at $c_2 < c_{\text{clust}}$. In this case the following flux equations for an infinite system are correct, formally setting $c_2 = c_{\text{clust}}$. Thus, taking into account the above discussed solution for $P(n)$, we get the following flux-density relation

$$j(c) = \begin{cases} \dfrac{bc(1-c)^2}{(cd)^2 + (1-c)^2} & : c \in [0; c_1] \cup]c_2; c_{\text{clust}}] \\ 1 - c + c(bw_{\text{opt}}(\Delta y_{\text{clust}}) - \Delta y_{\text{clust}}) : c \in [c_1; c_2] \end{cases} \quad (10.89)$$

These equations represent an exact analytical solution for the fundamental diagram of traffic flow in the framework of our relatively simple model, calculated in the thermodynamic limit. Since the fundamental diagram represents one of the most important characteristics of traffic flow, this result has a fundamental significance and has to be compared with vehicular experiments. As can be seen from these equations, the fundamental diagram consists of fragments of a nonlinear curve and of a straight line. The nonlinear curve represented by the first formula of (10.89) corresponds to homogeneous flow, whereas the straight line corresponds to nonhomogeneous (or congested) flow. The fundamental diagram calculated for the special case $\Delta y_{\text{clust}} = 0$ is shown in Figure 10.20. In this diagram

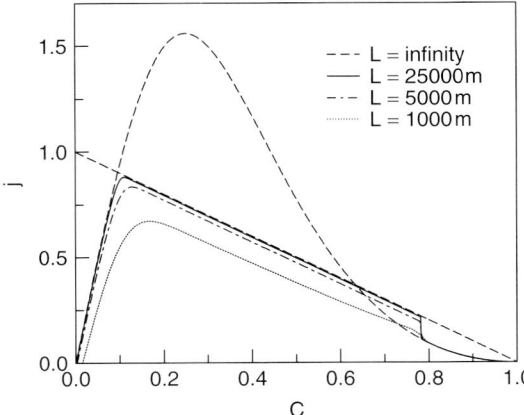

Figure 10.20 Based on the stationary solution of the stochastic master equation the fundamental diagram (dimensionless flow rate (flux) j vs. dimensionless car density c) is calculated. The dimensionless control parameters are $b = 10$, $d = 7/3$, and $\Delta y_{\text{clust}} = 0$. The length of road L varies, the effective length of a car being fixed, $\ell = 6$ m. For finite roads ($L < \infty$) and for infinitely long roads ($L \to \infty$) the flow j can be divided into homogeneous regimes (left: free flow as a gaseous phase, right: heavy traffic as a liquid phase) and a transition regime with free and congested vehicles (formation of a car cluster).

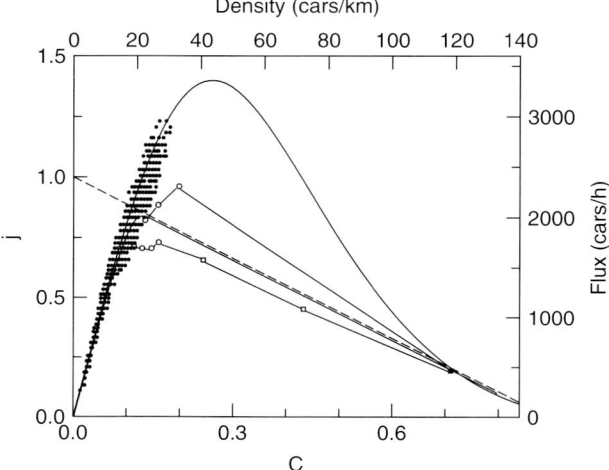

Figure 10.21 Comparison of the fundamental diagram of traffic flow calculated at fitted parameter values $\ell = 6$ m, $v_{max} = 34$ m/s, $\tau = 1.5$ s, $D = 13$ m, and $\Delta x_{clust} = 1$ m ($b = 8.5$, $d = 13/6$, and $\Delta y_{clust} = 1/6$) with experimental data of Ref. 101 (denoted by separate points and thin solid line connecting measured points). The thick solid line shows the solution for the finite road of length $L = 5000$ m, the theoretical curves for infinite system represented by the first and the second formula in Eq. (10.89) are shown by smooth thin solid line and dashed line, respectively.

three different regimes of the traffic flow can be distinguished: free flow as a gaseous phase at small densities ($c < c_1$), heavy traffic as a liquid phase at large densities ($c > c_2$), and a transition regime with free and congested vehicles at intermediate densities ($c_1 < c < c_2$).

In Figure 10.21 we show a comparison of our fundamental diagram to experimental traffic data [101]. In this case we have used realistic values of the control parameters found from experimental measurements.

10.4
Metastability Near Phase Transitions in Traffic Flow

In the previous section we have focused our attention on the stationary characteristics of traffic flow, such as the stationary probability distribution $p(n)$ over cluster (jam) sizes n and the fundamental diagram averaged over an infinite time interval. Here we discuss the time behavior of the probability distribution function $\mathbf{P}(t)$, introduced in Section 3.2, for our one-lane traffic model with the transition frequencies (10.62)–(10.64). We are particularly interested in the relaxation dynamics near the critical densities c_1 and c_2. By analogy with physical systems of many interacting particles, the relaxation behavior near criticality can be very slow (the critical slowing down), so that the long-time behavior of $\mathbf{P}(t)$ is important.

10.4 Metastability Near Phase Transitions in Traffic Flow

The general time-dependent solution of the master equation is given by (3.37). We need merely the eigenvalues λ_i for our specific case. In fact, the problem reduces to finding the eigenvalues of the transition matrix $\mathbf{W}$ (cf. (3.26) and (3.27)) which in our case is a three-diagonal matrix. Below we will describe the method of solution of this problem.

Consider the determinant Det(m) of a three-diagonal matrix of size m comprised of elements $a_{i,j}$. It is assumed that $a_{i,j}$ are given constants independent of the matrix size m. It follows from the structure of such a matrix that

$$\text{Det}(m) = a_{m,m}\text{Det}(m-1) - a_{m-1,m}a_{m,m-1}\text{Det}(m-2). \tag{10.90}$$

The initial condition for this recurrence relation is Det(0) = 1 and Det(1) = $a_{1,1}$.

The actual problem is to calculate determinant of a matrix of size $N+1$ (N is the total number of cars, $0 \leq n < N$) with diagonal elements $a_{i,i} = -w_+(i-1) - w_-(i-1) - \lambda$ (where $w_-(0) = w_+(N) = 0$, $1 \leq i \leq N+1$) and nondiagonal elements $a_{i-1,i} = w_-(i-1)$ and $a_{i,i-1} = w_+(i-2)$ (where $2 \leq i \leq N+1$), and to find the values of λ (eigenvalues) at which Det($N+1$) = 0. It follows from the specific properties of the matrix $\mathbf{W}$, discussed in Section 3.2, that one eigenvalue is $\lambda_0 = 0$ and all other eigenvalues $0 > \lambda_1 > \lambda_2 > \ldots > \lambda_N$ are negative. They can be calculated from the equation of Nth order

$$f(\lambda) = \sum_{n=0}^{N} B_n^{(N)} \lambda^n = 0, \tag{10.91}$$

where coefficients $B_n^{(N)}$ are found based on (10.90). Det(m) can be represented as

$$\text{Det}(m) = \sum_{n=0}^{m} A_n^{(m)} \lambda^n, \tag{10.92}$$

where coefficients $A_n^{(m)}$ with $0 \leq n \leq m$ satisfy the recurrence relation

$$A_n^{(m)} = -[w_+(m-1) + w_-(m-1)]A_n^{(m-1)} - A_{n-1}^{(m-1)}$$
$$- w_+(m-2)w_-(m-1)A_n^{(m-2)}. \tag{10.93}$$

We can subsequently calculate all the coefficients $A_n^{(m)}$ from (10.93) starting with $m = 1$ and finishing with $m = N+1$. In this case we formally set $A_0^{(0)} = 1$ and $A_n^{(m)} = 0$ if $n > m$. $A_0^{(N+1)} = 0$ holds since one eigenvalue is zero. According to this $B_n^{(N)} \equiv A_{n+1}^{(N+1)}$. All N roots of (10.91) are real and negative. This means that function $f(\lambda)$ has exactly $N-1$ extremum points (solutions of equation $df/d\lambda = 0$ of the $(N-1)$th order) located between zeros of this function. Thus, $f(\lambda)$ is monotonous within $\lambda_1 < \lambda < 0$ which allows us to find the eigenvalue λ_1 by solving (10.91) numerically (e.g, by the Newton linearization method) with the initial approximation $\lambda = 0$. After this we rewrite (10.91) in the form

$$(\lambda - \lambda_1) \sum_{n=0}^{N-1} B_n^{(N-1)} \lambda^n = 0 \tag{10.94}$$

based on the recurrence relation

$$B_{n-1}^{(N-1)} = B_n^{(N)} + \lambda_1 B_n^{(N-1)}, \tag{10.95}$$

where $B_N^{(N-1)} = 0$ and n subsequently takes the values $N, N-1, \ldots, 1$. Then we find the eigenvalue λ_2 from the equation

$$\sum_{n=0}^{N-1} B_n^{(N-1)} \lambda^n = 0 \tag{10.96}$$

with the initial approximation $\lambda = \lambda_1$. This procedure can be continued to find all the eigenvalues by subsequent reduction of the equation order.

In exact arithmetics this method allows us to find all eigenvalues for arbitrarily large matrix. However, due to numerical inaccuracy, in practice this can be done only for quite a small matrix (e.g. $N \leq 20$). Nevertheless, the eigenvalues which are the smallest in magnitude can be calculated for large enough N, which allows us to investigate the long-time behavior of the system. In particular, it follows from (3.31) that the probability distribution $\mathbf{P}(t)$ tends to the equilibrium distribution $\mathbf{P}^{\mathrm{eq}}$ as

$$\mathbf{P}(t) - \mathbf{P}^{\mathrm{eq}} \simeq c_1 \mathbf{u}_1 e^{\lambda_1 t} \tag{10.97}$$

at $t \to \infty$, where λ_1 is the eigenvalue closest to zero and $\mathbf{u}$ is the corresponding eigenvector of the transition matrix $\mathbf{W}$. In other words, $-1/\lambda_1$ is the relaxation time.

Results of the calculation for λ_1 (solid lines) and also for λ_2 (dotted lines) depending on the density of cars $c = N\ell/L$ for different lengths of the road L are shown in Figure 10.22. We have used the same set of control parameters:

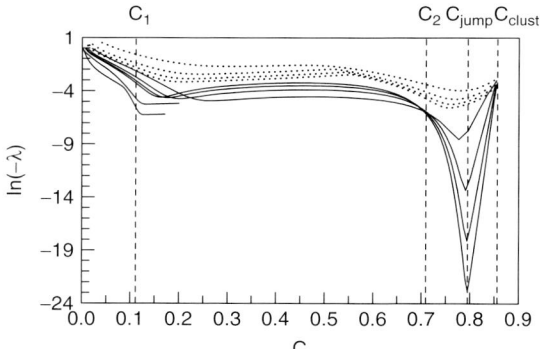

Figure 10.22 Representation of the eigenvalues as $\ln(-\lambda_1)$ (solid lines) and $\ln(-\lambda_2)$ (dotted lines) vs density of cars c at different sizes of the system L/ℓ. From the top to the bottom (if looking at $c = c_1$ and also at $c = c_{\mathrm{jump}}$) the sizes are $L/\ell = 51.46, 101.08, 151.03, 201.09, 1500,$ and 4000. The first four sizes are chosen such that the minima of solid curves at $c \approx c_{\mathrm{jump}}$ correspond to $N = 40, 80, 120,$ and 160. Parameters of the system are $b = 8.5$, $d = 13/6$, $\Delta y_{\mathrm{clust}} = 1/6$, $p = 0.001$. The time is assumed to be dimensionless, so $w_-(n) = 1$.

$b = 8.5$, $d = 13/6$, $\Delta y_{\text{clust}} = 1/6$, and $p = 0.001$, as in our previous calculation in Figure 10.19. It is evident from Figure 10.22 that $-\lambda_1$ has a sharp minimum at $c \approx c_{\text{jump}}$, and the density at the minimum tends to c_{jump} with increasing L. In this case we have denoted by c_{jump} the second critical value c_2 of the density in (10.89) which corresponds to a jump in the maximum of the stationary probability distribution $p(n) \equiv p^{\text{eq}}(n)$ (at N between 663 and 664 in Figure 10.19) as well as in the flux j (Figure 10.20). As distinct from (10.89), here we denote by c_2 the minimal density at which the stationary probability distribution $p(n)$ has two maxima. In general, two maxima exist at $c_2 < c < c_{\text{clust}}$ where c_{clust} is the car density inside a cluster. The value of c_{jump} about 0.796 corresponds to the jump-like first-order phase transition, as is evident from Figure 10.19. The value of $-\lambda_1$ in Figure 10.22 decreases exponentially with increasing system size (linearly in the logarithmic scale used in Figure 10.22) at densities $c_2 < c < c_{\text{clust}}$. This result has a simple physical interpretation. At these densities the equilibrium probability distribution $p^{\text{eq}}(n)$ has two maxima, one of which corresponds to the stable, and another to the metastable state of the system. So, $-1/\lambda_1$ represents the time in which the system finds the stable state if one starts from the metastable state. At $c = c_{\text{jump}}$ both states have the same probability, therefore the relaxation is the slowest. In general, the increase in the relaxation time is exponential because of the necessity to overcome a 'potential barrier' when switching from one state to another. The 'height of the potential barrier' is proportional to the number of unstable intermediate states (proportional to N) which need to be overcome.

It is easy to construct the analytical asymptotic solution at $t \to \infty$ in large system size limit $L \to \infty$ in the case when the time-dependent probability distribution $p(n, t)$ meets the initial condition $p(n, 0) \approx p^{\text{me}}(n)$, where $p^{\text{me}}(n)$ is the probability distribution of the quasi-equilibrium or metastable state. The system exhibits relaxation to the metastable state within a time interval of about $-1/\lambda_2$. This time interval of does not diverge exponentially at $L \to \infty$, as can be seen from Figure 10.22 (dotted lines). Thus, for $-1/\lambda_2 \ll t \ll -1/\lambda_1$ we have $p(n, t) \simeq p^{\text{me}}(n)$, which means that the term $c_1 \mathbf{u}_1$ in (10.97) is (approximately) $\mathbf{P}^{\text{me}} - \mathbf{P}^{\text{eq}}$. In this way, we obtain

$$p(n, t) \simeq p^{\text{eq}}(n) + \left(p^{\text{me}}(n) - p^{\text{eq}}(n)\right) e^{\lambda_1 t} \quad \text{at } t \gg -1/\lambda_2. \tag{10.98}$$

This is an asymptotically exact equation in the limit $L \to \infty$ and $t\lambda_2 \to \infty$ for the values of n around both maxima of $p^{\text{eq}}(n)$ (except the irrelevant intermediate states where $p^{\text{me}}(n)$ is not precisely defined but, in any case, is vanishingly small). In this case $p^{\text{me}}(n)$ can be calculated in the same way as $p^{\text{eq}}(n)$, but neglecting those states which practically cannot be reached at $t\lambda_2 \to \infty$ and $t\lambda_1 \to 0$. The correct result is always ensured by taking into account the states either with $n < n_{\text{unst}}$ (at $c_2 < c < c_{\text{jump}}$) or with $n > n_{\text{unst}}$ (at $c_{\text{jump}} < c < c_{\text{clust}}$) to include the maximum of $p(n, 0)$. Here n_{unst} is the unstable cluster size corresponding to the minimum of $p^{\text{eq}}(n)$.

It is also interesting to investigate the relaxation behavior of the system near the first phase transition point $c = c_1$ where the spontaneous formation of a large car cluster starts when increasing the density. Some singularity of the relaxation time is expected at $c = c_1$ in the limit $L \to \infty$. Our calculations at the larger system sizes $L/\ell = 1500$ and $L/\ell = 4000$ indicate the existence of a breakpoint in $\ln(-\lambda_1)$ vs c plot at $c = c_1$ in the limit $L \to \infty$. In this case $-\lambda_1$ has an almost constant value at densities somewhat above c_1. This value behaves as $\sim 1/L$ (within the estimation accuracy allowed by our calculations at finite L), which means that the relaxation time increases proportionally with the system size L. The fact that $-\lambda_1 \sim 1/L$ can be understood as follows. The relaxation time at the densities actually considered is proportional to the time of stable (stationary) cluster formation, but the latter is proportional to L at a given c. Both the mean velocity of cluster growth and the stable cluster size are proportional to $c - c_1$; therefore, the mean time of the cluster formation is practically independent of $c - c_1$, which explains the fact that $-\lambda_1$ as a function of c is almost constant.

10.5
Car Cluster Formation as First-Order Phase Transition

Here we propose an essential innovation in our traffic flow model: now the detachment frequency depends on the cluster size. By analogy with physical systems like droplets in supersaturated vapor, one can expect that smaller clusters dissolve easier, so $w_-(n)$ increases considerably when the cluster size becomes smaller than some characteristic value n_0. In traffic flow, this idea is based on the concept of higher irregularity of the cluster structure when the size is small. The increase in $w_-(n)$ might also be related to some multi-lane effect; due to several possible overtaking maneuvers, clusters consisting of a few cars are particularly unstable. In some approximation our one-lane model effectively describes a multi-lane freeway. In this case n and N are the number of congested cars and the total number of cars per lane. The above mentioned multi-lane effect is simulated by choosing an appropriate ansatz for $w_-(n)$. Our specific choice is to replace (10.62) by

$$w_-(n) = \frac{1}{\tau}\left[1 + \beta\left(\frac{n_0}{n + n_0}\right)^s\right] : \quad n \geq 1, \tag{10.99}$$

where β and s are positive constants. Assuming $n_0 \gg 1$, the parameter $\beta \approx (w_-(1) - w_-(\infty))/w_-(\infty)$ shows the relative increase in $w_-(n)$ when the cluster size n is reduced from large values to 1. The parameter s is responsible for the speed of $w_-(n)$ converging to its asymptotic value $w_-(\infty) = 1/\tau$ at $n \to \infty$. The specific form of $w_-(n)$ is chosen somewhat arbitrarily to get a simple, but still realistic, description.

The advanced traffic flow model, introduced in this section, exhibits similar features to those observed in supersaturated vapor as discussed in Chapter 9. The most essential distinguishing feature of the new model is the existence of metastability near the phase transition from the free flow to cluster phase. The latter, therefore, can be interpreted as a jump-like, that is first-order phase transition where the system goes over from one state to another by overcoming a potential barrier.

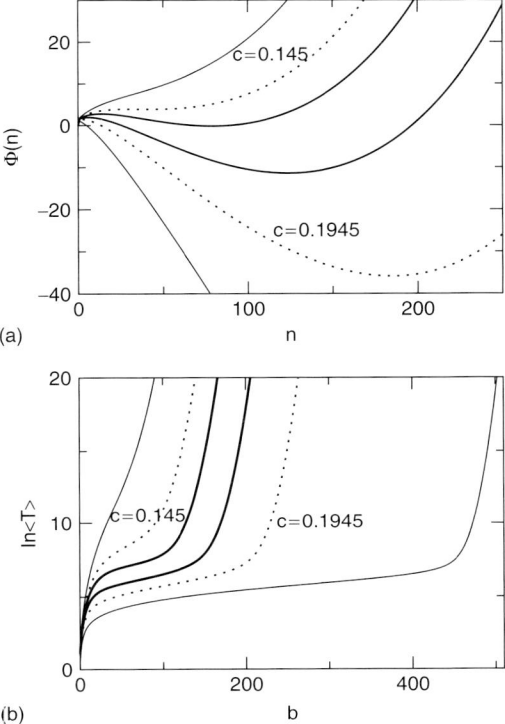

Figure 10.23 The potential of the circular one-lane-road model depending on the car cluster size n (a), and the logarithm of the mean first-passage time from the free-flow state $n = 0$ to the congested state with cluster of a certain size $n = b + 1$ (b). The control parameters of the model are $b = 8.5$, $d = 13/6$, $\Delta y_{\text{clust}} = 1/6$, $\beta = 0.8$, $s = 1$, and $n_0 = 10$ at the total dimensionless length of the road $L/\ell = 2000$. Curves from the top to the bottom correspond to increasing densities $c = 0.13, 0.145, 0.154, 0.17, 0.1945$, and 0.3. The dotted curves at $c = 0.145$ and $c = 0.1945$ indicate the range of bistability with the double-well potential shown in the upper picture by thick solid lines.

The potential (3.48) with the specific attachment frequencies (10.63), (10.64) and the detachment frequencies (10.99) of our advanced model is shown in Figure 10.23 (a) for different total densities of cars at realistic values of the dimensionless control parameters $b = 8.5$, $d = 13/6$, and $\Delta y_{\text{clust}} = 1/6$ estimated in [147]. Other parameters have been chosen: $\beta = 0.8$, $L/\ell = 2000$, $s = 1$, and $n_0 = 10$. In a certain range of concentrations from $c = 0.145$ to $c = 0.1945$, indicated by dotted lines, the potential has two minima (one at $n = 0$, another at $n > 1$) separated by a smooth maximum. This is the so-called double-well potential. The stable state of a large system corresponds to the absolute minimum of the potential. However, the system needs considerable time to go over from one minimum to another by climbing over the potential barrier. Thus, the switching from a free-flow state with $n = 0$ to the congested state occurs due to rear fluctuations of large amplitude, which means that the system exhibits metastability and first-order phase transition.

This behavior is completely analogous to nucleation in supersaturated vapor (see Chapter 9), where the growth of liquid droplets of overcritical size occurs, the critical size being reached due to stochastic fluctuations caused by overcoming the free energy barrier.

An essential parameter characterizing the switching process is the mean first-passage time $\langle T \rangle$ (3.74) from the free-flow state with $n = 0$ to the jammed states beyond the maximum of potential $\Phi(n)$. Its logarithm $\ln\langle T \rangle$ depending on the cluster size $b + 1$ which has to be reached for the first time, is shown in Figure 10.23(b) at the same overall car densities c as the potential depicted in Figure 10.23(a) [99]. The time is dimensionless, measured in units of τ. The thick solid lines at $c = 0.154$ (upper line) and $c = 0.17$ (lower line) correspond to the bistable situation with the double-well potential. Assuming $\tau = 1.5$ s, in agreement with the estimation in [147], we can evaluate from Figure 10.23, the mean time during which the system overcomes the potential barrier starting from the free flow state $n = 0$. In our example the maximum of the potential is located at $n = 14$ at the density $c = 0.154$ and $n = 7$ at the density $c = 0.17$. The corresponding mean first-passage times are about 6 min and 80 s, respectively. However, the system still needs some time to reach the stable cluster size $n = 79$ at $c = 0.154$ and $n = 124$ at $c = 0.17$ which coincides with the minimum of the potential. In our example, the corresponding mean times are 40 min and 25 min. The local maximum of the potential corresponds to the unstable or critical cluster size n_crit, since smaller clusters with $n < n_\text{crit}$ tend, on average, to shrink, while larger clusters with $n > n_\text{crit}$ tend to grow up to the stable cluster size n_stable. The mean first-passage time increases relatively slightly within $b \in [n_\text{crit}; n_\text{stable}]$. This corresponds to the middle part of the thick solid lines in Figure 10.23(b). In principle, the car cluster can also somewhat exceed the stable size due to stochastic fluctuations. However, the mean first-passage time increases dramatically in this case. It corresponds to the sharply increasing r. h. s. part of the curves in Figure 10.23(b). The sharp increase in the mean first-passage time is observed for all concentrations shown in Figure 10.23 at large enough values of b, which can be reached by moving against the driving force $-\partial\Phi(n)/\partial n$ of the increasing potential. At small densities (the upper curve at $c = 0.13$) the formation of a remarkably large car cluster is highly improbable, therefore the mean first-passage time increases sharply starting from $b = 0$. At intermediate densities, represented by the lower curve at $c = 0.3$, the formation of a large cluster proceeds without overcoming a potential barrier. There is only a relatively small delay at the beginning of the process, where the driving force $-\partial\Phi(n)/\partial n$ is smaller.

Note that the upper cut-off of the $\ln\langle T \rangle$ vs b curves corresponds to the time $4.85 \times 10^8 \tau$ which is about 23 years. This means that the states with cluster sizes which are considerably larger than the stable stationary cluster size (which is zero at small densities and nonzero at larger densities) are practically never reached.

One must mention that the bistability exists also at large densities of cars, that is, at $c > 0.5$ in our example. In this case the system can switch from a dense homogeneous state to a heterogeneous cluster state by overcoming a potential barrier which is much higher than in the actual range of concentrations $c \in [0.145; 0.1945]$ indicated in Figure 10.23.

In our traffic-flow model the inverse time constant $1/\tau$ plays the role of temperature. The detachment frequency $w_-(n)$ (10.99) diverges as $1/\tau$ at $\tau \to 0$, whereas the attachment frequency (10.64) remains finite in this case. This means that, for all vehicle densities, the car cluster tends to dissolve at $\tau \to 0$ or at high temperatures. In other words, the formation of a stable car cluster as well as bistability is possible only at undercritical values of the temperature $1/\tau$. In the simplest case of $\beta = 0$ the critical temperature corresponds to a vanishing parameter σ in (10.85), as consistent with the fact that the balance condition $w_+(n)/w_-(n) = 1$ for the average number of attached and detached cars can be fulfilled only at $\sigma > 0$.

To show further analogy with a supersaturated vapor, we consider the mean cluster size as described by the deterministic equation (10.70). The stationary equation in our advanced model becomes

$$b \frac{w_{\mathrm{opt}}(\Delta y_{\mathrm{free}}(\langle n \rangle)) - w_{\mathrm{opt}}(\Delta y_{\mathrm{clust}})}{\Delta y_{\mathrm{free}}(\langle n \rangle) - \Delta y_{\mathrm{clust}}} = 1 + \beta \left(\frac{n_0}{n + n_0} \right)^s \qquad (10.100)$$

consistent with the ansatz for transition probabilities (10.64) and (10.99). The stationary mean cluster sizes $\langle n \rangle_{\mathrm{st}}$ correspond, in general, to different stationary solutions of these equations. Also the stable free-flow state with $\langle n \rangle_{\mathrm{st}} = 0$ has to be identified as a state with $d \langle n \rangle / dT < 0$ at $\langle n \rangle \to 0$. The corresponding dimensionless fluxes

$$j = J\tau = b \left[w_{\mathrm{opt}}(\Delta y_{\mathrm{clust}}) \frac{\ell \langle n \rangle_{\mathrm{st}}}{L} + w_{\mathrm{opt}}(\Delta y_{\mathrm{free}}(\langle n \rangle)) \left(c - \frac{\ell \langle n \rangle_{\mathrm{st}}}{L} \right) \right] \qquad (10.101)$$

can be calculated as well, where J is the flux averaged over the whole road or, which is the same, over an infinitely long time interval at a fixed coordinate. In this case we have neglected the stochastic fluctuations taking $n \equiv \langle n \rangle_{\mathrm{st}}$.

The equation for the stationary cluster size on the road, (10.100), has been solved together with (10.61) and (10.68) in the approximation $S(n) \approx n$,

$$\Delta y_{\mathrm{free}}(n) = \frac{(L/\ell)(1-c) - n \Delta y_{\mathrm{clust}}}{c(L/\ell) - n}, \qquad (10.102)$$

at the same realistic values of the dimensionless control parameters $b = 8.5$, $d = 13/6$, $\Delta y_{\mathrm{clust}} = 1/6$ as we used in Figure 10.23. Three different values of the parameter β, $\beta = 0$, 0.8, and 1.6, have been considered. Other parameters have been chosen $L/\ell = 500$, $s = 1$, and $n_0 = 10$.

The results are shown in Figures 10.24 and 10.25. In Figure 10.24 we have illustrated the behavior of the stationary cluster size $\langle n \rangle_{\mathrm{st}}$ (a) and the corresponding flux (b) in a special case of parameter $\beta = 0$ where our advanced model reduces to the previous one discussed in Section 10.3. As a result, the stationary cluster size behaves in a similar way as in Figure 10.17, with the only essential difference that $\Delta y_{\mathrm{clust}} > 0$. In this case the free flow with $\langle n \rangle_{\mathrm{st}} = 0$ is stable up to some critical density $c = c_1$. Then a car cluster appears which grows linearly in size with increasing density c from c_1 to the maximum value $c_{\mathrm{clust}} = 1/(1 + \Delta y_{\mathrm{clust}})$ (the density in the cluster), as shown in Figure 10.24 by a thin solid line. This corresponds to one of the stationary solutions $\langle n \rangle = n_1(c)$ of (10.100). At densities

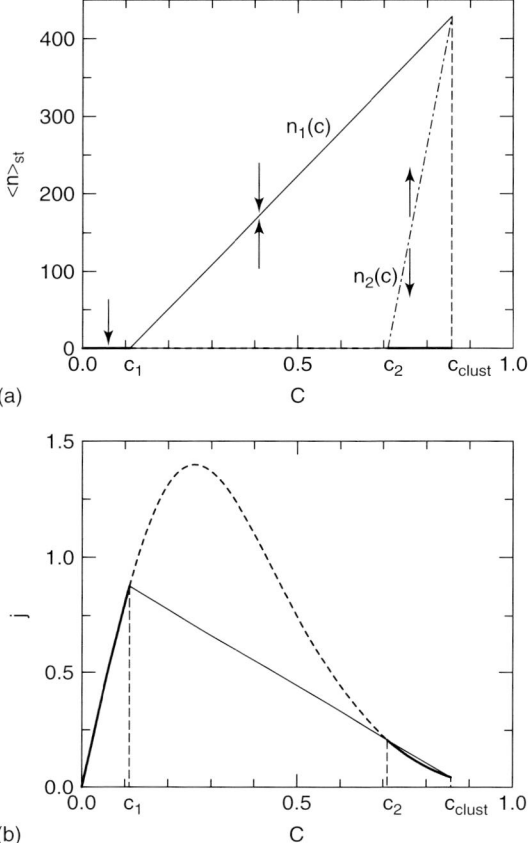

Figure 10.24 Stationary cluster size (a) and flux (b) vs concentration calculated for a finite road at dimensionless control parameters $\beta = 0, b = 8.5, d = 13/6, \Delta\gamma_{clust} = 1/6, L/\ell = 500$. The thick and the thin solid lines correspond to stable free and congested states, respectively. Dashed lines represent an unstable free flow and the dot-dashed line shows the unstable (critical) cluster size.

$c_2 < c < c_{clust}$ another positive solution $\langle n \rangle = n_2(c)$ exists (dot-dashed line) which corresponds to an unstable or critical cluster size. According to (10.70), the cluster tends to grow ($dn/dt > 0$) if $n > n_2(c)$, and it tends to dissolve ($dn/dt < 0$) if $n < n_1(c)$. The most probable time evolution of n, given by the deterministic equation (10.70), depending on the density c and initial conditions, is indicated in Figure 10.24 by arrows. As in the supersaturated vapor, the growth of a cluster starting from undercritical size $n < n_2(c)$ and finishing with the stable cluster size $n = n_1(c)$ is also possible, but only due to stochastic fluctuations.

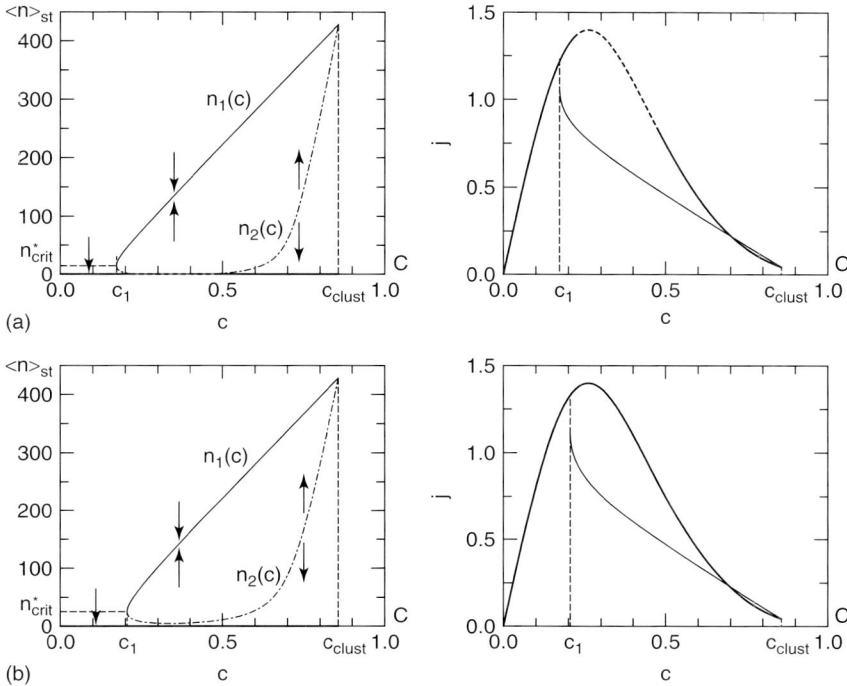

Figure 10.25 Stationary cluster size (left) and flux (right) vs concentration calculated for finite road at $\beta = 0.8$ (a) and $\beta = 1.6$ (b). Other dimensionless control parameters are $n_0 = 10, s = 1, b = 8.5, d = 13/6, \Delta y_{clust} = 1/6, L/\ell = 500$. The meaning of solid, dashed, and dot-dashed lines is the same as in Figure 10.24.

In Figure 10.25 we have shown the same diagrams at $\beta > 0$. A distinguishing feature is that a nonzero critical cluster size already appears at $c = c_-$ where a phase transition takes place with a jump-like increase in the stable cluster size $\langle n \rangle_{st}$ from 0 to n^*_{crit}. This is a first-order phase transition, a behavior observed in physical systems like supersaturated vapor. At small values of β, however, a density region with vanishing critical cluster size still exists ($\beta = 0.8$, Figure 10.25(a)) which disappears at larger β values ($\beta = 1.6$, Figure 10.25(b)). Note also that at $\beta = 0.8$ the free flow becomes unstable somewhat above c_1 (see $j(c)$ plots in Figure 10.25), while both free and congested states of the system are stable at $\beta = 1.6$. The bifurcation diagram at $\beta = 1.6$ is qualitatively similar to that of the supersaturated vapor in Figure 9.4. At positive β, a jump-like decrease in the flux j is observed at the critical point $c = c_1$ when switching from the free flow (thick solid line) to the congested flow (thin solid line). This, however, is a finite-size effect which disappears in the limit $N \to \infty$, since $n^*_{crit}/N \to 0$.

10.6
Thermodynamics of Traffic Flow

An extension of thermodynamic concepts from equilibrium to nonequilibrium or driven systems is one of the fundamental problems in physics. It refers also to so-called nonphysical systems, such as traffic or granular flow, economics, biological systems, etc., where the laws of microscopic interaction and motion differ from those known in physics. Different approaches have been developed previously. In the geometrical formulation of thermodynamics [62], the latter is regarded as a theory arising in the analysis of dynamics. In this concept, equilibrium thermodynamics is represented by a manifold of time-independent equilibrium states, whereas the thermodynamics of a driven system is represented by a manifold of slowly evolving states. A k-component system undergoing chemical reaction is considered as an example in [62]. A more widely discussed approach is based on the introduction of entropy [188, 218] and usage of the entropy maximization principle in various applications, e.g. linear dissipative driven systems [218] and single-lane traffic [188]. An appropriate definition of temperature is a relevant question when we consider a nonphysical system. In [188] the temperature T and pressure p of traffic flow have been introduced via derivatives of certain thermodynamic functions, and it has been found that T is negative at typical velocities. In another approach [118] similarities between traffic and granular flow have been discussed proposing two effective temperatures: one characterizing fast or single-car dynamics, and another – slow or collective dynamics of traffic flow.

As mentioned in [188], entropy need not occupy a position of primacy in a general theory beyond the classical equilibrium thermodynamics. We have found that, cases where the stationary state of a driven system has the property of detailed balance in the space of a suitable stochastic variable, the thermodynamic potential can be easily introduced based on this property in complete analogy with equilibrium systems. This approach can prove to be useful in many applications due to its relative simplicity. As an example we consider the formation of a car cluster in one-lane traffic and show its analogy with the phase separation in supersaturated vapor–liquid system.

The aggregation of particles out of an initially homogeneous situation is well known in physics, as well as in other branches of natural sciences and engineering. The formation of bound states as an aggregation process is related to self-organization phenomena [203, 210, 229]. The formation of car clusters (jams) at overcritical densities in traffic flow is an analogous phenomenon in the sense that cars can be considered as interacting particles [107, 108, 184]. The development of traffic jams in vehicular flow is an everyday example of the occurrence of nucleation and aggregation in a system of many point-like cars. For previous work focusing on the description of jam formation as a nucleation process, see [112, 147, 149, 151]. This is related to phase separation and metastability in low-dimensional driven systems, a topic which has attracted much recent interest [43, 89, 97, 166, 209]. Metastability and hysteresis effects have been observed in real traffic, see, e.g., [31, 77, 100, 133, 239, 240] for a discussion of empirical data and the various different modeling approaches.

Here we focus on the application of thermodynamics to a many-particle system such as traffic flow. In a first step we do not consider real traffic with its very complicated behavior but instead limit our investigations to simple models of a directional one-lane vehicular flow. We hope this will trigger further development in the description of more realistic situations for multi-lane traffic as well as synchronized flow [100]. We have found a certain analogy with physical systems like supersaturated vapor-liquid, although there are also essential differences, since the traffic flow is a driven system. We would like to outline some basic ideas and concepts.

1. On a microscopic level, traffic flow can be described by the optimal velocity model (OVM). In this case the equations of motion can be written as Newton's law with accelerating and decelerating forces and one can define the potential V and the kinetic energy T of the car system, as well as the total energy $E = T + V$. The latter has a thermodynamic interpretation as $\langle E \rangle = U$, where U is the internal energy of the system.

2. Traffic flow is a dissipative system of driven or active particles. It means that the total energy is not conserved, but we have an energy balance equation

$$\frac{dE}{dt} + \Phi = 0$$

with the energy flux Φ following from the equations of motion and consisting of dissipation (due to friction) and energy input (due to the burning of petrol).

3. In the long-time limit the many-car system tends to a certain stationary state. In the microscopic description it is either the fixed-point or the limit cycle in the phase space of velocities and headways depending on the overall car density and control parameters. The stationary state is characterized by a certain internal energy.

4. On a mesoscopic level, traffic flow can be described by a stochastic master equation, where the stochastic variable is the number of congested cars n, that is, the size of car cluster. In this case the fixed-point solution corresponds to $n = 0$, and the limit cycle – to the coexistence of a car cluster with $n > 0$ and the free-flow phase.

5. In the space of the cluster size, the detailed balance holds for the stationary solution just as in equilibrium physical systems. It allows one to describe various properties of the stationary state by equilibrium thermodynamics. In particular, we calculate the free energy of the system and the chemical potentials of coexisting phases in complete analogy with the known treatment for a supersaturated liquid-gas system.

6. As distinct from equilibrium systems, the chemical potential of the cluster phase of traffic flow is not an internal property of this phase, since it depends on an outer parameter – the density of cars in the free-flow phase. It allows one to distinguish between the traffic flow as a driven system and purely equilibrium systems.

Traffic flow can be viewed as a random dynamical system [6, 7] of active or intelligent particles [38, 41, 211]. To describe it on a microscopic level, here we use the optimal velocity (OV) model for point-like cars, moving on a one-lane road with periodic boundary conditions, defined by (10.13)–(10.15). Equation (10.13) can be written as

$$m\frac{dv_i}{dt} = F_{acc}(v_i) + F_{dec}(\Delta x_i), \tag{10.103}$$

where

$$F_{acc}(v_i) = \frac{m}{\tau}(v_{max} - v_i) \geq 0 \tag{10.104}$$

$$F_{dec}(\Delta x_i) = \frac{m}{\tau}(v_{opt}(\Delta x_i) - v_{max}) \leq 0 \tag{10.105}$$

are the accelerating and decelerating forces, respectively. The coordinate-dependent force term is due to the interaction between cars

$$F_{dec}(\Delta x) = v_{max}\frac{m}{\tau}\left(\frac{(\Delta x)^2}{D^2 + (\Delta x)^2} - 1\right) \tag{10.106}$$

and is always negative, starting at $F_{dec}(\Delta x = 0) = -v_{max}m/\tau$, approaching zero at infinite distances. The potential energy of the car system can be defined as $V = \sum_{i=1}^{N} \phi(\Delta x_i)$, where $\phi(\Delta x_i)$ is the interaction potential of the ith car with the car ahead, which is given by

$$F_{dec}(\Delta x_i) = -\frac{\partial \phi(x_{i+1} - x_i)}{\partial x_i} = \frac{d\phi(\Delta x_i)}{d\Delta x_i} \tag{10.107}$$

By integrating this equation we get

$$\phi(\Delta x) = v_{max}\frac{Dm}{\tau}\left[\frac{\pi}{2} - \arctan\left(\frac{\Delta x}{D}\right)\right], \tag{10.108}$$

where the integration constant is chosen such that $\phi(\infty) = 0$. The potential energy depending on the headway is shown in Figure 10.26. Note that $F_{dec}(\Delta x_i)$ in this case is not given by $-\partial V/\partial x_i$, since the latter quantity includes an additional term $-\partial \phi(x_i - x_{i-1})/\partial x_i$. This term is absent in our definition of the force because the car behind does not influence the motion of the actual ith vehicle. It reflects the fact that, unlike in physical systems, Newton's third law does not hold here.

The total time derivative of the potential energy is

$$\frac{dV}{dt} = \sum_{i=1}^{N}\left[\frac{\partial \phi(\Delta x_i)}{\partial x_i}\frac{dx_i}{dt} + \frac{\partial \phi(\Delta x_i)}{\partial x_{i+1}}\frac{dx_{i+1}}{dt}\right]$$

$$= \sum_{i=1}^{N}(v_{i+1} - v_i)F_{dec}(\Delta x_i) \tag{10.109}$$

The total time derivative of the kinetic energy $T = \sum_{i=1}^{N} mv_i^2/2$ is obtained by multiplying both sides of (10.103) by v_i and summing over i. It leads to the following energy balance equation

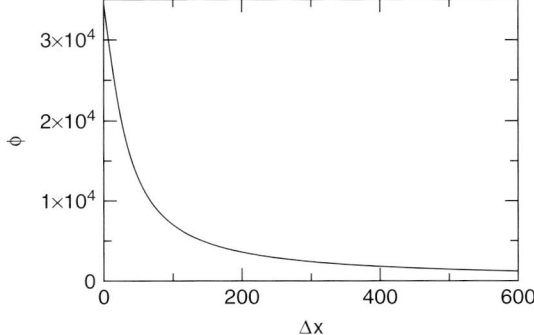

Figure 10.26 Graph of the potential energy function $\phi(\Delta x)$.

$$\frac{dE}{dt} + \Phi = 0 \qquad (10.110)$$

for the total energy $E = T + V$ of the car system, where

$$\Phi = -\sum_{i=1}^{N}\left[v_i F_{\text{acc}}(v_i) + v_{i+1} F_{\text{dec}}(\Delta x_i)\right] \qquad (10.111)$$

is the energy flux. It includes both energy dissipation due to friction and energy input from the engine. Equation (10.110) shows that, as distinct from closed mechanical systems, the total energy is not conserved in traffic flow. Nevertheless, it approaches a constant value in the long-time limit, where the system converges to one of two possible stationary states: either to the fixed point $\Delta x_i = \Delta x_{\text{hom}}$, $v_i = v_{\text{opt}}(\Delta x_{\text{hom}})$ (where $\Delta x_{\text{hom}} = L/N$ is the distance between N homogeneously distributed cars over the road of length L), or to the limit cycle in the phase space of headways and velocities. Both situations are illustrated in Figure 10.27. At a small enough density of cars there is a stable fixed point (solid circle), which lies on the optimal velocity curve (dotted line). An unstable fixed point (empty circle) exists at larger densities. In the latter case any small perturbation of the initially homogeneous fixed point situation leads to the limit cycle (solid line) in the long-time limit.

In the thermodynamic interpretation the mean energy $\langle E \rangle$ is the internal energy U of the system. The latter thus has a certain value in any one of the stationary states. The temporal behavior of E for the same sets of parameters as in Figure 10.27 is shown in Figure 10.28. In the case of the convergence to the limit cycle (solid line) for $\rho = 0.0606\ \text{m}^{-1}$, one can distinguish six plateaus in the energy curve. The first one represents the short-time behavior when starting from an almost homogeneous initial condition with zero velocities, and the second plateau is the unstable fixed-point situation. Further on, four car clusters have been formed in the actual simulation, and this temporal situation is represented by the third, relatively small, plateau. The next three plateaus with three, two and, finally, one car clusters reflect the coarse graining or Ostwald ripening process. The dashed line shows the convergence to the stable fixed point value at $\rho = 0.0303\ \text{m}^{-1}$.

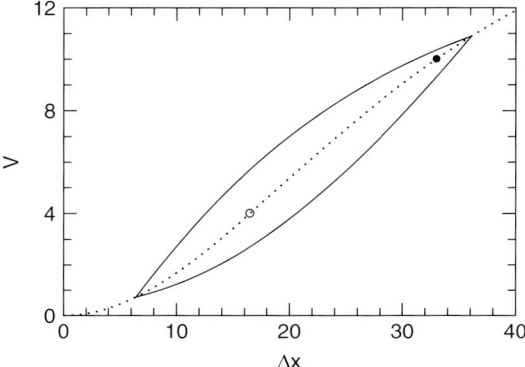

Figure 10.27 Fixed points (circles) and the limit cycle (solid line) in the space of headways Δx and velocities v of cars. The solid circle represents the stable fixed point at the car density $\rho = N/L = 0.0303$ m^{-1}. The empty circle is the unstable fixed point at a larger density $\rho = 0.0606$ m^{-1}, where the long-time trajectory for any car is the limit cycle shown. The fixed points lie on the optimal velocity curve (dotted line) given by (10.15). The parameters are chosen as $N = 60$, $D = 33$ m, $v_{max} = 20$ m s^{-1}, $\tau = 1.5$ s, and $m = 1000$ kg.

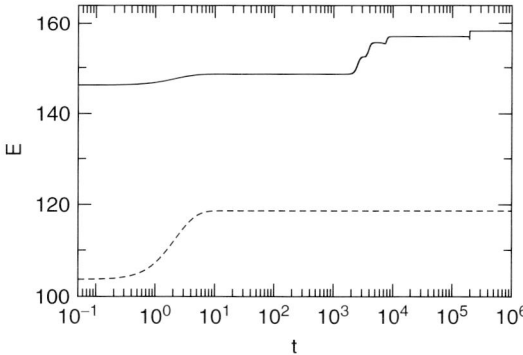

Figure 10.28 The total energy E of the car system, measured in units of $mv_{max}^2/2$, depending on the time t given in seconds. The same sets of parameters have been used as in Figure 10.27. The upper solid line corresponds to a larger density $\rho = 0.0606$ m^{-1} where the limit cycle forms, whereas the lower dashed line corresponds to a smaller density $\rho = 0.0303$ m^{-1} where the convergence to a stable fixed point is observed.

Apart from the internal energy, other thermodynamic functions can be introduced as well. In the following we will calculate the free energy F of the traffic flow. By using the known relation $F = U - T^*S$ we can also calculate the entropy S of traffic flow for a properly defined 'temperature' T^*.

Up to now we have considered purely deterministic equations of motion. Randomness can be included, e.g. by adding a multiplicative noise term to (10.103).

This leads to stochastic differential equations (cf. (10.45), (10.46))

$$m\,dv_i(t) = (F_{\text{acc}}(v_i) + F_{\text{dec}}(\Delta x_i))\,dt + \sigma\,v_i\,dW_i(t), \tag{10.112}$$

$$dx_i(t) = v_i\,dt \tag{10.113}$$

A similar equation with an additive noise term has been studied in [79, 108]. An advantage of the version with multiplicative noise is that it guarantees the positiveness of velocities v_i. In the deterministic model the departure (leaving a cluster) times are strongly correlated in such a way that, in the stationary regime, one car leaves the cluster after each time interval of a given length τ_1. The arrival (adding to a cluster) times also are strongly correlated due to the repulsive forces. The noise makes these correlations weaker. It allows one to apply the formalism of stochastic Markov processes in order to describe approximately the fluctuations of the cluster size, as discussed later.

It is easier to study the formation of car congestion on a mesoscopic level, as has been done in [123, 147–151, 153], where we do not follow each individual car, but only look for the number of congested cars n, that is, the size of car cluster. In this description it is also very easy to introduce the randomness, by considering n as a stochastic variable. Some properties of our stochastic traffic flow model, given by the master equation (10.49), can be described by equilibrium thermodynamics in analogy to the liquid-vapor system in spite of the fact that the traffic flow is a driven, namely, a nonequilibrium system. Here we consider the simplest version of the model, assuming that cars are point-like. In this case the transition rates (10.50) and (10.52) reduce to

$$w_-(n) = \frac{1}{\tau} \tag{10.114}$$

$$w_+(n) = \frac{v_{\text{opt}}(\Delta x_{\text{free}})}{\Delta x_{\text{free}}}, \tag{10.115}$$

where τ is a reaction time constant and $\Delta x_{\text{free}}(n) = L/(N-n)$ is the mean headway distance in the free-flow phase. As we have discussed earlier, no large stable cluster forms at low car densities, whereas a macroscopic fraction of them are condensed (jammed) into the cluster above a certain critical density. The first situation corresponds to the fixed-point solution of the OVM (optimal velocity model, see Section 10.1), whereas the second corresponds to the limit cycle. The stationary solution $p^{\text{st}}(n) = \lim_{t\to\infty} p(n,t)$ obeys the detailed balance condition $p^{\text{st}}(n)\,w_+(n) = p^{\text{st}}(n+1)\,w_-(n+1)$ (cf. (3.49)). It is a very remarkable property of the actual model, which allows one to make a connection to thermodynamics.

In equilibrium the detailed balance holds in a physical system like, e.g. the supersaturated vapor-liquid discussed in Chapter 9. The free energy and chemical potentials can be derived based on this principle in a physical system, as well as in the actual traffic flow model. By analogy with (9.65), the detailed balance for a system containing a cluster of size n can be written as

$$\frac{w_+(n-1)}{w_-(n)} = \exp\left(-\frac{F(n) - F(n-1)}{T^*}\right), \tag{10.116}$$

where T^* is a parameter with energy dimension, which corresponds to $k_B T$ in (9.65) or to the temperature measured in energy units, and $F(n)$ is the free energy of state (including all possible microscopic distributions of coordinates and momenta) with cluster size n. Also, in traffic flow, T^* can be interpreted as some 'temperature'.

Following (9.66)–(9.68), also in traffic flow, the free energy of a system with large cluster size n can be represented by the transition rates as

$$\ln\left[\frac{w_+(n)}{w_-(n)}\right] = -\frac{1}{T^*}\frac{\partial F}{\partial n}. \tag{10.117}$$

or

$$F = F_0 - T^* \int_0^n \ln\left[\frac{w_+(n')}{w_-(n')}\right] dn', \tag{10.118}$$

where $F_0 = F(n=0)$ does not depend on the cluster size n. It is the free energy of the system without cluster.

According to (10.114) and (10.115), the ratio of transition rates reads

$$\frac{w_+(n)}{w_-(n)} = \tau \frac{v_{\text{opt}}(\Delta x_{\text{free}})}{\Delta x_{\text{free}}}, \tag{10.119}$$

where $v_{\text{opt}}(\Delta x)$ is the optimal velocity function given by (10.15). It yields

$$\frac{w_+(n)}{w_-(n)} = v_{\max}\tau\rho\frac{1 - n/N}{1 + \rho^2 D^2(1 - n/N)^2}, \tag{10.120}$$

where $\rho = N/L$ is the car density. Introducing the dimensionless density $\tilde{\rho} = \rho D$ and a dimensionless control parameter $\tilde{b} = D/(v_{\max}\tau)$, this becomes

$$\frac{w_+(n)}{w_-(n)} = \frac{1}{\tilde{b}}\frac{\tilde{\rho}(1 - n/N)}{1 + \tilde{\rho}^2(1 - n/N)^2}, \tag{10.121}$$

or

$$\ln\left[\frac{w_+(n)}{w_-(n)}\right] = \ln\left(\frac{\tilde{\rho}}{\tilde{b}}\right) + \ln\left[1 - \frac{n}{N}\right] - \ln\left[1 + \tilde{\rho}^2\left(1 - \frac{n}{N}\right)^2\right]. \tag{10.122}$$

By inserting the latter relation into (10.118), the integration using $\int \ln(1+x^2)\,dx = 2\arctan x + x\ln(1+x^2) - 2x$ yields

$$\frac{F - F_0}{\tilde{L} T^*} = \tilde{\rho}\left\{\left(1 - \frac{n}{N}\right)\ln\left(1 - \frac{n}{N}\right) - \frac{n}{N} - \frac{n}{N}\ln\left(\frac{\tilde{\rho}}{\tilde{b}}\right)\right.$$
$$\left. - \left(1 - \frac{n}{N}\right)\ln\left(1 + \tilde{\rho}^2\left[1 - \frac{n}{N}\right]^2\right) + \ln(1 + \tilde{\rho}^2)\right\}$$
$$+ 2\arctan\tilde{\rho} - 2\arctan\left(\tilde{\rho}\left[1 - \frac{n}{N}\right]\right), \tag{10.123}$$

where $\tilde{L} = L/D$ is the dimensionless length of the road.

The results for $w_+(n)/w_-(n)$ and $(F-F_0)/(\tilde{L}T^*)$ depending on the fraction of congested cars n/N at four different densities are shown in Figures 10.29 and 10.30. The value of the dimensionless control parameter has been chosen as $\tilde{b} = 2/7 \approx 0.2857$. It corresponds, e.g., to $D = 24$ m, $v_{max} = 42$ m s^{-1}, and $\tau = 2$ s. The ratio $w_+(n)/w_-(n)$ is never 1 and no stable car cluster forms at small densities (dotted line). These plots have to be compared with those for the liquid-gas system shown in Figures 9.7 and 9.8. In distinction to the liquid-gas system, the cluster appears without a nucleation barrier in the actual traffic-flow model at somewhat larger densities $\tilde{\rho}$ (dot-dashed line), whereas the nucleation barrier (free energy maximum) shows up only at even larger $\tilde{\rho}$ values (solid line).

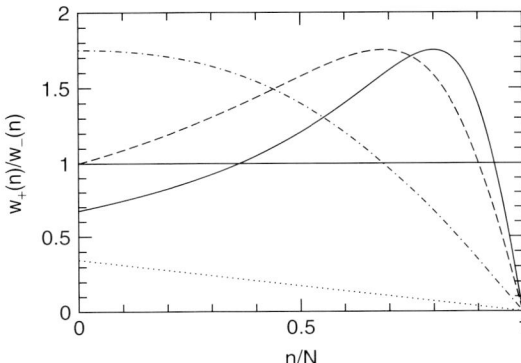

Figure 10.29 The ratio of transition rates $w_+(n)/w_-(n)$ depending on the fraction of congested cars n/N for four dimensionless densities $\tilde{\rho} = 0.1$ (dotted line), $\tilde{\rho} = 1$ (dot-dashed line), $\tilde{\rho} = 3.186$ (dashed line), and $\tilde{\rho} = 5$ (solid line).

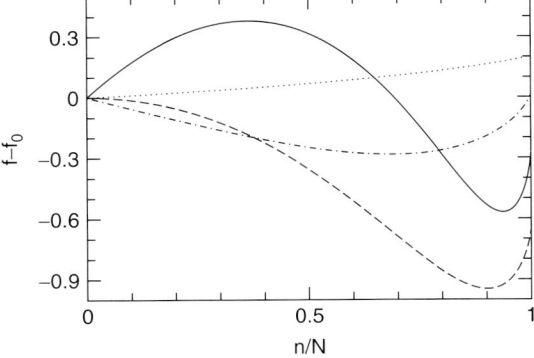

Figure 10.30 Normalized free energy difference $(F-F_0)/(\tilde{L}T^*) = f - f_0$ depending on the fraction of congested cars n/N for four dimensionless densities $\tilde{\rho} = 0.1$ (dotted line), $\tilde{\rho} = 1$ (dot-dashed line), $\tilde{\rho} = 3.186$ (dashed line), and $\tilde{\rho} = 5$ (solid line).

In the above calculation we have determined only the difference $F - F_0$, but not the free energy F_0 of the ideal system without car cluster. As in physical systems, e.g. the supersaturated vapor discussed in Chapter 9, the latter cannot be derived from the detailed balance, but should be calculated from a microscopic model. We should take into account that the distribution over momenta for cars is not the same as that for molecules in an ideal gas. As a first approximation we may assume a similar Gaussian distribution with only shifted mean value $\langle p \rangle = m \langle v \rangle = m v_{\text{opt}} (\Delta x_{\text{hom}})$ in accordance with the optimal velocity $v_{\text{opt}} (\Delta x_{\text{hom}})$ in the homogeneous flow of cars with the mean headway distance $\Delta x_{\text{hom}} = L/N = 1/\rho$. The Gaussian form of the distribution is consistent with the simulation results for the stochastic car-following models [79, 108, 151]. We should take into account also that cars are always moving in one direction, so momentum $p > 0$ always holds. Finally, as distinct from the ideal gas of molecules, the coordinates and momenta of cars are one-dimensional. Hence, by integration over coordinates and momenta of all cars, for the ideal part of the partition function we can write

$$Z_{\text{ideal}} = \frac{1}{N!} \prod_{\alpha=1}^{N} \frac{1}{h} \int_0^L dx_\alpha \int_0^\infty dp_\alpha \, \exp\left(-\frac{(p_\alpha - \langle p \rangle)^2}{2mT^*}\right)$$
$$\approx \frac{1}{N!} \left(\frac{L}{\lambda_0(T^*)}\right)^N, \qquad (10.124)$$

where $\lambda_0(T^*) = h/(2\pi m T^*)^{1/2}$. The latter approximate equality in (10.124) holds when $\langle p \rangle^2/(2mT^*) \gg 1$ or, in other words, when the width of the velocity distribution is narrow as compared to the mean velocity. The latter condition is satisfied for the model (10.112) and (10.113) with a certain set of control parameters used in the simulations (see Figure 10.12). The distribution width, however, increases with the noise amplitude. In fact, the approximation (10.124) is good enough when the distribution function has a small value at zero momentum $p = 0$, as in the simulation results of [79, 108]. According to the above consideration, the temperature in traffic flow is a parameter which controls this distribution width or the amplitude of the velocity and momentum fluctuations, as in the ideal gas of molecules. According to (10.124), the ideal part of the free energy reads

$$F_0 = -T^* \left[N \ln \left(L/\lambda_0(T^*)\right) - \ln N!\right] \simeq T^* N \left[\ln \left(\rho \lambda_0(T^*)\right) - 1\right]. \qquad (10.125)$$

As in liquid-gas system, the congested traffic can be considered as consisting of two phases: the cluster or jam (like a liquid) phase with n particles (vehicles) and free energy $F_{\text{clust}}(n)$, and the free flow (like an ideal gas) phase with $N_{\text{ideal}} = N - n$ particles and free energy $F_{\text{ideal}}(N_{\text{ideal}})$. The total free energy is $F = F_{\text{clust}} + F_{\text{ideal}}$, and we can define the same basic relations for the chemical potentials of these phases as in physical systems, so $\mu_{\text{clust}} = \partial F_{\text{clust}}/\partial n$ and $\mu_{\text{ideal}} = \partial F_{\text{ideal}}/\partial N_{\text{ideal}} = -\partial F_{\text{ideal}}/\partial n$. Hence, we also have (cf. (9.86))

$$\frac{\partial F}{\partial n} = \mu_{\text{clust}} - \mu_{\text{ideal}}. \qquad (10.126)$$

The free energy of the free-flow phase in traffic is

$$F_{\text{ideal}}(T^*, L, N, n) = T^*(N-n)\left[\ln\left(\lambda_0(T^*)\frac{N-n}{L}\right) - 1\right], \quad (10.127)$$

as consistent with (10.125) where we put $N \to N_{\text{ideal}} = N - n$ and $\rho \to N_{\text{ideal}}/L$. From (10.127) we get

$$\mu_{\text{ideal}} = -\frac{\partial F_{\text{ideal}}}{\partial n} = T^* \ln\left(\lambda_0(T^*)\frac{N-n}{L}\right) = T^* \ln\left(\frac{\lambda_0(T^*)}{D}\tilde{\rho}_{\text{free}}\right), \quad (10.128)$$

where $\tilde{\rho}_{\text{free}} = D(N-n)/L$ is the dimensionless density of cars in the free-flow phase. The chemical potential of the cluster phase can be easily calculated from (10.117), (10.122), and (10.126). It yields

$$\mu_{\text{clust}} = -T^* \left\{\ln\left(\frac{D}{\lambda_0 \tilde{b}}\right) - \ln\left[1 + \tilde{\rho}^2\left(1 - \frac{n}{N}\right)^2\right]\right\}$$

$$= -T^* \left\{\ln\left(\frac{D}{\lambda_0 \tilde{b}}\right) - \ln\left[1 + \tilde{\rho}_{\text{free}}^2\right]\right\}. \quad (10.129)$$

It is remarkable that, as distinct from the liquid-gas system, the chemical potential of the cluster phase is not an internal property of this phase, since it depends on the outer parameter – the density of the surrounding free-flow phase $\tilde{\rho}_{\text{free}}$. The physical interpretation of this fact is that the traffic flow is a driven system, which approaches a stationary rather than equilibrium state in the usual sense. However, as we have shown here, various properties of this stationary state can be described by equilibrium thermodynamics.

The free energy of the cluster phase can be calculated consistently from (10.123), (10.125), and (10.127) according to $F = F_{\text{clust}} + F_{\text{ideal}}$. The result is

$$F_{\text{clust}}(T^*, L, N, n) = T^* N \left\{-\frac{n}{N}\left(2 + \ln\left(\frac{D}{\lambda_0 \tilde{b}}\right)\right)\right.$$

$$+ \frac{2}{\tilde{\rho}}\left[\arctan\tilde{\rho} - \arctan\left(\tilde{\rho}\left[1 - \frac{n}{N}\right]\right)\right] \quad (10.130)$$

$$\left. - \left(1 - \frac{n}{N}\right)\ln\left(1 + \tilde{\rho}^2\left[1 - \frac{n}{N}\right]^2\right) + \ln\left(1 + \tilde{\rho}^2\right)\right\}.$$

It is consistent with $\mu_{\text{clust}} = \partial F_{\text{clust}}/\partial n$.

As a summary of this chapter, we have shown how thermodynamics can be applied to a many-particle system such as traffic flow, based on a microscopic (car-following) as well as mesoscopic (stochastic) cluster description. By analogy with equilibrium physical systems like supersaturated vapor-forming liquid droplets (see Chapter 9), we have derived the free energy function and chemical potentials by using the detailed balance in the space of car cluster sizes. Distinguishing features between the traffic flow as a driven system and equilibrium physical systems have also been discussed.

10.7
Exercises

E 10.1 Optimal velocity model I

Formulate the optimal velocity model, considered in Section 10.6, writing Newton's equation of motion in the dimensionless form

$$\frac{d}{d\tilde{t}}\tilde{v}_i = \tilde{F}_{acc}(\tilde{v}_i) + \tilde{F}_{dec}(\Delta\tilde{x}_i)$$

by introducing dimensionless variables, time $\tilde{t} = t/\tau$, coordinate $\tilde{x} = x/D$, velocity $\tilde{v} = v/v_{max}$, and the kinetic and potential energy of the ith car given in dimensionless form as $\tilde{E}_{kin}(\tilde{v}_i) = \tilde{v}_i^2/2$ and $\tilde{E}_{pot}(\Delta\tilde{x}_i) = -\int \tilde{F}_{dec}(\tilde{x}_i)d\tilde{x}_i$. Here D and v_{max} are the interaction distance and the maximum velocity entering the optimal velocity function $v_{opt}(\Delta x) = v_{max}(\Delta x)^2/(D^2 + (\Delta x)^2)$.

Derive expressions for dimensionless quantities: the acceleration and deceleration forces $\tilde{F}_{acc}(\tilde{v}_i)$ and $\tilde{F}_{dec}(\Delta\tilde{x}_i)$, the dimensionless kinetic and potential energy of the car system, and the total energy flux. Find the dimensionless control parameters which determine the behavior of the model.

E 10.2 Optimal velocity model II

Having the energy balance in mind (see Figure 10.31) make numerical calculations of the energy and energy flux for the optimal velocity model in analogy with those presented in Section 10.6, but, using dimensionless variables as defined in the previous exercise. Find how the systems behavior changes depending on the values of the dimensionless control parameters and calculate the limit cycle in the headway – velocity phase space.

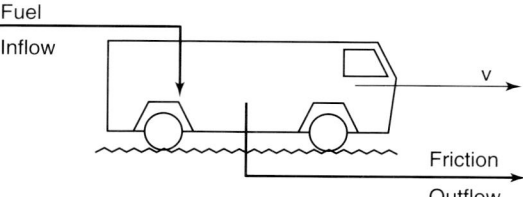

Figure 10.31 Energy flow in steady state vehicular traffic.

E 10.3 Optimal velocity model III

Consider the optimal velocity model introduced in Section 10.2. In Figure 10.32 a phase diagram is depicted which shows different states of the car system on a circular road depending on the average headway ($\tilde{L}/N$) and the dimensionless control parameter b. Find the answers to the following questions.

- What is the meaning of the spinodal function?
- What is the meaning of the binodal function?
- What does b_{crit} mean?
- What does b_{coll} mean?

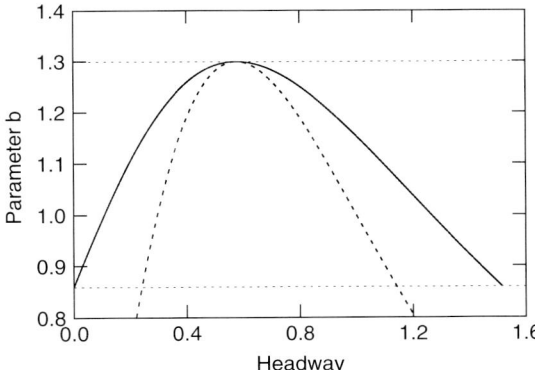

Figure 10.32 Binodal (solid) and spinodal function (dashed) in phase diagram of traffic. The upper dotted line is b_{crit}. The lower dotted line is b_{coll}.

- Identify five different regions in the phase diagram (only homogeneous solution, only limit-cycle solution, homogeneous and limit-cycle solution possible, homogeneous solution with collisions in the limit-cycle, no stable solution) and discuss their meaning in detail.
- Think about the hysteresis effect in this diagram. What is the meaning of this effect in real traffic?

11
Noise-Induced Phase Transitions

11.1
Equilibrium and Nonequilibrium Phase Transitions

This chapter is devoted to a novel type of cooperative phenomenon that can be observed in systems far from their equilibrium state or in systems where the notion of thermal equilibrium is not applied at all. These are called phase transitions induced by noise.

The ability of noise to produce order under certain conditions, in particular, to give rise to various cooperative phenomena and the corresponding phase transitions, is now a well established fact [6, 54, 86, 102, 114, 200]. Such phase transitions manifest themselves in the phase-space distribution changing its structure, for example, the number of maxima. Popular examples are stochastic resonance [36, 52, 66], coherence resonance [114, 183], noise-induced transport [158], and a number of noise-induced phase transitions of the diffusion type [54, 232, 233]. Typically, the latter phenomena are due to multiplicative noise. However, additive noise in the presence of another multiplicative noise can also induce phase transitions [115, 245, 246] or individually cause a hidden phase to become visible [116]. Systems of elements with motivated behavior, e.g. fish and bird swarms, car ensembles on highways, stock markets, etc., also can display a wide variety of cooperative effects caused by noise action [77]. However, the theory of these phenomena is far from being well understood.

To elucidate the main features of the nonequilibrium phase transitions under consideration, let us first discuss briefly the basic notions of the equilibrium phase transitions of second order. We make use of the Landau order parameter theory and for the sake of simplicity consider a one-dimensional system with a nonconservative scalar order parameter $h(x, t)$ with dynamics governed by the equation [137]

$$\Gamma \frac{\partial h}{\partial t} = -\frac{\delta \mathcal{H}\{h\}}{\delta h} + g\xi(t), \tag{11.1}$$

where Γ is the kinetic coefficient, g is the amplitude of the random Langevin force proportional to white noise $\xi(t)$ whose averages satisfy the equalities

$$\langle \xi(t) \rangle = 0 \quad \text{and} \quad \langle \xi(t)\xi(t') \rangle = \delta(t - t'), \tag{11.2}$$

and the functional $\mathcal{H}\{h\}$ is the free energy, typically written in the form

$$\mathcal{H}\{h\} = \int_{-\infty}^{+\infty} dx \left\{ \frac{1}{2} \ell^2 (\nabla h)^2 + H(h) \right\}$$

with the free energy density $H(h)$ specified by the expression

$$H(h) = \tfrac{1}{2} \alpha(T) h^2 + \tfrac{1}{4} \beta h^4.$$

Here, by definition, the symbol $\nabla := \partial/\partial x$, the characteristic spatial scale ℓ, the coefficient $\beta > 0$, the kinetic coefficient $\Gamma > 0$ and the noise amplitude g are certain constant parameters of the given system, whereas the coefficient $\alpha(T)$ depends on the temperature T. In the case of $\ell = 0$, it changes the sign at the critical temperature $T = T_c$. The following ansatz with some constant α_0

$$\alpha(T) = \alpha_0 \frac{(T - T_c)}{T_c}$$

is usually adopted. In these terms the governing equation (11.1) becomes

$$\Gamma \frac{\partial h}{\partial t} = \ell^2 \nabla^2 h - \alpha(T) h - \beta h^3 + g\xi(t). \tag{11.3}$$

For $T > T_c$ the system possesses only one phase state matching the stable minimum of the free energy $h(x) = 0$ (Figure 11.1). When the temperature drops below the critical value, $T < T_c$, the free energy changes in form. Two new additional extrema appear which are stable, whereas the previous one becomes unstable. As a result, the homogeneous state decays, giving rise to the composition of domains with $h_\pm = \pm\sqrt{|\alpha|/\beta}$. This bifurcation is shown in Figure 11.1 which also exhibits the resulting evolution of the probability function $P(h)$ of the order parameter. It should be pointed out that the width of the distribution $P(h)$ is determined by the intensity g of the Langevin forces. Within the framework of the given model the temperature dependence of the coefficient $\alpha(T)$ and the intensity g of the Langevin forces are independent characteristics of the system properties. This is quite a formal statement because, on the microscopic level, both of them are related to the stochastic motion of the particles forming this system. Nevertheless, when dealing with equilibrium phase transitions, one can regard the random forces as just a source of disorder and the appearance of new phases becomes pronounced when the difference between the new phase branches, $(h_+ - h_-)$, essentially exceeds the Langevin force intensity, g, in magnitude.

In some sense the nonequilibrium phase transitions are distinguished from the equilibrium ones by the fact that the phenomena causing them cannot be described within models similar to (11.1) without substantial modification.

Below, we will consider two types of nonequilibrium phase transitions that are due to the creative action of noise. The former appears in systems where the amplitude $g(h)$ of the random Langevin forces depends essentially on the order parameter h, for example, tends to zero as $h \to 0$. The latter comes into being via the anomalous behavior of the kinetic coefficient $\Gamma(\mathbf{h})$ depending on the order parameter now being some vector $\mathbf{h} = \{h_1, h_2, \ldots\}$. Namely, the phase space $\{h_1, h_2, \ldots\}$ of such

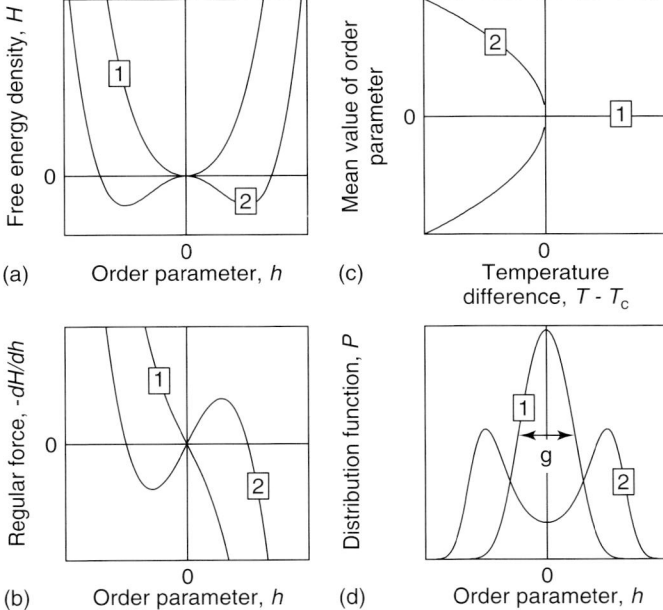

Figure 11.1 Schematic illustration of the mechanism of the second-order phase transitions in equilibrium systems. (a) shows the free energy density above and below the critical value of the temperature, which is labeled with numbers 1 and 2, respectively. The resulting regular force is shown in (b). The averaged order parameter vs the temperature is demonstrated in (c), whereas (d) depicts the distribution function of the order parameter again above and below the phase transition.

systems contains a narrow layer, a 'low-dimensional' domain called the dynamical trap region, where the kinetic coefficient $\Gamma(\mathbf{h})$ takes extremely large values. So, when the system enters this region its dynamics becomes stagnated, which gives rise to long-lived states which are treated as new dynamical phases. In all of these cases, noise gives rise to a certain ordering in addition to random perturbations in the system motion. As a result, after the system has undergone phase transition, the difference $(h_+ - h_-)$ between the mean values of the order parameter h in the new phases and the width of local extrema of the distribution function $P(h)$ are of the same order of magnitude. In some sense they are due to the same phenomena and, thus, cannot be analyzed independently of each other. So, such transitions are typically detected by analyzing the distribution function and fixing the transition between, e.g. the unimodal distribution to the bimodal one (Figure 11.2) as the control parameter changes.

Before passing directly to the nonequilibrium phase transition it is necessary to outline the mathematical concepts to be used in describing nonlinear stochastic systems.

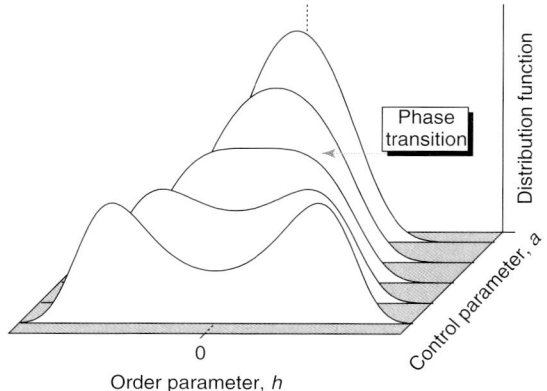

Figure 11.2 Illustration of the approach to detecting phase transitions in onequilibrium systems which are caused by the creative action of noise.

11.2
Types of Stochastic Differential Equations

An introduction to the rigorous description of stochastic processes is given in Chapter 1. Here we apply a more informal construction which is actually justified by the results presented within the chapter.

For an ordinary dynamical system with, for example, a one-dimensional phase space $x \in \mathbb{R}$, the governing equation is written as

$$\frac{dx}{dt} = f(x,t), \tag{11.4}$$

where $f(x,t)$ is a force acting on the system. From a naive point of view one might just replace the force $f(x,t)$ by the sum of the regular component $f(x,t)$ and the random Langevin source $g(x,t)\xi(t)$ to describe a similar stochastic system, rewriting the governing equation as

$$\frac{dx}{dt} = f(x,t) + g(x,t)\xi(t). \tag{11.5}$$

Unfortunately, such a generalization is justified only in the case when the intensity of the Langevin sources does not depend on the phase variable, so that $g(x,t) = g(t)$. To understand this fault we consider the model where

$$f(x,t) = 0, \quad g(x,t) = 1 + \epsilon x, \tag{11.6}$$

and $\epsilon \ll 1$ is a small parameter. Initially, the system is assumed to be located at the origin $x(t=0) = 0$. Then, within an accuracy up to the first order in ϵ, the solution of (11.5) can be obtained by iteration yielding

$$x(t) = \int_0^t dt'\, \xi(t') + \epsilon \int_0^t dt' \int_0^{t'} dt''\, \xi(t')\xi(t''). \tag{11.7}$$

By virtue of (11.2) the value of $x(t)$ averaged over all the realizations of white noise $x(t)$ is given by the expression

$$\langle x(t) \rangle = \epsilon \int_0^t dt' \int_0^{t'} dt'' \, \delta(t' - t''). \tag{11.8}$$

It is exactly these expressions which are responsible for ambiguities in the formal use of differential equations for describing nonlinear stochastic systems. The problem is that the point where the Dirac δ-function differs from zero lies at the boundary $t' = t''$ of the integration region (Figure 11.3). In this case it is not clear which part of the δ-function should be ascribed to the integration region or, what amounts to the same thing, which part is outside it.

In order to overcome this problem, one can apply to the notion of infinitesimals, namely, the hyperreal numbers. Their rigorous description can be found in the literature [1, 3]. Here we will confine our consideration to the physical level of rigor. Dealing again with a standard dynamical system, its governing equation (11.4) can be rewritten using infinitesimals as

$$dx = f(x, t) \, dt, \tag{11.9}$$

where dx and dt are infinitesimally small increments of variable x and a time step. Roughly speaking, in standard dynamical systems there is only one infinitesimally small variable, the time step dt. All the other infinitesimals are derived from it by multiplying dt by some smooth function. From the standpoint of such a microscopic level the distinction between systems with regular and stochastic dynamics is mainly due to the stochastic ones possessing two or more really independent infinitesimals, the time step dt and the infinitesimal moments $\{dW(t)\}$ of the random Langevin forces. In the case under consideration there is only one moment of the random force, $dW(t)$. Here, the time t is used with the

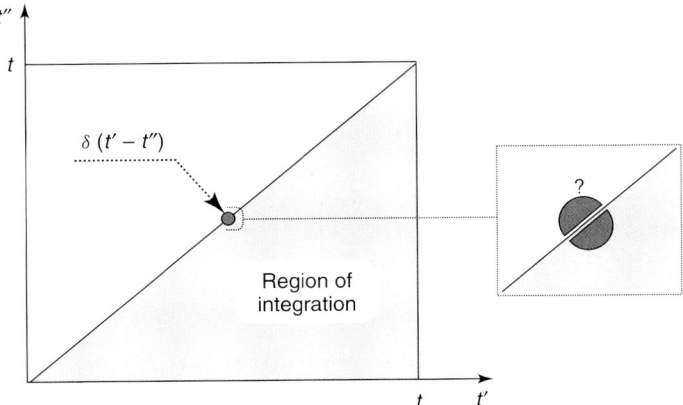

Figure 11.3 Illustration of the fault encountered in describing stochastic systems by nonlinear Langevin forces. See the discussion preceding expression (11.8).

symbol $dW(t)$ in order to relate the random force moment to the current time. At different moments t, t' of time the quantities $dW(t)$ and $dW(t')$ are supposed to be mutually independent. It is possible to consider $dW(t)$ as the integral of white noise $\xi(t)$

$$dW(t) = \int_{t}^{t+dt} dt'\xi(t') \tag{11.10}$$

which, however, is no more than a qualitative explanation of the properties ascribed to the random force moment $dW(t)$. Namely, they are

$$\langle dW(t)\rangle = 0 \quad \text{and} \quad \langle [dW(t)]^2\rangle = dt. \tag{11.11}$$

The latter equality enables us to estimate the amplitude of the random force moment as $dW(t) \sim \sqrt{dt}$. In the following it will be seen that a stochastic infinitesimal quantity affects the system dynamics only if its amplitude scales with the time step as $\sqrt{dt}$, whereas all the essential regular infinitesimals scale as dt. Therefore, when dealing with $[dW(t)]^2$ we can ignore its random component and regard it as a regular infinitesimal. In other words, the following relationship

$$[dW(t)]^2 = dt \tag{11.12}$$

is adopted. In some sense it specifies the algebraic manipulations with infinitesimals of the stochastic systems (see [55] for the details).

Now we are ready to write down the governing equation for the given stochastic system in infinitesimals. However, before doing this it is worthwhile to note that the notion of infinitesimals enables one to pose a question about the point x_θ at which the force $f(x_\theta, t)$ in (11.9) should be calculated. In particular, it is naturally to set (Figure 11.4)

$$x_\theta = x + \theta\,dx \tag{11.13}$$

where $0 \leq \theta \leq 1$ is a certain constant or a smooth function of x. For standard dynamical systems, however, this shift in (11.10) from the point x to point x_θ makes no sense because it gives rise to addition terms on the left-hand side of (11.10) having no effect on the system dynamics. Namely,

$$f(x, t)\,dt \Rightarrow f(x_\theta, t) \approx f(x, t)\,dt + \frac{\partial f}{\partial x}\theta\,dx\,dt, \tag{11.14}$$

and since $dx \propto dt$ in the given case, the last term scales as $(dt)^2$ which is negligible. This is not the case if the Langevin force undergoes a shift $x \to x_\theta$ as will be seen below.

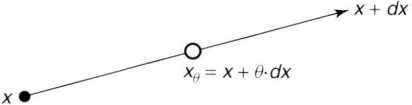

Figure 11.4 Intermediate point x_θ which determines the values of the forces acting on the system during the time interval $(t, t + dt)$.

Keeping in mind this possibility of dealing with the intermediate points $\{x_\theta\}$ we apply to the following equation in infinitesimals (cf. [55])

$$dx = f_\theta(x_\theta, t)\, dt + g(x_\theta, t)\, dW(t) \tag{11.15a}$$

to describe the stochastic system with a nonlinear Langevin force. This equation actually inherits the structure of the formal equation (11.5) after its 'integration' from t to $t + dt$ and the formal relationship (11.10) between the random infinitesimal $dW(t)$ and white noise $\xi(t)$. In other words, the given stochastic system is characterized by the regular force $f_\theta(x, t)$, the intensity $g(x, t)$ of the random Langevin force, and the parameter θ specifying the intermediate point x_θ which determines the magnitudes of these forces.

Concluding the construction of the given governing equation we note the following. First, the introduced basic parameter θ has no analogy in standard dynamical systems. Moreover, it characterizes the physical properties of a given system reflecting its features at the microscopic level. The subscript θ for the function $f_\theta(x, t)$ is used to underline the fact that it belongs to the triplet $\{f, g, \theta\}$. Second, in the general case, $f_\theta(x, t)$ is a vector and $g(x, t)$ as well as $\theta(x, t)$ is some tensor. Third, as it follows from (11.15a) that $dx \propto dW \propto \sqrt{dt}$ holds at the leading order in dt. Therefore, the difference

$$f_\theta(x_\theta, t)\, dt - f_\theta(x, t)\, dt \approx \frac{df}{dx} \theta (dt)^{3/2}$$

can be ignored because it scales as $(dt)^{3/2}$ with time step dt and, therefore, is an infinitesimal of higher order than dt. So the replacement

$$f_\theta(x_\theta, t) = f_\theta(x, t)$$

in equation (11.15a) will be further adopted. In this case, (11.15a) reads

$$dx = f_\theta(x, t)\, dt + g(x_\theta, t)\, dW(t). \tag{11.15b}$$

Fourth, special types of stochastic equations have individual names (Table 11.1). The reasons for this will become clear.

As we have mentioned above, a stochastic system is described by the triplet $\{f_\theta, g, \theta\}$ which aggregates the system properties on the microscopic level, and the question concerning the specific value of θ should correspond to physics rather than mathematics. For example, diffusion in solids is the Hänggi–Klimontovich process, whereas current induced by temperature gradient is the Ito process.

Table 11.1 Basic types of stochastic systems.

θ	Name of stochastic process
0	Ito process
1/2	Stratonovich process
1	Hänggi–Klimontovich process

Nevertheless, for a given system it is possible to make use of the description $\{f_\theta, g, \theta\}$, where the value of the parameter θ is chosen for specific purposes. This is due to the fact that the collection of triplets $[\{f_\theta, g, \theta\}]_\theta$, where the form of the function $g(x, t)$ is fixed, gives the equivalent description of stochastic dynamics, provided that the regular forces $\{f_\theta\}$ meet some relationship. To show this we expand the function $g(x_\theta, t) = g(x + \theta\, dx, t)$ in the Taylor series with respect to the infinitesimal quantity dx and cut it off at the first-order term

$$g(x_\theta, t) = g(x, t) + \theta \frac{\partial g(x, t)}{\partial x}\, dx.$$

As follows from (11.15), at the leading order in dt, that is, within the accuracy of $\sqrt{dt}$, the infinitesimals dx and dW are related by the expression $dx = g(x, t)\, dW$, so

$$g(x_\theta, t) = g(x, t) + \frac{\theta}{2} \frac{\partial g^2(x, t)}{\partial x}\, dW.$$

Substituting the latter expression into (11.15b) and taking into account (11.12), this reduces to the following Ito-type stochastic governing equation

$$dx = \left[f_\theta(x, t) + \frac{\theta}{2} \frac{\partial g^2(x, t)}{\partial x}\right] dt + g(x, t)\, dW(t). \tag{11.16}$$

Therefore, all the triplets $\{f_\theta, g, \theta\}$ with a given intensity $g(x, t)$ of Langevin forces actually describe the same stochastic system if the regular forces $\{f_\theta(x, t)\}$ belong to a family of functions such that

$$f_0(x, t) = f_\theta(x, t) + \frac{\theta}{2} \frac{\partial g^2(x, t)}{\partial x}, \tag{11.17}$$

where $f_0(x, t)$ is the regular component of the Ito-type triplet. A similar relationship for the regular force of the Hänggi–Klimontovich type is written as

$$f_1(x, t) = f_\theta(x, t) - \frac{(1 - \theta)}{2} \frac{\partial g^2(x, t)}{\partial x}. \tag{11.18}$$

As a closing remark in this section, we note that the Ito representation of a stochastic system has an advantage in describing the system dynamics as an array of succeeding steps $\{dx\}$ because the probability of the system transition from the state x to the state $x + dx$ is specified completely by its properties at the state x only. The properties distinguishing the other two types of stochastic process will be discussed below.

11.3
Transformation of Random Variables

Conversion from a random variable, x, to a new one, y, poses the problem of finding the governing equation for the random process $y(t) \Leftarrow x(t)$ provided that the governing equation of the system dynamics described in terms of the variable x is known, for example, it is (11.15). It is normal to consider this problem under the assumption that the old and new random processes are characterized by the same value of the parameter θ. Unfortunately, the standard rules valid for the ordinary differential equations do not hold, in general, for stochastic systems. The fact is that, even for a smooth function $x = \varphi(y)$, the expansion

$$\varphi(y + dy) = \varphi(y) + \varphi'(y)\,dy + \tfrac{1}{2}\varphi''(y)(dy^2) + \cdots$$

contains terms proportional to dt from two sources. The former is the second term in this expansion due to the presence of the regular force; the latter is the third term due to the contribution of the random force moment, namely, $(dW)^2 = dt$.

To derive the transformation rule for the stochastic system under consideration we make use of the following equalities

$$\begin{aligned} x_\theta &= \theta\varphi[y_\theta + (1-\theta)\,dy] + (1-\theta)\varphi[y_\theta - \theta\,dy] \\ &= \varphi[y_\theta] + \tfrac{1}{2}\theta(1-\theta)\varphi''[y_\theta](dy)^2 \end{aligned} \quad (11.19)$$

and

$$\begin{aligned} dx &= \varphi[y_\theta + (1-\theta)\,dy] - \varphi[y_\theta - \theta\,dy] \\ &= \varphi'[y_\theta]\,dy + (\tfrac{1}{2}-\theta)\varphi''[y_\theta](dy)^2, \end{aligned} \quad (11.20)$$

where $y_\theta = y + \theta\,dy$. In (11.15a) the argument x_θ of the functions $f_\theta(x_\theta, t)$ and $g(x_\theta, t)$ can be replaced by another quantity deviating from x_θ by an infinitesimal of order less than $\sqrt{dt}$. Thus, by virtue of (11.20) the transformation of the variables $x = \varphi(y)$ is implemented, first using the direct replacement

$$f_\theta(x_\theta, t),\ g(x_\theta, t) \mapsto f_\theta[\varphi(y_\theta), t],\ g[\varphi(y_\theta), t].$$

Second, the substitution of (11.19) into (11.15a) converts it to

$$\varphi'[y_\theta]\,dy + (\tfrac{1}{2}-\theta)\varphi''[y_\theta](dy)^2 = f_\theta[\varphi(y_\theta), t]\,dt + g[\varphi(y_\theta), t]\,dW(t). \quad (11.21)$$

Whence it follows that, at the leading order in dt, that is, within the accuracy $\sqrt{dt}$ the infinitesimals dy and dW are related as

$$dy = \frac{g[\varphi(y_\theta), t]}{\varphi'[y_\theta]}\,dW(t).$$

Taking into account rule (11.12), formula (11.21) leads us to the required governing equation written for the random variable y related to x as $x = \varphi(y)$ (cf. [55])

$$dy = \frac{1}{\varphi'[y_\theta]} \left\{ f_\theta[\varphi(y_\theta), t] + \left(\theta - \frac{1}{2}\right) g^2[\varphi(y_\theta), t] \frac{\varphi''[y_\theta]}{(\varphi'[y_\theta])^2} \right\} dt$$

$$+ \frac{g[\varphi(y_\theta), t]}{\varphi'[y_\theta]}\,dW(t). \quad (11.22)$$

It is the second term in the braces that makes the transformation of random variables distinct from one for ordinary dynamical systems. However, if the random process is of the Stratonovich type, so that $\theta = 1/2$, this term vanishes and the transformation of random variables takes the standard form [55]

$$dy = \frac{1}{\varphi'[\gamma_{1/2}]} f_{1/2}[\varphi(\gamma_{1/2}), t] \, dt + \frac{g[\varphi(\gamma_{1/2}), t]}{\varphi'[\gamma_{1/2}]} \, dW(t). \tag{11.23}$$

This property endows the Stratonovich representation of stochastic processes with an advantage in dealing with a description based on various transformations of the system phase space.

11.4
Forms of the Fokker–Planck Equation

The relationship between the stochastic differential equations dealing with individual realizations of stochastic trajectories and the Fokker–Planck equation describing the dynamics of the distribution function $P(x,t)$ was considered in detail in Chapter 5. Here we will make use of the results presented there.

For the systems under consideration the diffusion coefficient is calculated as follows

$$D(x,t) = \frac{\langle (dx)^2 \rangle}{2 \, dt} = \frac{g^2(x,t)}{2}.$$

Thus, for a stochastic system governed by the Ito-type triplet $\{f_0, g, 0\}$ the Fokker–Planck equation takes the form

$$\frac{\partial P}{\partial t} = \frac{\partial}{\partial x}\left[\frac{1}{2}\frac{\partial (g^2 P)}{\partial x} - f_0 P\right]. \tag{11.24}$$

In order to write the corresponding equation for a system with a general triplet $\{f_\theta, g, \theta\}$, we can make use of the Fokker–Planck equation (11.24) after passing to the equivalent Ito representation according to (11.17). In this way we get

$$\frac{\partial P}{\partial t} = \frac{\partial}{\partial x}\left[\frac{g^{2\theta}}{2}\frac{\partial (g^{2(1-\theta)} P)}{\partial x} - f_\theta P\right]. \tag{11.25}$$

In particular, for the Stratonovich processes, where $\theta = 1/2$, the Fokker–Planck equation is of the form

$$\frac{\partial P}{\partial t} = \frac{\partial}{\partial x}\left[\frac{g}{2}\frac{\partial (gP)}{\partial x} - f_{1/2} P\right], \tag{11.26}$$

whereas for the Hänggi–Klimontovich processes it becomes

$$\frac{\partial P}{\partial t} = \frac{\partial}{\partial x}\left[\frac{g^2}{2}\frac{\partial P}{\partial x} - f_1 P\right]. \tag{11.27}$$

The Hänggi–Klimontovich representation of stochastic dynamics is singled out by the relation between the steady-state distribution $P^{st}(x)$, the regular force $f(x)$, and the intensity $g(x)$ of the Langevin source. Namely, for an autonomous system, that is, when its properties do not depend on time, the steady-state solution $P^{st}(x)$

of (11.27) obeys the equality

$$\frac{g^2(x)}{2}\frac{dP^{st}}{dx} - f_1(x)P^{st} = 0.$$

This equality is quite natural for unbounded one-dimensional systems, whereas for multidimensional systems it also is the case when the detailed balance holds. Direct integration of the latter equality gives us the expression for the steady-state distribution

$$P^{st}(x) = \frac{1}{Z}\exp\left\{\int^x \frac{2f_1(x')}{g^2(x')}\,dx'\right\}. \tag{11.28}$$

Dealing with the initial triplet $\{f_\theta, g, \theta\}$ of the general type, formula (11.28) can be rewritten as

$$P^{st}(x) = \frac{1}{Zg^{2(1-\theta)}(x)}\exp\left\{\int^x \frac{2f_\theta(x')}{g^2(x')}\,dx'\right\} \tag{11.29}$$

by virtue of (11.18). Here the coefficient Z is specified by the normalization condition

$$\int_{-\infty}^{+\infty} P^{st}(x)\,dx = 1. \tag{11.30}$$

Naturally, all these integrals should exist. Expression (11.28) will be used in the analysis of the noised-induced phase transitions in the next section.

In the following two sections we consider typical examples of the stochastic systems undergoing nonequilibrium phase transition caused by nonlinear Langevin sources.

11.5
The Verhulst Model of Third Order

This model is related to one-dimensional systems with fading dynamics near the stationary point $x = 0$ or between two stable points $x = \pm\sqrt{\lambda_s}$ (when $\lambda_s > 0$), where the returning force contains linear and nonlinear component like this

$$\frac{dx}{dt} = -\lambda(t)x - x^3, \tag{11.31}$$

with the coefficient $\lambda(t) = \lambda_s + \sigma\xi(t)$ not being a constant but exhibiting random fluctuations in time. The intensity of these fluctuations is quantified by the parameter σ. When the amplitude of these fluctuations exceeds some critical value, the coefficient $\lambda(t)$ changes the sign and during a short time the system behavior becomes anomalous, for example, the point $x = 0$ becomes unstable. The Verhulst model actually describes the competition between these random events of the system instability and regular fading. The corresponding governing equation is of the form [86]

$$dx = \lambda_s x - x^3 + \sigma x_\theta\,dW(t) \tag{11.32}$$

and the parameter θ is assumed to be given in the interval $0 < \theta < 1$. Therefore $f_\theta(x) = \lambda_s x - x^3$ and $g(x) = \sigma x$ hold for this system. In the following we will analyze only the properties of the stationary distribution $P^{st}(x)$.

Formula (11.29) together with the normalization condition (11.30) immediately gives us the desired function

$$P^{st}(x) = \frac{1}{\Gamma(\beta)\sigma} \left(\frac{|x|}{\sigma}\right)^{2\beta-1} \exp\left\{-\frac{x^2}{\sigma^2}\right\} \quad \text{(for } \beta > 0\text{)}, \tag{11.33}$$

where the parameter $\beta = \beta(\lambda_s, \sigma, \theta)$ is

$$\beta = \frac{\lambda_s}{\sigma^2} + \theta - \frac{1}{2} \tag{11.34}$$

and $\Gamma(p)$ is the gamma function.

Let us consider two possible cases with $\lambda_s > 0$ or $\lambda_s < 0$ individually. When $\lambda_s > 0$ and $\sigma = 0$ the system has one unstable stationary point $x = 0$ and two stable ones $x = \pm\sqrt{\lambda_s}$ according to (11.31). The random Langevin force of low intensity, $\sigma \ll 1$, can only disturb the system motion near these points. In the given limit $\beta > \frac{1}{2}$ holds and the distribution function is bimodal (Figure 11.5). This bimodality, however, is not due to the noise effect but is a result of the regular

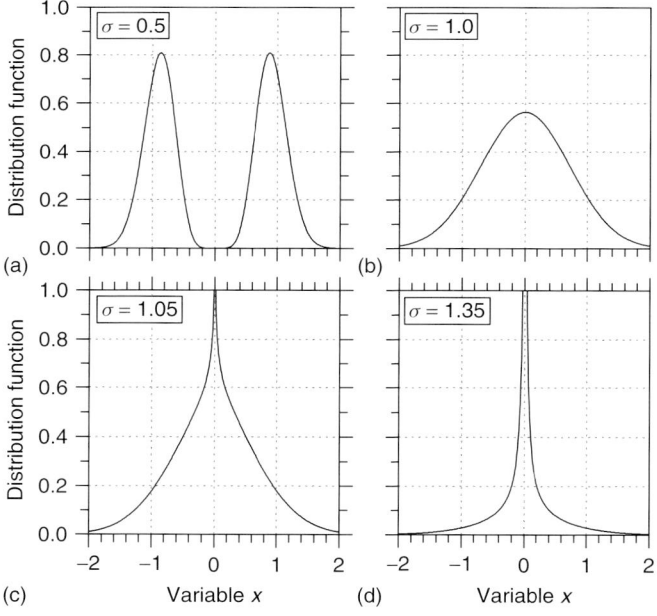

Figure 11.5 The stationary distribution function of the Ito-type Verhulst random process for different values of the noise intensity σ. (a) depicts the distribution function for $\sigma < \sigma_{c1}$, (b) shows it at the critical value $\sigma = \sigma_{c1}$, whereas (c) and (d) exhibits the distribution function in the case $\sigma_{c1} < \sigma < \sigma_{c2}$ for two values of σ located near the boundaries of this interval. In drawing the curves the parameter $\lambda_s = 1$ was used for which $\sigma_{c1} = 1$ and $\sigma_{c2} = \sqrt{2}$.

force structure. As the noise intensity grows, the Langevin force destroys it, and for $0 < \beta < \frac{1}{2}$ the stationary distribution $P^{st}(x)$ becomes unimodal. For large values of σ, corresponding to $\beta < 0$, the system undergoes collapse indicated formally by the fact that function (11.33) has a nonintegrable singularity at $x = 0$. Accordingly, when the system wandering in space reaches the origin, it will never leave it. This 'ordering' is really due to the noise effect. In the given case the stationary distribution takes the form of the Dirac δ-function

$$P^{st}(x) = \delta(x) \quad (\text{for } \beta > 0). \tag{11.35}$$

Figure 11.5 visualizes this evolution of the stationary distribution as the intensity of the random Langevin forces increases. According to (11.33), the system is characterized by the bimodal stationary distribution if

$$\sigma < \sigma_{c1} := \sqrt{\frac{\lambda_s}{(1-\theta)}}. \tag{11.36}$$

For $\sigma \ll 1$ its extrema are located near $x \pm \sqrt{\lambda_s}$. In the given case, the random Langevin force affecting the system dynamics mainly disturbs its motion. We note that for the Hänggi–Klimontovich process, regarded here as the limit $\theta \to 1 - 0$, it is the only possible behavior of the system. For the intermediate values of the noise intensity σ, namely,

$$\sigma_{c1} < \sigma < \sigma_{c2} := \sqrt{\frac{\lambda_s}{\left(\frac{1}{2} - \theta\right)}} \tag{11.37}$$

the Langevin force destroys the ordering by a regular force but cannot order the system motion itself. For stochastic processes with $\theta \geq \frac{1}{2}$ the critical value σ_{c2} does not exist and the noise effect is purely destructive. This is true, in particular, for the Stratonovich-type processes. When $\theta < \frac{1}{2}$ and the noise intensity is quite high,

$$\sigma > \sigma_{c2}, \tag{11.38}$$

the Langevin force causes the system to collapse at the origin, which means that, after some finite time, the system will inevitably be trapped at the point $x = 0$.

If the parameter $\lambda_s < 0$, the Verhulst system without noise admits only one stationary point $x = 0$ being stable. In this case a noise of low intensity cannot destroy the system collapse at the origin and the stationary distribution function is the Dirac δ-function. It is justified by the divergence of integral (11.30) for the solution (11.33) with $\beta < 0$. The latter inequality holds for small values of σ, as follows from (11.34). For a stochastic system with $\theta > \frac{1}{2}$ the Langevin force, however, destroys this collapse when its intensity σ exceeds some critical value

$$\sigma > \sigma_c^* := \sqrt{\frac{|\lambda_s|}{\left(\theta - \frac{1}{2}\right)}}. \tag{11.39}$$

Here again the Langevin force plays a destructive role.

We note that the phenomena described by the given Verhulst model are mainly the effects of the order breakdown proceeding via phase transitions caused by the nonlinearity of the Langevin force. In the next section a system with the creative role of nonlinear Langevin forces will be considered.

11.6
The Genetic Model

This model reflects the characteristic features in the description of population genetics. By way of an example we analyze it within a somewhat simplified formulation dealing with the variable X determined inside the interval $(0, 1)$ with dynamics governed by the equation [86]

$$\frac{dX}{dt} = \frac{1}{2} - X + \lambda(t)X(1 - X),$$

where the parameter $\lambda(t)$ undergoes random fluctuations in time. For the sake of simplicity we convert to the new variable $x = (2X - 1)$ and assume the process under consideration to be of the Stratonovich type. In this case the governing equation in infinitesimals is written as

$$dx = -x\, dt + \sigma \left(1 - x_0^2\right) dW, \tag{11.40}$$

where, as previously, the parameter σ characterizes the noise intensity. Also, the system is assumed to be localized initially at some point of the interval $-1 < x < 1$. Here the regular force $f_{1/2}(x)$ describes pure damping relaxation toward the origin, $x = 0$, the unique stationary point being stable without the noise effect. The intensity of the random Langevin forces, $g(x) = \sigma(1 - x^2)$, drops essentially at the points $x = \pm 1$, which is responsible for the anomalous system behavior.

In this case, (11.29) together with (11.30) gives

$$P^{st}(x) = \exp\left\{\frac{1}{2\sigma^2}\right\} K_0^{-1}\left(\frac{1}{2\sigma^2}\right) \frac{1}{(1 - x^2)} \exp\left\{-\frac{1}{\sigma^2(1 - x^2)}\right\}, \tag{11.41}$$

where $K_0(\ldots)$ is the modified Bessel function of the second kind of order 0. Analyzing this expression directly, we conclude that the distribution function $P^{st}(x)$ is of unimodal form when the noise intensity is quite low, namely, $\sigma < 1$. For $\sigma > 1$, the Langevin force induces essential deviation of the system from the origin and the distribution function become biomodal (Figure 11.6). In particular, the function $P^{st}(x)$ attains its maximum at the points

$$x_m = \pm \frac{\sqrt{1 - \sigma^2}}{\sigma}.$$

In the given system it is the nonlinear Langevin force that causes the formation of two new phases, which is a good example of the constructive action of nonlinear random forces.

11.7
Noise-Induced Instability in Geometric Brownian Motion

Both of the examples considered in the previous two sections demonstrate the fact that the system distribution can change its form depending on the noise intensity,

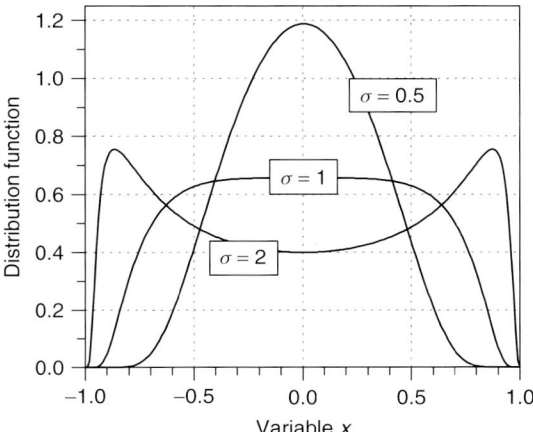

Figure 11.6 A stationary distribution function of the Stratonovich type. Genetic random process for different values of the noise intensity σ.

when the Langevin force is essentially nonlinear. At first glance it might be possible to assume that the transitions found in the distribution form are no more than 'artifacts' and can be eliminated by converting to the appropriate variables $y = \psi(x)$. Indeed, this transformation of variables leads to the conversion of the probability density $P(x) \mapsto p(y)$, namely,

$$p(y) = \frac{1}{\psi'(x)} P(x), \tag{11.42}$$

and the nonlinearity of the transformation $y = \psi(x)$ can eliminate the distribution bimodality.

The remaining part of this section is a counter-example of this statement. It is provided by geometric random walks already introduced in Section 5.9. The process is governed by the following equation in infinitesimals (cf. [206])

$$dx = \beta x\, dt + \sigma x_\theta\, dW(t), \tag{11.43}$$

where the system motion is considered in the half-space $x > 0$ with the coefficient $\beta = \pm 1$, and, as previously, the parameter σ quantifies the noise intensity. The parameter θ of this process is assumed to be a given constant not equal to $1/2$, so $0 \leq \theta < 1/2$ or $1/2 < \theta \leq 1$. This model provides a rather simplified description of the birth–death processes in many biological and ecological objects.

Let us convert to the new variable

$$x = \exp(y). \tag{11.44}$$

Then, by virtue of (11.22), equation (11.43) becomes

$$dy = \left[\beta + \left(\theta - \frac{1}{2}\right)\sigma^2\right] dt + \sigma\, dW(t). \tag{11.45}$$

Whence it follows that, if the Langevin source is rather weak, namely,

$$\sigma < \sigma_c := \sqrt{\frac{2}{|2\theta - 1|}}, \qquad (11.46)$$

then its presence cannot essentially affect the system dynamics. In other words, in the y-space the system drifts either to $-\infty$ or $+\infty$, depending on the sign of the parameter β. In the x-space this is seen in the system tending to the origin x or to infinity with probability equal to unity as time increases. When the noise intensity increases the critical value, $\sigma > \sigma_c$, the system motion can change direction. This occurs when $\beta = -1$ and $\theta > 1/2$ or $\beta = 1$ and $\theta < 1/2$. This actually demonstrates the instability of the point $x = 0$ caused by multiplicative noise.

Finalizing this section, we present the distribution function $p(y, t)$ and $P(x, t)$ written for the random variables y and x governed by (11.45). The latter describes the standard diffusion process with constant drift, so

$$p(y, t) = \frac{1}{\sqrt{2\pi\sigma^2 t}} \exp\left\{-\frac{[(y - y_0) - (\beta + (\theta - \tfrac{1}{2})\sigma^2)t]^2}{2\sigma^2 t}\right\}, \qquad (11.47a)$$

$$P(x, t) = \frac{1}{x\sqrt{2\pi\sigma^2 t}} \exp\left\{-\frac{[\ln(x/x_0) - (\beta + (\theta - \tfrac{1}{2})\sigma^2)t]^2}{2\sigma^2 t}\right\}, \qquad (11.47b)$$

where $x_0 = \exp(y_0)$ are the coordinates of the initial position of the particle and (11.42) has also been taken into account. Figure 11.7 visualizes these functions for different moments of time. It also should be noted that the geometric random walks have quite anomalous properties. To demonstrate this fact, let us analyze the time dependence of the moments of the variable x

$$M_p(t) := \int_0^\infty x^p P(x, t)\, dx \quad \text{for } p = 1, 2, \ldots \qquad (11.48)$$

Substituting (11.47) into (11.48), direct integration yields

$$M_p(t) = \exp\left\{\left[\beta + \left(\theta + \frac{p-1}{2}\right)\sigma^2\right]pt\right\}. \qquad (11.49)$$

So, when the effect of noise is ignorable, $\sigma \to 0$, the system either goes to zero or to infinity, depending on the sign of β. For the Langevin force of finite intensity, $\sigma > 0$, there is always an order p so that

$$p > 1 - 2\left(\frac{\beta}{\sigma^2} + \theta\right)$$

and the moment $M_p(t)$ diverges as $t \to \infty$. In particular, even for $\theta = 0$ and $\beta = 0$, the first-order moment M_1 does not depend on time, whereas the second moment $M_2(t)$ goes to infinity as $\exp\{\sigma^2 t\}$. In other words, even if the mean value of the random variable x is fixed, its random fluctuations increase without bounds as time increases.

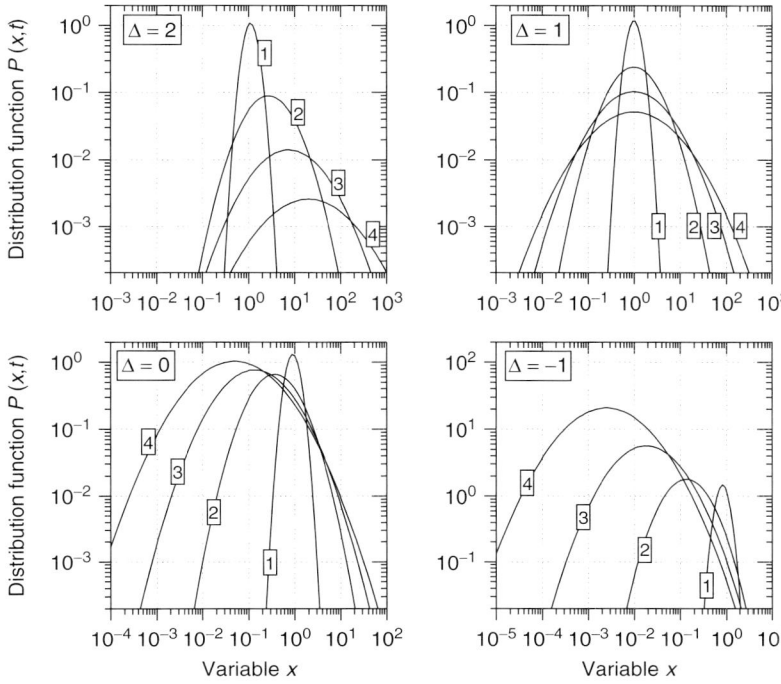

Figure 11.7 The distribution function for geometric random walks governed by (11.43) for different times $\{t_i\}$ such that $\sigma^2 t_1 = 0.1$, $\sigma^2 t_2 = 1$, $\sigma^2 t_3 = 2$, and $\sigma^2 t_4 = 3$. The curves in the figure are labeled with the corresponding numbers and the values used $\Delta = \beta/\sigma^2 + \theta - 1/2$ are also shown. The system was initially located at $x_0 = 1$.

11.8
System Dynamics with Stagnation

Now we pass on to the consideration of another type of phase transition induced by the action of noise, namely, phase transitions caused by dynamical traps. These occur in systems where the kinetic coefficient $\Gamma(h)$, see (11.1), depends essentially on the system state h.

Originally, the development of the dynamical trap concept [247] or, more precisely, the dynamical traps of stagnation type were stimulated by a wide class of intricate cooperative phenomena found in the dynamics of various systems, e.g. vehicle ensembles moving on highways, fish and bird swarms, stock markets, etc. (see [77] for a review). The background of the models to be developed is as follows. People, as elements of a certain system, cannot individually control all the governing parameters. Therefore one chooses a few crucial parameters and mainly focuses attention on them. When equilibrium with respect to these crucial parameters is attained, human activity slows down, in turn retarding the system dynamics as a whole. For example, when driving a car, the control over the relative velocity is of prime importance in comparison with the correction of the headway distance. So, under normal conditions a driver should first reduce the relative velocity between his car and the car ahead and only then optimize the headway. In markets, the deviation from the supply-demand equilibrium, reflected in price changes, also has to exhibit faster time variations than, e.g. the production cost determined by technological capabilities. In physical systems this situation can also be found, e.g. in Pd-metal alloys charged with hydrogen where the structure relaxation exhibits non-monotonic dynamics [8,92]. In these alloys hydrogen atoms and nonequilibrium vacancies form long-lived complexes essentially affecting the structure relaxation. Their generation and disappearance governed, in turn, by the structure evolution causes the non-monotonic dynamics which can be described in terms of different time scales.

These speculations have led us to the concept of dynamical traps, that is, a certain 'low' dimensional region (trap region) in the phase space where all the main kinetic coefficients exhibit anomalous behavior [129, 130, 132, 133]. As a result, all the time scales of the system dynamics in the trap region become large in comparison with their values outside it. The latter effect, in turn, causes long-lived states to appear in the system. In time patterns these states manifest themselves as a sequence of fragments within which at least one of the phase variables remains approximately constant. These fragments are continuously connected by sharp jumps of the given variable. Paper [133] demonstrated that such long-lived states do exist in dense traffic flow and proposed some model of dynamical traps to explain the observed features of the car velocity time series (Figure 11.8). Papers [129] and [132] simplified this model to single out the dynamical trap effect on its own. Paper [130] studied a single oscillator with dynamical traps and demonstrated numerically that the white noise can cause the distribution function of the oscillator position to convert from the unimodal form to the bimodal one. This is due to the fact that, inside the trap region, the regular 'force' is only depressed, rather than changing

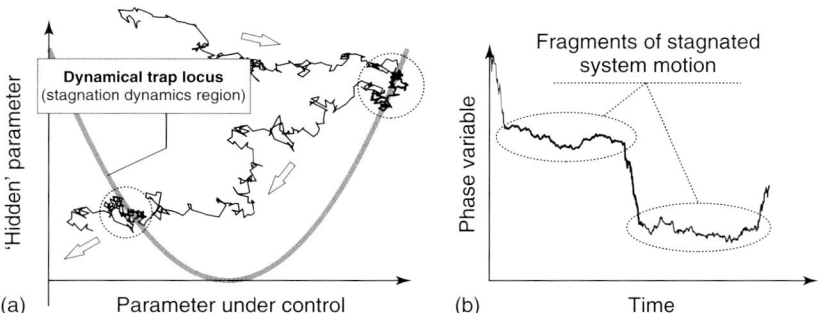

Figure 11.8 Illustration of the dynamical trap effect: (a) depicts the phase space with the region of the dynamical traps where the system motion is stagnated; (b) shows the time pattern for one of the phase state variables.

the sign, and the system motion is governed by a random Langevin force. A first step towards this effect in oscillator ensembles was made in [129].

11.9
Oscillator with Dynamical Traps

In order to elucidate the mechanism of such nonequilibrium phase transitions, this section analyzes a model derived from the damping harmonic oscillator where the dynamical trap region is a narrow layer in the phase space, inside of which the particle velocity is equal to zero. Namely, the following system is under consideration

$$\frac{dx}{dt} = v, \qquad (11.50)$$

$$\frac{dv}{dt} = -\omega_0^2 \Omega(v) \left[x + \frac{\sigma}{\omega_0} v \right] + \epsilon_0 \xi_v(t). \qquad (11.51)$$

Here x and v are the dynamical variables treated as the coordinate and velocity of a certain particle, ω_0 is the circular frequency of oscillations provided the system is not affected by other factors, σ is the damping decrement, and the term $\epsilon_0 \xi_v(t)$ in (11.51) is a random Langevin force of intensity ϵ_0 proportional to the white noise $\xi_v(t)$,

$$\langle \xi_v(t) \rangle = 0, \quad \langle \xi_v(t) \xi_v(t') \rangle = \delta(t - t'), \qquad (11.52)$$

with unit amplitude. At $\Omega(v) = 1$ and $\sigma = 0$ the system of equations (11.50)–(11.51) corresponds to the classical harmonic oscillator. Here we treat another case, where the function $\Omega(v)$ describes the dynamical-trap effect in the vicinity of $v = 0$. The following simple ansatz

$$\Omega(v) = \frac{v^2 + \Delta^2 \vartheta_t^2}{v^2 + \vartheta_t^2}, \qquad (11.53)$$

is adopted, where the parameter ϑ_t characterizes the thickness of the trap region and the parameter $\Delta \leq 1$ measures the trapping efficacy. When $\Delta = 1$ the dynamical-trap effect is negligible and for $\Delta = 0$ it is most effective. It should be pointed out that the governing equations (11.50) and (11.51) are written in terms of time derivatives because in the given case the noise is additive and the result is independent of the random process parameter θ.

The characteristic features of the given system are illustrated in Figure 11.9. The shaded area shows the trap region where the regular force, the former term in (11.51), is depressed. The latter is described by the factor $\Omega(v)$ taking small values in the trap region (for $\Delta \ll 1$). Inside the trap region the system is mainly governed by the random Langevin force. Outside the trap region it is approximately harmonic.

In order to analyze the system dynamics, a dimensionless time t and the dynamical variables η and u are used. Namely, the time t is measured in units of

11 Noise-Induced Phase Transitions

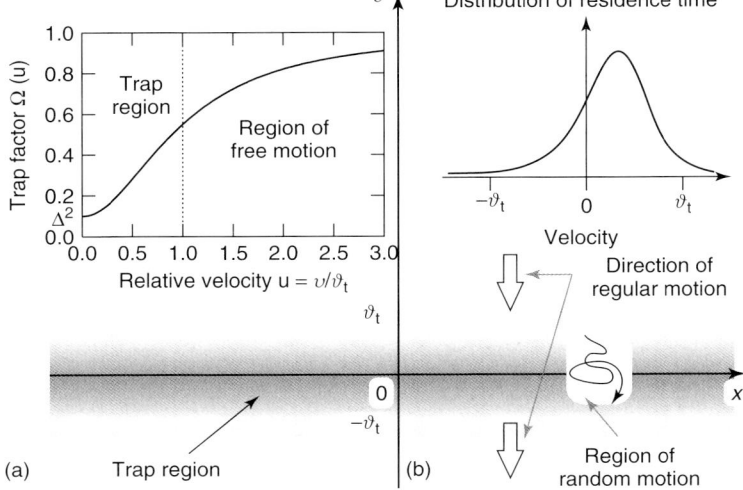

Figure 11.9 Characteristic structure of the phase space $\{x, v\}$. The shaded area represents the trap region where the regular force is depressed and the system motion is random. The regular force depression is described by the factor $\Omega(v)$ illustrated in (a). The essence of the trap effect on the system dynamics is shown in (b). Outside the trap region the system dynamics is mainly regular.

$1/\omega_0$, that is, $t \to t/\omega_0$ and the units of the coordinate x and the velocity v are ϑ_t/ω_0 and ϑ_t, respectively. So, by introducing the new variables

$$\eta = \frac{x\omega_0}{\vartheta_t} \quad \text{and} \quad u = \frac{v}{\vartheta_t},$$

the dynamical equations (11.50), (11.51) read (for the dimensionless time t)

$$\frac{d\eta}{dt} = u, \quad \frac{du}{dt} = -\Omega[u](\eta + \sigma u) + \epsilon \xi(t), \tag{11.54}$$

where the noise $\xi(t)$ obeys conditions like equalities (11.52), the parameter ϵ is $\epsilon = \epsilon_0/(\sqrt{\omega_0}\vartheta_t)$, and the function $\Omega[u]$ is given by

$$\Omega[u] = \frac{u^2 + \Delta^2}{u^2 + 1}.$$

Without noise, this system has only one stationary point $\{\eta = 0, u = 0\}$ being stable because it possesses a Lyapunov function

$$\mathcal{H}(\eta, u) = \frac{\eta^2}{2} + \frac{u^2}{2} + \frac{1 - \Delta^2}{2} \ln\left(\frac{u^2 + \Delta^2}{\Delta^2}\right). \tag{11.55}$$

This Lyapunov function attains the absolute minimum at the point $\{\eta = 0, u = 0\}$ and obeys the inequality

$$\frac{d\mathcal{H}(\eta, u)}{dt} = -\sigma u^2 < 0 \quad \text{for} \quad u \neq 0. \tag{11.56}$$

In particular, if $\sigma = 0$ and $\epsilon = 0$, then function (11.55) is the first integral of the system. In the following, the values σ and ϵ will be treated as small parameters.

The dynamics of system (11.54) was analyzed numerically using a high-order stochastic Runge–Kutta method [23] (see also [24]). The distribution function $\mathcal{P}(\eta, u)$ was calculated numerically by finding the cumulative time during which the system is located inside a given mesh on the (η, u)-plane for a path of a sufficiently long time of motion, $t \approx 500\,000$. The size of mesh was chosen to be about 1% of the dimension characterizing the system location on the (η, u) plane.

The evolution of the distribution function $\mathcal{P}(\eta, u)$ is shown in Figure 11.10 in the form of the level contours dividing the variation scale into ten equal parts. Part (a) corresponds to the case of $\triangle = 1$ where the trap effect is absent and the distribution function is unimodal; (c) illustrates the case when the distribution function has a well pronounced bimodal shape, shown also in Figure 11.11. Comparing (a), (b) and (c) in Figure 11.10, it becomes evident that there is a certain relation $\Phi_c(\triangle, \sigma, \epsilon) = 0$ between the parameters $\triangle$, σ, and ϵ when the system undergoes a second-order phase transition, which manifests itself as a change in the shape of the phase space density $\mathcal{P}(\eta, u)$ from unimodal to bimodal. In particular, for $\sigma = 0.1$ and $\epsilon = 0.1$ the critical value of the parameter $\triangle$ is $\triangle_c(\sigma, \epsilon) \approx 0.5$, as can be seen in part (b).

To understand the mechanism of the noise-induced phase transition observed numerically in the given system, consider a typical fragment of the system motion through the trap region for $\triangle \ll 1$ as shown in Figure 11.12. When it goes into the trap region $\mathcal{Q}_t$, $-\vartheta_t \ll v \ll \vartheta_t$, the regular force $\Omega[u](\eta + \sigma u)$ containing the trap factor $\Omega[u]$ and governing the regular motion becomes small. Hence, inside this region the system dynamics becomes random due to the remaining weak Langevin force $\epsilon \xi(t)$. However, the boundaries $\partial_+ \mathcal{Q}_t$ (where $v \sim \vartheta_t$) and $\partial_- \mathcal{Q}_t$ (where $v \sim -\vartheta_t$) are not identical in properties with respect to the system motion. At the boundary $\partial_+ \mathcal{Q}_t$ the regular force leads the system inwards to the trap region $\mathcal{Q}_t$, whereas at the boundary $\partial_- \mathcal{Q}_t$ it causes the system to leave the region $\mathcal{Q}_t$. Outside the trap region $\mathcal{Q}_t$ the regular force is dominant. So, from the standpoint of the system motion inside the region $\mathcal{Q}_t$, the boundary $\partial_+ \mathcal{Q}_t$ is 'reflecting' whereas the boundary $\partial_- \mathcal{Q}_t$ is 'absorbing'.

As a result, the distribution of the residence time at different points in the region $\mathcal{Q}_t$ should be asymmetric, as schematically shown in Figure 11.9(b). This asymmetry is also seen in the distribution function $\mathcal{P}(\eta, u)$ obtained numerically. Its maxima are located at the points with nonzero values of the velocity, clearly visible in Figure 11.10(d). Therefore, during location inside the trap region the mean velocity of the system must be positive and tends to go away from the origin. This effect gives rise to an increase in the 'energy' $\mathcal{H}(\eta, u)$. Outside the trap region the 'energy' $\mathcal{H}(\eta, u)$ decreases according to (11.56). So, when the former effect becomes sufficiently strong, that is, the random force intensity ϵ exceeds a certain critical value, $\epsilon > \epsilon_c(\triangle, \sigma)$, the distribution function $\mathcal{P}(\eta, u)$ becomes bimodal.

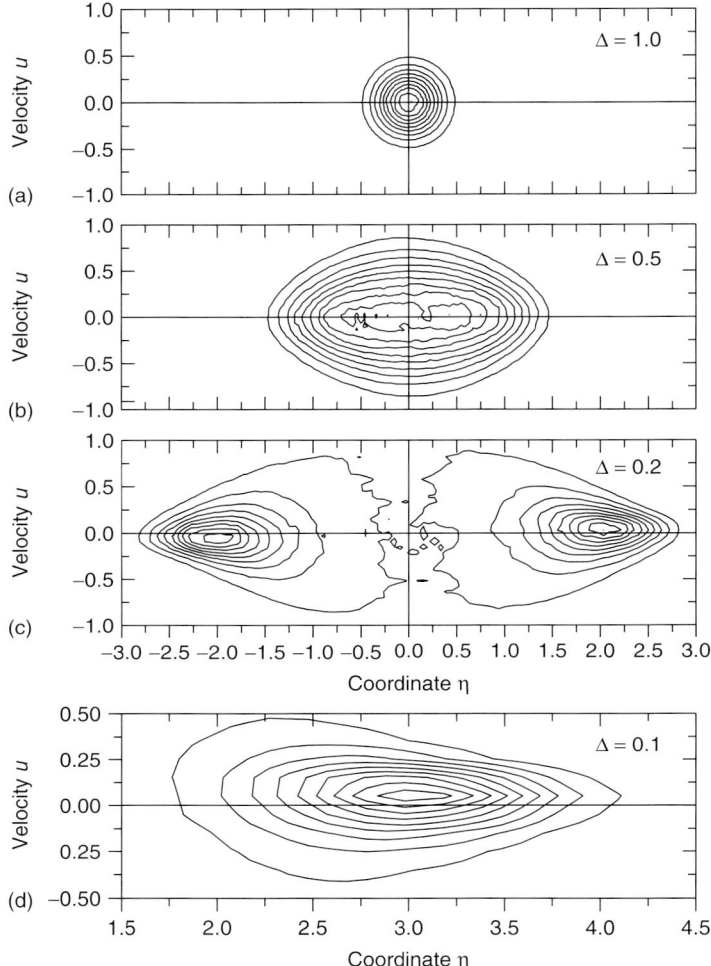

Figure 11.10 Evolution of the distribution function $\mathcal{P}(\eta, u)$ (shown by level contours) as the parameter $\triangle$ decreases. In numerical calculations the values $\sigma = 0.1$ and $\epsilon = 0.1$ were used; (d) depicts only one maximum of the whole distribution function.

11.10
Dynamics with Traps in a Chain of Oscillators

The previous section was devoted to the mechanism via which phase transitions induced by dynamical traps arise. This section demonstrates that dynamical traps do in fact give rise to cooperative phenomena that can be regarded as the formation of new phases.

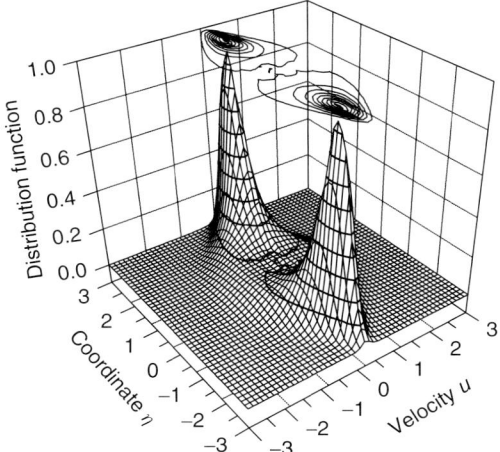

Figure 11.11 The form of the distribution function $\mathcal{P}(\eta, u)$ for the parameters $\sigma = 0.1$, $\epsilon = 0.1$, and $\triangle = 0.2$.

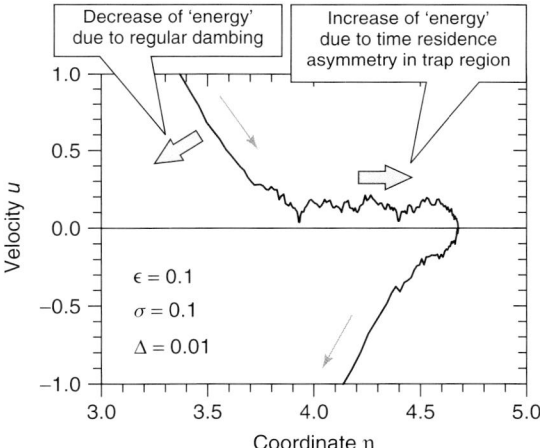

Figure 11.12 A typical fragment of the system path going through the trap region. The parameters $\sigma = 0.1$, $\epsilon = 0.1$, and $\triangle = 0.01$ were used in numerical simulations in order to make the trap effect more pronounced.

We analyze a one-dimensional ensemble of 'lazy' particles. These particles are characterized by their positions and velocities $\{x_i, v_i\}$ as well as possessing some motives for active behavior. Particle i 'wishes' to get the 'optimal' middle position between the nearest neighbors. Thus, one of the stimuli causing it to accelerate or decelerate is the difference $\eta_i = x_i - \frac{1}{2}(x_{i-1} + x_{i+1})$ provided its relative velocity $\vartheta_i = v_i - \frac{1}{2}(v_{i-1} + v_{i+1})$ with respect to the pair of nearest neighbors is sufficiently low. Otherwise, especially if particle i is currently located near the optimal position, it has to eliminate the relative velocity ϑ_i, being the other stimulus for particle i to change its state of motion. Since a particle cannot predict the dynamics of its neighbors, it has to regard them as moving uniformly with the current velocities. The acceleration dv_i/dt is determined directly by both stimuli. The model to be formulated combines both of these stimuli within a linear approximation $(\eta_i + \sigma\vartheta_i)$, where σ is the relative weight of the second stimulus.

When, however, the relative velocity ϑ_i attains sufficiently low values, the current situation for particle i cannot become worse, at least, it cannot occur soon. In this case particle i 'prefers' not to change the state of motion and to retard the correction of its relative position. This assumption leads to the appearance of some common cofactor $\Omega(\vartheta_i)$ in the corresponding governing equation as

$$\frac{dv_i}{dt} \propto -\Omega(\vartheta_i)(\eta_i + \sigma\vartheta_i).$$

The cofactor $\Omega(\vartheta)$ has to satisfy the inequality $\Omega(\vartheta) \ll 1$ for $\vartheta \ll \vartheta_c$ and, $\Omega(\vartheta) \approx 1$ when $\vartheta \gg \vartheta_c$, where ϑ_c is a critical value quantifying the particle's 'perception' of speed. The inclusion of such a factor is the implementation of the dynamical-trap effect.

Now we will describe the model. The following linear chain of N point-like particles is considered (Figure 11.13). Each internal particle $i \neq 1, N$ can freely move along the x-axis interacting with the nearest neighbors, namely, particles $i-1$ and $i+1$ interact via ideal elastic springs with some quasi-viscous friction. The dynamics of this particle ensemble is governed by the collection of coupled equations

$$\frac{dx_i}{dt} = v_i, \tag{11.57}$$

$$\frac{dv_i}{dt} = -\Omega(\vartheta_i, h_i)[\eta_i + \sigma\vartheta_i + \sigma_0 v_i] + \epsilon\xi_i(t). \tag{11.58}$$

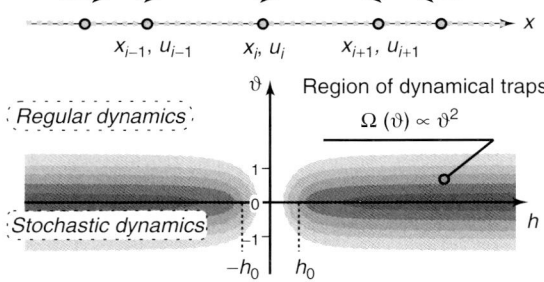

Figure 11.13 The particle ensemble under consideration and the structure of the phase space. The darkened region depicts the points where the dynamical-trap effect is pronounced. For the relationship between the variables x_i, v_i, h_i, and ϑ_i see (11.60) and (11.61).

Here for $i = 2, 3, \ldots, N-1$ the variables η_i and ϑ_i to be called the symmetry distortion and the distortion rate, respectively, are specified as

$$\eta_i = x_i - \tfrac{1}{2}(x_{i-1} + x_{i+1}), \tag{11.59}$$

$$\vartheta_i = v_i - \tfrac{1}{2}(v_{i-1} + v_{i+1}), \tag{11.60}$$

the mean distance h_i between the particles at the point x_i, by definition, is

$$h_i = \tfrac{1}{2}(x_{i+1} - x_{i-1}), \tag{11.61}$$

and $\{\xi_i(t)\}$ is the collection of mutually independent white-noise sources of unit amplitude, so

$$\langle \xi_i(t) \rangle = 0, \quad \langle \xi_i(t)\xi_{i'}(t') \rangle = \delta_{ii'}\delta(t - t'). \tag{11.62}$$

Also, the parameter ϵ is the noise amplitude, σ is the viscous friction coefficient of the springs, σ_0 is a small parameter that can be treated as some viscous friction related to the particle motion with respect to the given physical frame. It is introduced to prevent the system motion as a whole reaching an infinitely high velocity. The symbol $\langle \ldots \rangle$ denotes averaging over all the noise realizations; $\delta_{ii'}$ and $\delta(t - t')$ are the Kronecker symbol and the Dirac δ-function. The factor $\Omega(\vartheta_i, h_i)$ is due to the effect of dynamical traps and, following our previous ansatz, we write

$$\Omega(\vartheta, h) = \frac{\vartheta^2 + \triangle^2(h)}{\vartheta^2 + 1}, \tag{11.63}$$

where the function $\triangle(h)$ of the form

$$\triangle^2(h) = \triangle^2 + (1 - \triangle^2)\frac{h_0^2}{h^2 + h_0^2} \tag{11.64}$$

is used. The parameter $\triangle \in [0,1]$ quantifies the dynamical trap influence and the spatial scale h_0 specifies the small distances within which the trap effect is depressed, so for $h \ll h_0$ its value is $\triangle(h) \approx 1$, whereas for $h \gg h_0/\triangle$ it is $\triangle(h) \approx \triangle$. If this parameter is $\triangle = 1$, then the dynamical traps do not exist at all. In the opposite case, $\triangle \ll 1$, their influence is pronounced inside a certain neighborhood of the h-axis (trap region) whose thickness is about unity (Figure 11.17). The temporal and spatial scales have been chosen so that the thickness of the trap region is about unity, and the oscillation circular frequency is also equal to unity outside the trap region. The terminal particles, $i = 1$ and $i = N$, are assumed to be fixed, so

$$x_1(t) = 0, \quad x_N(t) = (N-1)l, \tag{11.65}$$

where l is the particle spacing in the homogeneous chain. The particles are treated as mutually impermeable ones. Therefore, when the coordinates x_i and x_{i+1} of an internal particle pair become identical, an absolutely elastic collision is assumed to happen, so if $x_i(t) = x_{i+i}(t)$ at a certain time t, then the timeless velocity exchange

$$v_i(t+0) = v_{i+1}(t-0),$$
$$v_{i+1}(t+0) = v_i(t-0) \tag{11.66}$$

comes into being. Multiparticle collisions are ignored. The system of equations (4.155)–(11.66) forms the model under consideration. We note again that the Langevin sources enter this model linearly, so the governing equations admit the representation with time derivatives.

The stationary point $x_i^{st} = (i-1)l$ is stable with respect to small perturbations; it stems from the linear stability analysis with respect to perturbations of the form

$$\delta x_i(t) \propto \exp\{\gamma t + ikl(i-1)\}, \tag{11.67}$$

where γ is the instability increment, k is the wave number, and the symbol **i** denotes the imaginary unit. The boundary conditions (11.65) are fulfilled by assuming the wave number k to take the values $k_m = \pi m/[(N-1)l]$ for $m = \pm 1, \pm 2, \ldots, \pm(N-2)$. For large values of the particle number N the parameter k can be treated as a continuous variable. Using the standard technique, the system of equations (11.57), (11.58) for perturbation (11.67) leads us to the following relation between the instability increment $\gamma(k)$ and the wave number k:

$$\gamma = -\Omega_0 \left[\frac{1}{2}\sigma_0 + \sigma \sin^2\left(\frac{kl}{2}\right)\right]$$
$$+ \mathbf{i}\sqrt{2\Omega_0 \sin^2\left(\frac{kl}{2}\right) - \Omega_0^2\left[\frac{1}{2}\sigma_0 + \sigma \sin^2\left(\frac{kl}{2}\right)\right]^2}. \tag{11.68}$$

Ansatz (11.63) has been used in deriving (11.68), enabling us to set $\Omega_0 = \Omega(0, l) = \Delta^2(l)$. Whence it follows that $\text{Re}\,\gamma(k) > 0$ for $k > 0$, so the homogeneous state of the chain is stable with respect to infinitely small perturbations of the particle arrangement.

The nonlinear dynamics of the given system has been analyzed numerically. Integration of the stochastic differential equations (4.155) and (11.58) was performed using the E2 high-order stochastic Runge–Kutta method [23,24]. Particle collisions were implemented analyzing a linear approximation of the system dynamics within *one* elementary step of the numerical procedure and finding the time at which a collision has happened. Then this step, treated as a complex one, was repeated. The integration time step of 0.02 was used, the results obtained were checked to be stable with respect to decreasing integration time step. An ensemble of 1000 particles was studied in order to make the statistics sufficient and to avoid the strong effect of the boundary conditions. The integration time T was chosen from 5000 to 8000 time units in order to make the calculated distributions stable. At the initial stage all the particles were distributed uniformly in space, whereas their velocities were randomly and uniformly distributed within the unit interval. The results of numerical simulation were used to evaluate the following partial distributions

$$\mathcal{P}(z) = \frac{1}{(N-2M)(T-T_0)} \sum_{i=M}^{N-M} \int_{T_0}^{T} dt\, \delta(z - z_i(t)), \tag{11.69}$$

where the time dependence $z_i(t)$ describes the dynamics of one of the variables $\eta_i(t)$, $\vartheta_i(t)$, and $v_i(t)$ ascribed to particle i. Here z is a given point of the space $\mathbb{R}_z$ describing the symmetry distortion η, the distortion rate ϑ, and the particle velocity v, respectively. The variables $\{\eta, \vartheta, v\}$ enable one to represent the system dynamics portrait within the space $\mathbb{R}_\eta \times \mathbb{R}_\vartheta \times \mathbb{R}_v$ or its subspace; N is the total number of particles in the ensemble, and M is the number of particles located near each of its boundaries. These are excluded from the consideration in order to weaken the possible effect of the specific boundary conditions. The same is true for the lower boundary of time integration T_0; its value is chosen to eliminate the effect of the specific initial conditions. The numerical implementation of the integration over time in (11.69) was related to the direct summation of the time series obtained. The partition of the corresponding space $\mathbb{R}_z$ was chosen such that the results are practically independent of the cell size. The value of M was also chosen using the stability of the result with respect to the double increase in M. The values of $M \sim 50$ and $T_0 \sim 500\text{--}1000$ were used.

Let us first discuss local properties of these ensembles. The term 'local' means that the corresponding state variable can take practically independent values when the particle index i changes by one or two. The variable η_i (expression (11.59)) may be regarded in this manner as it describes the symmetry of the particle arrangement in space. When $\eta_i = 0$, particle i takes the middle position between the nearest neighbors, particles $i-1$ and $i+1$. A nonzero value of η_i denotes its deviation from this position; in other words, a local distortion of the ensemble symmetry. This was the reason for the name used for the variables η_i as well as the variables $\vartheta_i = d\eta_i/dt$.

Figure 11.14 shows the distribution of the variables η and ϑ depending on the dissipation rate σ and the initial distance l between particles, that is, their mean density.

In the case of weak dissipation, the distribution functions of the symmetry distortion $\mathcal{P}(\eta)$ possess two maxima, matching the effect described in the previous section. Noise makes the uniform particle distribution unstable and the particles spend the main time in the vicinity of one or other neighbors. This leads to the bimodal distribution of the symmetry distortion η. After entering the region of dynamical traps the particle motion is stagnated, whereas outside it particles move relatively fast. This fact is reflected in the distribution of the distortion rate ϑ actually found to contain two components of different scales. The narrow component is due to the particle motion inside the trap region. This should be practically independent of the mean distance between particles. By contrast, the wide one depends remarkably on the particle density because it matches the fast motion of particles outside the trap region and, thus, has to be affected by their relative dynamics. This effect is exactly demonstrated in Figure 11.14 which also depicts the corresponding properties of the particle paths.

11 Noise-Induced Phase Transitions

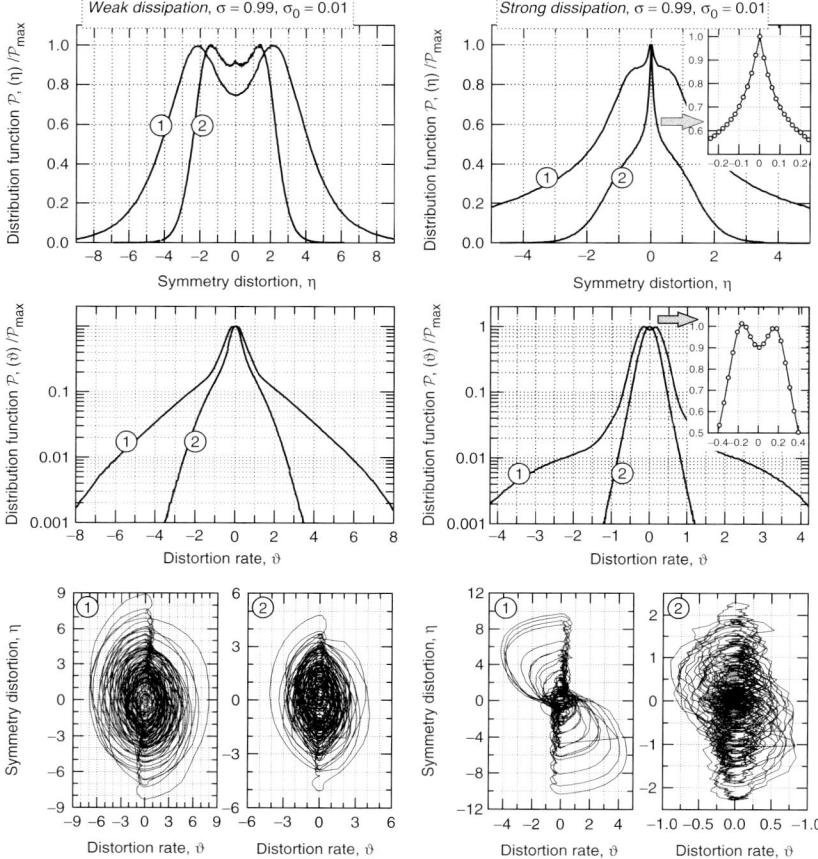

Figure 11.14 The distribution functions of the symmetry distortion η and the distortion rate ϑ for the 1000 particle ensemble with low ($l = 50$, label 1) and high ($l = 5$, label 2) density and weak ($\sigma \approx 0.1$) and strong ($\sigma \approx 1.0$) dissipation. The lower four windows depict characteristic path fragments of duration 1000 time units formed by a single particle with index $i = 500$ on the phase plane $\{\eta, \vartheta\}$ which was chosen due to its middle position in the given ensemble. The other parameters used are the noise amplitude $\epsilon = 0.1$, the trap-effect measure $\triangle = 0.1$, the small regularization friction coefficient $\sigma_0 = 0.01$ and the regularization spatial scale $h_0 = 0.25$. The time interval within which the data were averaged changed from 2000 to 5000 in order to make the obtained distributions stable.

In the case of strong dissipation, $\sigma \approx 1.0$, the situation changes dramatically, although the characteristic scales of the corresponding distributions turn out to be of the same order of magnitude. Now the distribution function $\mathcal{P}(\eta)$ of the symmetry distortion has only one maximum at $\eta = 0$; however, its form is characterized by two scales. In other words, it looks like a sum of two monoscale components. One of these is sufficiently wide, its thickness is about the same value as that obtained for the corresponding particle ensemble with weak dissipation. This component exhibits a remarkable dependence on the particle density, enabling us to relate it to the particle motion outside the trap region. The other is characterized by an extremely narrow and sharp form shown in detail in the inset of Figure 11.14 for the dense particle ensemble. Its sharpness leads us to the assumption that 'many–particle' effects in such systems with dynamical traps cause the symmetrical state to be singled out from the other possible states concerning the system properties. By contrast, the distortion rate behaves in a similar way to the previous case except for some details. When the mean particle density is high ($l = 5$) the wide component of the distortion rate distribution disappears and only the narrow one remains, with the latter having a quasi-cusp form $\propto \exp\{-|\vartheta|\}$. For the system with low density, the peak of the distortion rate distribution splits into two small spikes.

These features can be explained by referring to the frames in Figure 11.14, which exhibit typical path fragments formed by the motion of a single particle on the $\{\eta\vartheta\}$ plane. Roughly speaking, three motion types can be singled out: some stagnation inside a narrow neighborhood of the origin $\{\eta = 0, \vartheta = 0\}$; slow wandering inside the trap region that, on average, follows a line with a finite positive slope; and the fast motion outside the trap region. The fast motion fragments typically stem from an arbitrary point of the low motion region and lead to a certain neighborhood of the origin. It seems that the systems with a low density of particles have the possibility of going sufficiently far from the origin and, during the fast motion, rarely come into the stagnation region. As a result, first, the distortion rate distribution function is of a two-scale form and contains two spikes on the peak. In the case of high density, the fast motion is depressed substantially and the system migrates mainly in the slow-motion region, entering the stagnation region many times. Thus, the distortion-rate distribution converts into a single-scale function and the symmetric state occurs often, giving rise to a significant sharp component of the distortion distribution located near the point $\eta = 0$.

Now let us discuss the nonlocal characteristics of the 1000-particle ensembles. Figure 11.15 depicts the velocity distributions. As we can see, it depends essentially on both parameters; the mean particle density and the dissipation rate. When the mean particle density is low and the dissipation is weak ($l = 50$ and $\sigma \approx 0.1$), the velocity distribution is practically of Gaussian form, however, its width has extremely large values about 10. The tenfold increase in the particle density, $l : 50 \mapsto 5$, shrinks the velocity distribution to the same order and its scale becomes similar to that of the distortion rate distribution in magnitude. However, in this case the form of the velocity distribution is a monoscale function of the well pronounced cusp form $\propto \exp\{-|v|\}$. In the case of strong dissipation ($\sigma \approx 1.0$) the situation is the opposite. The system with low density ($l = 50$), as previously, is characterized by an extremely wide velocity distribution, its width is about 10. However, now its form deviates substantially from the Gaussian one. For the corresponding ensemble with high density ($l = 5$) the velocity distribution is Gaussian with width about 1. The latter, nevertheless, is much larger than the same width in the absence of dynamical traps.

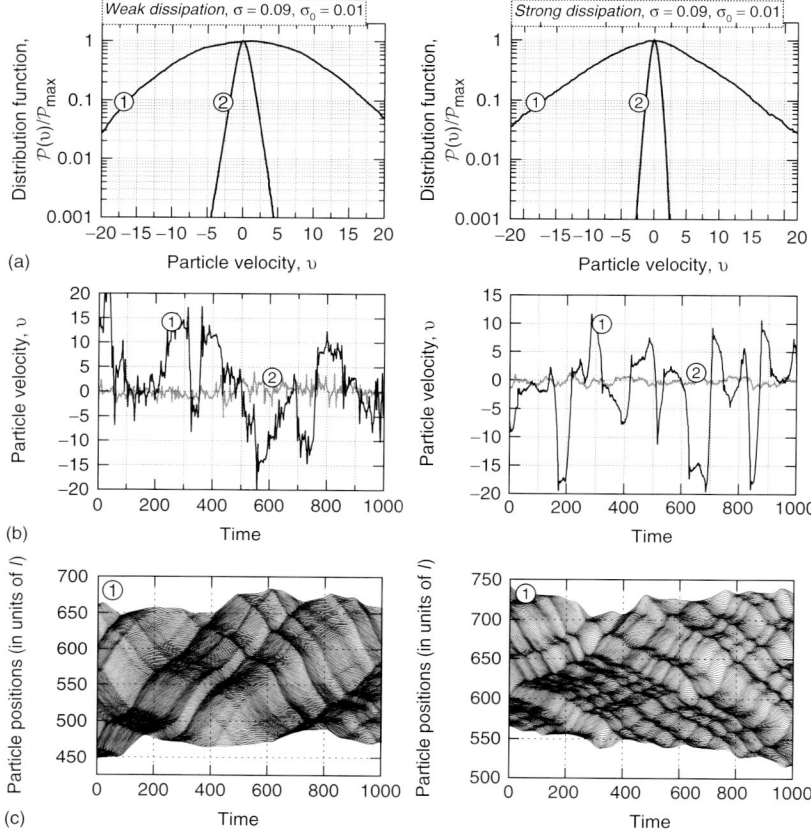

Figure 11.15 The distribution functions of the particle velocities and the characteristic time patterns formed by the velocity variations of the 500 th particle. Dynamics of the 1000-particle ensemble with low ($l = 50$, label 1) and high ($l = 5$, label 2) mean density and weak ($\sigma \approx 0.1$) and strong ($\sigma \approx 1.0$) dissipation was implemented for the calculation time up to 8000 time units to make the obtained distributions stable with respect to a time increase. (c) shows the time patterns formed by 200 paths of particle motion during 1000 time units and chosen in the middle of the given ensemble. Here the curve thickness has been chosen so that the difference in brightness can depict local variations in the path spacing due to changes either in the particle density or in the velocities of cooperative particle motion (in this way the different long-lived states of the given particle ensemble become apparent). The other parameters used are the noise amplitude $\epsilon = 0.1$, the trap-effect measure $\Delta = 0.1$, the small regularization friction coefficient $\sigma_0 = 0.01$ and the regularization spatial scale $h_0 = 0.25$.

These features of the velocity distribution characterize the cooperative behavior of particles rather than their individual dynamics. In other words, there should be strong correlations in the motion of not only neighboring particles but also distant ones. Therefore, the velocity variations responsible for the formation of such distributions in fact describe the motion of multiparticle clusters. To justify this, we refer to Figure 11.15(b) which demonstrates some typical fragments of the time patterns formed by the velocities of individual particles. When the mean particle density is low ($l = 50$), these patterns look like a sequence of fragments $\{v_\alpha\}$ inside which the particle velocity varies in the vicinity of some level v_α. The values $\{v_\alpha\}$ are rather randomly distributed inside a certain region of thickness $V \sim 10$ in the vicinity of $v = 0$. The continuous transitions between these fragments occur via sharp jumps. The typical duration of these fragments is about $T \sim 100$, which enables us to regard them as long-lived states because the temporal scales of individual particle dynamics are about several units. Moreover, these long-lived states can persist only if a group of many particles moves as a whole because the characteristic distance L individually traveled by a particle involved in such state is about $L \sim VT \sim 1000 \gg l$.

The spatial structure of these cooperative states is depicted in Figure 11.15 which also shows time patterns formed by paths $\{x_i(t)\}$ for 200 particles of duration about 1000 time units. These particles were chosen in the middle part of the 1000-particle ensembles with low density. For high-density ensembles such patterns also develop, but are not so pronounced. As we can see, a large number of different mesoscopic states are formed in these systems. They differ from one another in size, direction of motion, speed, life-time, etc. Moreover, the life-time of such a state can be much longer than the characteristic time interval during which particles forming it will currently belong to this state individually. Besides, the patterns found could be classified as hierarchical structures. Some relatively small domains formed by cooperative motion of individual particles, in their turn, together make up larger superstructures. In other words, the observed long-lived cooperative states have their 'own' life independent, in some sense, of the individual particle dynamics. The latter properties are the reason for regarding them as certain dynamical phases arising in the systems under consideration due to the dynamical traps affecting the individual particle motion. The term 'dynamical' has been used to underline the fact that the complex cooperative motion of particles is responsible for these long-lived states; without the continuous particle motion such states cannot exist.

11.11
Self-Freezing Model for Multi-Lane Traffic

This section is devoted to a model for multi-lane congested traffic flow, where the dynamical traps give rise to the continuum of long-lived states observed in real traffic. When vehicles move on a multi-lane highway without changing lanes, they interact practically only with the nearest neighbors ahead. The more frequently lane changing occurs, the more correlated is the traffic flow on a multi-lane highway. Therefore, to characterize traffic flow on multi-lane highways, it is reasonable to introduce an additional state variable, the order parameter η [131]. In this case the mean velocity v of multi-lane traffic flow is determined by both the vehicle density ρ and the order parameter η, namely, $v = \vartheta(\eta, \rho)$.

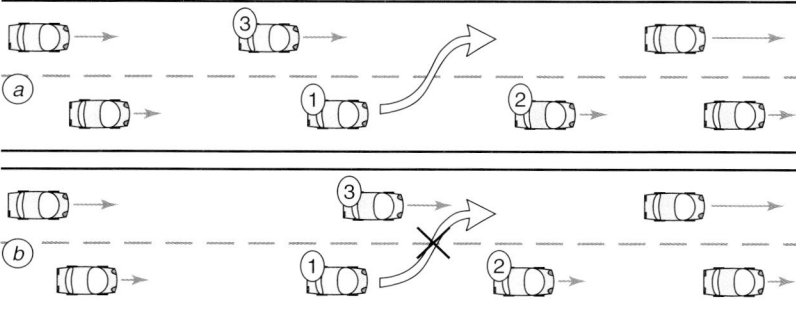

Figure 11.16 Schematic illustration of the car arrangement in the neighboring lane in the synchronized mode that enables overtaking (a) or hinders it (b).

However, to describe phase transitions in cooperative vehicle motion, we have to treat the multi-lane car interaction in more detail. The fact is that, for a car to change lane, the local vehicle arrangement at the neighboring lanes should be of a special form, otherwise it will be prevented for a certain time, as illustrated in Figure 11.16. For car 1 to be able to overtake car 2 the neighboring car 3 should provide room for this maneuver. In the opposite case the driver of car 1 has to wait and the local car arrangement will not vary substantially. In other words, changes in the particular realizations of the local car arrangement can be frozen for a certain time although the globally optimal car configuration is not attained at the current moment of time. Due to this self-freezing effect, the synchronized mode can comprise a great number of locally metastable states and correspond to a certain two-dimensional region in the ρq plane rather than to a line $q = \vartheta(\rho)\rho$. This feature seems to be similar to that met in physical media with local order; for example, in glasses where phase transitions are characterized by a wide range of controlling parameters (temperature, pressure, etc.) rather than their fixed values (see, e.g. [252]).

We should specify the evolution of the order parameter a to complete the description of the long-lived state continuum of the cooperative car motion. Since the order parameter a allows for microscopic details of the fundamental cluster structure, its fluctuations will be treated as a random noise whose amplitude depends on the vehicle density only. In contrast, the rate da/dt of time variations in the order parameter a has to be affected substantially by the current value of the order parameter η. In fact, as the order parameter η tends to the local optimum value $\eta_0(a, \rho)$ for the given a, the rate da/dt should be depressed because all the drivers forming the fundamental cluster prefer to wait until a more comfortable car configuration arises, therefore inhibiting the evolution of the fundamental cluster structure.

11.11 Self-Freezing Model for Multi-Lane Traffic

The dimensionless model describing these effects in the congested multi-lane traffic takes the form [132]

$$d\eta = -(\eta - a^2) dt + \epsilon dW_\eta(t), \tag{11.70}$$

$$da = -\Omega_0 \varpi(\eta, a) a \, dt + \Omega_0^{1/2} \varpi^{1/2}(\eta, a) \, dW_a(t), \tag{11.71}$$

where the infinitesimal moments of the random Langevin forces dW_η and dW_a are mutually independent and the random process is of the Ito type. Finally, the function $\varpi(\eta, a)$ may be specified as:

$$\varpi(\eta, a) = \begin{cases} (\eta - a^2)^2/\phi_0^2 & \text{if } |\eta - a^2| \leq \phi_0 \\ 1 & \text{if } |\eta - a^2| > \phi_0 \end{cases}. \tag{11.72}$$

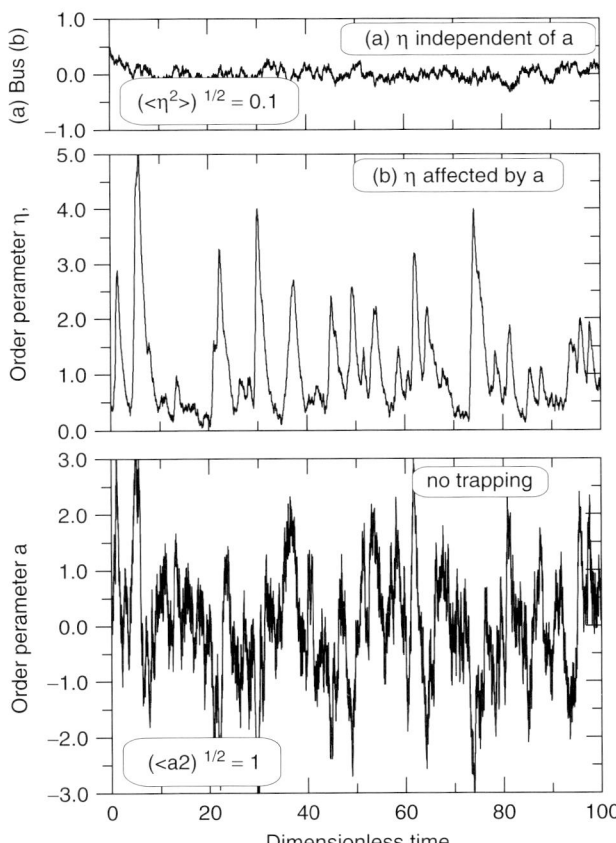

Figure 11.17 The time pattern of the order parameters η and a if the self-freezing effect is suppressed: (a) exhibits time variations in the order parameter η in the case where the influence of the order parameter a has been ignored; (b) presents the dynamics of η affected by a.

Here $\eta = a^2$ is the optimal value of the order parameter η attained for the given value of the car ensemble arrangement, and $a = 0$ matches the totally optimal structure of the car ensemble. When the car arrangement attains a local optimum for a fixed value of a, the drivers 'do not know what to do' and the system dynamics stagnates. In other words, the curve $\eta = a^2$ is the locus of dynamical traps and factor (11.72) describes them, with ϕ_0 being the threshold of driver perception.

Let us now discuss the results obtained by numerically simulating the system of (11.70), and (11.71) for $\Omega_0 = 1$, $\epsilon = 0.1$, and $\phi_0 = 0.5$. First of all, Figure 11.17 illustrates the evolution of the order parameters η and a when there is no self-freezing. In this case the order parameter a exhibits the standard random pattern of Brownian movement inside a region of unit width. We see a collection of practically independent spikes of unit width. A similar pattern (see Figure 11.17(a)) is demonstrated by the dynamics of the order parameter η provided the interaction of the parameters η and a have been ignored, so $(\eta - a^2)$ is replaced by η. Naturally, in this case the synchronized mode matches a line on the ρq plane for $\epsilon \ll 1$.

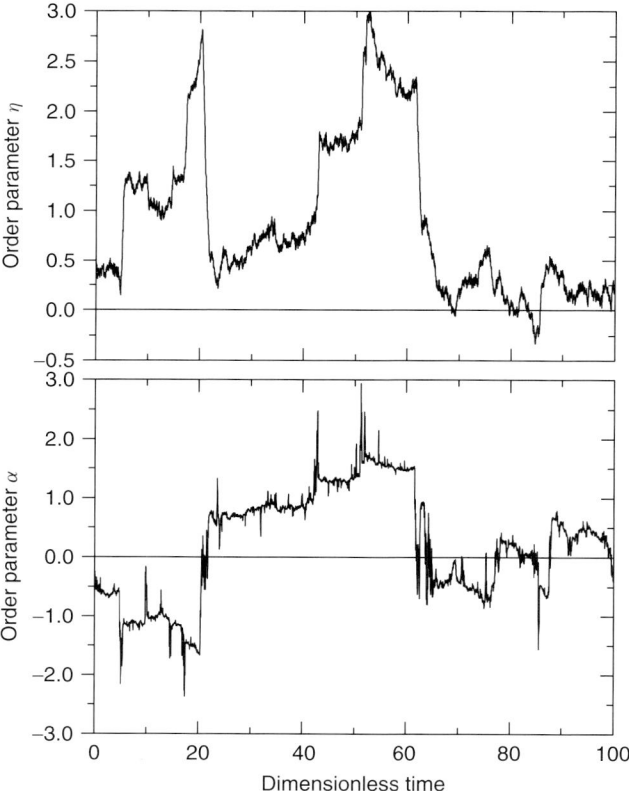

Figure 11.18 The time pattern of the order parameters η and a when the self-freezing effect is considerable.

The dynamics of the order parameter η, affected by the variable a but without the reciprocal influence, is shown in Figure 11.17(b). Again a collection of spikes can be seen, whose amplitude as well as width has increased tenfold.

The dynamics changes dramatically for the full problem, see Figure 11.18. The time pattern takes a form corresponding to the long-lived state continuum. When the point $\{a(t), \eta(t)\}$ representing the current state of the synchronized mode wanders on the $a\eta$ plane and reaches the curve $\eta = a^2$ at any point, it will be trapped for a certain time until it finally escapes from the trap due to the noise $\epsilon \, dW_\eta(t)$. After this the system again wanders in the $a\eta$ plane during a time interval about unity before being trapped for the next time. Since the characteristic duration of the trapping is much longer than unity, the pattern looks like a certain collection of local metastable states of the synchronized mode. However, such prolonged stays of the system are not metastable in a rigorous sense and we prefer to call them simply long-lived states. Since each point of the curve $\eta = a^2$ is a trap, the long-lived states make up a certain continuum.

The problem of a car following a lead car driven with constant velocity is considered in [134, 135]. To derive the governing equations for the dynamics of the following car a cost functional is constructed. This functional ranks the outcomes of different driving strategies. Assuming rational driver behavior, the existence of the Nash equilibrium is proved.

11.12
Exercises

E 11.1 Stochastic differential equations and stochastic processes
Verify whether the following stochastic differential equations of different types

$$dx = x \, dt + x_0^2 \, dW, \qquad (11.73a)$$

$$dx = (x - x^3) \, dt + x_{1/2}^2 \, dW, \qquad (11.73b)$$

$$dx = (x - 2x^3) \, dt + x_1^2 \, dW \qquad (11.73c)$$

describe the same process.

E 11.2 Transformation of variables in stochastic differential equations
Find the transformation of the variables $y = \phi(x)$ that converts the following stochastic differential equation of the Hänggi–Klimontovich type

$$dx = \frac{x}{(1 + x^2)^3} \, dt + \frac{1}{1 + x_1^2} \, dW \qquad (11.74a)$$

to the stochastic differential equation

$$dy = dW \qquad (11.74b)$$

describing the standard diffusion process.

E 11.3 Noise-induced phase transitions in a genetic processes

Let us consider a genetic model of stochastic processes specified by the equation

$$dx = -(\beta x + x^3)\,dt + \sigma(1 - x_\theta^2)\,dW \tag{11.75}$$

for $x \in (-1, 1)$. Find the threshold value of the noise intensity $\sigma = \sigma_c$ causing the unimodal–bimodal transition depending on the parameter $\beta > 0$ and the process type, that is, on the value of θ.

E 11.4 Anomalous properties of stochastic processes with nonlinear noise

Let a random process in $\mathbb{R}$ be governed by the following stochastic differential equation of the Hänggi–Klimontovich type

$$dx = -\alpha x\,dt + \sqrt{1 + x_1^2}\,dW, \tag{11.76}$$

where $\alpha > 0$ is a given positive constant. Find the stationary distribution $P^{\mathrm{st}}(x)$ of the random variable x and find the conditions when the moments

$$M_p := \int_{-\infty}^{+\infty} |x|^p P^{\mathrm{st}}(x)\,dx, \quad p = 1, 2, \ldots \tag{11.77}$$

diverge.

E 11.5 Geometric random walk without crossing boundaries

Let one-dimensional geometric random walks be governed by the equation

$$dx = x\,dt + x_{1/2}\,dW \tag{11.78}$$

of the Stratonovich type and let the walker be absorbed immediately when it gets to the point $x = x_{\mathrm{tr}}$. Find the probability for the walker to survive depending on its initial position $x_0 > x_{\mathrm{tr}}$ as time goes to infinity, $t \to \infty$.

12
Many-Particle Systems

12.1
Hopping Models with Zero-Range Interaction

In Chapters 9 and 10 in which we discussed the nucleation in supersaturated vapor, as well as the jam formation in traffic flow, we have mainly focused on the applications of stochastic Markov processes to one-cluster models. Here we will consider in some detail more complex many-particle models starting with a particular totally asymmetric particle-hopping model, where many clusters can coexist, as described by the so-called zero-range stochastic process as a system of interacting random walks [43, 67–69, 89, 97, 166, 209, 221].

In general, the particle-hopping models are those where the spatial coordinates of particles are discretized or split into cells. Each cell can be either empty or occupied by a particle. It can also contain many particles depending on the specific definition of the model. If the time is also discretized, then at each time step particles can jump to other cells. Different updating rules are possible such as parallel, sequential, or random. Each transition to the new discrete state of the particle system is characterized by certain probability. If in a one-dimensional model the probabilities for a particle to hop to the right and to the left are different, then the model is called asymmetric. The hopping models can also be defined in continuous time. In this case the hopping probabilities are replaced by transition rates, as is usual in the master equation approach discussed in Chapter 3. In a totally asymmetric particle-hopping model, particles can jump in only one direction. The stochastic particle-hopping process is called the exclusion process, if a cell can contain no more than one particle. The totally asymmetric simple exclusion process (TASEP) may be suitable for describing the traffic flow, as the motion of cars is indeed totally asymmetric (in one direction), and two vehicles cannot occupy the same position on the road. The Nagel–Schreckenberg cellular automaton model [170] with the maximum velocity $v_{\max} = 1$, where at each time step a car can move forwards to the next empty cell with certain probability, represents a simple TASEP. Characteristic plots showing sets of trajectories generated by the Nagel–Schreckenberg model are shown in Figure 12.1.

An extension of the TASEP to two-lane traffic has been made in [187]. In this case two possible internal states, like spin-up and spin-down in the Ising model,

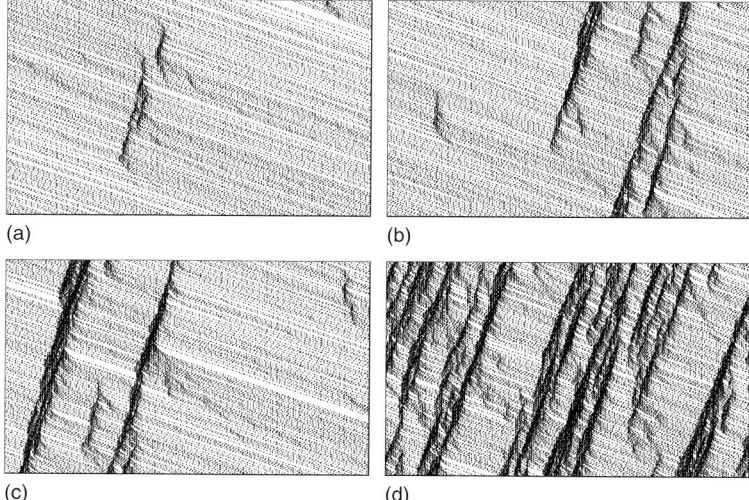

Figure 12.1 Spatio-temporal diagram of road traffic simulated by the Nagel–Schreckenberg model with $v_{max} = 5$. The results for four different car densities: (a) 8 %, (b) 12 %, (c) 15 %, (d) 30 %, are presented.

are assigned to each particle. These internal states allow one to distinguish between two lanes. The exclusion principle in TASEP is equivalent to the Pauli exclusion principle for electrons in this case. Another application of an asymmetric simple exclusion process was considered in [81] by mapping it onto an *XXZ* quantum chain.

A widely studied particle-hopping process is the so-called zero-range process, which could be viewed as a generic model for domain dynamics in one dimension [89]. In this model, each cell or box can contain an arbitrary number of particles n, whereas the process is defined by the hopping rates to the neighboring boxes, and these rates depend only on the number of particles in the departure box. Due to the latter property, it is called the zero-range process (ZRP).

The zero-range model is attractive from the point of view that many particle-hopping processes such as the TASEP can be easily mapped to the ZRP [43,89]. The mapping can be done in such a way that the number of particles in a box in ZRP corresponds to the size of a particle cluster or, alternatively, to the size of a gap between clusters in TASEP. The mapping of a traffic model to ZRP will be discussed in detail in the next section. The zero-range model is also attractive as the stationary probability distribution can be represented exactly as a product measure [42,221]. Also, it provides exact criteria for the phase separation [89] in a one-dimensional driven system. In the following section we will discuss these points in relation to the model, which is designed to describe some features of traffic flow such as phase separation and metastability [97].

12.2
The Zero-Range Model of Traffic Flow

The development of traffic jams in vehicular flow is an everyday example of the occurrence of phase separation in low-dimensional driven systems, a topic which has attracted much recent interest (see, e.g., [166, 209] and references therein). In [89] the existence of phase separation is related to the size-dependence of domain currents and a quantitative criterion is obtained by considering the zero-range process (ZRP) as a generic model for domain dynamics. Phase separation corresponds to the phenomenon of condensation in the ZRP (see [43] for a recent review) in which a macroscopic proportion of particles accumulate on a single site.

In the following we will use the zero-range picture to study the phase separation in traffic flow. We consider a model of traffic flow, where cars are moving along a circular road. Each car occupies a certain length of road ℓ. We divide the whole road of total length L into cells of size ℓ. Each cell can be either empty or occupied by a car, just as in cellular automaton traffic models (see, e.g., [31, 77, 151] and references therein). Most of these models use a discrete-time update rule; for example, see [120] for a class of traffic models related to a parallel updating version of the ZRP. In contrast, we consider the development of our system in continuous time. The probability per unit time for each car to move to the next cell is given by a certain transition rate, which depends on the actual configuration of empty and occupied cells. This configuration is characterized by the cluster distribution. An uninterrupted string of n occupied cells, bounded by two empty cells, is called a cluster of size n. The clusters of size $n = 1$ are associated with freely moving cars. The first car in each cluster is allowed to move forward by one cell. The transition rate w_n of this stochastic event depends on the size n of the cluster to which the car belongs. In this case w_1 is the mean of the inverse time necessary for a free car to move forward by one cell. The transition rate w_1 is related to the distribution of velocities in the free-flow regime or phase, which is characterized by a certain car density c_{free}. For small densities, expected in the free-flow phase in real traffic, the interaction between cars is weak and, therefore, the transition rate w_1 depends only weakly on the density c_{free}. Hence, in the first approximation we may assume that w_1 is a constant.

This model can be directly mapped to the zero-range process. Each vacancy (empty cell) in the original model is related to a box in the zero-range model. The number of boxes is fixed, and each box can contain an arbitrary number of particles (cars), which is equal to the size of the cluster located to the left (if cars are moving to the right) of the corresponding vacancy in the original model. If this vacancy has another vacancy to the left, then it means that the box is empty. Since the boundary conditions are periodic in the original model, they remain periodic also in the zero-range model. In this representation, one particle moves from a given box to the right with transition rate w_n, which depends only on the number of particles n in this box. In the grand canonical ensemble, where the total number of particles is allowed to fluctuate, the stationary distribution over the cluster-size configurations is the product of independent distributions for individual boxes. The probability that there are just n particles in a box in a homogeneous phase is [42, 221] $P(n) \propto z^n / \prod_{m=1}^{n} w_m$ for $n > 0$, $P(0)$ being given by the normalization condition. Here $z = e^{\mu/k_B T}$ is the fugacity a parameter which controls the mean number of particles in the system.

This result can be obtained and interpreted within the stochastic master equation approach [43]. Assuming the statistical independence of the distributions in different boxes, we have a multiplicative ansatz

$$P_2(k, m, t) = P(k, t) P(m, t) \qquad (12.1)$$

for the joint probability $P_2(k, m, t)$ that there are m particles in one box and k particles in the neighboring box on the left at time t. This approximation leads to the mean-field dynamics described by the master equations [43]

$$\frac{\partial P(n, t)}{\partial t} = \langle w \rangle P(n-1, t) + w_{n+1} P(n+1, t)$$
$$- [\langle w \rangle + w_n] P(n, t) \quad : \quad n \geq 1, \qquad (12.2)$$

$$\frac{\partial P(0, t)}{\partial t} = w_1 P(1, t) - \langle w \rangle P(0, t), \qquad (12.3)$$

where

$$\langle w \rangle(t) = \sum_{k=1}^{\infty} w_k P(k, t) \qquad (12.4)$$

is the mean inflow rate in a box. The ansatz (12.1) is an exact property of the stationary state of the grand canonical ensemble or, alternatively, of an infinitely large system [221]. Hence, in these cases, the master equations (12.2) and (12.3) give the exact stationary state while providing a mean-field approximation to the dynamics of reaching it.

The stationary solution $P(n)$ corresponding to $\partial P(n, t)/\partial t = 0$ can be found recursively, starting from $n = 0$. It yields the known result [42, 43, 221]

$$P(n) = P(0) \langle w \rangle^n \prod_{m=1}^{n} \frac{1}{w_m} \qquad (12.5)$$

for $n > 0$, where $P(0)$ is found from the normalization condition.

Denoting the number of boxes by M, which corresponds to the number of vacancies in the original model, the mean number of cars on the road is given by $\langle N \rangle = M \langle n \rangle$, where

$$\langle n \rangle = \sum_{n=1}^{\infty} n P(n) \qquad (12.6)$$

is the average number of particles in a box. Note that, in the grand canonical ensemble, the total number of cars and the length of the road L fluctuate. For the mean value, measured in units of ℓ, we have $\langle L \rangle = M + \langle N \rangle$. Hence, the average density of cars is

$$c = \frac{\langle N \rangle}{\langle L \rangle} = \frac{\langle n \rangle}{1 + \langle n \rangle}. \qquad (12.7)$$

According to (12.7), (12.6), and (12.5), we have the following relation

$$\frac{c}{1-c} = \frac{\sum_{n=1}^{\infty} n \langle w \rangle^n \prod_{m=1}^{n} \frac{1}{w_m}}{1 + \sum_{n=1}^{\infty} \langle w \rangle^n \prod_{m=1}^{n} \frac{1}{w_m}} \qquad (12.8)$$

from which the stationary mean inflow rate $\langle w \rangle$ can be calculated at a given average density c.

12.3
Transition Rates and Phase Separation

Now we make the following choice for the transition rate dependence on the cluster size n:

$$w_n = w_\infty \left(1 + \frac{b}{n^\sigma}\right) \quad \text{for} \quad n \geq 2, \qquad (12.9)$$

the value of w_1 being given separately, since it is related to the motion of uncongested cars, whereas w_n with $n \geq 2$ represents escape from a jam of size n. Although an individual driver does not know how many cars are jammed behind him, the effective current of cars from a jam, represented by w_n, is a collective effect which is expected to depend on the correlations and internal structure (e.g., distribution of headways) within the cluster [89]. A monotonously decreasing dependence on cluster size, such as (12.9), can be considered as a type of slow-to-start rule: the longer a car has been stationary the larger the probability of a delay when starting (cf. [14, 19, 127, 225]).

We now explore the consequences of the choice (12.9) in terms of the ZRP phase behavior and its implications for the description of traffic flow. In numerical calculations we have assumed $w_\infty = 1/\tau_\infty = 1$ and $w_1 = 5$, by choosing the time constant τ_∞ as a time unit, whereas the control parameters b and σ have been varied.

If $\sigma > 1$, as well as for $b \leq 2$ at $\sigma = 1$, then (12.8) has a solution for any density $0 < c < 1$ (see dashed and dotted curves in Figure 12.2). This implies that the homogeneous phase is stable in the whole range of densities, so there is no phase transition in a strict sense. If $\sigma < 1$ (solid curve in Figure 12.2), as well as for $b > 2$ at $\sigma = 1$ (see Figure 12.3), $\langle w \rangle/w_\infty$ reaches 1 at a critical density $0 < c_{\mathrm{cr}} < 1$, and there is no physical solution of (12.8) for $c > c_{\mathrm{cr}}$. This means that the homogeneous phase cannot accommodate a larger density of particles and condensation takes place at $c > c_{\mathrm{cr}}$.

This behavior underlies the known criterion for phase separation in one-dimensional driven systems [89]. For illustration, we comment that, in the multi-cluster model considered in [96], the transition rates do not depend on

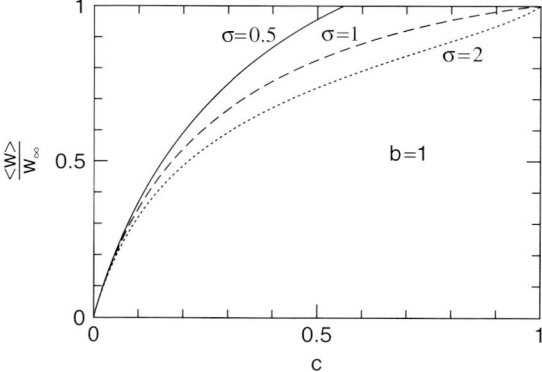

Figure 12.2 $\langle w \rangle / w_\infty$ versus density c at $b = 1$ for different σ.

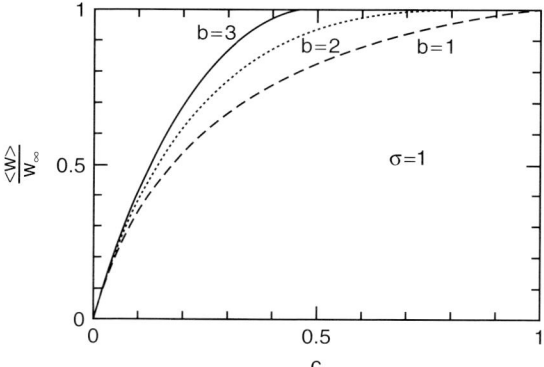

Figure 12.3 $\langle w \rangle / w_\infty$ versus density c at $\sigma = 1$ for different b.

the cluster sizes, only the inflow rate in a cluster depends on the overall car density and fraction of congested cars. This corresponds to the case $b = 0$, where, according to the criterion, no macroscopic phase separation takes place in agreement with the theoretical conclusions and simulation results of [96]. In contrast, a class of microscopic models was introduced in [90] where correlations within the domain (jam) give rise to currents of the form (12.9) with $\sigma = 1$ and $b > 2$; phase separation is then observed.

At $c < c_{\text{cr}}$ in our model the cluster distribution function $P(n)$ decays exponentially fast for large n (dashed and dotted curves in Figures 12.4 and 12.5), whereas the decay is slower at $c = c_{\text{cr}}$ (solid curves in Figures 12.4 and 12.5). It is well known that the decay in this case is power-like for $\sigma = 1$, so that $P(n) \sim n^{-b}$ [89].

The large n asymptotics for $0 < \sigma < 1$ at the critical density $c = c_{\text{cr}}$ is calculated as follows [97]. According to (12.5), at $c = c_{\text{cr}}$ where $\langle w \rangle = w_\infty$, we have

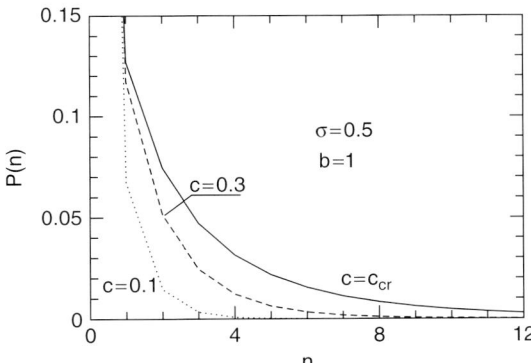

Figure 12.4 Probability distribution function $P(n)$ over cluster sizes for different densities c at $\sigma = 0.5$ and $b = 1$.

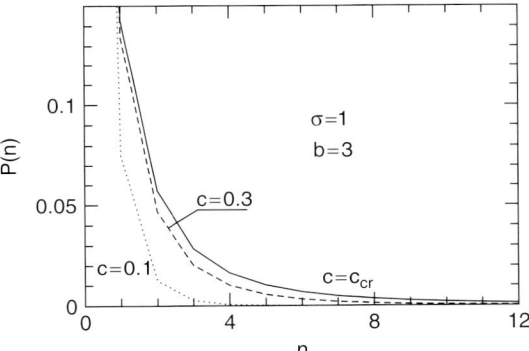

Figure 12.5 Probability distribution function $P(n)$ over cluster sizes for different densities c at $\sigma = 1$ and $b = 3$.

$$\ln P(n) = \ln\left(\frac{P(0)w_\infty}{w_1}\right) - \sum_{m=2}^{n} \ln\left(1 + \frac{b}{m^\sigma}\right) \quad (12.10)$$

$$= \ln\left(\frac{P(0)w_\infty}{w_1}\right) - \int_1^n \ln\left(1 + \frac{b}{m^\sigma}\right) dm + C + \delta(n),$$

where C is a constant, and $\delta(n) \to 0$ at $n \to \infty$. The latter follows from the fact that each term with $m = k$ in the sum generates terms $\propto k^{-\gamma}$ with $\gamma > 0$ when the logarithm is expanded in a Taylor series, and for each of these terms we have

$$k^{-\gamma} = \int_{k-1}^{k} m^{-\gamma}\, dm + O\left(k^{-\gamma-1}\right) \text{ at } k \to \infty.$$

This ensures that the difference between the integral and the sum in (12.10) is finite and tends to some constant C at $n \to \infty$ for $\sigma > 0$. The remainder term $\delta(n)$ is irrelevant for the leading asymptotic behavior of $P(n)$. By expanding the logarithm and integrating term by term, for $0 < \sigma < 1$ and $\sigma \neq 1/2, 1/3, 1/4, \ldots$ we obtain

$$P(n) \propto \prod_{k=1}^{[1/\sigma]} \exp\left\{\frac{(-b)^k}{k} \frac{n^{1-k\sigma}}{1-k\sigma}\right\}, \qquad (12.11)$$

where $[1/\sigma]$ denotes the integer part of $1/\sigma$. The cases where σ is an inverse integer are special, since a term $\propto 1/m$ appears in the expansion of the logarithm, giving rise to a power-like correction to the stretched exponential behavior, namely

$$P(n) \propto n^{\sigma(-b)^{1/\sigma}} \prod_{k=1}^{\sigma^{-1}-1} \exp\left\{\frac{(-b)^k}{k} \frac{n^{1-k\sigma}}{1-k\sigma}\right\} \qquad (12.12)$$

for $\sigma = 1/2, 1/3, 1/4, \ldots$. The known result for $\sigma = 1$ can also be obtained by this method: it corresponds to the power-like prefactor in (12.12). Only the linear expansion term of the logarithm is relevant at $1/2 < \sigma < 1$, so we find $P(n) \propto \exp\left[-b n^{1-\sigma}/(1-\sigma)\right]$ in the limit of large n. The first two terms are relevant for $1/3 < \sigma \leq 1/2$, the third one becomes important for $1/4 < \sigma \leq 1/3$, and so on. Equations (12.11) and (12.12) represent an exact analytical result at $n \to \infty$ which we have also verified numerically at different values of σ and b. In this form, where the proportionality coefficient is not specified, Equations (12.11) and (12.12) are universal, that is, they do not depend on the choice of w_1.

At $\langle w \rangle / w_\infty = 1$ the inflow $\langle w \rangle$ in a macroscopic cluster of size $n \to \infty$ is balanced by the outflow w_∞. This means that at overall density $c > c_{cr}$ the homogeneous phase with density c_{cr} is in equilibrium with a macroscopic cluster, represented by one of the boxes containing a non-vanishing fraction of all particles in the thermodynamic limit [63, 88]. Hence, $\langle w \rangle / w_\infty = 1$ holds in the phase coexistence regime at $c > c_{cr}$.

According to (12.7), the critical density c_{cr} is given by

$$c_{cr} = \frac{\langle n \rangle_{cr}}{1 + \langle n \rangle_{cr}}, \qquad (12.13)$$

where $\langle n \rangle_{cr}$ is the mean cluster size at the critical density. Since $\langle w \rangle = w_\infty$ holds in this case, we have

$$\langle n \rangle_{cr} = \frac{\sum_{n=1}^{\infty} n\, w_\infty^n \prod_{m=1}^{n} \frac{1}{w_m}}{1 + \sum_{n=1}^{\infty} w_\infty^n \prod_{m=1}^{n} \frac{1}{w_m}}. \qquad (12.14)$$

The critical density, calculated numerically from (12.13) and (12.14) as a function of parameters σ and b, is shown in Figures 12.6 and 12.7, respectively. In contrast to the situations discussed previously in the literature, in our model w_1 is not given by the general formula (12.9) but is an independent parameter. This distinction leads to quantitatively different results, for example, the critical density for $\sigma = 1$ is analytically shown to be

$$c_{cr} = \frac{b(b+1)}{(b-1)[2(b+1) + w_1(b-2)]}. \qquad (12.15)$$

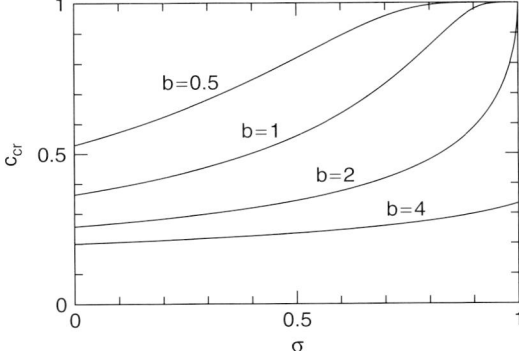

Figure 12.6 Critical density as a function of control parameter σ for different values of b.

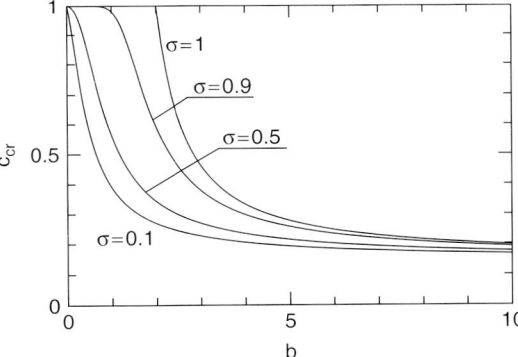

Figure 12.7 Critical density as a function of control parameter b for different values of σ.

12.4 Metastability

Suppose that, at the initial time moment $t = 0$, the system is in a homogeneous state with overall density slightly larger than c_{cr}. Here we study the development of such a state in the mean-field approximation provided by (12.2) and (12.3). With this initial condition, the mean inflow rate in a box $\langle w \rangle$ is slightly larger than that at $c = c_{cr}$, so $\langle w \rangle = w_\infty + \varepsilon$ holds with small and positive ε. Hence, only large clusters with $w_n < w_\infty + \varepsilon$ have a stable tendency to grow, whereas any smaller cluster typically (except a rare case) fluctuates until it finally dissolves. In other words, the initially homogeneous system with no large clusters can stay in this metastable supersaturated state for a long time until a large stable cluster appears due to a rare fluctuation.

Neglecting the fluctuations, the time development of the size n of a cluster is described by the deterministic equation

$$\frac{dn}{dt} = \langle w \rangle - w_n. \tag{12.16}$$

According to this equation, the undercritical clusters with $n < n_{cr}$ tend to dissolve, whereas the overcritical ones with $n > n_{cr}$ tend to grow, where the critical cluster size n_{cr} is given by the condition

$$\langle w \rangle = w_{n_{cr}}. \tag{12.17}$$

Using (12.9) yields

$$n_{cr} \simeq \left(\frac{b}{\langle w \rangle/w_\infty - 1}\right)^{1/\sigma}. \tag{12.18}$$

In this case n_{cr} is rounded to an integer value.

This deterministic approach describes only the most probable scenario for an arbitrarily chosen cluster of a given size. It does not allow one to obtain the distribution over cluster sizes: the deterministic equation (12.16) suggests that all clusters shrink to zero size if they are smaller than n_{cr} at the beginning, whereas the real size distribution arises from the competition between opposite stochastic events of shrinking and growing. Assuming that the distribution of relatively small clusters contributing to $\langle n \rangle$ is quasi-stationary, that is, the detailed balance (equality of the terms in (12.2) and (12.3) describing opposite stochastic events) for these clusters is almost reached before any cluster with $n > n_{cr}$ has appeared, we have

$$\langle n \rangle \simeq \sum_{n=1}^{n_{cr}} n\, P(n) \tag{12.19}$$

for such a metastable state. In this case, from (12.7), we obtain

$$\frac{c}{1-c} \simeq \frac{\sum_{n=1}^{n_{cr}} n \langle w \rangle^n \prod_{m=1}^{n} \frac{1}{w_m}}{1 + \sum_{n=1}^{n_{cr}} \langle w \rangle^n \prod_{m=1}^{n} \frac{1}{w_m}}. \tag{12.20}$$

instead of (12.8) for calculation of $\langle w \rangle$ in this homogeneous metastable state. The critical cluster size is found self-consistently by solving (12.18) and (12.20) as a system of equations. From (12.18) we can see that the critical cluster size n_{cr} diverges at $c \to c_{cr}$, since $\langle w \rangle \to w_\infty$. The results of the calculation of n_{cr} (rounding down to an integer value) at $\sigma = 0.5$, $b = 1$ and at $\sigma = 1$, $b = 3$ are shown in Figure 12.8.

Within the framework of mean-field dynamics, the mean nucleation time in our model can be evaluated as follows. Let $\mathcal{P}(t)$ be the probability density of the first-passage time of exceeding the critical number of particles n_{cr} in a single box. By our

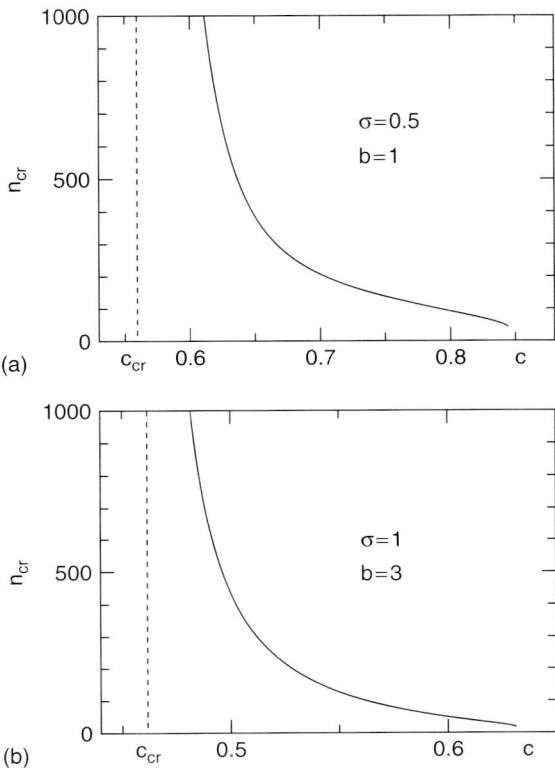

Figure 12.8 Critical cluster size vs density at $\sigma = 0.5$, $b = 1$ (a) and $\sigma = 1$, $b = 3$ (b). The critical density is indicated by a vertical dashed line.

definition, the nucleation occurs when one of the M boxes reaches the cluster size $n_{cr} + 1$. The probability that it occurs first in a given box within a small time interval $[t; t + dt]$ is thus $\mathcal{P}(t)\,dt \times \left[1 - \int_0^t \mathcal{P}(t')\,dt'\right]^{M-1}$ according to our assumption that the boxes are statistically independent. The term $\left[1 - \int_0^t \mathcal{P}(t')\,dt'\right]^{M-1}$ is the probability that in all other boxes, except the given one, the overcritical cluster size $n_{cr} + 1$ has still not been reached. Since the nucleation can occur in any of M boxes, the nucleation probability density $\mathcal{P}_{M(t)}$ for the system of M boxes is given by

$$\mathcal{P}_{M(t)} = M\mathcal{P}(t) \times \left[1 - \int_0^t \mathcal{P}(t')\,dt'\right]^{M-1}$$
$$\simeq M\mathcal{P}(t) \exp\left(-M \int_0^t \mathcal{P}(t')\,dt'\right). \tag{12.21}$$

The latter equality holds for large M, since all M boxes are equivalent, and therefore the probability $\int_0^t \mathcal{P}(t')\,dt'$ that the nucleation occurs in a given box within a characteristic time interval $t \sim \langle T \rangle_M$ is a small quantity of order $1/M$. The mean nucleation time for the system of M boxes is

$$\langle T \rangle_M = \int_0^\infty t\, \mathcal{P}_M(t)\, dt. \tag{12.22}$$

Here $\langle T \rangle_1$ is the mean first-passage time for a single box.

In order to estimate $\langle T \rangle_M$ according to (12.21) and (12.22), one needs some idea about the first-passage time probability density for one box $\mathcal{P}(t)$. This is actually the problem of a particle escaping from a potential well. Since we start with an almost homogeneous state of the system, we may assume zero cluster size $n = 0$ as the initial condition. The first-passage time probability density can be calculated as the probability per unit time of reaching the state $n_{\rm cr} + 1$, assuming that the particle is absorbed there. It is reasonable to assume that after a certain equilibration time $t_{\rm eq}$, when a quasi-stationary distribution of the cluster sizes within $n \leq n_{\rm cr}$ is reached, the escaping from this region is characterized by a certain transition rate $w_{\rm esc}$. Hence, for $t > t_{\rm eq}$ we have

$$\mathcal{P}(t) \simeq w_{\rm esc} \times \left[1 - \int_0^t \mathcal{P}(t')\, dt' \right], \tag{12.23}$$

where the expression in square brackets is the probability that the absorption at $n_{\rm cr} + 1$ has still not occurred up to the time t. At high enough potential barriers (large mean first-passage times) the short-time contribution to the integral is irrelevant and, by means of (12.22), the solution of (12.23) can be written as

$$\mathcal{P}(t) = \frac{1}{\langle T \rangle_1} \exp\left(-\frac{t}{\langle T \rangle_1} \right), \tag{12.24}$$

where $\langle T \rangle_1 = w_{\rm esc}^{-1}$. Obviously, this approximate solution of the first-passage problem is not valid for very short times $t \ll t_{\rm eq}$, since the short-time solution should explicitly depend on the initial condition. In particular, if we start at $n = 0$, then the state $n_{\rm cr} + 1$ cannot be reached immediately, so that $\mathcal{P}(0) = 0$. Nevertheless (12.24) can be used to estimate the mean nucleation time $\langle T \rangle_M$ provided that $\langle T \rangle_M > t_{\rm eq}$.

We have checked the correctness of these theoretical expectations within the mean-field dynamics represented by (12.2) and (12.3) by comparing them with the results of the simulation of stochastic trajectories generated according to these equations. The simulation curves for $\mathcal{P}(t)$ at two different sets of parameters: $\sigma = 0.5$, $b = 1$, $c = 0.84$ (with $n_{\rm cr} = 48$), and $\sigma = 1$, $b = 3$, $c = 0.61$ (with $n_{\rm cr} = 35$) are shown in Figure 12.9 in two different time scales. As we can see, (12.24) is a good approximation for large enough times $t > t_{\rm eq}$. For definiteness, we have identified the equilibration time $t_{\rm eq}$ with the crossing point of the theoretical and simulated curves. An interesting additional feature is the presence of an apparent nucleation time lag, which is about $t_{\rm lag} \approx 60$ for the first set of parameters and about $t_{\rm lag} \approx 30$ for the second one. Evidently, the first-passage time probability density $\mathcal{P}(t)$ tends to zero very rapidly when t decreases below $t_{\rm lag}$.

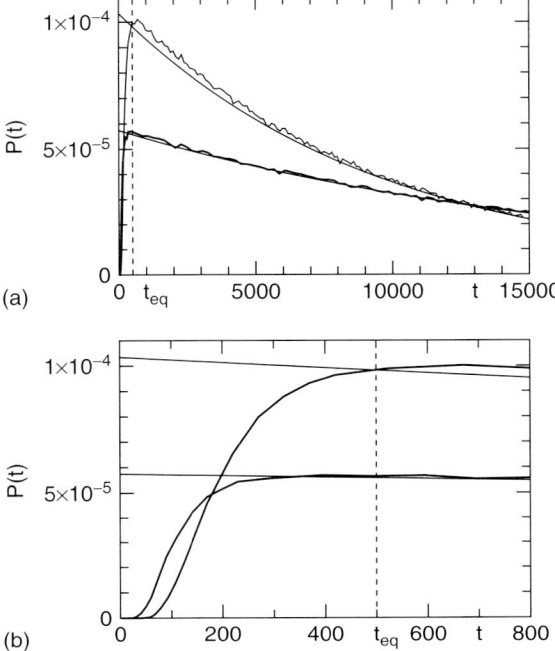

Figure 12.9 Comparison between the theoretical approximation (12.24) for $\mathcal{P}(t)$ (smooth curves) and mean-field simulation results (fluctuating curves) shown in a longer (a) and in a shorter (b) time scale. The vertical dashed line indicates the equilibration time $t_{eq} \approx 500$ for the set of parameters $\sigma = 0.5$, $b = 1$, and $c = 0.84$ represented by the upper curves in both pictures. The other curves correspond to $\sigma = 1$, $b = 3$, and $c = 0.61$.

By inserting (12.24) in (12.22) we obtain

$$\langle T \rangle_M \simeq M \langle T \rangle_1 \int_0^\infty x e^{-x} \exp\left(-M\left[1 - e^{-x}\right]\right) dx \qquad (12.25)$$

after changing the integration variable $t/\langle T \rangle_1 \to x$. Taking into account that only the region $x \sim 1/M$ contributes to the integral at large M, we arrive at a very simple expression

$$\langle T \rangle_M \simeq \frac{\langle T \rangle_1}{M} \qquad (12.26)$$

relating the mean first-passage time or nucleation time in a system of M boxes to that of one box. The latter can be calculated easily by the known formula [55]

$$\langle T \rangle_1 = \sum_{n=0}^{n_{cr}} \left[\langle w \rangle \tilde{P}(n)\right]^{-1} \sum_{m=0}^{n} \tilde{P}(m), \qquad (12.27)$$

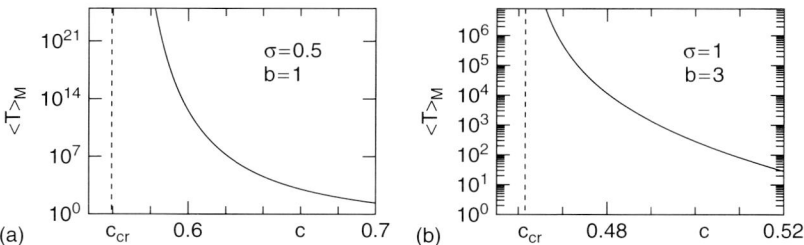

Figure 12.10 Mean nucleation time versus density at $\sigma = 0.5$, $b = 1$ (a) and $\sigma = 1$, $b = 3$ (b). In both cases $M = 10^6$. The critical density is indicated by a vertical dashed line.

where $\tilde{P}(0) = 1$ and $\tilde{P}(n) = \prod_{k=1}^{n}(\langle w \rangle / w_k)$ with $n > 1$ represent the unnormalized stationary probability distribution.

The mean nucleation time versus the density c, calculated from (12.26) and (12.27) at $M = 10^6$ is shown in Figure 12.10. These figures show that the mean nucleation time increases dramatically as the critical cluster size n_{cr} increases (see the corresponding plots in Figure 12.8) approaching the critical density c_{cr}.

According to our previous discussion, estimate (12.26) is valid for large enough mean nucleation times $\langle T \rangle_M > t_{eq}$; in particular, when approaching the critical density $c \searrow c_{cr}$ at any large but fixed M. It is not valid in the thermodynamic limit $M \to \infty$ at a fixed density c. Namely, (12.26) suggests that $\langle T \rangle_M$ decreases as $\sim 1/M$, whereas in reality the decrease must be slower for small nucleation times (large M) since $\mathcal{P}(t) \to 0$ as $t \to 0$. In particular, the mean-field dynamics suggests that, for a wide range of M values, $\langle T \rangle_M$ quasi-saturates at $\langle T \rangle_M \approx t_{lag}$, since the critical cluster size is almost never reached before $t = t_{lag}$.

12.5
Monte Carlo Simulations of the Hopping Model

Numerical simulations of the zero-range model show clear evidence for the existence of a metastable state prior to condensation In Figure 12.11 we show the largest cluster size as a function of time for three separate Monte Carlo runs in the case $\sigma = 0.5$, $b = 1$, $w_1 = 5$, $M = 10^5$. For each run the system was started in a random uniform initial condition with density $c = 0.66$ (for these parameters $c_{cr} \simeq 0.56$). It can be clearly seen that, after a short equilibration period, the system fluctuates in a metastable state before a condensate appears. The critical cluster size is observed to be around 400 in good agreement with the prediction $n_{cr} \simeq 330$ from (12.18) and (12.20) (see Figure 12.8). However, the metastable time is about an order of magnitude larger than predicted.

In Figure 12.12 we show the distribution of cluster sizes (for small clusters) averaged over the metastable state of one such run. The distribution is in good agreement with (12.5) with $\langle w \rangle = w_{n_{cr}}$, thus supporting the assumption of quasi-stationarity.

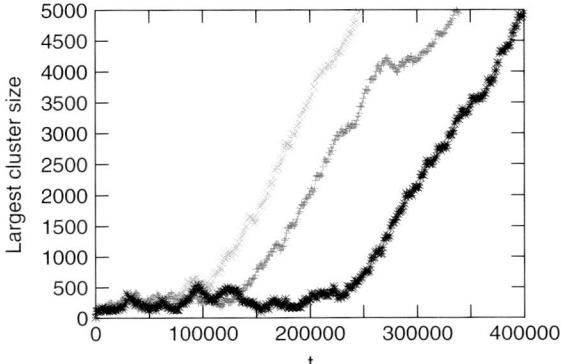

Figure 12.11 Largest cluster size versus time for $\sigma = 0.5$, $b = 1$, $w_1 = 5$, $c = 0.66$, $M = 10^5$. Results from three independent Monte Carlo runs are shown.

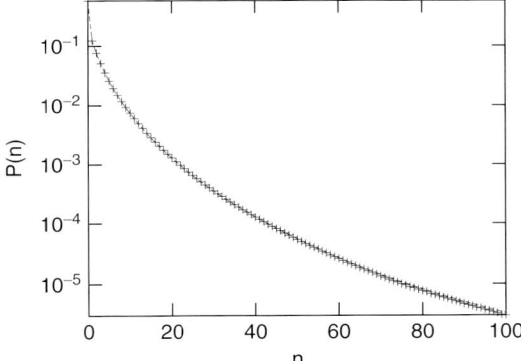

Figure 12.12 Distribution over small cluster sizes in the metastable state. Results for Monte Carlo simulation (crosses) compared to prediction of (12.5) with $\langle w \rangle = w_{n_{cr}}$ and n_{cr} calculated numerically from (12.18) and (12.20) (dashed line).

In the analytical treatment of the previous section we calculated the mean time for the maximum cluster size to exceed n_{cr} under the assumption that the current in the metastable state is constant. In practice, of course, the metastable current also fluctuates (and the fluctuations are greater when w_n depends more strongly on cluster size, e.g. for the $\sigma = 1$ case compared with, say, $\sigma = 0.5$). Simulations suggest that these fluctuations can destroy the metastable state in cases where the metastable current $w_{n_{cr}}$ is close to the current of the condensed phase w_∞.

In contrast, for parameters where the metastable state is well separated from the condensed state we find relatively good quantitative agreement between theory and simulation. For example, in Figures 12.13 and 12.14 we compare the average

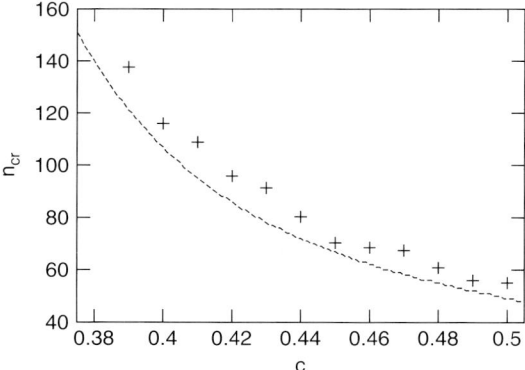

Figure 12.13 Critical cluster size versus density for $\sigma = 0.5$, $b = 3$, $w_1 = 0.5$, $M = 10^5$ ($c_{cr} \simeq 0.27$). Crosses show simulation data (averaged over 10 Monte Carlo histories), the dashed line is a prediction of (12.18) and (12.20).

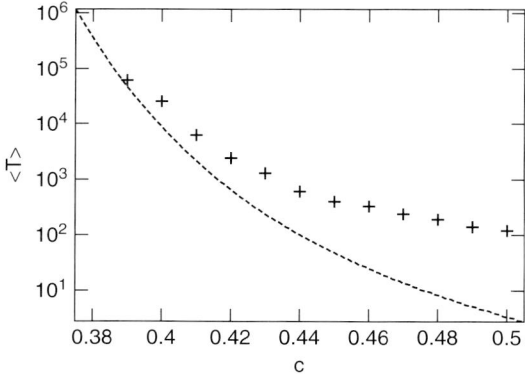

Figure 12.14 Nucleation time versus density for $\sigma = 0.5$, $b = 3$, $w_1 = 0.5$, $M = 10^5$ ($c_{cr} \simeq 0.27$). Crosses show simulation data (averaged over 10 Monte Carlo histories), the dashed line is a prediction of (12.26) and (12.27).

simulation values of the critical cluster size and nucleation time with the theoretical predictions for a range of densities in the case $\sigma = 0.5$ and $b = 3$.

In these simulations we crudely identified the end of the metastable state as the point when the current out of the largest cluster had been less than the average system current for 50 consecutive Monte Carlo time steps.

We find that our mean-field theory fairly accurately reproduces the critical cluster size, but systematically underestimates the nucleation time. This discrepancy may be partly due to the presence of (weak) dynamical correlations between the numbers of particles in the boxes in the fluctuating metastable state. Namely, the appearance of a large cluster with $n \simeq n_{cr}$ is likely to be accompanied by a slight depletion of the surrounding medium. Furthermore, we only calculated the mean first-passage time and ignored the probability that a cluster reaches $n_{cr} + 1$ and is immediately driven by a fluctuation back below n_{cr}. Monte Carlo histories which involve such a fluctuation back into the metastable state before a condensate is established, would increase the average simulation nucleation time above the theoretical prediction.

Despite the neglect of current fluctuations, dynamical correlations, etc., our simulations show that the simple mean-field approach provides a good qualitative description of the metastable state and its dependence on density. It thus represents an important first step towards more refined theories.

12.6
Fundamental Diagram of the Zero-Range Model

The relation between the density c and flux j of cars is known as the fundamental diagram of traffic flow. The average stationary flux can be calculated as follows

$$j = \sum_{n=1}^{\infty} Q(n) \, w_n, \tag{12.28}$$

where $Q(n)$ is the probability that there is a car in a given cell (in the original model) which can move forwards at a rate w_n. Note that only those cars contribute to the flux, which are the first in some cluster. Hence, $Q(n) = \varphi P(n) / \sum_{m=1}^{\infty} P(m)$, where φ is the fraction of cells which contain such cars. This fraction can be calculated easily as the number of clusters divided by the total number of cells. These quantities fluctuate in our model. For large systems, however, they can be replaced by the mean values. The mean number of clusters is equal to the mean number of non-empty boxes $M \sum_{n=1}^{\infty} P(n)$ in the zero-range model, whereas the mean number of cells, that is, the mean length of the road is $\langle L \rangle = M + \langle N \rangle = M(1 + \langle n \rangle) = M/(1-c)$, as we have already discussed in Section 12.2. Hence, $Q(n) = (1-c) P(n)$ and (12.28) reduces to

$$j = (1 - c) \langle w \rangle. \tag{12.29}$$

The mean stationary transition rate $\langle w \rangle$ depends on the car density c. For undercritical densities $c < c_{cr}$, this quantity is the solution of (12.8). For overcritical densities we have $\langle w \rangle = w_{\infty}$ in the phase coexistence regime, as discussed in Section 12.3; therefore, in this case the fundamental diagram reduces to a straight line

$$j = (1 - c) w_{\infty} : \quad c \geq c_{cr}. \tag{12.30}$$

In the metastable homogeneous state at $c > c_{cr}$ the mean transition rate $\langle w \rangle$ together with the critical cluster size n_{cr} can be found from the system of equations (12.18) and (12.20), which allows calculation of the metastable branch of flux j.

The resulting fundamental diagrams for $\sigma = 0.5$, $w_1 = 5$ and two values of parameter b are shown in Figure 12.15. As we can see, the shape of the fundamental

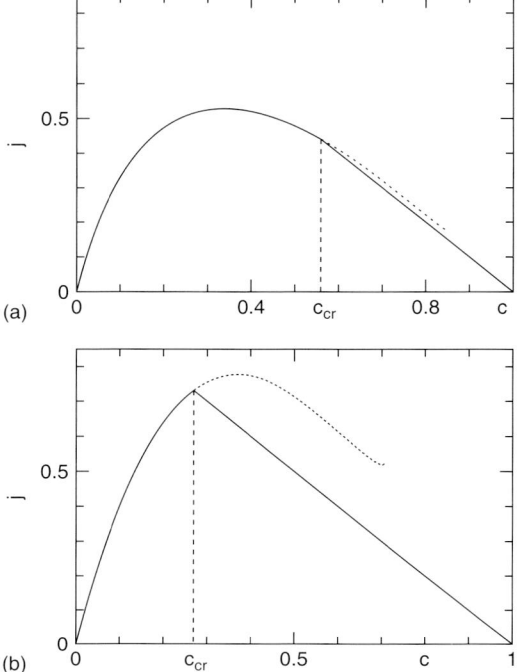

Figure 12.15 The fundamental (flux–density) diagram for two different sets of control parameters: $\sigma = 0.5$, $b = 1$, $w_1 = 5$ (a); $\sigma = 0.5$, $b = 3$, $w_1 = 5$ (b). The branches of metastable homogeneous state are shown by dotted lines, the critical densities c_{cr} are indicated by vertical dashed lines.

diagram, as well as the critical density and location of the metastable branch, depend remarkably on the value of b. These features will also depend on the values of σ and w_1. The metastable branch ends abruptly at certain density above which (12.18) and (12.20) have no real solution. This corresponds to a relatively small, but finite value of the critical cluster size n_{cr}.

Note that a metastable branch is also observed in simulations of cellular automata with slow-to-start rules [14, 19, 225]. In our examples, however, the metastable branch is located at larger densities and decreases with increasing c over a certain wide range of values depending on b, σ and w_1. The simulations of the previous section suggest that, when this metastable branch is well separated from the condensed section of the fundamental diagram, our picture is robust even in the presence of fluctuations.

In summary, therefore, we believe that by suitable variation of parameters our simple model can reproduce some important features of real traffic flow [97].

12.7
Polarization Kinetics in Ferroelectrics with Fluctuations

Here we would like to complete our discussion of many-particle systems with one more application, which is more nontrivial in the sense that a multidimensional state space is considered to describe collective phenomena in a ferroelectric.

The stochastic description of collective phenomena like phase transitions is a key destination in solid-state physics. The problems of this kind are nontrivial and usually are solved by means of the perturbation theory [51, 94, 137, 253] with some exceptions like the mean-field model considered in [196]. Here we study the kinetics of polarization switching in ferroelectrics, taking into account the spatio–temporal fluctuations of the polarization field, given by the Langevin and multidimensional Fokker–Planck equations [64, 179]. The problem was studied earlier [93, 95] by means of the Feynman diagram technique. Here we derive the Fokker–Planck equation in the Fourier representation, which is suitable for a numerical approach.

We consider a ferroelectric with the Landau–Ginzburg Hamiltonian

$$H = \int \left(\frac{\alpha}{2} P^2(\mathbf{x}) + \frac{\beta}{4} P^4(\mathbf{x}) + \frac{c}{2} (\nabla P(\mathbf{x}))^2 - \lambda(\mathbf{x}, t) P(\mathbf{x}, t) \right) d\mathbf{x}, \qquad (12.31)$$

where $P(\mathbf{x}, t)$ is the local polarization and $\lambda(\mathbf{x}, t)$ is the time-dependent external field. Only those configurations of the polarization are allowed which correspond to the cut-off $k < \Lambda$ in the Fourier space with $\Lambda = \pi/a$, where a is the lattice constant. Hamiltonian (12.31) can be approximated by a sum over discrete cells, where the size of one cell can be even larger than the lattice constant. It is a small domain with almost constant polarization. Thus, the Hamiltonian of a system consisting of N cells with total volume V reads

$$H = \frac{V}{N} \sum_{\mathbf{x}} \left(\frac{\alpha}{2} P^2(\mathbf{x}) + \frac{\beta}{4} P^4(\mathbf{x}) + \frac{c}{2} (\nabla P(\mathbf{x}))^2 - \lambda(\mathbf{x}, t) P(\mathbf{x}, t) \right), \qquad (12.32)$$

where $V/N = \Delta V$ is the volume of one cell, and the coordinates of their centers are given by the set of discrete d-dimensional vectors $\mathbf{x} \in \mathbb{R}^d$. The stochastic dynamics of the system is described by the Langevin equation

$$\dot{P}(\mathbf{x}, t) = -\gamma \frac{\partial H}{\partial P(\mathbf{x}, t)} + \xi(\mathbf{x}, t), \qquad (12.33)$$

where $\xi(\mathbf{x}, t)$ is the white noise, that is,

$$\langle \xi(\mathbf{x}, t) \xi(\mathbf{x}', t') \rangle = 2\gamma \theta \, \delta_{\mathbf{x}, \mathbf{x}'} \delta(t - t'). \qquad (12.34)$$

In the case of Gaussian white noise, the probability distribution function

$$f = f\left(P(\mathbf{x}_1), P(\mathbf{x}_2), \ldots, P(\mathbf{x}_N), t \right)$$

is given by the Fokker–Planck equation

$$\frac{1}{\gamma}\frac{\partial f}{\partial t} = \sum_{\mathbf{x}} \frac{\partial}{\partial P(\mathbf{x})}\left(\frac{\partial H}{\partial P(\mathbf{x})}f + \theta\frac{\partial f}{\partial P(\mathbf{x})}\right), \qquad (12.35)$$

as consistent with (5.4) for the N-dimensional state vector with components $P(\mathbf{x}_i)$, where $i = 1, 2, \ldots, N$. At equilibrium we have a vanishing flux which corresponds to Boltzmann's distribution $f \propto \exp(-H/\theta)$ with $\theta = k_B T$.

Assuming periodic boundary conditions, we consider the Fourier transformation

$$P(\mathbf{x}) = N^{-1/2} \sum_{\mathbf{k}} P_{\mathbf{k}} e^{i\mathbf{k}\mathbf{x}}$$

$$P_{\mathbf{k}} = N^{-1/2} \sum_{\mathbf{x}} P(\mathbf{x}) e^{-i\mathbf{k}\mathbf{x}}. \qquad (12.36)$$

The Fourier amplitudes are complex numbers $P_{\mathbf{k}} = P'_{\mathbf{k}} + iP''_{\mathbf{k}}$. Since $P(\mathbf{x})$ is real, $P'_{-\mathbf{k}} = P'_{\mathbf{k}}$ and $P''_{-\mathbf{k}} = -P''_{\mathbf{k}}$ hold. It is assumed that the total number of modes N is an odd number. This means that there is a mode with $\mathbf{k} = 0$ and the modes with $\pm\mathbf{k}_1, \pm\mathbf{k}_2, \ldots, \pm\mathbf{k}_m$, where $m = (N-1)/2$ is the number of independent nonzero modes.

The Fokker–Planck equation for the probability distribution function

$$f = f\left(P_0, P'_{\mathbf{k}_1}, P'_{\mathbf{k}_2}, \ldots, P'_{\mathbf{k}_m}, P''_{\mathbf{k}_1}, P''_{\mathbf{k}_2}, \ldots, P''_{\mathbf{k}_m}, t\right)$$

reads

$$\frac{1}{\gamma}\frac{\partial f}{\partial t} = \sum_{\mathbf{k}\in\Omega} \frac{\partial}{\partial P'_{\mathbf{k}}}\left\{\frac{1}{2}(1+\delta_{\mathbf{k},0})\left[\frac{\partial H}{\partial P'_{\mathbf{k}}}f + \theta\frac{\partial f}{\partial P'_{\mathbf{k}}}\right]\right\}$$
$$+ \sum_{\mathbf{k}\in\overline{\Omega}} \frac{\partial}{\partial P''_{\mathbf{k}}}\left\{\frac{1}{2}\left[\frac{\partial H}{\partial P''_{\mathbf{k}}}f + \theta\frac{\partial f}{\partial P''_{\mathbf{k}}}\right]\right\}, \qquad (12.37)$$

where $P'_0 \equiv P_0$, $\overline{\Omega}$ is the set of m independent nonzero wave vectors, and Ω also includes $\mathbf{k} = 0$. Here the Fourier-transformed Hamiltonian is given by

$$H = \Delta V \left(\frac{1}{2}\sum_{\mathbf{k}}(\alpha + c\mathbf{k}^2)|P_{\mathbf{k}}|^2 - \sum_{\mathbf{k}}\lambda_{-\mathbf{k}}(t)P_{\mathbf{k}}\right.$$
$$\left. + \frac{\beta}{4}N^{-1} \sum_{\mathbf{k}_1+\mathbf{k}_2+\mathbf{k}_3+\mathbf{k}_4=0} P_{\mathbf{k}_1}P_{\mathbf{k}_2}P_{\mathbf{k}_3}P_{\mathbf{k}_4}\right). \qquad (12.38)$$

Some of the variables in (12.38) are dependent according to $P'_{-\mathbf{k}} \equiv P'_{\mathbf{k}}$ and $P''_{-\mathbf{k}} \equiv -P''_{\mathbf{k}}$, and $\lambda_{\mathbf{k}}(t) = \lambda'_{\mathbf{k}}(t) + i\lambda''_{\mathbf{k}}(t)$ is the Fourier transform of $\lambda(\mathbf{x}, t)$. The

Fokker–Planck equation (12.37) can be written explicitly as

$$\frac{1}{\gamma}\frac{\partial f}{\partial t} = \sum_{\mathbf{k}\in\Omega} \frac{\partial}{\partial P'_\mathbf{k}} \left\{ \Delta V f \left[(\alpha + c\mathbf{k}^2) P'_\mathbf{k} + \beta S'_\mathbf{k} - \lambda'_\mathbf{k}(t) \right] + \frac{\theta}{2}(1+\delta_{\mathbf{k},0}) \frac{\partial f}{\partial P'_\mathbf{k}} \right\}$$
$$+ \sum_{\mathbf{k}\in\bar{\Omega}} \frac{\partial}{\partial P''_\mathbf{k}} \left\{ \Delta V f \left[(\alpha + c\mathbf{k}^2) P''_\mathbf{k} + \beta S''_\mathbf{k} - \lambda''_\mathbf{k}(t) \right] + \frac{\theta}{2} \frac{\partial f}{\partial P''_\mathbf{k}} \right\}, \quad (12.39)$$

where

$$S'_\mathbf{k} = N^{-1} \sum_{\mathbf{k}_1+\mathbf{k}_2+\mathbf{k}_3=\mathbf{k}} \left\{ P'_{\mathbf{k}_1} P'_{\mathbf{k}_2} P'_{\mathbf{k}_3} - 3 P'_{\mathbf{k}_1} P''_{\mathbf{k}_2} P''_{\mathbf{k}_3} \right\}, \quad (12.40)$$

$$S''_\mathbf{k} = N^{-1} \sum_{\mathbf{k}_1+\mathbf{k}_2+\mathbf{k}_3=\mathbf{k}} \left\{ -P''_{\mathbf{k}_1} P''_{\mathbf{k}_2} P''_{\mathbf{k}_3} + 3 P''_{\mathbf{k}_1} P'_{\mathbf{k}_2} P'_{\mathbf{k}_3} \right\}. \quad (12.41)$$

The simplest case is the spatially homogeneous polarization when only the $\mathbf{k}=\mathbf{0}$ mode is retained in (12.39) with a spatially homogeneous external field $\lambda(\mathbf{x},t) = \lambda_0(t) = A\sin(\omega t)$. In this case we have

$$\frac{1}{\gamma}\frac{\partial f}{\partial t} = \frac{\partial}{\partial P_0} \left\{ V f \left[\alpha P_0 + \beta P_0^3 - A\sin(\omega t) \right] + \theta \frac{\partial f}{\partial P_0} \right\}. \quad (12.42)$$

This equation has been solved numerically by using a difference scheme with special exponential-type substitution described in [98]. The numerical solution has been found within $P_0 \in [-2; 2]$ at the values of parameters $\gamma = V = \beta = 1, \alpha = -1$, $\theta = 0.05, A = 0.309$, and $\omega = 10^{-3}$. The boundary conditions $f(\pm 2, t) = 0$ and the initial condition

$$f(P_0, 0) = \frac{1}{\sqrt{2\pi}\sigma} \exp\left(-\frac{(P_0 - \tilde{P})^2}{2\sigma^2}\right) \quad (12.43)$$

have been used with $\sigma = 0.3$, $\tilde{P} = -1$ for $P_0 < 0$, and $\tilde{P} = 1$ for $P_0 > 0$. The calculated mean polarization $\bar{P}_0$ depending on the external field $\lambda_0(t)$ forms a hysteresis loop shown in Figure 12.16.

Further we shall consider a quasi one-dimensional case, where a three-dimensional ferroelectric sample is stretched out in x the direction, so $L_x \gg L_y$ and $L_x \gg L_z$ hold for the linear sizes. In this case we assume that the polarization as well as the external field depend only on the coordinate x. This means that the wave vectors also have only one nonvanishing component, which is a scalar quantity $k = (2\pi/L_x) \cdot n$, where $n = 0, \pm 1, \pm 2, \ldots, \pm m$.

As the first step, we include only one ($m = 1$) independent nonzero wave vector $k_1 = 2\pi/L_x$ (totally $N = 3$ wave vectors $k = -k_1, 0, k_1$) and homogeneous external field $\lambda(x,t) = 1/\sqrt{3}\lambda_0(t) = 1/\sqrt{3}A\sin(\omega t)$. Furthermore, we assume that the probability distribution function in real space is translation invariant at the initial time. Due to the translation symmetry of the model, it holds also at later

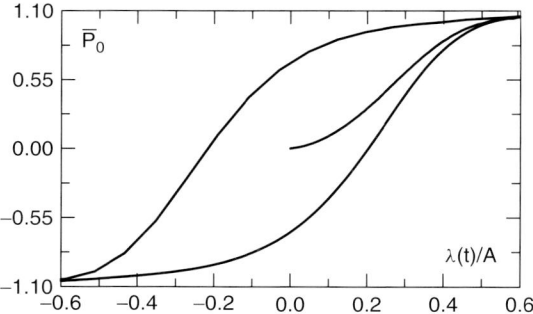

Figure 12.16 The polarization hysteresis: the mean polarization $\bar{P}_0$ versus normalized external field $\lambda_0(t)/A$ calculated numerically at the values of parameters $\gamma = V = \beta = 1$, $\alpha = -1$, $\theta = 0.05$, $A = 0.309$, and $\omega = 10^{-3}$.

times. In the Fourier representation, this means that the probability distribution function depends on the modulus of P_{k_1}, but not on its phase. Thus, we have

$$f = f\left(P_0, P'_{k_1}, P''_{k_1}, t\right) = \frac{\hat{f}\left(P_0, |P_{k_1}|, t\right)}{2\pi |P_{k_1}|},$$

where $\hat{f}\left(P_0, |P_{k_1}|, t\right)$ is the probability density in the $(P_0, |P_{k_1}|)$ space. It obeys the Fokker–Planck equation

$$\frac{1}{\gamma}\frac{\partial \hat{f}}{\partial t} = \frac{\partial}{\partial P_0}\left\{\Delta V \hat{f}\left[\alpha P_0 + \beta\left(\frac{1}{3}P_0^3 + 2P_0 |P_{k_1}|^2\right) - A\sin(\omega t)\right] + \theta \frac{\partial \hat{f}}{\partial P_0}\right\}$$
$$+ \frac{\partial}{\partial |P_{k_1}|}\left\{\Delta V \hat{f}\left[(\alpha + ck_1^2) |P_{k_1}| + \beta\left(|P_{k_1}|^3 + P_0^2 |P_{k_1}|\right)\right]\right.$$
$$\left. + \frac{\theta}{2}\left[\frac{\partial \hat{f}}{\partial |P_{k_1}|} - \frac{\hat{f}}{|P_{k_1}|}\right]\right\}. \tag{12.44}$$

Since f is finite, $\hat{f}$ vanishes at $|P_{k_1}| = 0$. The physical boundary conditions correspond to zero flux at the boundaries $P_0 = \pm \infty$, $|P_{k_1}| = 0$, and $|P_{k_1}| = \infty$. An appropriate initial condition has to be chosen which fulfills these relations, e.g. $\hat{f}\left(P_0, |P_{k_1}|, 0\right) \propto |P_{k_1}| \exp\left(-a_0 P_0^2 - a_1 |P_{k_1}|^2\right)$.

In the present section a multidimensional Fokker–Planck equation has been derived, which describes the polarization-switching kinetics in a ferroelectric in the presence of an external field. The probability distribution function entering this equation, depends on a set of Fourier amplitudes. An example calculation has been performed in a spatially homogeneous approximation, retaining only the zero mode $\mathbf{k} = 0$. The calculated mean polarization versus external field forms a hysteresis, as observed in real ferroelectrics.

12.8 Exercises

E 12.1 *Generalization of zero-range model*

Consider a generalization of the zero-range model, where one particle from each of N boxes on a ring can hop forwards with the transition rate $u(n, m)$, where n is the number of particles in the actual box and m the number in the destination box. By analogy with the zero-range model, formulate the master equation describing, in the mean-field approximation, the dynamics of the probability $p(n, t)$ of having just n particles in a box at time t.

E 12.2 *Particle-hopping model*

Consider a string of M subsequent boxes in the particle-hopping model defined in the previous exercise and formulate the master equation for the probability $p_M(n_1, n_2, \ldots, n_M; t)$ of having n_1 particles in box 1, n_2 particles in box 2, and so on at time t. Consider the thermodynamic limit of a large total number of boxes $N \to \infty$ with periodic boundary conditions and use the Mth order cluster approximation

$$P_{M+1}(n_1, n_2, \ldots, n_{M+1}; t) \simeq \frac{P_M(n_1, \ldots, n_M; t)\, P_M(n_2, \ldots, n_{M+1}; t)}{P_{M-1}(n_2, \ldots, n_M; t)} \qquad (12.45)$$

to obtain a closed equation for any M. Find the condition at which the stationary probability distribution calculated from this master equation factorizes (for any M) as in the case of the zero-range model. If the factorization takes place for the transition rates which depend both on n and m, then we have the so-called misanthrope process.

E 12.3 *Flux-density fundamental diagram*

Consider the particle-hopping model defined in previous two exercises with transition rates

$$u(n, m) = f(n)\, g(m), \qquad (12.46)$$

where

$$\begin{aligned} &f(0) = 0 \\ &f(1) = \text{independent constant} \\ &f(n) = f_\infty \left(1 + \frac{b_1}{n^{\sigma_1}}\right): \quad n \geq 2 \\ &g(n) = g_\infty \left(1 + \frac{b_2}{(n+1)^{\sigma_2}}\right). \end{aligned} \qquad (12.47)$$

Calculate the flux-density fundamental diagram based on the master equation in the mean-field approximation. Consider different values of the parameters b_1, b_2, σ_1, σ_2 and compare the results with those of the zero-range model, corresponding to $b_2 = 0$.

E 12.4 *Equilibrium fluctuations*

Consider the Fokker–Planck equation (12.42) describing the spatially homogeneous polarization fluctuations. Find the condition at which this equation has the equilibrium solution in the form of the Boltzmann distribution.

Epilogue

Finally we would like to classify this textbook within the general framework of the natural and engineering sciences. One should note that, at a very basic or fundamental level, nature is described by the laws of quantum mechanics and elementary particle physics. The explanation of macroscopic physical phenomena, especially in nonequilibrium, based on a microscopic description of the underlying processes is the subject of many scientific papers and books which have been written on this topic. For any interested reader we recommend the monograph *Statistical Mechanics of Nonequilibrium Processes* by our colleagues Dmitrii Zubarev, Vladimir Morozov and Gerd Röpke [254] which studies physical processes on a microscopic level. There are also fundamental mathematical studies of stochastic processes [57,91]. An exact application of the fundamental laws to complex systems, however, is too complicated, as has already been pointed out by Paul Dirac in 1929:

> The underlying physical laws necessary for the mathematical theory of a large part of physics and the whole of chemistry are thus completely known, and the difficulty is only that the exact application of these laws leads to equations much too complicated to be soluble. It therefore becomes desirable that approximate practical methods of applying quantum mechanics should be developed, which can lead to an explanation of the main features of complex atomic systems without too much computation.

This book tends towards practical applications, starting from classical physical systems like supersaturated vapors, and extending the known methods to other complex systems such as traffic flow and financial markets. Taking into account the difficulties mentioned we begin from a mesoscopic or a macroscopic level of description, where the microscopic fundamental laws are taken into account in an averaged or integrated way.

Since the pioneering works on Brownian motion and its interpretation at the molecular level by Albert Einstein, the study of open systems at a phenomenological level, based on coarse-grained kinetic equations, is still crucial in classical statistical mechanics. The stochastic theory of nonequilibrium systems allows one to connect isolated pure-state dynamics (also without noise as deterministic chaos [208]) with open mixed-state dynamics in order to study active propelling or motor-driven particles. For further reading we recommend the book *Statistical Thermodynamics and Stochastic Theory of Nonequilibrium Systems* by Werner Ebeling (my former encouraging supervisor at Rostock University, R. M.) and Igor Sokolov [38] as well as many other important references [6, 86, 211, 220].

The authors thank all readers who have persevered to the end of the whole text including 150 figures!

References

1. R. Abraham: *Non-standard Analysis*, (Princeton University Press, New York 1996).
2. A.Z. Akcasu, J.P. Holloway: Fokker–Planck description of particle transport in finite media: Boundary conditions, Phys. Rev. E **58**, 4321 (1998).
3. S. Albeverio, J.E. Fenstad, R. Hoegh-Krohn, T. Lindstrøm: *Nonstandard Methods in Stochastic Analysis and Mathematical Physics*, (Academic Press, New York 1986).
4. E. Allen: *Modeling with Itô Stochastic Differential Equations*, (Springer, Dordrecht 2007).
5. D.J. Amit: *Field Theory, the Renormalization Group, and Critical Phenomena*, (World Scientific, Singapore 1984).
6. V.S. Anishchenko, V. Astakhov, A. Neimann, T. Vadivasova, L. Schimansky–Geier: *Nonlinear Dynamics of Chaotic and Stochastic Systems*, 2002, 2nd ed. (Springer, Berlin 2007).
7. L. Arnold: *Random Dynamical Systems*, (Springer, Berlin 1998).
8. V.M. Avdjukhina, A.A. Anishchenko, A.A. Katsnelson, G.P. Revkevich: Nonmonotonic relaxation in hydrogen-saturated alloys Pd-Mo, Perspekt. Materials, n. 4, 5 (2002) (in Russian).
9. V. Balakrishnan, C. Van den Broeck, P. Hänggi: First-passage times of non-Markovian processes: The case of a reflecting boundary, Phys. Rev. A **38**, 4213 (1988).
10. M. Bando, K. Hasebe, A. Nakayama, A. Shibata, Y. Sugiyama: Structure stability of congestion in traffic dynamics, Japan. J. Indust. and Appl. Math. **11**, 203 (1994).
11. M. Bando, K. Hasebe, A. Nakayama, A. Shibata, Y. Sugiyama: Dynamical model of traffic congestion and numerical simulation, Phys. Rev. E **51**, 1035 (1995).
12. M. Bando, K. Hasebe, K. Nakanishi, A. Nakayama, A. Shibata, Y. Sugiyama: Phenomenological study of dynamical model of traffic flow, J. Phys. I France **5**, 1389 (1995).
13. R.B. Banks: *Growth and diffusion phenomena*, (Springer, Berlin 1994).
14. R. Barlovic, L. Santen, A. Schadschneider, M. Schreckenberg: Metastable states in cellular automata for traffic flow, Eur. Phys. J. B **5**, 793 (1998).
15. R.J. Baxter: *Exactly Solved Models in Statistical Mechanics*, (Academic Press, London 1989).
16. R. Becker, W. Döring: Kinetische Behandlung der Keimbildung in übersättigten Dämpfen, Ann. Phys. **24**, 719 (1935).
17. J. Bect: A unifying formulation of the Fokker–Planck–Kolmogorov equation for general stochastic hybrid systems, IFAC World Congress, July 6–11, 2008, Seoul, e-print arXiv:0801.3725v2, 26 Feb 2008.
18. J. Bect, H. Baili, G. Fleury: Fokker–Planck–Kolmogorov equation for stochastic differential equations

with boundary hitting resets, e-print arXiv:math/0504583v1, 28 Apr 2005.
19. S.C. Benjamin, N.F. Johnson, P.M. Hui: Cellular automata models of traffic flow along a highway containing a junction, J. Phys. A **29**, 3119 (1996).
20. J. Bernardini: Grain boundary diffusion in metallic nano and polycrystals, Interface Science **5**, 55 (1997).
21. R.V. Bobryk, A. Chrzeszczyk: Transitions in a Duffing oscillator excited by random noise, Nonlin. Dyn. **51**, 541 (2008).
22. A.N. Borodin, P. Salminen: *Handbook of Brownian Motion – Facts and Formulae*, 2nd ed., (Birkhäuser, Basel 2002).
23. K. Burrage, P.M. Burrage: High strong order explicit Runge–Kutta methods for stochastic ordinary differential equations, App. Num. Math. **22**, 81 (1996).
24. P.M. Burrage: *Numerical methods for stochastic differential equations*, Ph.D. Thesis (University of Queensland, Brisbane, Queensland, Australia, 1999).
25. V. Capasso, D. Bakstein: *An Introduction to Continuous–Time Stochastic Processes. Theory, Models, and Applications to Finance, Biology, and Medicine*, (Birkhäuser, Berlin 2004).
26. B. K. Chakrabarti, L. G. Benguigui: *Statistical Physics of Fracture and Breakdown in Disordered Systems*, (Clarendon Press, Oxford 1997).
27. P.H. Chavanis: Exact diffusion coefficient of self-graviting Brownian particles in two dimensions, Eur. Phys. J. B **57**, 391 (2007).
28. A. Chetverikov, J. Dunkel: Phase behavior and collective excitations of the Morse ring chain, Eur. Phys. J. B **35**, 239 (2003).
29. M.H. Choi, R.F. Fox: Evolution of escape processes with a time-varying load, Phys. Rev. E **66**, 031103, 2002.
30. A.J. Chorin, O.H. Hald: *Stochastic Tools in Mathematics and Science*, (Springer, New York 2006).
31. D. Chowdhury, L. Santen, A. Schadschneider: Statistical physics of vehicular traffic and some related systems, Phys. Reports **329**, 199, 2000.
32. D. Chowdhury, A. Schadschneider, K. Nishinari: Physics of transport and traffic phenomena in biology: From molecular motors and cells to organisms, Phys. Life Rev. **2**, 318 (2005).
33. K. Christensen, N.R. Moloney: *Complexity and Criticality*, (Imperial College Press 2005).
34. W.T. Coffey, Yu.P. Kalmykov, J.T. Waldron: *The Langevin Equation* (World Scientific, New Jersey 2004).
35. B. Dybiec, L. Schimansky-Geier: Emergence of bistability in noisy systems with single-well potential, Eur. Phys. J. B, **57**, 313 (2007).
36. M.I. Dykman, P.V.E. McClintock: What can stochastic resonance do? Nature **391**, 344 (1998).
37. W. Ebeling, P.S. Landa, V.G. Ushakov: Self-oscillations in ring Toda chains with negative friction, Phys. Rev. E **63**, 046601 (2001).
38. W. Ebeling, I.S. Sokolov: *Statistical Thermodynamics and Stochastic Theory of Nonequilibrium Systems* (World Scientific, New Jersey 2005).
39. P. & T. Ehrenfest: Über zwei bekannte Einwände gegen das Boltzmannsche H-Theorem Physikalische Zeitschrift **8**, 311 (1907).
40. A. Einstein: Über die von der molekularkinetischen Theorie der Wärme geförderte Bewegung in ruhenden Flüssigkeiten suspendierten Teilchen, Annalen der Physik, **17**, 549 (1905).
41. U. Erdmann, W. Ebeling, L. Schimansky-Geier, F. Schweitzer: Brownian particles far from equilibrium, Eur. Phys. J. B **15**, 105 (2000).
42. M.R. Evans: Phase transitions in one-dimensional nonequilibrium systems, Braz. J. Phys. **30**, 42 (2000).
43. M.R. Evans, T. Hanney: Nonequilibrium statistical mechanics of the zero-range process and related models, J. Phys. A: Math. Gen. **38**, R195 (2005).
44. H. Ez-Zahraouy, Z. Benrihane, A. Benyoussef: The optimal velocity traffic flow models with open boundaries, Eur. Phys. J. B **36**, 289 (2003).

45 L. Farkas: The velocity of nucleus formation in supersaturated vapours, Z. Phys. Chemie **125**, 236 (1927).
46 W. Feller: *An Introduction to Probability Theory and its Applications*, Vol. I (John Wiley & Sons, New York 1968).
47 W. Feller: *An Introduction to Probability Theory and its Applications*, Vol. II, 2nd ed. (John Wiley & Sons, New York 1971).
48 R.F. Fox, M.H. Choi: Rectified Brownian motion and kinesin motion along microtubules, Phys. Rev. E **63**, 051901, 2001.
49 T.D. Frank: *Nonlinear Fokker–Planck Equations*, (Springer, Berlin 2005).
50 G. Gallavotti: Heat and fluctuations from order to chaos, Eur. Phys. J. B **61**, 1 (2008).
51 A. Gambassi: Relaxation phenomena at criticality, Eur. Phys. J. B **64**, 379 (2008).
52 L. Gammaitoni, P. Hänggi, P. Jung, F. Marchesoni: Stochastic resonance, Rev. Mod. Phys. **70**, 223 (1998).
53 F.R. Gantmacher: *The Theory of Matrices*, (Chelsea, New York 1959); Russian ed (Nauka, Moscow 1967); German ed (Deutscher Verlag der Wissenschaften, Berlin 1986).
54 J. García-Ojalvo, J.M. Sancho: *Noise in Spatially Extended Systems*, Institute for Nonlinear Science (Springer, New York 1999).
55 C.W. Gardiner: *Handbook of Stochastic Methods for Physics, Chemistry and the Natural Sciences*, 1985, 3rd ed. (Springer, Berlin 2004).
56 N.M. Ghoniem: Stochastic theory of diffusional planar-atomic clustering and its application to dislocation loops, Phys. Rev. B **39**, 11810 (1989).
57 I.I. Gikhman, A.V. Skorokhod: *Introduction to the Theory of Random Processes*, (Dover, New York 1996).
58 D. Gillespie: Exact stochastic simulation of coupled chemical reactions, J. Phys. Chem. **81**, 2340 (1977).
59 C. Godrèche: Dynamics of condensation in zero-range process, J. Phys. A **36**, 6313 (2003).
60 I.S. Gradshteyn, I.M. Ryzhik, Alan Jeffrey (ed.), Daniel Zwillinger (ed.): *Table of Integrals, Series, and Products*, 6th ed. (Academic Press, San Diego 2000).
61 M. Grigoriu: *Stochastic Calculus. Applications in Science and Engineering*, (Birkhäuser, Boston 2002).
62 M. Grmela: Thermodynamics of a driven system, Phys. Rev. E **48**, 919 (1993).
63 S. Grosskinsky, G.M. Schütz, H. Spohn: Condensation in the zero-range process: Stationary and dynamical properties, J. Stat. Phys. **113**, 389 (2003).
64 H. Haken: *Synergetics. Introduction and Advanced Topics*, (Springer, Berlin 2004).
65 P. Hänggi, H. Grabert, P. Talkner, H. Thomas: Bistable systems: master equation versus Fokker–Planck modeling, Phys. Rev. A **29**, 371 (2002).
66 P. Hänggi, P. Talkner, M. Borkovec: Reaction-rate theory: fifty years after Kramers, Rev. Mod. Phys. **62**, 251 (1990).
67 R.J. Harris, A. Rákos, G.M. Schütz: Current fluctuations in the zero-range process with open boundaries, J. Stat. Mech., P08003 (2005).
68 R.J. Harris, A. Rákos, G.M. Schütz: Breakdown of Gallavotti–Cohen symmetry for stochastic dynamics, Europhys. Lett. **75**, 227 (2006).
69 R.J. Harris, R.B. Stinchombe: Scaling approach to related disordered stochastic and free-fermion models, Phys. Rev. E **75**, 031104 (2007).
70 S. Harris: Absorbing-boundary limit for Brownian motion: Demonstration for a model, Phys. Rev. A **36**, 3392 (1987).
71 D. Helbing: Improved fluid-dynamic model for vehicular traffic, Phys. Rev. E **51**, 3164 (1995).
72 D. Helbing: Theoretical foundation of macroscopic traffic models, Physica A **219**, 375 (1995).
73 D. Helbing: High-fidelity macroscopic traffic equations, Physica A **219**, 391 (1995).
74 D. Helbing: Derivation and empirical validation of a refined traffic flow model, Physica A **233**, 253 (1996).

75. D. Helbing: Empirical traffic data and their implications for traffic modelling, Phys. Rev. E **55**, R25 (1997).
76. D. Helbing: *Verkehrsdynamik. Neue physikalische Modellierungskonzepte*, (Springer, Berlin 1997).
77. D. Helbing: Traffic and related self-driven many-particle systems, Rev. Mod. Phys., **73**, 1067 (2001).
78. D. Helbing, B. Tilch: Generalized force model of traffic dynamics, Phys. Rev. E **58**, 133 (1998).
79. D. Helbing, M. Treiber, A. Kesting, Understanding interarrival and interdeparture time statistics from interactions in queuing systems, Physica A **363**, 62 (2006).
80. D. Helbing, M. Treiber: Gas-kinetic-based traffic model explaining observed hysteretic phase transitions, Phys. Rev. Lett. **81**, 3042 (1998).
81. M. Henkel, G. Schütz: Boundary-induced phase transitions in equilibrium and non-equilibrium systems, Physica A **206**, 187 (1994).
82. J. Hinkel: *Applications of Physics of Stochastic Processes to Vehicular Traffic Problems*, (Dissertation, Univ. Rostock 2007).
83. J. Hinkel, R. Mahnke: Outflow probability for drift–diffusion dynamics, Int. J. Theor. Phys. **46**, 1542 (2007).
84. J. Honerkamp: *Stochastische Dynamische Systeme*, (VCH, Weinheim 1990); English ed *Stochastic Dynamical Systems*, (VCH, New York 1994).
85. J. Honerkamp: *Statistical Physics. An Advanced Approach with Applications*, (Springer, Berlin 1998).
86. W. Horsthemke, R. Lefever: *Noise-Induced Transitions. Theory and Applications in Physics, Chemistry, and Biology*, 1984, 2nd ed. (Springer, Berlin 2006).
87. H. Huang, N.M. Ghoniem: Formulation of a moment method for multidimensional Fokker–Planck equations, Phys. Rev. E **51**, 5251 (1995).
88. I. Jeon, P. March: Condensation transition for zero range invariant measures, Can. Math. Soc. Conf. Proc. **26**, 233 (2000).
89. Y. Kafri, E. Levine, D. Mukamel, G.M. Schütz, J. Török: Criterion for phase separation in one-dimensional driven systems, Phys. Rev. Lett. **89**, 035702 (2002).
90. Y. Kafri, E. Levine, D. Mukamel, G.M. Schütz, R.D. Willmann: Phase-separation transition in one-dimensional driven models, Phys. Rev. E **68**, 035101(R) (2003).
91. I. Karatzas, St. E. Shreve: *Brownian Motion and Stochastic Calculus*, 1988, 4-th ed. (Springer, New York 1997).
92. A.A. Katsnelson, A.I. Olemskoi, I.V. Sukhorukova, G.P. Revkevich: Self–oscillation processes during the structure relaxation of palladium–metal alloys (Pd–W) saturated with hydrogen, Physics-Uspekhi, **165**, 331 (1995).
93. J. Kaupužs: Fluctuations in ferroelectrics and dielectric properties from the Fokker–Planck equation, physica status solidi (b) **195**, 325 (1996).
94. J. Kaupužs: Critical exponents predicted by grouping of Feynman diagrams in φ^4 model, Ann. Phys. (Leipzig) **10**, 299 (2001).
95. J. Kaupužs, E. Klotins: Spatio-temporal correlations of local polarization in ferroelectrics, Ferroelectrics **296**, 239 (2003).
96. J. Kaupužs, R. Mahnke: A stochastic multi-cluster model of freeway traffic, Eur. Phys. J. B **14**, 793 (2000).
97. J. Kaupužs, R. Mahnke, R.J. Harris: Zero-range model of traffic flow, Phys. Rev. E **72**, 056125 (2005).
98. J. Kaupužs, J. Rimshans: Numerical solution of semiconductor Fokker–Planck kinetic equations, In: Proc. of the European Congress ECCOMAS 2000, Barcelona, Spain, pp. 1–18, 2000.
99. J. Kaupužs, H. Weber, J. Tolmacheva, R. Mahnke: Applications to traffic breakdown on highways, In: *Progress in Industrial Mathematics at ECMI 2002*, A. Buikis, R. Ciegis, A.D. Fitt, eds., pp. 133–138, (Springer, Berlin 2004).
100. B.S. Kerner: *The Physics of Traffic*, (Springer, Berlin 2004).

101 B. S. Kerner, H. Rehborn: Experimental properties of complexity in traffic flow, Phys. Rev. E **53**, R4275 (1996).

102 D.O. Kharchenko, A.V. Dvornichenko: Phase transitions induced by thermal fluctuations, Eur. Phys. J. B **61**, 95 (2008).

103 M. Kiessling, C. Lancellotti: The linear Fokker–Planck equation for the Ornstein–Uhlenbeck process as an (almost) nonlinear kinetic equation for an isolated N-particle system, J. Stat. Phys. **123**, 525 (2006).

104 P.E. Kloeden, E. Platen: *Numerical Solution of Stochastic Differential Equations*, 1992, Corr. 3rd ed. (Springer, Berlin 1999).

105 W. Knospe, L. Santen, A. Schadschneider, M. Schreckenberg: Towards a realistic microscopic description of highway traffic, J. Phys. A **33**, L477 (2000).

106 W. Knospe, L. Santen, A. Schadschneider, M. Schreckenberg: Human behavior as origin of traffic phases, Phys. Rev. E **65**, 015101 (2002).

107 M. Krbalek: Equilibrium distribution in a thermodynamical traffic gas, J. Phys. A: Math. Theor. **40**, 5813 (2007).

108 M. Krbalek, D. Helbing: Determination of interaction potentials in freeway traffic from steady-state statistics, Physica A **333**, 370 (2004).

109 N. Krepysheva, L. Di Pietro, M.-C. Néel: Fractional diffusion and reflective boundary condition, Physica A **368**, 355 (2006).

110 R. Kühne, R. Mahnke: Controlling traffic breakdowns, In: *Transportation and Traffic Theory* (Ed.: H.S. Mahmassani), pp. 229–244 (Elsevier Ltd., Oxford 2005).

111 R. Kühne, R. Mahnke, J. Hinkel: Understanding traffic breakdown: A stochastic approach, In: *Transportation and Traffic Theory* (Eds.: R.E. Allsop, M.G.H. Bell, B.G. Heydecker), pp. 777–790 (Elsevier Ltd., Oxford 2007).

112 R. Kühne, R. Mahnke, I. Lubashevsky, J. Kaupužs: Probabilistic description of traffic breakdowns, Phys. Rev. E **65**, 066125 (2002).

113 D. Labudde, R. Mahnke, V. Frischfeld: Monte Carlo simulation of thermodynamic systems with cluster formation under different boundary conditions, Comp. Phys. Comm. **106**, 181 (1997).

114 P. Landa: *Nonlinear Oscillations and Waves in Dynamical Systems*, (Kluwer Academic Publ., Dordrecht 1996).

115 P.S. Landa, A.A. Zaikin, L. Schimansky-Geier: Influence of additive noise on noise-induced phase transitions in nonlinear chains, Chaos, Solitons and Fractals **9**, 1367 (1998).

116 P.S. Landa, A.A. Zaikin, V.G. Ushakov, J. Kurths: Influence of additive noise on transitions in nonlinear systems, Phys. Rev. E **61**, 4809 (2000).

117 G. Lamm, K. Schulten: Extended Brownian dynamics. II. Reactive, nonlinear diffusion, J. Chem. Phys. **78**, 2713 (1983).

118 M.E. Lárraga, J.A. del Río, A. Mehta: Two effective temperatures in traffic flow models: Analogies with granular flow, Physica A **307**, 527 (2002).

119 M. Lax, W. Cai, M. Xu: *Random Processes in Physics and Finance*, (Oxford University Press, New York, 2006).

120 E. Levine, G. Ziv, L. Gray, D. Mukamel: Traffic jams and ordering far from thermal equilibrium, Physica A **340**, 636 (2004).

121 A.J. Lichtenberg, M.A. Lieberman: *Regular and Stochastic Motion*, (Springer, New York 1983).

122 Ch. Liebe: *Stochastik der Verkehrsdynamik: Von Zeitreihen-Analysen zu Verkehrsmodellen*, (Diplom, Univ. Rostock, 2006).

123 Ch. Liebe, R. Mahnke, J. Kaupužs, H. Weber: Vehicular motion and traffic breakdown: Evaluation of energy balance, In: *Traffic and Granular Flow '07* (Eds.: C. Appert-Rolland, F. Chevoir, Ph. Gondret, S. Lassarre,

J.-P. Lebacque, M. Schreckenberg), (Springer–Verlag, Berlin 2008).

124 I.M. Lifshitz, V.V. Slyozov: The kinetics of precipitation from supersaturated solid solutions, J. Phys. Chem. Solids **19**, 35 (1961).

125 V. Linetsky: The spectral representation of Bessel processes with constant drift: Applications in queueing and finance, J. Appl. Prob. **41**, 327, 2004.

126 V. Linetsky: On the transition densities for reflected diffusions, Adv. Appl. Prob. **37**, 435, 2005.

127 O' Loan, M.R. Evans, M.E. Cates: Jamming transition in a homogeneous one-dimensional system: the bus route model, Phys. Rev. E **58**, 1404 (1998).

128 I. Lubashevsky, R. Friedrich, R. Mahnke, A. Ushakov, N. Kubrakov: Boundary singularities and boundary conditions for the Fokker–Planck equations, e-print arXiv:math-ph/0612037v1, 12 Dec 2006.

129 I. Lubashevsky, M. Hajimahmoodzadeh, A. Katsnelson, P. Wagner: Noise–induced phase transition in an oscillatory system with dynamical traps, Eur. Phys. J. B **36**, 115 (2003).

130 I. Lubashevsky, S. Kalenkov, R. Mahnke: Towards a variational principle for motivated vehicle motion, Phys. Rev. E **65**, 036140 (2002).

131 I. Lubashevsky, R. Mahnke: Order-parameter model for unstable multilane traffic flow, Phys. Rev. E **62**, 6082 (2000).

132 I. Lubashevsky, R. Mahnke, M. Hajimahmoodzadeh, A. Katsnelson: Long-lived states of oscillator chains with dynamical traps, Eur. Phys. J. B **44**, 63 (2005).

133 I. Lubashevsky, R. Mahnke, P. Wagner, S. Kalenkov: Long-lived states in synchronized traffic flow: Empirical prompt and dynamical trap model, Phys. Rev. E **66**, 016117 (2002).

134 I. Lubashevsky, P. Wagner, R. Mahnke: Rational-driver approximation in car-following theory, Phys. Rev. E **68**, 056109 (2003).

135 I. Lubashevsky, P. Wagner, R. Mahnke: Bounded rational driver models, Eur. Phys. J. B **32**, 243 (2003).

136 J. Luczka, M. Niemiec, P. Hänggi: First-passage time for randomly flashing diffusion, Phys. Rev. E **52**, 5810 (1995).

137 Shang-Keng Ma: *Modern Theory of Critical Phenomena*, (W.A. Benjamin, New York 1976).

138 A.J. MacConnell: *Applications of Tensor Analysis*, (Dover Publications, New York 1957).

139 R. Mahnke: *Zur Evolution in nichtlinearen dynamischen Systemen*, (Habilitation, Univ. Rostock 1990).

140 R. Mahnke: *Nichtlineare Physik in Aufgaben*, Teubner Studienbücher Physik, (Teubner, Stuttgart 1994).

141 R. Mahnke: Aggregation phenomena to a single cluster regime under different boundary conditions, Zeitschr. f. Phys. Chem. (Leipzig) **204**, 85 (1998).

142 R. Mahnke: Probabilistic description of nucleation in vapours and on roads, In: *Interface and Transport Dynamics – Computational Modelling* (Eds.: H. Emmerich, B. Nestler, M. Schreckenberg), pp. 361–389, (Springer, Berlin 2003).

143 R. Mahnke, A. Budde: A new formula for the binding energy of clusters, Zeitschr. f. Phys. Chem. (Leipzig) **271**, 857 (1990).

144 R. Mahnke, A. Budde: Monte-Carlo-experiments to aggregation phenomena in complex systems, In: *Models of Selforganisation in Complex Systems MOSES* (Eds.: W. Ebeling, M. Peschel, W. Weid-lich), Mathematical Research, Vol. 64, pp. 164–171, (Akademie-Verlag, Berlin, 1991).

145 R. Mahnke, A. Budde: Pattern formation by cellular automata, Journal of Mathematical Modelling and Simulation in System Analysis (Syst. Anal. Model. Simul.) **10**, 133 (1992).

146 R. Mahnke, H. Hartmann: Keimbildung in übersättigten Gasen und auf überfüllten Autobahnen, In:

147. R. Mahnke, J. Kaupužs: Stochastic theory of freeway traffic, Phys. Rev. E **59**, 117 (1999).
148. R. Mahnke, J. Kaupužs: Probabilistic description of traffic flow, Networks and Spatial Economics **1**, 103 (2001).
149. R. Mahnke, J. Kaupužs, V. Frishfelds: Nucleation in physical and nonphysical systems, Atmospheric Research **65**, 261 (2003).
150. R. Mahnke, J. Kaupužs, J. Hinkel, H. Weber: Applications of thermodynamics to driven systems, Eur. Phys. J. B **57**, 463 (2007).
151. R. Mahnke, J. Kaupužs, I. Lubashevsky: Probabilistic description of traffic flow, Physics Reports **408**, 1–130 (2005).
152. R. Mahnke, R. Kühne, J. Kaupužs, I. Lubashevsky, R. Remer: Stochastic description of traffic breakdown. In: *Noise in Complex Systems and Stochastic Dynamics*, L. Schimansky-Geier, D. Abbott, A. Neiman, Ch. Van den Broeck, eds., Proc. SPIE **5114**, 126 (2003).
153. R. Mahnke, N. Pieret: Stochastic master-equation approach to aggregation in freeway traffic, Phys. Rev. E **56**, 2666 (1997).
154. R. Mahnke, J. Schmelzer, G. Röpke: *Nichtlineare Phänomene und Selbstorganisation*, Teubner Studienbücher Physik, (Teubner, Stuttgart 1992).
155. R. Mahnke, H. Urbschat, A. Budde: Nucleation and condensation to a single equilibrium cluster regime in a Monte Carlo experiment, Zeitsch. f. Physik D **20**, 399 (1991).
156. H. Malchow, L. Schimansky-Geier: *Noise and Diffusion in Bistable Nonequilibrium Systems*, Teubner-Texte zur Physik, Vol. 5, (Teubner, Leipzig 1986).
157. R. Mannella: Integration of stochastic differential equations on a computer, Int. J. Mod. Phys. **13**, 1117 (2004).
158. F. Marchesoni: Conceptual design of a molecular shuttle, Phys. Lett. A **237**, 126 (1998).
159. M. Martin: The source solution for diffusion with a linearly position dependent diffusion coefficient, Zeitschrift für Physikalische Chemie, NF, **162**, 245 (1989).
160. R.M. Mazo: *Brownian motion: Fluctuations, Dynamics, and Applications*, (Clarendon Press, Oxford 2006).
161. B. McCoy, T.T. Wu: *The Two-Dimensional Ising Model*, (Harvard University Press 1973).
162. S.V.G. Menon, D.C. Sahni: Derivation of the diffusion equation and radiation boundary condition from the Fokker–Planck equation, Phys. Rev. A **32**, 3832 (1985).
163. R. Metzler: Non-homogeneous random walks, generalized master equations, fractional Fokker–Planck equations, and the generalized Kramers–Moyal expansion, Eur. Phys. J. B **19**, 249 (2001).
164. M.A. Miller, J.P.K. Doye, D.J. Wales: Structural relaxation in atomic clusters: Master equation dynamics, Phys. Rev. E **60**, 3701 (1999).
165. E.W. Montroll, B.J. West: On an enriched collection of stochastic processes. In: *Studies in Statistical Mechanics*, vol. VII: Fluctuation Phenomena, ed by E.W. Montroll and J.L. Lebowitz, (North Holland Publ, Amsterdam 1979).
166. D. Mukamel: Phase transitions in nonequilibrium systems, In: *Soft and Fragile Matter. Nonequilibrium Dynamics, Metastability and Flow*, (Eds.: M.E. Cates and M.R. Evans), p. 205, (Institute of Physics Publishing, Bristol 2000).
167. A. Münster: *Statistical Thermodynamics*, vol. I, (Springer, Berlin 1969).
168. K. Nagel: Particle hopping models and traffic flow theory, Phys. Rev. E **53**, 4655 (1996).
169. D. Jost, K. Nagel: Probabilistic traffic flow breakdown in stochastic car following models, In: *Traffic and Granular Flow '03* (Eds.: S.P. Hoogendoorn, S. Luding, P.H.L. Bovy, M. Schreckenberg, D.E. Wolf), pp. 87–103, (Springer, Berlin 2005).

170 K. Nagel, M. Schreckenberg: A cellular automaton model for freeway traffic, J. Phys. I France **2**, 2221 (1992).

171 K.R. Naqvi, K.J. Mork, S. Waldenstrøm: Reduction of the Fokker–Planck equation with an absorbing or reflecting boundary to the diffusion equation and the radiation boundary condition, Phys. Rev. Lett. **49**, 304 (1982).

172 E. Nelson: *Dynamical Theories of Brownian Motion*, (Princeton University Press 1967).

173 G.F. Newell: Mathematical models of freely moving traffic, Oper. Res. **9**, 209 (1961).

174 H. Nobach: *Vorteile Klassischer Signal–und Datenverarbeitungsverfahren in der Optischen Strömungsmesstechnik*, (Habilitation, Univ. Darmstadt 2007)

175 B. Øksendal: *Stochastic Differential Equations*, 1985, 6th ed. (Springer, Berlin 2003).

176 B. Øksendal, A. Sulem: *Applied Stochastic Control of Jump Diffusion*, (Springer, Berlin 2005).

177 L. Onsager: Crystal Statistics. I. A two-dimensional model with an order–disorder transition, Phys. Rev. **65**, 117 (1944).

178 H. Öttinger: Computer simulation of reptation theories. I. Doi-Edwards and Curtiss-Bird models, J. Chem. Phys. **91**, 6455 (1989).

179 G. Parisi, N. Sourlas: Random magnetic fields, supersymmetry, and negative dimensions, Phys. Rev. Lett. **43**, 744 (1979).

180 W. Paul, J. Baschnagel: *Stochastic Processes. From Physics to Finance*, (Springer, Berlin 1999).

181 A. Pelissetto, E. Vicari: Critical phenomena and renormalization-group theory, Physics Reports **368**, 549 (2002).

182 E.A.J.F. Peters, Th.M.A.O.M. Barenbrug: Efficient Brownian dynamics simulation of particles near walls. I. Reflecting and absorbing walls, II. Sticky walls, Phys. Rev. E **66**, 056701, 056702 (2002).

183 A.S. Pikovsky, J. Kurths: Coherence Resonance in a noise-driven excitable system, Phys. Rev. Lett. **78**, 775 (1997).

184 I. Prigogine, R. Herman: *Kinematic Theory of Vehicular Traffic*, (Elsevier, New York 1971).

185 S. Redner: *A Guide to First-Passage Processes*, (Cambridge University Press, New York 2001).

186 P. Réfrégier: *Noise Theory and Application to Physics. From Fluctuations to Information*, (Springer, New York 2004).

187 T. Reichenbach, E. Frey, T. Franosch: Traffic jams induced by rare switching events in two-lane transport, New J. Phys. **9**, 159 (2007).

188 H. Reiss, A.D. Hammerich, E.W. Montroll: Thermodynamic treatment of nonphysical systems: Formalism and an example (single-lane traffic), J. Stat. Phys. **42**, 647 (1986).

189 R. Remer: *Theorie und Simulation von Zeitreihen mit Anwendungen auf die Aktienkursdynamik*, (Dissertation, Univ. Rostock 2005).

190 R. Remer, R. Mahnke: Application of Heston model and its solution to German DAX data, Physica A **344**, 236 (2004).

191 R. Remer, R. Mahnke: Stochastic volatility models and their application to German DAX data, Fluctuation and Noise Letters **4**, R67 (2004).

192 R. Remer, R. Mahnke: Application of the Heston and Hull–White models to German DAX data, Quantitative Finance **4**, 685 (2004).

193 H. Risken: *The Fokker–Planck Equation: Methods of Solutions and Applications*, 1984, 3rd ed (Springer, Berlin 1996).

194 G. Röpke: *Statistische Mechanik für das Nichtgleichgewicht*, (Deutscher Verlag der Wissenschaften, Berlin 1987).

195 J. Rudnik, G. Gaspari: *Elements of the Random Walk. An Introduction for Advanced Students and Researchers*, (Cambridge University Press, 2004).

196 S. Ruffo: Equilibrium and nonequilibrium properties of systems

with long-range interactions, Eur. Phys. J. B **64**, 355 (2008).

197 Yu.B. Rumer, M.Sh. Ryvkin: *Thermodynamics, Statistical Physics and Kinetics*, (Mir Publishers, Moscow, 1980).

198 A. Schadschneider: The Nagel–Schreckenberg model revisited, Eur. Phys. J. B **10**, 573 (1999).

199 A. Schadschneider, M. Schreckenberg: Traffic flow models with 'slow-to-start' rules, Ann. Phys. (Leipzig) **6**, 541 (1997).

200 L. Schimansky-Geier, Th. Pöschel (eds.): *Stochastic Dynamics*, Lecture Notes in Physics, Vol. 484 (Springer, Berlin 1998).

201 R.B. Schinazi: *Classical and Spatial Stochastic Processes*, (Birkhäuser, Boston 1999).

202 J. Schmelzer (ed.): *Nucleation Theory and Applications*, (Wiley-VCH, Weinheim 2005).

203 J. Schmelzer, G. Röpke, R. Mahnke: *Aggregation Phenomena in Complex Systems*, (Wiley-VCH, Weinheim 1999).

204 J. Schnackenberg: Network theory of microscopic and macroscopic behavior of master equation systems, Rev. Mod. Phys. **48**, 571 (1976).

205 M. Schreckenberg, A. Schadschneider, K. Nagel, N. Ito: Discrete stochastic models for traffic flow, Phys. Rev. E **51**, 2939 (1995).

206 M. Schulz: *Statistical Physics and Economics*, (Springer, New York 2003).

207 M.F. Schumaker: Boundary conditions and trajectories of diffusion processes, J. Chem. Phys. **117**, 2469 (2002).

208 H.G. Schuster: *Deterministic Chaos. An Introduction*, 2nd ed. (VCH, Weinheim 1989).

209 G.M. Schütz: Critical phenomena and universal dynamics in one-dimensional driven diffusive systems with two species of particles, J. Phys. A: Math. Gen. **36**, R339 (2003).

210 F. Schweitzer (ed.): *Self-Organization of Complex Structures*, (Gordon and Breach Science Publ., Amsterdam 1977).

211 F. Schweitzer: *Brownian Agents and Active Particles. Collective Dynamics in the Natural and Social Sciences*, (Springer, Berlin 2003).

212 U. Seifert: Stochastic thermodynamics: Principles and perspectives, Eur. Phys. J. B **64**, 423 (2008).

213 J.P. Sethna: *Statistical Mechanics: Entropy, Order Parameters and Complexity*, (Oxford University Press, New York 2006).

214 R. Seydel: *Tools for Computational Finance*, (Springer, Berlin 2004).

215 A. Siegert: On the first passage time probability, Phys. Rev. **81**, 617 (1951).

216 R. da Silveira: An introduction to breakdown phenomena in disordered systems, Am. J. Phys. **67**, 1177 (1999).

217 A. Singer, Z. Schuss, D. Holcman: Narrow escape, Part I –III, J. Stat. Phys. **122**, 437; 465; 491 (2006).

218 E. Smith, Thermodynamic dual structure of linear-dissipative driven system, Phys. Rev. E **72**, 036130 (2005).

219 I.M. Sokolov: Solution of a class of non-Markovian Fokker–Planck equations, Phys. Rev. E **66** 041101 (2002).

220 D. Sornette: *Critical Phenomena in Natural Sciences*, (Springer, Berlin 2004).

221 F. Spitzer: Interaction of Markov processes, Adv. Math. **5**, 246 (1970).

222 D. Stirzaker: *Stochastic Processes and Models*, (Oxford University Press, New York 2005).

223 S.X. Sun: Path summation formulation of the master equation, Phys. Rev. Lett. **96**, 210602 (2006).

224 P. Szymczak, A.J.C. Ladd: Boundary conditions for stochastic solutions of the convection–diffusion equation, Phys. Rev. E **68**, 036704 (2003).

225 M. Takayasu, H. Takayasu: $1/f$ noise in a traffic model, Fractals **1**, 860 (1993).

226 T. Taniguchi, E.G.D. Cohen: Nonequilibrium steady state thermodynamics and fluctuations for stochastic systems, J. Stat. Phys. **130**, 633 (2008).

227 M. Treiber, A. Hennecke, D. Helbing: Derivation, properties, and simulation of a gas-kinetic-based nonlocal traffic model, Phys. Rev. E **59**, 239 (1999).

228 St. Trimper: Master equation and two heat reservoirs, Phys. Rev. E **74**, 051121 (2006).

229 H. Ulbricht, J. Schmelzer, R. Mahnke, F. Schweitzer: *Thermodynamics of Finite Systems and the Kinetics of First-Order Phase Transitions*, Teubner-Texte zur Physik, vol. 17, (Teubner, Leipzig 1988).

230 H. Ulbricht, F. Schweitzer, R. Mahnke: Nucleation theory and dynamics of first-order phase transitions in finite systems, In: *Selforganisation by Nonlinear Irreversible Processes* (Eds.: W. Ebeling, H. Ulbricht), pp. 23–36, Springer–Series in Synergetics, vol. 33 (Springer, Berlin 1986).

231 G.E. Uhlenbeck, L.S. Ornstein: On the theory of the Brownian motion, Phys. Rev. **36**, 823, 1930.

232 C. Van den Broeck, J.M.R. Parrondo, R. Toral: Noise-induced nonequilibrium phase transition, Phys. Rev. Lett. **73**, 3395 (1994).

233 C. Van den Broeck, J.M.R. Parrondo, R. Toral, R. Kawai: Nonequilibrium phase transitions induced by multiplicative noise, Phys. Rev. E **55**, 4084 (1997).

234 N.G. Van Kampen: *Stochastic Processes in Physics and Chemistry*, 1981, 2nd edn (North Holland Publ, Amsterdam 1992).

235 J. Voit: *The Statistical Mechanics of Financial Markets*, (Springer, Berlin 2001).

236 M. Volmer: *Kinetik der Phasenbildung*, (Th. Steinkopff, Dresden 1939).

237 M. Volmer, A. Weber: Nuclei formation in supersaturated states, Z. Phys. Chemie **119**, 227 (1926).

238 C. Wagner: Theory of precipitate change by redissolution, Z. Elektrochemie **65**, 581 (1961).

239 P. Wagner, How human drivers control their vehicle Eur. Phys. J. B **52**, 427 (2006).

240 P. Wagner, K. Nagel: Comparing traffic flow models with different number of phases, Eur. Phys. J. B **63**, 315 (2008).

241 H. Weber, R. Mahnke, Ch. Liebe, J. Kaupužs: Dynamics and thermodynamics of traffic flow, In: *Traffic and Granular Flow '07* (Eds.: C. Appert-Rolland, F. Chevoir, Ph. Gondret, S. Lassarre, J.-P. Lebacque, M. Schreckenberg), (Springer, Berlin 2008).

242 M.F. Wehner, W.G. Wolfer: Numerical evaluation of path-integral solutions to Fokker–Planck equations. II. Restricted stochastic processes, Phys. Rev. A **28**, 3003 (1983).

243 G.B. Whitham: Exact solution for a discrete system arising in taffic flow, Proc. R. Soc. London, Ser. A **428**, 49 (1990).

244 D.T. Wu: Nucleation theory, In: *Solid State Physics*, vol. 50, (Eds.: H. Ehrenreich, F. Spaepen), p. 37 (Academic Press, San Diego 1997).

245 A.A. Zaikin, J. Garcia-Ojalvo, L. Schimansky-Geier: Nonequilibrium first-order phase transitions induced by additive noise, Phys. Rev. E **60**, R6275 (1999).

246 A.A. Zaikin, L. Schimansky-Geier: Spatial patterns induced by additive noise, Phys. Rev. E **58**, 4355 (1998).

247 G.M. Zaslavsky: Dynamical traps, Physica D, **168–169**, 292 (2002).

248 Ya. B. Zeldovich: Theory of the formation of a new phase. Cavitation, Sov. Phys. JETP **12**, 525 (1942).

249 A. Zettl: *Sturm–Liouville Theory*, (Providence, American Mathematical Society, 2005).

250 J.W. Zhang, Y. Zou, L. Ge: A force model for single-line traffic, Physica A **376**, 628 (2007).

251 X. Zhang, G. Hu: $1/f$ noise in a two-lane highway traffic model, Phys. Rev. E **52**, 4664 (1995).

252 J.M. Ziman: *Models of Disorder*, (Cambridge University Press, 1979).

253 J. Zinn-Justin: *Quantum Field Theory and Critical Phenomena*, (Clarendon Press, Oxford 1996).

254 D. Zubarev, V. Morozov, G. Röpke: *Statistical Mechanics of Nonequilibrium Processes*, Vol. 1 and 2 (Wiley-VCH, Berlin 1997).

Index

a
absorbing boundary, 44, 90, 199, 216
active Brownian particle, 149
active particle, 306, 343
adjoint operator, 119
advection–diffusion problem, 32
aggregation, XV, 85, 342
  –reaction limited, 284
anomalous behavior, 357
arithmetric Brownian motion, 173
Arrhenius ansatz, 104
Arrhenius, Svante, 104
attachment, 281, 290, 322

b
Bachelier, Louis, XIV, 5
backward Fokker–Planck equation, 26, 36, 39, 240
backward Kolmogorov equation, 27
backward transition, 92
balance equation, 82, 86, 216
Bessel equation, 135
Bessel function, 136, 368
Bethe–Weizsäcker formula, 283
bifurcation, 287, 356
  –subcritical, 308
  –supercritical, 308
bifurcation diagram, 154
binding energy, 282, 283
binomial distribution, 183, 185, 212
birth-and-death process, 85
bistability, 99
Black–Scholes equation, 177
Boltzmann distribution, 111, 410
Boltzmann, Ludwig, 115
Boltzmann–Gibbs distribution, 147
boundary condition, 91

boundary layer, 56
  –scaling, 61
boundary singularity, 37, 40
  –vector, 38
boundary trap, 34
Box–Muller method, 171
breakdown function, 217
breakdown probability
  –cumulative, 228
breakdown probability density, 197, 226
breakdown probability distribution, 204
breakdown rate, 92
Brown, Robert, XII, 4, 77, 212
Brownian motion, XIV, XV, 4, 5, 77, 212
  –3d velocity space, 160
  –direct integration, 161
  –first moment, 161
  –harmonic analysis, 163
  –second moment, 162
  –variance, 163
Brownian particle, 123, 146, 176
Brownian path, 29
Brusselator, 116
bumper-to-bumper distance, 306

c
canonical ensemble, 148
capillary length, 285
car cluster, 321
car cluster model, 336
car dynamics, 309
  –acceleration, 312
  –deceleration, 312
  –vector field, 310
car ensemble, 320
car following model, 303

car system
  –total energy, 345
Carus, Titus Lucretius   XII
Cauchy problem, 256
Cauchy sequence, 8, 17
cellular automata, XIV, 391, 408
chaos, XII
Chapman–Kolmogorov equation, 23, 31, 32, 34, 35, 79, 81, 89, 120
Chebyshev inequality, 5
chemical potential, 284, 298
circular road
  –one-lane, 320
circular traffic, 306
cluster
  –binding energy, 283
  –critical, 277
cluster dissolution, 93
cluster distribution, 281
cluster distribution function, 279
cluster size
  –average, 325
  –stationary, 326
  –time evolution, 325
collective dynamics, 342
colored noise, 159
complex number, 249
condensation, 85, 395, 404
conditional probability density, 77
congested flow, 331
congested traffic, 317
continuity equation, 117, 146, 187
cooperative behavior, 306
cooperative phenomenon, 355
correlation function, 158, 165, 273
covariance, 273
creative action, 357
critical cluster
  –curvature, 287
critical cluster size, 286, 292
critical exponent, 308, 309
critical phenomena, XV
critical point, 308
criticality, 332
cumulative breakdown probability, 232
cumulative life-time distribution, 217
cumulative probability, 217
current fluctuations, 407
curvature, 283

d
Darwin, Charles, 211
de Broglie length, 279, 282

decay process, 92
deceleration force, 314
depletion, 278
detachment, 290
detachment rate, 283, 284
detailed balance, 83, 85, 88, 111, 113, 281, 289, 293, 347, 400
dichotomic process, 99
dichotomous noise, 151
differential equation
  –first-order, 306
  –second-order, 238
differential operator, second-order, 119
diffusion, XV, 31
  –absorbing boundaries, 210
  –boundaries, 208
  –finite interval, 208
  –finite intervall, 193
  –mixed boundaries, 193
  –natural boundaries, 186
  –semi-open, 209
  –separation ansatz, 135, 194
diffusion coefficient, 133, 146, 186, 364
diffusion equation, XIV
  –one-dimensional, 186
  –superposition, 200
diffusion layer, 31
diffusion length, 295
diffusion process, 19, 22–24, 27, 32
diffusion tensor, 38, 46, 47
diffusional growth, 278
Dirac function, 37
Dirac, Paul, 415
discrete random walk, 211
displacement, 35
dissipative system, 343
dissolution, 93
double-well potential, 87, 154
drift field, 46
drift velocity, 54
drift–diffusion dynamics
  –eigenfunction, 219
drift–diffusion motion
  –first moment, 125
drift–diffusion problem
  –bounded, 119
  –dimensionless, 216
  –one-dimensional, 119
driven system, 342, 395
drunkard's walk, 181
dynamical trap, XV, 372
dynamics, deterministic, 154

e

Ebeling, Werner,   416
econophysics,   85, 177, 269
Ehrenfest, Paul and Tatiana,   115
eigenvalue equation,   84
eigenvalue problem,   195
Einstein formula,   163, 258
Einstein relation,   147, 148
Einstein, Albert,   XIV, 5, 212, 260, 415
energy balance,   343, 352
energy dissipation,   149
energy exchange,   XI
energy flux,   345
energy input,   343
ensemble-averaging,   165
equilibrium distribution,   83
equilibrium phase transition,   355
equilibrium state,   163
erratic motion,   XI
Euler discretization,   29
Euler formula,   157
evaporation,   281
evolution,   7
exponential decay,   218

f

fast diffusion boundary,   44
fat tails,   269
Fermi distribution,   104
Fermi's golden rule,   81
ferroelectric,   409
Feynman diagram,   409
Fick's law,   187
Fick, Adolf,   187
filtration,   7
financial market,   177
finite-size effect,   308, 341
first moment,   124
first-passage time,   126, 242
first-passage time distribution,   197, 217
first-passage time problem
  –absorbing boundary,   252
  –mixed boundaries,   252
first-order phase transition,   277, 336
first-passage problem,   90
first-passage time,   89, 402
first-passage time distribution
  –semi-finite interval,   199
flow vector,   310
fluctuation-dissipation relation,   147
fluctuation-dissipation theorem,   163, 165, 258

flux-density relation,   329
Fokker, Adriaan,   XIV
Fokker–Planck dynamics,   124
  –V-shaped potential,   252
  –linear potential,   251
Fokker–Planck equation,   XIV, 27, 31, 32, 35, 117, 148, 409
  –backward,   118, 128, 240
  –boundary conditions,   31, 32
  –conservation form,   43
  –derivation,   126
  –forward,   117, 127, 128
  –mixed boundaries,   215
  –multidimensional,   117, 118
  –multidimensional,   145
  –natural boundaries,   213
  –one-dimensional,   118
  –stationary solution,   167
  –Sturm–Liouville type,   131
Fokker–Planck operator,   37, 118, 121
Fokker–Planck dynamics
  –comparison,   130
force
  –accelerating,   344
  –decelerating,   344
forward Fokker–Planck equation,   26, 42
Fourier representation,   157, 409
Fourier space,   189, 409
Fourier transformation,   187, 255, 410
  –inverse,   261
fractal media,   XV
free energy,   282, 294, 348, 356
  –ideal gas,   294
free flow solution,   307
friction
  –nonlinear,   149
friction force,   149
fundamental diagram,   329, 407

g

Galton board,   183, 184, 211
Galton, Francis,   183, 211
Gauss theorem,   42
Gaussian distribution,   58, 166. 173, 186, 190, 206, 214, 269
  –moments,   214
Gaussian profile,   187
Gaussian white noise,   78, 146, 151, 168, 409
  –multiplicative,   316
generating function,   36, 98
Genetic model,   368

geometric Brownian motion, 28, 173, 177, 269, 369
geometric random walks, 369
German stock index, 270
Girsanov transformation, 23
grand canonical ensemble, 393
graph theory, 109
gravitational potential, 148
Green function, 33, 34, 37, 39
ground state, 220

h

Hamiltonian dynamics, XVI
Hamiltonian
  –many-particle system, 281
Hänggi–Klimontovich process, 166, 361
harmonic oscillator, 176
  –dynamical trap, 373
harmonic potential, 143
Hausdorff dimension, 29
heat bath, XI, 104, 149
Heston model, 270, 274
homogeneous flow, 308
Hull–White model, 270
hyperbolic tangens, 304
hysteresis, 308, 342

i

ideal gas model, 284
impermeability, 37
impermeable boundary, 44
Ingenhousz, Jan, XII
initial-boundary-value problem, 215, 233
interaction potential, 150, 344
internal energy, 345
Ising model, 99
Ito diffusion, 29
Ito formula, 20, 27
Ito integral, 15, 17, 28
Ito process, 18, 361
  –diffusion, 19, 20
  –drift, 19, 20
  –increment, 19
  –path, 19
  –quadratic variation, 18
Ito stochastic calculus, 167
Ito stochastic integral, 167
Ito stochastic process, 166

j

jam formation, 320
jam shrinkage, 94
joint probability density, 77
Julien Bect, 31

k

kinetic coefficient, 357
kinetic equation, 297
Kirchhoff diagram, 112
Kirchhoff's method, 110
Kirchhoff, Gustav, 109
Kolmogorov backward equation, 25
Kolmogorov forward equation, 25
Kramers–Moyal approach, 40
Kramers–Moyal expansion, XIV, 31, 127
Kronecker delta, 47

l

Landau theory, 297
Landau–Ginzburg Hamiltonian, 409
Langevin equation, XV, 19, 128, 316, 409
  –additive noise, 152
  –multidimensional, 145
  –one-dimensional, 151
  –overdamped limit, 148
Langevin force, 38, 54, 145, 151, 355
Langevin, Paul, XIV
Laplace operator, 147
Laplace transformation, 141, 242
  –inverse, 243
lattice random walk, 54, 58, 62
law of large numbers, 36, 169
Lennard–Jones potential, 304
Levy walk, 212
Levy, Paul, 212
life-time, 217
Lifshitz–Slyozov–Wagner theory, 278
limit cycle, 308, 312
Liouville, Joseph, 238
liquid–gas interface, 295
liquid–gas system, 293, 295, 298
local homogeneity, 35
log-normal distribution, 175
logarithmic return, 271
long-lived state, 372, 386
long-range interaction, 148
Lyapunov exponent, 307
Lyapunov function, 374

m

many-particle system, XI, 99
 – vehicular traffic, 351
 – Hamiltonian, 147
Markov chain, 183
Markov dichotomic system
 – diagonalization method, 101
 – master equation, 100
 – time evolution matrix, 103
Markov process, 14, 20, 23, 79, 167, 322
 – history, 80
Markov property, 20, 79, 120, 212
Martin, Manfred, 133
martingale, 18
master equation, XV, 81, 82, 288
 – boundaries, 86
 – matrix form, 83
 – multidimensional, 290
 – one-step, 322
 – three-level system, 105
 – transition rate, 81
 – two heat reservoirs, 104
maximal tree, 110
maximum value distribution, 206
Maxwell distribution, 142, 258
mean first-passage time, 123, 124, 338
mean reverting process, 270
mean value, 191, 273
mean-field approximation, 99, 394
mean-field dynamics, 394, 400
mean-reverting process, 253
metastability, XVI, 332, 392, 399
metastable state, 405
metric tensor, 47
mirror method, 200
molecular dynamics, 305
moment, nth order, 191
Monte Carlo method, 288
Monte Carlo simulation, 95, 404
Morse potential, 149, 150
Morse ring chain, 149
multi-lane effect, 336
multi-lane traffic flow, 385

n

Nagel–Schreckenberg model, 317, 391
narrow escape problem, 123
Newton's third law, 305
Nobach, Holger, 157
noise, 316
 – $1/f$ noise, 159
 – spectral density, 164

noise amplitude, 317
noise-induced transition, 151
noise-induced transport, 355
nonequilibrium phase transition, 357
non-Gaussian behavior, XVI
nonlinear Langevin force, 361
normal distribution, 27, 28, 78
normalization, 191
normalized eigenfunctions, 197
nucleation, 277, 342
 – free energy, 282
 – isothermal–isochoric, 280
 – multi-droplet case, 289
 – on roads, 287
 – single-droplet case, 279
 – supersaturated vapor, 286
nucleation theory, 290
nucleation time, 402

o

one-step master equation, 86
one-step process, 85, 88, 90
one-step processes, 328
optimal velocity, 322
optimal velocity function, 306
optimal velocity model, 304, 306, 312, 344, 352
order parameter, 357, 385
order parameter theory, 355
Ornstein, Leonard Salomon, 273
Ornstein–Uhlenbeck process, 27, 29, 160, 255, 273
orthogonality, 196
 – Bessel functions, 137
oscillator ensemble, 373
Ostwald ripening, 277, 278, 292
outflow probability, 124

p

particle conservation, 279
particle-hopping model, 391
 – asymmetric, 391
 – totally asymmetric, 391
partition function, 281
Pauli exclusion principle, 392
Pearson, Karl, 181
periodic boundary conditions, 410
periodic perturbation, 307
phase equilibrium, 298
phase separation, XVI, 392
phase transition, XV, 148, 308, 409
 – noise-induced, 355

phase-space distribution, 355
Planck, Max, XIV
Poisson distribution, 98
Poisson integral, 190
Poisson process, 93, 114
polar coordinates, 190
Polar method, 171
polarization, 409
polarization hysteresis, 411
Polya, George, 184
Pontryagin technique, 40
population genetics, 368
potential energy, 344
power-like singularity, 308
precluster, 324
Prigogine, Ilya, 116
probability current, 86, 124, 271
probability density
   –first-passage time, 171
probability density distribution, 271
probability density function, 148
probability distribution, 78, 324
probability flux, 216
probability flux operator, 43
probability outflow, 203
probability theory, XII, 18

q
quadratic variation, 5, 6

r
Röpke, Gerd, 415
random force, XII, 360
random number, 171
   –uniform distribution, 171
random path, XII
random process, XIV
random variable, 4, 8, 16, 17, 22
random walk, XIV, 46, 54, 56, 85, 181, 391
   –boundary condition, 184
   –moments, 184
   –one-dimensional, 182
   –recurrence, 184
   –symmetric, 182
random walk experiment, 182
randomness, 347
Rayleigh formula, 123
reaction limited aggregation, 290
reaction–diffusion equation, 82
recurrence relation, 87
reduced Fokker–Planck equation, 218
reflected diffusion, 208
reflecting boundar, 215
reflection principle, 170
relaxation, 297
relaxation dynamics, 332
relaxation time, 334
Remer, Ralf, 255
root-mean-square, 182
Runge–Kutta method, 375

s
scale invariance, 181
scaling property, 169
Schlögl model, 115
Schlögl, Friedrich, 115
Schmelzer, Jürn, 278
Schrödinger equation, 143
self-adjoint operator, 119
self-driven motion, 145
self-gravity, 149
self-organization, XV, 342
semi–infinite interval, 200
semigroup, 23, 24
separation ansatz, 218
shadow path, 170
Siegert, Arnold, 124
sigmoidal function, 304
single-lane traffic, 342
slow-to-start rule, 395, 408
soft matter physics, 149
soliton, 150
spectral analysis, 157
spectral density, 158
spectral representation, 189
spinodal decomposition, 278
Spitzer, Frank, XVI
stability analysis, 307, 380
stability region, 307
stagnation, 371
standard deviation, 192
stationarity, 82
stationary distribution, 82
steady state, 307
steady-state distribution, 365
Stirling formula, 185, 282, 294
stochastic collision, 160
stochastic differential equation, 23, 27, 152, 347
   –drift–diffusion, 173
   –multiplicative noise, 173
   –one-dimensional, 166
stochastic dynamics, XVI, 33
stochastic equation, XV
stochastic force, properties, 160

stochastic hybrid systems, 31
stochastic integration
    –intermediate point, 361
stochastic modeling, 254
stochastic motion, 31, 38
stochastic process, XIV, 4, 78, 80
    –exclusion process, 391
    –Markov approximation, XIV, 80
    –Stratonovich type, 364
    –totally asymmetric, 391
    –zero-range, 391
stochastic realization, 153
stochastic system
    –Ito type, 362
    –Langevin source, 358
stochastic tool, XVI
stochastic trajectory, 327
    –maximum, 205
stochastic volatility, 269
stochasticity, XII
stock market, XIV, XVI, 269
stock price dynamics, 269
stopping time, 169
Stratonovich process, 166, 361
Stratonovich stochastic calculus, 167
Sturm, Jacques Ch. F., 238
Sturm–Liouville operator, 132, 238
    –self-adjoint, 132, 239
Sturm–Liouville problem, 117, 134
Sturm–Liouville theory, 238
sub-diffusion, 32
subcritical bifurcation, 155
subgraph
    –connected, 110
super-diffusion, 32
supercritical bifurcation, 154
superposition, 223
supersaturated vapor, 289
supersaturation, 278, 292
survival function
    –open system, 211
    –semi-open system, 209
symmetry breakdown, 32
synchronized flow, 343
Szilard, Leo, 277, 299

t
Taylor expansion, 21, 186, 297
thermodynamic equilibrium, 83, 147
thermodynamic limit, 282, 329, 398

thermodynamic potential, 88, 289
    –free energy, 281
Thiele, Thorvald N., XIV
time confinement, 35, 72
time delay, 303
time lag, 199, 204, 402
    –tangent construction, 205
time-series analysis, XII
time-averaging, 165
Toda chain, 150
Toda potential, 149
traffic
    –flux-density relation, 331
    –fundamental diagram, 329
traffic breakdown, 215
traffic breakdown probability, 89
traffic flow, XVI, 93, 303
    –chemical potential, 351
    –force model, 304
    –free energy, 351
    –phase diagram, 308
    –temperature, 348
    –thermodynamics, 343
traffic jam, 342
traffic model
    –optimal velocity, 303
    –time lag, 303
transcendental equation, 219, 229, 245
    –roots, 249
transition frequency, 324
transition matrix, 83, 334
transition probability, 324
translation symmetry, 411
transport, 32
transportation, 293
Trimper, Steffen, 104
two-level system, 99

u
Uhlenbeck, George Eugene, 273
uncorrelated process, 78
universality, 181
urn model, 115

v
vector field, 310
vehicular flux, 329
vehicular interaction, 344
vehicular traffic, XV
Verhulst model, 365
volatility, 270

W

wave equation, 194, 218
wave number, 195, 218
  –ground state, 221
white noise, 152, 409
  –additive, 145
Wiener increment, 128
Wiener process, 4, 6, 15, 22, 27, 32, 152, 168, 317
  –boundaries, 171
  –finite interval, 171
  –increment, 153, 169
  –moments, 168
  –trajectory, 172
  –variance, 168
Wiener trail, 29
Wiener, Norbert, 5, 7
Wiener–Khinchin theorem, 159

Z

zero-flux relationship, 87
zero-range model, 393, 404
zero-range process, XVI, 391